Level 2

PRINCIPLES OF LIGHT VEHICLE MAINTENANCE & REPAIR

DIPLOMA

Graham Stoakes
Eric Sykes
Catherine Whittaker

www.pearsonschoolsandfecolleges.co.uk

✓ Free online support
✓ Useful weblinks
✓ 24 hour online ordering

0845 630 44 44

babcock
trusted to deliver™

Heinemann
Part of Pearson

Heinemann is an imprint of Pearson Education Limited, Edinburgh Gate, Harlow, Essex, CM20 2JE.

www.pearsonschoolsandfecolleges.co.uk

Heinemann is a registered trademark of Pearson Education Limited

Text © Babcock International Group and Graham Stoakes 2011

Edited by Melanie Birdsall, Caroline Low and Liz Evans
Designed by Pearson Education Limited
Typeset by Phoenix Photosetting
Original illustrations © Pearson Education Limited 2011
Illustrated by KJA Artists
Cover design by Woodenark
Cover photo/illustration © Shutterstock/Steve Mann

The rights of Babcock International Group and Graham Stoakes to be identified as authors of this work have been asserted by them in accordance with the Copyright, Designs and Patents Act 1988.

First published 2011

15 14 13 12
10 9 8 7 6 5 4 3 2

British Library Cataloguing in Publication Data
A catalogue record for this book is available from the British Library

ISBN 978 0 435 04816 7

Copyright notice
All rights reserved. No part of this publication may be reproduced in any form or by any means (including photocopying or storing it in any medium by electronic means and whether or not transiently or incidentally to some other use of this publication) without the written permission of the copyright owner, except in accordance with the provisions of the Copyright, Designs and Patents Act 1988 or under the terms of a licence issued by the Copyright Licensing Agency, Saffron House, 6–10 Kirby Street, London EC1N 8TS (www.cla.co.uk). Applications for the copyright owner's written permission should be addressed to the publisher.

Printed in Spain by Grafos S.A.

Websites
There are links to relevant websites in this book. In order to ensure that the links are up to date and that the links work we have made the links available on our website at www.pearsonhotlinks.co.uk. Search for this title Level 2 Diploma in Light Vehicle Maintenance & Repair or ISBN 9780435048167.

Acknowledgements

Pearson Education Limited would like to thank the following people for providing technical feedback: Ian Gillgrass and Beverley Lilley of the Institute of the Motor Industry (IMI) and Stephen Mitchell of Whitstable Community College for his invaluable help in the development of this title.

Babcock International Group would like to thank Eric Sykes for his technical direction and Catherine Whittaker, Sarah Harrison and Christine Potts of Babcock International Group for their invaluable help in the development of this material. We would also like to thank Derrick Insley for photography direction and Zoe Fielding and Jake Davenport for modelling during the photo shoot.

Graham Stoakes would like to thank Stella Mbubaegbu CBE, Highbury College Principal and Chief Executive, for the use of the college workshops during the photo shoot. Graham Stoakes would also like to thank Holly Stoakes and Jonny Walker for modelling during the photo shoot.

The author and publisher would like to thank the following individuals and organisations for permission to reproduce photographs:

(Key: b-bottom; c-centre; l-left; r-right; t-top)

Alamy Images: Blend Images 70tl, Flake 99tc, Frank Naylor 18-19 (welding), Hympi 70cr, Juice Images 50cr, Leslie Garland Picture Library 410, Mike Harrington 50br, Pashkov Andrey 18-19 (chemical-resistant gloves), petpics 365/5, Philip Willis 428, Radius Images 50cl, Simon Clay 18-19 (roadside attendance), UK Stock Images Ltd 70bl; **Corbis:** Jonathan Bielaski / First Light 18-19 (grinding/sanding); **Getty Images:** Stone 257; **iStockphoto:** Alistair Forrester Shankie 46t, 46c, Arne Thaysen / Goldhafen 119, Pixonaut 201; **Masterfile UK Ltd:** Peter Griffith 361; **Pearson Education Ltd:** David Sanderson 18-19 (high visibility jacket), 54, 80, Gareth Boden 37, 45/1, 45/2, 45/3, 45/4, 45/5, 86c, 94c, 94b, 95c, 95b, 99tl, 99tr, 100c, 100b, 101, 319tr, Naki Photography 36, Trevor Clifford 86b, 374, 375bl, 386tl, 388; **Photolibrary.com:** Aflo FotoAgency 221, Cultura RM 222; **Shutterstock.com:** a_v_d 381, Alaettin YILDIRIM 18-19 (welding mask), 99cl, Alexey Nikolaew 18-19 (full face mask), Alkristeena 50tl, ching 55c, Cico 111, Dario Sabljak 1, Denis Dryash 90cr, Eduard Ionescu 366, 375br, Four Oaks 93b, Gualberto Becerra 98cl, Gunpreet 187t, IDAL 18-19 (removal of hypodermic needles), IRC 18-19 (filter respirator), Jan de Wild 123bl, Jeff Gynane 18-19 (removal of broken glass), John Kasawa 105cl, John Schiffer 105cr, 146/3, Karam Miri 18-19 (Kevlar® gloves), KTOMKRATHOG 188c, Lukusz Kwiatkowski 390, Luzi 315, Monkey Business Images 49, 69, 71, Olmar 421, Rafa Irusta 123tl, Rob Kemp 88br, STILLFX 47, StockHouse 55t, TerryM 105tr, Tratong 187b, Vladislav Gajic 46b, Yanas 96/1.88; **SuperStock:** imagebroker.net 188t

Cover image: *Front:* **Shutterstock.com:** Steve Mann

All other images © **Pearson Education Ltd**: Clark Wiseman, Studio 8

The authors and publishers would like to thank the following individuals and organisations for permission to reproduce material:

p. 71 British Standards Institution (BSI) Kitemark®

p. 149 Images in Table 2.4 based on illustrations from Renault, used with permission of Renault.

Every effort has been made to trace the copyright holders and we apologise in advance for any unintentional omissions. We would be pleased to insert the appropriate acknowledgement in any subsequent edition of this publication.

Contents

	Introduction	iv
	Features of this book	vii
1	General workshop operation principles for light vehicles	1
	1. Health and safety	2
	2. Job roles and communication	67
	3. Materials and tools	84
2	Light vehicle chassis system units & components	119
3	Light vehicle engine mechanical, lubrication & cooling system units & components	201
4	Light vehicle fuel, ignition, air & exhaust system units & components	257
5	Light vehicle transmission & driveline units & components	315
6	Light vehicle electrical units & components	361
7	Light vehicle inspection & servicing	421
	Index	481

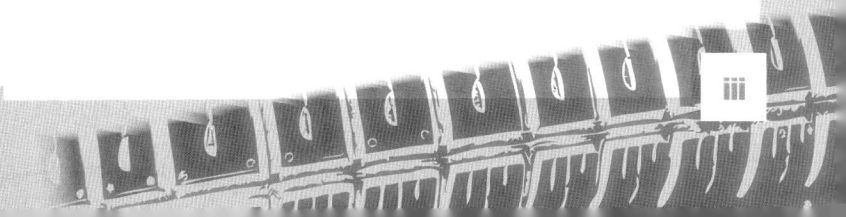

Introduction

Welcome to Principles of Light Vehicle Maintenance & Repair!

Working in the automotive industry is going to give you a great opportunity to work here in the UK or overseas: it's a challenging, stimulating and fulfilling career.

Working in this sector combines many different practical skills with a knowledge of specialised materials and techniques. It also requires good people skills: customers are as much a part of the sector as the light vehicles. This book will introduce you to the automotive industry and all the important systems of light vehicles from engines, chassis, transmission and electrics through to inspection and servicing.

About this book

This book has been produced to help you build a sound knowledge and understanding of all aspects of the Diploma and NVQ requirements associated with the automotive industry.

The topics in this book cover all the information you will need to attain your Level 2 qualification in Light Vehicle Maintenance and Repair Principles. Each chapter relates to particular units of the Diploma and provides the information needed to form the required knowledge and understanding of that area. It can be used for any awarding organisations' qualification including IMI Awards Limited and City & Guilds.

This book is just one part of a series of publications aimed at light vehicle maintenance and repair learners. The first in the series is aimed at Level 1 and this covers an introduction to the industry and how the light vehicle is constructed and maintained. This Level 2 book covers the knowledge and skills required for the learner to service and replace vehicle systems in order to progress to Level 3. The Level 3 book (published in Spring 2012) covers in depth diagnosis and repair for current and new vehicle technology.

Although there are many routes into a career in the automotive industry, most will involve undertaking an apprenticeship. An apprenticeship means you will probably be employed by a garage and study at a college one to two days a week. At college you will learn the theory needed to enable you to gain a 'technical certificate' which includes the knowledge and skills requirements of your qualification. At work you will practice the skills and gather evidence to show that you are able to undertake the required practical tasks competently. This will enable you to gain the NVQ or SVQ part of your apprenticeship qualification.

This book has been written by experienced trainers who have many years of experience within the sector. They believe in providing you with all the necessary information you need to support your studies and ensuring it is presented in a style which is both manageable and relevant.

This book will also be a useful reference tool for you in your professional life once you have gained your qualifications and are working in the sector.

About the Automotive industry

75 per cent of households in the UK have access to at least one car and 30 per cent own two or more cars. This means that there are more than 33 million cars licensed for use in the UK.

The automotive sector has a wide variety of career opportunities ranging from design through to manufacture, sales, maintenance and repair. All of which require specialist knowledge and skills which in turn bring the associated rewards. In the UK, the automotive industry directly employs more than 600,000 people and supports a further 1.9 million jobs (around 6 per cent of the UK workforce) – almost more than in any other – and it is constantly expanding and developing. There are more choices and opportunities than ever before. Your career doesn't have to end in the UK either – what about taking the skills and experience you are developing abroad? Having these automotive skills will give you a career you can take with you wherever you go. There's always going to be a car that needs your attention.

Qualifications for the automotive industry

There are many ways of entering the automotive industry, but the most common method is as an apprentice.

Apprenticeships

You can become an apprentice by being employed, usually be working for a garage or a vehicle manufacturer.

The Institute of the Motor Industry (IMI) operates the sector skills council which is responsible for setting the standards of automotive training in the UK.

The framework of an apprenticeship is based around an NVQ (or SVQ in Scotland). These qualifications are developed and approved by industry experts and will measure your practical skills and job knowledge on-site.

You will also need to achieve:

- a technical certificate
- the appropriate level of Functional Skills assessment
- an Employees Rights and Responsibilities briefing.

Diploma

The Level 2 Diploma in Light Vehicle Maintenance and Repair Principles VRQ (Vocational Related Qualification) provides the knowledge requirements for its related VCQ (Vocational Competence Qualification) and forms the knowledge component of the IMI SSC Maintenance and Repair Apprenticeship framework (for Light Vehicle). The assessment of this VRQ is made up of two components:

- practical tasks
- online testing.

The Level 2 Diploma meets the requirements of the new Qualifications and Credit Framework (QCF) which bases a qualification on the number of credits.

In order to pass the qualification, you must achieve a total of 78 credits derived from mandatory and optional units.

As part of the Level 2 Diploma you will gain the skills needed for the NVQ as well as the Functional Skills knowledge you will need to complete your qualification.

National Vocational Qualifications (NVQs)

NVQs are available to anyone, with no restrictions on age, length or type of training. There are different levels of NVQ (for example 1, 2, 3), which in turn are broken down into units of competence. NVQs are not like traditional examinations in which someone sits an exam paper. An NVQ is a 'doing' qualification, which means it lets the industry know that you have the knowledge, skills and ability to actually 'do' something.

NVQs are made up of both mandatory and optional units and the number of units that you need to complete for an NVQ depends on the level and vehicle area.

Authors

Graham Stoakes MIMI is a lecturer in automotive engineering for light vehicles and motorcycles at a large college of further education. With his background as a qualified master technician, senior automotive manager and specialist diagnostic trainer, he brings 28 years of technical industry experience to this title.

Eric Sykes B.Ed, AAE, MIMI, MSOE, MIFL, Eng.Tech – Automotive Quality Manager for Babcock International Group – has 28 years of experience in the motor industry from delivery of training through to management roles both nationally and internationally and has brought his wealth of technical knowledge and experience to this title.

Catherine Whittaker BA (Hons) – Babcock International Group – used her 12 years experience in learning resource development, specifically in the area of vocational training, to assist in the development of the title.

About the Institute of the Motor Industry (IMI)

The IMI is here to help you progress throughout your career and give you the recognition you deserve.

If you're looking to start a career in the sector, **1st Gear** is here to help you do just that. With free advice from industry experts, 1st Gear aims to help point you in the right direction to get started. To join for free, visit www.1stgear.org.uk. For those of you just starting your career in the sector, **Accelerate** is the IMI's fast paced online community for students in training who are at the beginning of their career journey into the automotive sector. It's where, from the start of your career, you can get help and information to assist you in your formal training. Accelerate is here to plug you into up to date information and guidance. It hooks you up with fellow industry apprentices and trainees. It provides the support that will help you achieve your goals and make your career a success.

Not only does Accelerate help your career, it also gives your toolbox and social life a real boost. To take a look at the benefits and join, visit www.motor.org.uk/accelerate.

For more information on the work of the IMI, visit: www.motor.org.uk.

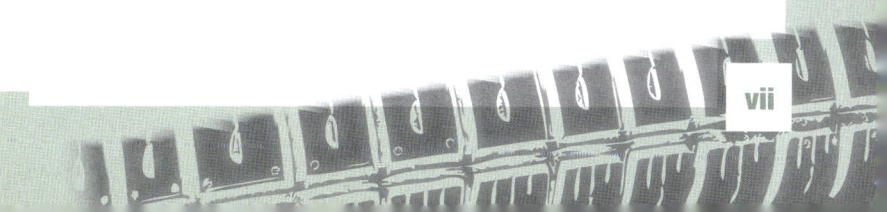

Qualification mapping grid

Chapter	QCF unit reference	IMIAL	City & Guilds
1 General workshop operation principles for light vehicle	D/601/6171 Y/601/7254 K/601/6237 Y/601/6279 T/601/6175 J/601/6262	G0102K and S G3 K and S G4 K and S (These are the mandatory units)	Unit 001 Unit 003 Unit 004 Unit 051 Unit 053 Unit 054
2 Light vehicle chassis units & components	A/601/3732 F/601/3876 D/601/3738 H/601/3885	LV04 K and S and some of LV11.3 (Knowledge of overhauling light vehicle steering and suspension units)	Unit 104 Unit 131 Unit 154 Unit 181
3 Light vehicle engine mechanical, lubrication & cooling system units & components	R/601/3719 K/601/3872	LV02.1 K LV02S	Unit 152 Unit 102
4 Light vehicle fuel, ignition, air & exhaust system units & components	H/601/3725 K/601/3872	LV02.2 K LV02S (Skills in removing and replacing light vehicle engine units and components)	Unit 172 Unit 102
5 Light vehicle transmission & driveline units & components	Y/601/3740 K/601/3886	LV12 K and S	Unit 112 Unit 162
6 Light vehicle electrical units & components	T/601/3731 T/601/3874	LV03 K and S	Unit 103 Unit 153
7 Light vehicle inspection & servicing	F/601/3716 H/601/3871 H/601/3742 A/601/3889	LV01 K and S LV0506 K and S (Knowledge of inspecting light vehicles using prescribed methods)	Unit 101 Unit 105 Unit 151 Unit 155

Introduction

Features of this book

This book has been fully illustrated with artworks and photographs. These will help to give you more information about a concept or a procedure, as well as helping you to follow a step by step technical skill procedure or identify a particular tool or material.

This book also contains a number of different features to help your learning and development.

Working practice pages

These pages pick up the key health and safety areas you need to be aware of as you work on particular light vehicle systems.

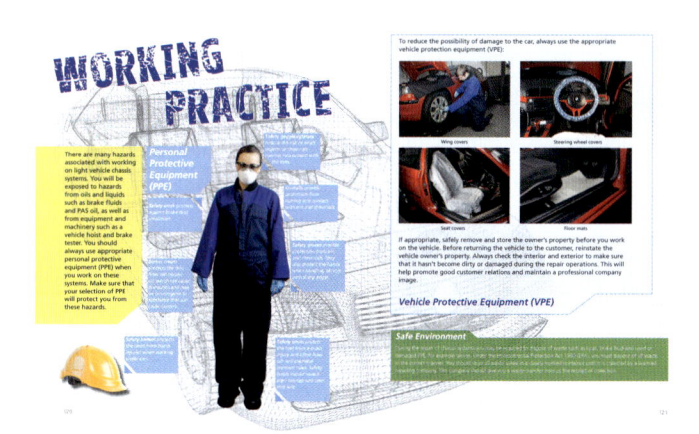

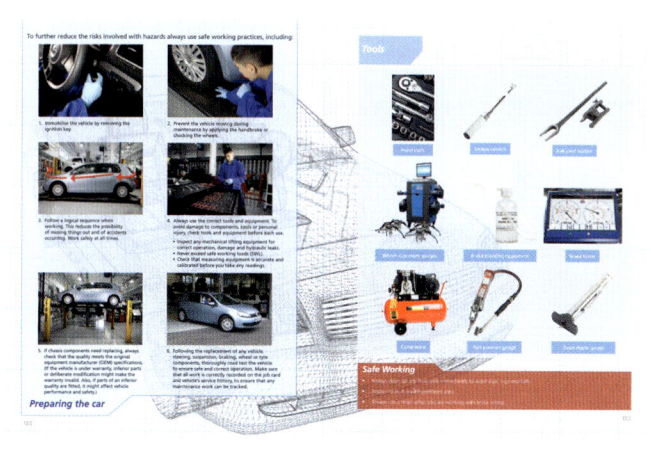

Technical skills

Throughout the book you will find step by step procedures to help you practice the technical skills you need to complete be successful in your studies (see the example to the right).

Safety tips

These four features give you guidance for working safely on cars and in the workshop.

Emergency
Green safety tips provide useful information about SAFE conditions in emergency situations and your personal safety.

Safe working
Red safety tips indicate a PROHIBITION (something you **must not** do).

Safe working
Blue safety tips indicate a MANDATORY instruction (something that you must do).

Safe working
Yellow safety tips indicate a WARNING (hazard or danger).

Level 2 Light Vehicle Maintenance & Repair

Other features

Key term
These are new or difficult words. They are picked out in **bold** in the text and then defined in the margin.

Did you know?
This feature gives you interesting facts about the automotive industry.

Find out
These are short activities and research opportunities, designed to help you gain further information about, and understanding of, a topic area.

Working life
This feature gives you a chance to read about and debate a real life work scenario or problem. Why has the situation occurred? What would you do?

CHECK YOUR PROGRESS

These are a few questions at the end of each section, usually related to a learning outcome to see how you are getting along.

FINAL CHECK

This is a series of multiple choice questions at the end of each chapter, in the style of the end of unit tests.

GETTING READY FOR ASSESSMENT

This feature provides guidance for preparing for the practical assessment. It will give you advice on using the theory you have learnt about in a practical way.

1 General workshop operation principles for light vehicles

Health and safety is a vital part of vehicle maintenance. This chapter covers the knowledge and skills you need to work safely. You will learn about the correct personal and vehicle protection you need to use and the importance of keeping the workplace clean. You will learn about the legislation covering working practices and the hazards and risks in your working environment.

One of the key skills in all workplaces is the ability to share knowledge and communicate effectively. Communicating clearly, both face to face and in writing, will enable you to work effectively as part of a team.

In this chapter you will find out about all the different tools and measuring devices you will be using in your work on light vehicles. The final section in this chapter covers the materials used in vehicle manufacture and maintenance and how to work with these materials. This information will help you to apply engineering, fabrication and fitting principles when modifying and repairing vehicles and components.

This chapter covers:

1. Health and safety
   - Health and safety legislation
   - Appropriate personal and vehicle protective equipment
   - Hazards and risks
   - Accident prevention
   - Good housekeeping practices

2. Job roles and communication
   - Organisational structures, functions and roles
   - Communicating with colleagues and customers

3. Materials and tools
   - Hand tools and measuring devices
   - Materials used in fabricating, modifying and repairing vehicles and fitting components

1. Health and safety

Health and safety legislation

This section is about the way you can contribute to making your workplace a safe, secure and healthy place for the people who use it. It will look at the laws that govern working practices and what they mean in reality.

Your workshop will have in place health and safety policies that have been developed to make sure that government **legislation** is observed. The legal side of things may seem like nothing to do with you and something that your employer will deal with. It is important that you are aware of the legislation and your rights and responsibilities, as well as those of your employer. It is your right to expect your employer to fulfil their responsibilities and it is your employer's right to expect you to fulfil yours.

Legislation is the law and, if you do not observe it, you are committing an offence (breaking the law).

The laws that are relevant to the automotive industry are covered below. The sections that follow will show you how these laws and regulations relate to your everyday working practice and to that of your colleagues. There will be cross-references throughout the chapter so you can access the practical information you need on safe working and the relevant laws and regulations.

The Health and Safety at Work Act 1974 (HASAWA)

The health and safety of those in your workplace is protected by the Health and Safety at Work Act. This law protects you, your employer and all employees while at work. It also protects your customers and the general public when visiting your workplace. Table 1.1 lists what you and your employer are responsible for under this law.

> **Did you know?**
>
> Your employer can be prosecuted under the terms of the Health and Safety at Work Act 1974 if you do not use the PPE provided when operating machinery.
>
> It is your duty to wear the PPE supplied by your employer when you operate machinery.

Table 1.1 Who's responsible for what under the Health and Safety at Work Act 1974?

You are responsible for ...	Your employer is responsible for ...
<ul><li>taking care that you do not endanger yourself or others who may be affected by your work</li><li>following the health and safety polices and procedures at your workplace</li><li>not damaging any machinery or PPE provided for your safety</li><li>taking care of machine guards and reporting any problems.</li></ul>	<ul><li>providing you with safe equipment and safe ways of carrying out work tasks</li><li>making sure that the equipment you use when handling, storing and transporting materials and substances is safe and without health risks</li><li>providing information, instruction, training and supervision to guarantee health and safety</li><li>maintaining the workplace, including means of entry and exit, in a safe condition</li><li>providing a safe and healthy working environment with adequate facilities and arrangements for employees' welfare</li><li>providing for employees a written statement of policy, organisation and arrangements for health and safety (employers with less than five employees are exempt from this).</li></ul>

Who checks that health and safety laws are enforced?

There is a government body called the Health and Safety Executive (HSE) which enforces the Health and Safety at Work Act. HSE inspectors have powers to issue **improvement notices** and **prohibition notices** if they believe that there are any poor health and safety practices in a workplace they are inspecting.

> **Working life**
>
> Ted has been servicing vehicles in a rented garage for many months and has been disposing of oil down the drain.
>
> 1 What would happen if a Health and Safety Executive inspector visited Ted's garage?

Consequences of non-compliance

If an employer does not comply with an improvement or prohibition notice, they can be fined or even imprisoned.

If employers do not properly look after the safety of their employees, they can be prosecuted. This could result in:

- charges of corporate manslaughter
- personal fines
- corporate fines
- imprisonment
- loss of production
- loss of income
- bad publicity in the media
- being served with a HSE prohibition notice.

Health and safety regulations

There are six sets of regulations which come under the Health and Safety at Work Act 1974 (HASAWA). Each of these covers specific areas of health and safety. They are:

1. Provision and Use of Work Equipment Regulations 1998 (PUWER)
2. Health and Safety (Display Screen Equipment) Regulations 1992
3. Personal Protective Equipment at Work Regulations 1992
4. The Management of Health and Safety at Work Regulations 1999
5. Workplace (Health, Safety and Welfare) Regulations 1992
6. Manual Handling Operations Regulations 1992

1. Provision and Use of Work Equipment Regulations 1998 (PUWER)

The equipment used in your workshop needs to be:

- safe to use
- maintained correctly
- inspected regularly
- only used by people who have received appropriate training.

> **Key terms**
>
> **Legislation** – law that has been passed by a government.
>
> **Improvement notice** – notification that the employer must eliminate a risk, for example a bad working practice such as draining petrol into open tanks. The improvement notice gives the employer a specific period of time to eliminate the risk.
>
> **Prohibition notice** – notification that the employer has to immediately stop all work until the safety risk is eliminated. A prohibition notice is only issued for serious safety risks, such as a damaged building that may collapse.

> **Did you know?**
>
> The six regulations which are contained in the HASAWA 1974 are often referred to as the 'Six Pack' regulations.

If these rules are not followed, you, your colleagues or your customers could be hurt or injured.

The Provision and Use of Work Equipment Regulations 1998 (PUWER) places the responsibility for the safety of workplace equipment on anyone who has control over the use of work equipment, including your employer, you and your colleagues.

Warning labels

On the equipment and machinery that you use at work, you will find a warning label informing you of the dangers it can pose. The labels might be:

- Warnings – make sure you follow the instructions to reduce the risk of damage to the machine and operator.
- Restrictions – follow these instructions so that you do not go into prohibited areas.
- Protective devices – ensure all guards are in place before use.

See the section on *Safety signs* on pages 58–60 for examples of the safety signs in use.

Part IV of the PUWER regulations contains specific requirements regarding power presses. These are:

- The press must be examined periodically so it is safe for use.
- All guards must be used.
- Only persons authorised to use the press can do so.
- All authorised persons must be listed in the employer's records.

> **Did you know?**
> Although employees do not have duties under PUWER, they do have general duties under the HASAWA and the Management of Health and Safety at Work Regulations 1999. These duties include taking reasonable care of themselves and others who may be affected by their actions, and co-operating with others.

> **Did you know?**
> The PUWER regulations do not apply to equipment used by the public, for example compressed-air equipment used in a garage forecourt. However, such circumstances are covered by the Health and Safety at Work Act 1974 (HASAWA).

Table 1.2 Your employer's responsibilities under the Provision and Use of Work Equipment Regulations 1998 (PUWER)

Your employer is responsible for ...
Making sure the equipment used is: • suitable for the intended use • safe for use, maintained in a safe condition and, in certain circumstances, inspected to ensure this remains the case • used only by people who have received adequate information, instruction and training • accompanied by suitable safety measures, e.g. protective devices, markings and warnings.

2. Health and Safety (Display Screen Equipment) Regulations 1992

There will be occasions when you need to use a computer or display screen, for example when searching parts catalogues or updating customer details.

In line with the Health and Safety (Display Screen Equipment) Regulations 1992, your employer or manager must take steps to reduce possible risks related to using a computer or display screen. Health risks include:

- eye fatigue
- mental stress
- muscle problems – in particular repetitive strain injury (RSI).

If you use a computer or display screen for a significant part of your normal working day, then your employer will complete risk assessments, make provision for eye tests and arrange instruction and training. Your employer must also make sure that:

- workstations provide sufficient space for you to change position and vary your movement
- lighting, humidity and heat produced by the workstation do not cause you any discomfort
- the noise from the workstation does not distract attention or disturb speech
- eye fatigue is minimal and that regular breaks are taken
- the chair that you use allows you freedom of movement and has adjustable seat height and back
- the software you use is suited to the tasks and adapted according to your knowledge.

Table 1.3 lists what you and your employer are responsible for under these regulations.

Table 1.3 Who's responsible for what under the Health and Safety (Display Screen Equipment) Regulations 1992?

You are responsible for ...	Your employer is responsible for ...
• taking regular breaks • learning how to use the systems to minimise injury through incorrect posture • making your employer aware of any problems with the equipment or software.	• scheduling sufficient breaks • providing effective training • regularly checking the equipment and updating the software.

Did you know?

Display screens are also referred to as visual display units (VDUs).

3. Personal Protective Equipment at Work Regulations 1992

Personal protective equipment (PPE) is equipment that you wear or hold at work, such as eye protectors and overalls to protect you against risks to health and safety. (See pages 17–21 for more detailed information on PPE.) PPE is covered by the Personal Protective Equipment at Work Regulations 1992.

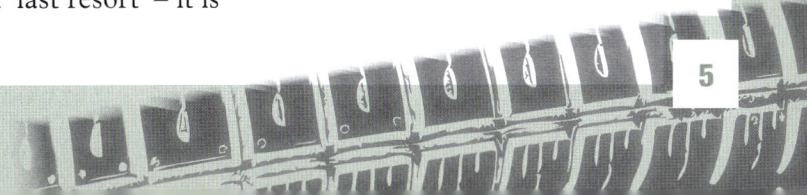

You must always wear the correct PPE for the task, since by wearing the correct PPE you are protecting yourself from injury. However, you should also be aware that PPE is considered to be a 'last resort' – it is just as important to work safely with the tools and equipment that are appropriate to carry out the job.

Table 1.4 lists what you and your employer are responsible for under these regulations.

Table 1.4 Who's responsible for what under the Personal Protective Equipment at Work Regulations 1992?

You are responsible for …	Your employer is responsible for …
• wearing the correct PPE for the task • keeping the PPE in good condition and clean • reporting any defects in PPE.	• supplying the PPE and providing training • providing adequate storage for the PPE • replacing any PPE that is damaged through normal wear and tear.

4. The Management of Health and Safety at Work Regulations 1999

These regulations apply mainly to employers. Your employer's main requirement under these regulations is to carry out **risk assessments** on tasks and duties within the workplace. If your employer employs five or more employees, they are required to record the findings of their risk assessments. (See the section on risk assessment which starts on page 55 for more detailed information.)

Table 1.5 lists what you and your employer are responsible for under these regulations.

Table 1.5 Who's responsible for what under the Management of Health and Safety at Work Regulations 1999?

You are responsible for…	Your employer is responsible for …
• as an appointed competent person, understanding the limits of your own knowledge and skills in your job role and working within these • following all necessary measures put in place to minimise risk • using the skills and knowledge you have gained from the training you have received.	• carrying out risk assessments • putting in place any necessary measures to minimise the risks • appointing people who are competent in their job role • arranging for suitable training to enable you to do your job safely and efficiently.

5. Workplace (Health, Safety and Welfare) Regulations 1992

The Workplace (Health, Safety and Welfare) Regulations 1992 were introduced to protect the health, safety and welfare of the workforce (you and your colleagues) when at work. This includes people with limited mobility, or with impaired sight or hearing.

> **Key term**
>
> **Risk assessment** – this is done to identify potential health and safety risks to staff and other people (e.g. visitors) in the workplace. Once the risk assessment has been done, it is the employer's responsibility to minimise the risks that they have identified.

> **Did you know?**
>
> A risk assessment should be straightforward and simple. It should only be complicated if it deals with serious hazards such as those at a nuclear power station, chemical plant, laboratory or on an oil rig.

These regulations make sure that your employer provides adequate welfare facilities for the staff. These include toilets, rest areas, washing facilities, changing areas, drinking water and eating facilities.

Table 1.6 lists what your employer's responsibilities are under these regulations.

Table 1.6 Your employer's responsibilities under the Workplace (Health, Safety and Welfare) Regulations 1992

Your employer is responsible for …
• providing adequate welfare, changing and rest room facilities, including separate facilities for men and women, running water, soap and towels • providing a place to sit and take meals in comfort • keeping the workshop at the correct temperature (the legal minimum indoor temperature is 13°C for people doing strenuous work and 16°C for people sitting at a desk) • ensuring that the workplace is well ventilated with fresh air; windows should be able to be opened safely • ensuring adequate lighting for safe work and access • ensuring a clean environment and adequate facilities for waste disposal.

6. Manual Handling Operations Regulations 1992

At some point in your working week you will be required to lift or move an object or load. When you do this, you will need to follow the Manual Handling Operations Regulations 1992.

The regulations cover any time that human effort (with or without mechanical assistance) is used to transport or support a load – including lifting, putting down, pushing, carrying or moving. If the regulations are followed correctly, they will help to prevent injury. Preventable injuries include sprains and strains to the back, lower limbs, hands, arms and fingers, as well as fractures and cuts.

The regulations state that all hazardous manual handling operations should be avoided, as far as is reasonably possible.

Any manual handling operation that cannot be avoided must be assessed and the risk of injury must be reduced.

Table 1.7 lists what you and your employer are responsible for under these regulations.

Table 1.7 Who's responsible for what under the Manual Handling Operations Regulations 1992?

You are responsible for …	Your employer is responsible for …
• attending training and lifting objects as instructed • not exceeding the manual lifting weight limits • using lifting devices where possible rather than lifting manually.	• providing instruction and training • asking employees only to lift weights that they feel comfortable lifting (and never more than the legal maximum) • supplying lifting devices.

Did you know?

The maximum weight for manual lifting is 25 kg for a male and 16 kg for a female. However, the actual amounts will vary between individuals. You must never lift a weight that is not comfortable for you.

(For detailed information on safe manual handling procedures, see the manual handling section on pages 42–45.)

Other health and safety regulations

There are several other health and safety regulations which apply to the automotive workshop, and which you need to know about.

Health and Safety (First-Aid) Regulations 1981

Your employer will do all they can to reduce the risk of accidents occurring in your workshop. However, accidents do occasionally happen. To comply with the Health and Safety (First-Aid) Regulations 1981, your workshop must have enough trained first aiders available to assist in the event of an accident.

The number of trained first aiders and the amount of training they need depends on how many people are employed in the workplace and the level of hazard associated with the working activities. There should be at least one trained first aider available at all times.

(For detailed information on first aid, see page 61.)

Health and Safety (Consultation with Employees) Regulations 1996

Under these regulations, your employer must consult with you on health and safety matters in certain situations. Table 1.8 lists the situations when your employer must consult with you or your appointed representative.

Table 1.8 When your employer needs to consult with you under the Health and Safety (Consultation with Employees) Regulations 1996

Your employer needs to consult with you ...
When they introduce any measure which may substantially affect your health and safety at work, for example the introduction of new equipment or new systems of work.
When arranging to get **competent people** to assist with complying with safety laws.
To give you information on: • any risks and dangers arising from your work • measures to reduce or get rid of these risks • what you should do if you are exposed to risk.
About planning and organising health and safety training.
About the health and safety consequences of introducing new technology.

Consultation involves employers not only giving information to you but also listening to you and taking account of what you say before making any health and safety decisions.

Health and Safety (Safety Signs and Signals) Regulations 1996

These regulations require that whenever there is a risk that has not been avoided or controlled by other means, your employer must display a safety sign to indicate the type of hazard.

There are many occasions where a safety sign will need to be displayed in your workshop, for example where floors are being cleaned or electrical maintenance is taking place. Hand signals may be used to manoeuvre vehicles. Acoustic signals, such as a siren, bell or fire alarm, might also be used as a way to communicate hazards and danger, such as extreme noise or fire in your workshop.

For more on safety signs, see pages 58–60.

Reporting of Injuries, Diseases and Dangerous Occurrences Regulations 1995 (RIDDOR)

Your employer must report any serious work-related health and safety incidents or accidents in your workplace to the Health and Safety Executive (HSE) enforcement authorities. A serious incident or accident is one which results in an injury or disease lasting three days or more, or in death.

- Cases of injuries lasting more than three days must be reported within ten days of the incident occurring.
- Cases of death, major injury or **dangerous occurrences** must be reported to the enforcing authority without delay. Your employer must keep records of these types of incidents for at least three years.

Table 1.9 lists what you and your employer are responsible for under the RIDDOR regulations.

Table 1.9 Who's responsible for what under the Reporting of Injuries, Diseases and Dangerous Occurrences Regulations 1995 (RIDDOR)?

You are responsible for ...	Your employer is responsible for ...
• making your employer aware of accidents, including near misses, and all incidents of ill health that have occurred • reporting all dangerous occurrences and accidents in the accident book provided by your employer.	• reporting serious accidents and dangerous occurrences to the HSE • keeping records of accidents and dangerous occurrences • investigating the cause of any accidents and dangerous occurrences and taking corrective action.

Environmental Protection Act 1990 (EPA)

Your daily tasks in the workshop, such as servicing, MOTs and replacing parts, could all potentially lead to polluting the environment. Obvious pollutants include hazardous substances, chemicals or materials. Other not so obvious pollutants include the emissions from exhaust fumes, vapours from welding and air-conditioning refrigerant leaks into the atmosphere.

Key terms

Competent person – someone who has sufficient training and experience or knowledge.

Dangerous occurrence – any incident that has a high potential to cause serious injury or death. For example, if a vehicle dropped off a jack in your workplace but your workmate managed to act quickly to avoid any serious injury occurring, there would be no injury to be reported. However, this near miss would be classed as a dangerous occurrence and it must be reported.

Find out

Ask your employer or have a look at the accident book in your workplace to see if there have been any injuries in the last 12 months and, if so, what they were. This will help to give you more understanding of the types of accidents which can happen in your workplace.

Damage to the environment may be caused by contaminating the atmosphere, water supply or drainage system. Under the Environmental Protection Act 1990 (EPA), your employer will have procedures in place for working with and disposing of any material which has potential to harm the environment.

Table 1.10 lists what you and your employer are responsible for under the Environmental Protection Act.

Table 1.10 Who's responsible for what under the Environmental Protection Act 1990 (EPA)?

You are responsible for …	Your employer is responsible for …
• following the procedures for disposing of waste; for example, don't tip waste oil or antifreeze down a drain or throw a used oil filter into a standard waste bin • informing your employer if waste oil drums are getting full.	• the correct disposal of any material which may pollute the environment • having procedures in place for the disposal of controlled waste • recording the disposal of any controlled waste.

Control of Substances Hazardous to Health Regulations 2002 (COSHH)

Not all the substances used within your workshop have the potential to be hazardous to health, but a large number of them do.

The legislation which you and your employer must observe when using hazardous substances in the workshop is the Control of Substances Hazardous to Health Regulations 2002 (COSHH).

Some types of hazardous substances are shown in Table 1.11.

Table 1.11 Types of hazardous substances and examples of each

Type of hazardous substance	Example
Natural/artificial	Brake fluids
Liquid	Used engine oil
Solid	Brake linings
Gas	Exhaust fumes
Vapour	Petrol fumes
Dust	Body filler

The degree of hazard for each substance is indicated by the hazard symbol on the label. For more information on COSHH symbols, see pages 28 and 29.

Table 1.12 lists what you and your employer are responsible for under the Control of Substances Hazardous to Health Regulations 2002 (COSHH).

> **Safe working**
>
> Never use any hazardous substance unless you have received COSHH training. Always follow the procedures closely on how to use a hazardous substance.

Table 1.12 Who's responsible for what under the Control of Substances Hazardous to Health Regulations 2002 (COSHH)?

You are responsible for …	Your employer is responsible for …
• following the procedures put in place • paying attention to training given • wearing the correct PPE.	• completing risk assessments of all hazardous chemicals and substances used by employees • providing training for employees • putting steps in place to prevent exposure to hazardous chemicals and substances.

There are eight steps that employers must take to protect employees from hazardous substances. These are:

Step 1: Find out what hazardous substances are used in the workplace and the risks these substances pose to people's health.

Step 2: Decide what precautions are needed before any work starts with hazardous substances.

Step 3: Prevent people being exposed to hazardous substances or, where this is not reasonably practicable, control the exposure.

Step 4: Make sure control measures are used and maintained properly and that safety procedures are followed.

Step 5: If required, monitor exposure of employees to hazardous substances.

Step 6: Carry out health surveillance where assessment has shown that this is necessary or where COSHH makes specific requirements.

Step 7: If required, prepare plans and procedures to deal with accidents, incidents and emergencies.

Step 8: Make sure employees are properly informed, trained and supervised.

For more information on the COSHH regulations, including information about different types of hazardous substances, risk assessments and disposing of hazardous wastes, see pages 23–32.

The Safety in Industry (Abrasive Wheels) Regulations 1982

In your workshop you may be asked to use an **abrasive wheel** to cut down a sharp metal component. The speed of the abrasive wheel will have been set to comply with the Safety in Industry (Abrasive Wheels) Regulations 1982. The speed will be marked on the abrasive wheel by the manufacturer.

Table 1.13 gives details of who is responsible for what under the Safety in Industry (Abrasive Wheels) Regulations 1982.

> **Did you know?**
> A chemical is a single fluid, gas or solid in its natural state.
> A substance is a mixture of chemicals.

> **Key term**
> **Abrasive wheel** – a power-driven wheel used for any grinding or cutting operation.

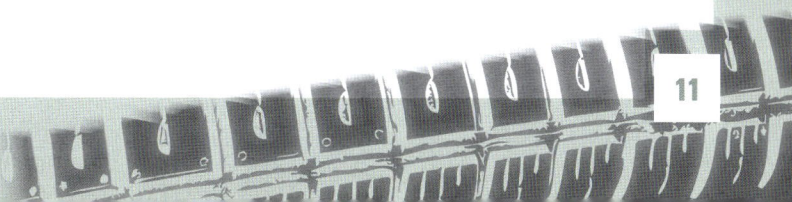

Level 2 Light Vehicle Maintenance & Repair

Table 1.13 Who's responsible for what under the Safety in Industry (Abrasive Wheels) Regulations 1982?

You are responsible for ...	Your employer is responsible for ...	The manufacturer is responsible for ...
• using the wheel correctly in line with training received • wearing appropriate PPE • reporting any damage to the machinery.	• displaying the rpm if the wheel is too small to do so • providing training to employees on how to use the wheel • providing the correct PPE.	• marking the maximum permissible speed in rpm (revolutions per minute) on the wheel.

Safe working

- When using a grinding machine, make sure that all guards are used, and that the work rest is as close to the wheel as possible (maximum 3 mm) without it touching.
- Always stand to the side when starting the grinder in case the wheel comes off under the intense acceleration.

Figure 1.1 Always stand to the side when starting the grinder in case the wheel comes off under the intense acceleration

Key term

Bar – a measure of pressure. 0.5 bar is equivalent to about 7 psi (pounds per square inch).

Pressure Systems and Transportable Gas Containers Regulations 1989

This legislation covers the safe manufacture, use, maintenance and repair of equipment and systems which use steam, gas or other fluids under pressure greater than 0.5 **bar** above atmospheric pressure – for example, air conditioning refrigerant.

Manufacturers and suppliers of pressure systems must follow safety guidelines and provide details of how the systems should be refilled and maintained. Table 1.14 lists what you and your employer are responsible for under these regulations.

Table 1.14 Who's responsible for what under the Pressure Systems and Transportable Gas Containers Regulations 1989?

You are responsible for ...	Your employer is responsible for ...
• following any training given to use the systems correctly.	• ensuring that the systems and equipment are in good working order • providing training in how to use the systems • ensuring that the systems are used correctly.

Electricity at Work Regulations 1989

As electricity forms the main energy supply for all we do, there needs to be specific legislation to cover safe practices in its installation, use and maintenance. This legislation is the Electricity at Work Regulations 1989, which cover any work involving electricity or electrical equipment.

Table 1.15 lists what you and your employer are responsible for under these regulations.

Table 1.15 Who's responsible for what under the Electricity at Work Regulations 1989?

You are responsible for …	Your employer is responsible for …
• reporting any faults in electrical equipment to your employer • following any procedures and safety measures you are given.	• keeping electrical systems safe and regularly maintained • putting in place procedures to reduce the risks of coming into contact with live electrical current. This applies to employees and people who are visiting the workplace.

Noise and Vibrations at Work Regulations 1989

Did you know that noise is generated by vibrations? There are different degrees of noise within a workshop depending on the type of work being carried out, for example the constant hum from a compressor when running.

The Noise and Vibrations at Work Regulations 1989 are in place to ensure that noise and the length of time you are exposed to noise are not at a level that will cause hearing damage or loss. Your employer will do a noise risk assessment and use the results to take action to protect you.

Table 1.16 lists what you and your employer are responsible for under these regulations.

Table 1.16 Who's responsible for what under the Noise and Vibrations at Work Regulations 1989?

You are responsible for …	Your employer is responsible for …
• wearing the ear protection supplied if you hear noise in the workshop • ensuring that all potentially noisy activities are necessary and cannot be done in another way • checking if there is any other way of carrying out the activity.	• carrying out a risk assessment on the level of noise and the length of time you are exposed to it • supplying adequate ear protection • taking action to reduce exposure to noise.

Working at Height Regulations 2005

For most of your working day, you will be working at ground level, but occasionally you may be required to work at height, for example when repairing the windscreen of a lorry cab or the sunroof of a 4 × 4. The risk associated with this kind of activity is falling from height.

The Working at Height Regulations 2005 were introduced to ensure that precautions are taken to reduce the risk of falling from height to the lowest possible level. These regulations have been updated many times since 2005, the latest being 2010.

Table 1.17 lists what you and your employer are responsible for under the Working at Height Regulations 2005.

Table 1.17 Who's responsible for what under the Working at Height Regulations 2005?

You are responsible for ...	Your employer is responsible for ...
• following training given for working at height • using the safety equipment provided • reporting any hazards to your employer.	• avoiding work at height where possible • providing training for working at height • providing equipment or safeguards to help prevent falls (for example, fencing should be erected around an inspection pit to prevent people falling in) • providing equipment and any other methods that will minimise a fall, the distance of any possible fall and the consequences of a fall (for example, a safety net).

Working life

Kuldeep has been asked to collect a part from the stores for a customer. The part is on the top shelf and is very heavy. The stepladder is kept at the back of the workshop but he can just reach the part if he stands on a swivel chair.

1. What could the consequences be if Kuldeep uses the chair rather than the ladder to reach the part?

Lifting Operations and Lifting Equipment Regulations 1998 (LOLER)

Different types of lifting equipment will be used in your workshop (these are described on pages 46–47). The Lifting Operations and Lifting Equipment Regulations 1998 (LOLER) require that any lifting equipment used at work for lifting or lowering loads is:

- strong and stable enough for its particular use

- marked to indicate its **safe working load (SWL)**
- positioned and installed to minimise any risks
- used safely – the work is planned, organised and performed by competent people
- regularly given a thorough examination and, where appropriate, a proper inspection by competent people.

The regulations cover:

- lifting equipment – any equipment used at work for lifting or lowering loads, for example an engine hoist
- any attachments used for anchoring, fixing or supporting the lifting equipment, for example D shackles
- lifting accessories such as chains, slings and eyebolts.

Table 1.18 lists what you and your employer are responsible for under the Lifting Operations and Lifting Equipment Regulations 1998 (LOLER).

> **Key term**
>
> **Safe working load (SWL)** – the maximum load which a lifting device, such as a crane or lifting arrangement, can safely lift, suspend or lower.

Table 1.18 Who's responsible for what under the Lifting Operations and Lifting Equipment Regulations 1998 (LOLER)?

You are responsible for …	Your employer is responsible for …
• carrying out any lifting work in line with set procedures • taking reasonable care of yourself and others who may be affected by your actions • co-operating with others.	• planning all lifting work properly • making sure that work is carried out by competent, trained people • putting safety measures in place • making sure that regular safety checks of equipment take place • assessing and minimising risks.

Other legislation to be aware of

- **Hazardous Waste (England and Wales) 2005** – This states that unsafe waste materials, such as explosive, chemical or toxic material, should be disposed of by your employer with special care.
- **Employers' Liability (Compulsory Insurance) Act 1969 and Regulations** – This requires employers to take out insurance in case of any work-related accidents or ill health involving employees or visitors to the premises.
- **Environment Management (Waste Disposal and Recycling) Regulations 2007** – Under this legislation, employers must ensure that the waste produced by the business is managed properly. They must control the safe retrieval and disposal of waste to prevent harm to human health or pollution of the environment. This includes ensuring that waste is only transferred to someone who is authorised to receive it.

- **Dangerous Substances and Explosive Atmospheres Regulations 2002 (DSEAR)** – This law requires the risks to safety from fire and explosions to be controlled by your employer. Your employer must take all necessary precautions against the risk of explosion from any potentially explosive materials found or used in the workplace.
- **Control of Asbestos at Work Regulations 2002** – Asbestos was a common material for brake and clutch friction linings in the past. However, manufacturers have now removed the asbestos content from these components. These regulations control the use, exposure to and disposal of all asbestos-based materials.
- **Confined Spaces Regulations 1997** – This law states that risk assessments must be done on areas where work needs to be carried out in confined spaces. Safety precautions, such as breathing apparatus and emergency arrangements, must also be in place if working in confined conditions is absolutely necessary.

> **Safe working**
>
> Brake lining manufacturers claim that asbestos is no longer used in the motor industry, but small amounts may still be found in friction linings of aftermarket suppliers. For this reason you must always wear a dust mask when working on brakes and clutches.

> **Working life**
>
> It is a Friday afternoon and the workshop is very busy. Emma has been asked to respray a wing mirror. The paint booth in the body shop is currently being used. She takes the portable compressor and the mirror to the back of the body shop where there are no windows. She takes only what she feels is essential: the portable compressor, spray gun and paint.
>
> 1. What could be the consequences of Emma's actions?
> 2. Is she correct to spray the wing mirror in this location?
> 3. Does she have all the essential equipment she needs?

CHECK YOUR PROGRESS

1. Name two of your responsibilities under the Health and Safety at Work Act 1974.
2. What is COSHH and what does it relate to?
3. Name three workshop activities that could potentially pollute your environment.
4. What is PPE and when should you use it?
5. What does PUWER refer to? Name one employer responsibility under this regulation.
6. What is a prohibition notice and who issues it?

1 General workshop operation principles for light vehicles

Appropriate personal and vehicle protective equipment

Personal protective equipment

Personal protective equipment (PPE) is the name for the clothes and other items that you wear in order to protect you against accidents or injury while you are carrying out your work. PPE is not the only way of preventing accidents or injury. It should be used together with all the other methods of staying healthy and safe in the workplace. These are:

- vehicle protective equipment (VPE)
- health and safety training
- following health and safety regulations and laws.

There are many dangers in the workshop, most of which can be prevented by using PPE.

> **Did you know?**
>
> 'There have been over 8000 injuries and 24 deaths in the motor vehicle repair (MVR) industry over the last 5 years.'
>
> Source: Health and Safety Executive website, 2010

> **Find out**
>
> Look at your hands. Do they feel rough and sore? Dermatitis is a skin condition caused by oil and grease getting into your skin. The oil and grease removes the natural moisturising oils from your hands.
>
> - What could you wear to prevent dermatitis? (You will find your answer in Table 1.19.)

What PPE to use and when to use it

You may be wearing some PPE at the moment. Understanding what to wear, when to wear it and how to care for this equipment is not an optional extra that you fit in if you have time. It is essential to keeping you and your colleagues safe.

Table 1.19 tells you what part of your body needs to be protected in a range of work situations, and what PPE to use. You will need to learn the PPE required for all of these situations.

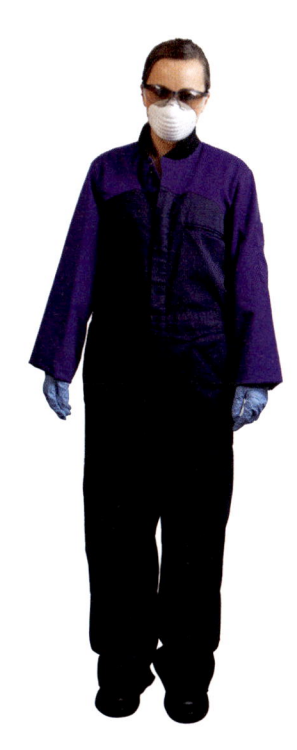

Figure 1.2 Personal protective equipment needed to work in an automotive environment

Level 2 Light Vehicle Maintenance & Repair

Table 1.19 The correct PPE to wear in different work situations

Activity	Head	Eye	Ears	Hands	Lungs	Body	Feet
Grinding/sanding	None	Full face mask	Defenders	Kevlar® gloves	Dust mask	Overalls	Steel toe-capped boots
Charging a battery	None	Safety goggles	None	Chemical-resistant gloves	Dust mask	Apron and overalls	Steel toe-capped boots
Brake lining replacement	None	Safety goggles	None	Latex gloves	Dust mask	Overalls	Steel toe-capped boots
Working under a ramp	Safety helmet	Safety glasses	None	Latex gloves	None	Overalls	Steel toe-capped boots
Welding	None	Welding mask	None	Heat-resistant gloves	None	Heat-resistant apron and overalls	Steel toe-capped boots
Cleaning parts	None	Safety goggles	None	Chemical-resistant gloves	Dust mask	Apron and overalls	Steel toe-capped boots

Parts of the body to protect and PPE to use

18

1 General workshop operation principles for light vehicles

Activity	Parts of the body to protect and PPE to use							
	Head	Eye	Ears	Hands	Lungs	Body	Feet
Oil draining	None	Safety goggles	None	Latex gloves	None	Overalls	Steel toe-capped boots
Chiselling	None	Safety goggles	Earplugs	Kevlar gloves	None	Overalls	Steel toe-capped boots
Roadside attendance	Safety helmet	None	Earplugs	Latex gloves	None	Overalls, high visibility jacket	Steel toe-capped boots
Removal of rubber seals from fire-damaged vehicles	None	Safety glasses	None	Chemical-resistant gloves	Filter respirator	Chemical-resistant apron and overalls	Steel toe-capped boots
Removal of broken glass	None	Safety goggles	None	Kevlar gloves	None	Overalls	Steel toe-capped boots
Removal of hypodermic needles	None	Safety glasses	None	Kevlar gloves	None	Overalls	Steel toe-capped boots

19

Level 2 Light Vehicle Maintenance & Repair

> **Working life**
>
> Ed was going to remove some fire-damaged rubber seals from a vehicle that had been in an accident. He knew that he should wear gloves. However, Ed was in a rush and didn't put on the gloves.
>
> 1   What PPE should Ed wear to do this task?
> 2   Why is this PPE necessary?

Checking and maintaining PPE and regulations

It is essential that your PPE is kept in good condition. If PPE is damaged, it will provide a lot less protection when you are wearing it.

Keep PPE in good condition by:

- **Cleaning** – for example, wash your overalls weekly or more often if working in dirty conditions.
- **Examining** – check for damage to equipment. For example, check that goggles fit well and that the lenses are clear.
- **Replacing** – if PPE becomes damaged, it must be replaced.

You can carry out simple maintenance of your PPE (for example, cleaning), but complicated repairs must be done only by an experienced person. It is your employer's responsibility to cover the cost of maintaining your PPE.

Your employer must provide suitable storage for your PPE when it is not being used, unless you take the PPE (e.g. footwear or clothing) away from the workplace. So it's your responsibility to use this storage space properly.

Storage may be simple, for example pegs for waterproof clothing or safety helmets. In some cases PPE may need to be kept in a cupboard or storage case.

The storage space should be sufficient to protect the PPE from contamination, loss, damage, damp or sunlight. If there is a risk that PPE may become contaminated during use, a separate storage area should be provided for non-PPE clothing.

Selecting the right PPE for the job

When selecting PPE, make sure that the equipment:

- is the right PPE for the job – ask for advice if you are not sure
- fits correctly – it needs to be adjustable so it fits you properly
- is properly looked after
- prevents or controls the risk for the job you are doing
- does not interfere with the job you are doing
- does not create a new risk, such as overheating
- is comfortable enough to wear for the length of time you need it

> **Safe working**
>
> PPE must be maintained regularly.

> **Did you know?**
>
> Your employer cannot ask you for money for your PPE. Your responsibility as an employee is to look after the PPE and use it when required. (See pages 5–6 for a detailed specification of the Personal Protective Equipment at Work Regulations 1992.) If you leave your place of work and you keep your PPE without your employer's permission, they might (depending on your contract) be able to deduct the cost of replacement PPE from your last pay.

- does not impair your sight, communication or movement
- is compatible with other PPE worn, for example your hard hat still fits correctly when worn with eye protection.

The CE mark found on PPE confirms that it has met the safety requirements of the Personal Protective Equipment at Work Regulations 1992. All PPE should have the CE mark (see Figure 1.3).

Risks of not using PPE

If you do not use the correct PPE you are potentially putting yourself at risk of severe and long-term health problems. Table 1.20 lists some examples of the risks of not using PPE.

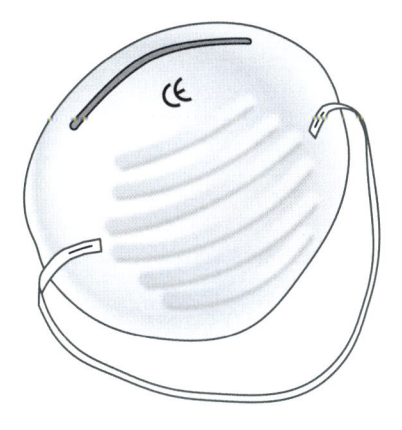

Figure 1.3 The CE mark

Table 1.20 Risks of not wearing appropriate PPE

Work carried out	PPE not worn	Exposure to …	Possibly resulting in …
Sanding and body filler	Mask Goggles	Dust (e.g. body filler)	Eye injuries Breathing problems Cancer
Brake cleaning	Gloves Mask Goggles	Solvents or chemicals	Dermatitis and other skin problems Inhalation of brake dust Cleaning fluid in eyes
Welding	Gloves Apron Helmet Mobile extraction unit with flexible hood and **trunking**	Flames, fumes, gases from primer and paint layers, underseal and lead in car bodies	Dryness of the throat/coughing Tightness of chest and difficulty breathing Long-term effects on the lungs

> **Key term**
>
> **Trunking** – a ventilation pipe used for the extraction of fumes.

> **Find out**
>
> PPE only works properly when it is worn and used correctly.
>
> - Choose two items of PPE from Table 1.19 on pages 18 and 19 and write down how to wear and use them correctly.
> - Swap the information you've found out with others in your group until you know about all the PPE items.
>
> Are you wearing the correct PPE for the tasks you are completing today?
>
> - Have a look at Table 1.19 on pages 18–19 and check.

> **Find out**
>
> Go to hotlinks and visit the Health and Safety Executive (HSE) website to read more about the latest PPE information (enter 'PPE' in the search box).

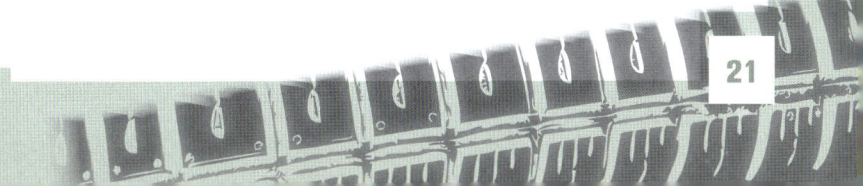

Level 2 Light Vehicle Maintenance & Repair

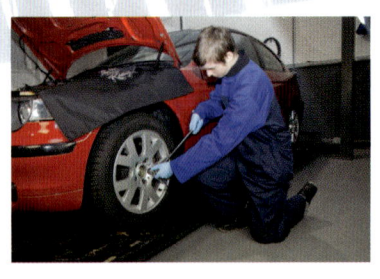

Figure 1.4 Wing covers

Figure 1.5 Steering wheel cover

Figure 1.6 Seat covers

Figure 1.7 Floor mat covers

Vehicle protective equipment

When carrying out any work on a vehicle, you need to make sure you do not damage the exterior bodywork or interior trim. Always use the vehicle protective equipment (VPE) shown in Figures 1.4 to 1.7.

- A wing cover will protect the vehicle body from scratches when you are leaning on the side of the vehicle.
- You should protect the steering wheel from oil and dirt left on hands.
- Use seat covers to prevent oil and dirt being left on the seats from dirty overalls.
- Use floor mat covers to prevent oil and dirt being left on the floor mats from dirty boots.

Meeting customer expectations

To meet the customer's expectations you should carry out vehicle maintenance to a high standard and return the vehicle in a clean condition. To do this you must carry out the following measures:

- Protect the inside of the vehicle with the covers intended for this purpose.
- Safely remove the owner's property and store if appropriate.
- Where possible, close the doors, windows and the boot during the repair process.
- Do not change the position of the seats or the mirrors or alter the radio programming.
- Never smoke in the vehicle.
- If you disconnect the battery, reprogram the systems.
- Clean the vehicle inside and out when you have finished your maintenance work.
- Return the customer's property to the car when you have completed the work.
- Leave the car locked.

Working life

Daniel finishes an oil change and picks up the keys for his next service job. He jumps into the car and drives it on to the ramp.

1. What VPE should Daniel have used before getting in the vehicle?
2. Do you think that Daniel has met his customer's expectations?

CHECK YOUR PROGRESS

1. When should you wear approved safety glasses?
2. Why are latex gloves used?
3. What type of mask must you use when replacing brake linings?
4. What three types of vehicle protective equipment (VPE) must you use when working inside a vehicle?

Hazards and risks

Although the automotive environment is an exciting and stimulating place to work in, it can also be dangerous. This section covers the **hazards** you may come across, including hazardous substances and how to dispose of them. You will also learn about assessing **risk** so that you understand how to manage your daily work in order to keep yourself, your colleagues and your customers safe.

> **Key terms**
>
> **Hazard** – something that has the potential to cause harm or damage.
>
> **Risk** – the likelihood of the harm or damage actually happening.

> **Find out**
>
> Copy the table on hazards and risks in the workplace and how to prevent them. Fill in the gaps and add another example of a hazard in the last row.
>
> | Hazard | Risk | Prevented by |
> |---|---|---|
> | Trailing wires | Falls or electrocution | |
> | Spilt oil | | |
> | | Crushed finger | |
> | | | |

Hazardous substances

Hazardous substances are the source of major health and safety risks in the workshop. Two particular dangers in the automotive workshop include asthma caused by breathing in brake lining dust or spray paint and dermatitis from engine oil.

Hazardous substances need to be handled, stored, transported and disposed of carefully.

Figure 1.8 Chemicals which are classed as hazardous substances

Routes of entry

Hazardous substances are dangerous because they can enter the human body and cause damage. There are various ways in which hazardous substances can enter the body, as shown in Table 1.21.

Table 1.21 Routes of entry for hazardous substances

Route of entry	Description	Example
Absorption	Absorbing a hazardous substance into your body through the skin. It then enters the bloodstream.	Using a compressed airline near to the skin when drying off parts
Inhalation	Breathing in toxic fumes or other hazardous substances	Brake dust breathed in during service or maintenance of the braking system
Ingestion	Swallowing a hazardous substance or having it splashed in or around the mouth	Working under a vehicle with fluids dripping on to the face
Injection	Being cut by a sharp object, such as a needle, that is contaminated with a hazardous substance	Cleaning out a vehicle and accidentally picking up a hypodermic syringe

Toxic substances

A **toxic** substance is one which, when inhaled, swallowed or absorbed through the skin, will act as a poison. If a toxic substance enters the body, it could result in illness, infection or death.

There are many toxic substances found in the automotive workshop which are hazardous to health. These include:

- paints, lacquers and underseals
- fuels, brake fluids and lubricants, including waste oil
- fumes and gases from welding and cutting
- dusts from abrasive wheels
- degreasing fluids and cleaning products, including strong hand cleaners
- adhesives and fillers
- battery acid.

Table 1.22 lists some types of toxic substances and gives an example of each type which can be found in a workshop.

> **Key term**
>
> **Toxic** – poisonous.

Table 1.22 Examples of toxic substances found in a workshop

Substance	In a workshop
Asbestos dust	Brake dust
Ethylene glycol or methanol	Antifreeze
Carbon monoxide	Exhaust fumes
Trichloroethylene	Brake cleaner
Sulphuric acid	Battery acid
Caustic soda	Radiator flush
Epoxy resins	Body filler

Some examples of areas in the workshop where you might encounter toxic substances are given in Table 1.23, along with the precautions you must take to minimise the risks from these hazards.

Table 1.23 Toxic hazards and precautions

Hazard area	Hazard	Precaution
Running an engine	Exhaust fumes	Use an extraction fan or pipe the gas outside.
Welding bench	Zinc oxide fumes given off when welding galvanised steel	Use a ventilated mask and extraction fan.
Body shop	Epoxy resin	Use a mask when sanding.
Paint shop	Water-based paint	Use fume extraction and suitable PPE (goggles, chemical-resistant apron, overalls and steel toe-capped boots).
Battery charging	Hydrogen gas and sulphuric acid	Do not smoke. Disconnect leads when charging. Wear the required PPE (chemical-resistant gloves, dust mask, apron, overalls, goggles, steel toe-capped boots).

COSHH risk assessment

Your employer must act in accordance with the Control of Substances Hazardous to Health 2002 (COSHH) regulations (see pages 10–11). Under the COSHH regulations, your employer must carry out a risk assessment for every hazardous substance that is present in the workplace.

Did you know?

Under the COSHH regulations, all employers are responsible for preventing employees from becoming exposed to hazardous chemicals and substances.

The safe use of chemicals and other substances						
Company Name:							
The findings should be recorded and the staff concerned informed as to their outcome and of the safe method of work. If you decide there are no risks to health, or risk is minimal this does not need to be recorded.							
STEP 1 – HAZARD	YES	NO	Comments/Action		Date Completed
Have all hazardous substances (marked with orange square with black symbol) been listed?			List substances:		
Have safety data sheets (supplier legally obliged to provide) being obtained?			Where are the data sheets located?		
Has the methods of use of the substance been determined?					
Have the number of people using the substance been determined?					
Has the affect of the substance on employees been assessed?			List affects:		
Has the risk been assessed using the above information?			List affects:		
STEP 2 – DECIDE ON PRECAUTIONS	YES	NO	Comments/Action		Date Completed
Has information from manufacturer and safety data sheets been used to determine suitable precautions?			What precautions needed?		
Has good working practices and standards recommended by trade association been used to determine suitable precautions?			What precautions needed?		
STEP 3 – PREVENT OR CONTROL EXPOSURE by making use of one or more of the following hierarchy of measures	YES	NO	Comments/Actions		Date Completed
Have you changed the process or activity (so that the subtance isn't needed)?			What was used? What is now used?		
Have you replaced the substance with a safer alternative (eg. water-based paints rather than solvent-based)?					
Have you made use of the substance in safer form (eg. pellets rather than loose powder)?					
Have you totally enclosed the process (removing all change of contact)?					
Have you partially enclosed the process and used local exhaust ventilation (eg. fume cabinets with extraction)?					
Have you provided adequate general ventilation (eg. in beauty salons, plenty of fresh air to avoid build-up of fumes from acetone etc)?					
Have you used methods of work which minimise handling or chance of spillage?					
Have you reduced the number of persons exposed or minimised time of exposure?					
Have you used personal protective equipment (eg. disposable gloves, overalls, safety goggles etc)?					
STEP 4 – CHECK YOUR ASSESSMENT AND PROCEDURES to assure the assessment's effectiveness	YES	NO	Comments/Actions		Date Completed
Have you checked that all staff are following procedures outlined in Step 2 and 3?					
Is the equipment used for controlling substances maintained and tested for effectiveness?					
Is air monitoring carried out if the assessment concludes there would be a serious risk to health if the controls should fail?					
Has health surveillance been carried out where specific health problems are associated with the substance (information on safety data sheet)? In most cases, this would be in the form of a simple question and answer or in cases such as potential dermatitis, regular examination of hands and other exposed areas.					

Figure 1.9 COSHH risk assessment checklist

COSHH risk assessments must be carried out by a competent person with the legal and technical knowledge required. They will record the procedure that must be followed when using hazardous substances. If there are any changes to the procedure or different substances are introduced, then a new risk assessment will be carried out. This will make sure that the precautions for each hazard still sufficiently control the risk.

The risk assessment must include:

- the information on the supplier's **safety data sheet**
- how hazardous the substance is
- how much and how often it is used
- who uses the substance.

On page 11, you saw the eight steps that your employer must follow when carrying out a COSHH risk assessment. Table 1.24 gives an example of how an employer might apply these steps to exposure to one hazardous substance – brake dust inhalation.

> **Key term**
>
> **Safety data sheet** – all the information required on a specific product to ensure it is used, stored and handled with care. The manufacturer must provide a safety data sheet with each product.

> **Safe working**
>
> If any health problems occur or there are any defects in the control measures or PPE, you must report this to your employer immediately.

Table 1.24 Applying a COSHH risk assessment to the control of brake dust inhalation

Steps	Example (control of brake dust inhalation)
Step 1 Assess the risks of hazardous substances in the workplace.	There is a risk of inhaling brake dust when removing brake linings.
Step 2 Decide what precautions are needed.	Supply the correct PPE, equipment and materials: • brake cleaner • vacuum extraction • mask, goggles, boots, overalls and latex gloves.
Step 3 Prevent or adequately control exposure to the hazardous substance.	Follow the manufacturer's directions for use, e.g. do not apply brake cleaner to hot brake or engine parts. Ensure vacuum extraction is used.
Step 4 Make sure control measures are used and maintained.	Monitor the use of brake cleaner and vacuum extraction.
Step 5 Monitor the exposure of employees to the hazardous substance.	Ensure that employees complete the activity as quickly as possible.
Step 6 Carry out health surveillance where necessary.	Observe the activity and feed back to employees.
Step 7 Prepare plans and procedures to deal with accidents, incidents and emergencies.	Provide a first aid box. Display a health and safety poster showing information in the event of an emergency, e.g. explosion/fire from the brake cleaner igniting as brake cleaner is extremely flammable.
Step 8 Make sure employees are properly informed, trained and supervised.	Provide training on: • PPE • dangers of brake dust • use of extraction equipment.

Level 2 Light Vehicle Maintenance & Repair

> **Safe working**
> - Do not store chemicals in open containers.
> - Do not transfer concentrated chemicals into an unlabelled container.
> - Do not reuse a concentrate container. Dispose of it safely or return it to the supplier.
> - Do not store more than 50 litres of flammable liquid indoors.

Safe storage of chemicals and hazardous substances

Correct storage of hazardous substances is essential and is covered by the COSHH regulations.

- Products containing hazardous chemicals and substances must be securely stored in a cool, dry, dark place.
- Do not store or stock more chemicals than you need.
- Make sure you keep apart chemicals that might react together.
- Control spills from burst containers. (Keep a spill clean-up kit nearby).
- Keep all hazardous chemicals and substances in locked storage containers which can only be accessed by COSHH-trained employees.

Figure 1.10 shows how to store hazardous chemicals and substances safely.

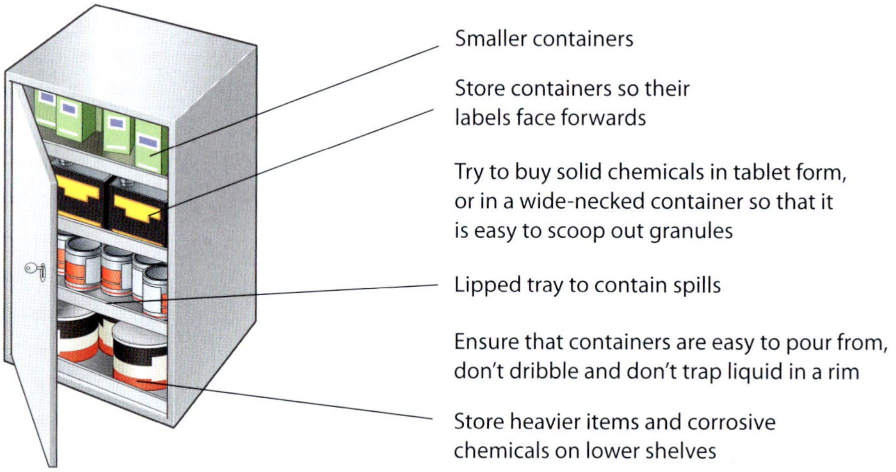

Labels on storage cabinet:
- Smaller containers
- Store containers so their labels face forwards
- Try to buy solid chemicals in tablet form, or in a wide-necked container so that it is easy to scoop out granules
- Lipped tray to contain spills
- Ensure that containers are easy to pour from, don't dribble and don't trap liquid in a rim
- Store heavier items and corrosive chemicals on lower shelves

Caution: Never store chemicals in open containers

Figure 1.10 How to store chemicals and hazardous substances safely

Figure 1.11 shows some warning labels that you will find on containers of hazardous chemicals and substances.

Dust | Toxic | Flammable | Irritant | Corrosive | Oxidising agent

Figure 1.11 Warning labels that you will find on containers of hazardous chemicals and substances

Using hazardous chemicals and substances

When using hazardous chemicals and substances:

- Always follow the instructions on product labels.
- You may need to use respiratory protective equipment (RPE) in case of a spill.
- Wear protective single-use nitrile gloves.
- If you must use latex gloves, use only 'low-protein, powderfree' gloves.
- Throw away single-use gloves every time you take them off.

> **Safe working**
> Use a flammables store for flammable liquids.

1 General workshop operation principles for light vehicles

Isocyanate-based paints

A hazardous substance which is widely used in light vehicle body shops is spray paint that contains the substance isocyanate. This type of paint is the number one cause of occupational asthma in the UK.

> **Working life**
>
> Edward is spray painting a damaged car wing using **isocyanate-based** two-pack top coat. He is aware that isocyanate can cause asthma and so he reduces his exposure to the paint in line with the recommended **WELs**.
>
> 1   Why does Edward's employer arrange regular health checks for him?

For further information on isocyanate-based paints, go to hotlinks to visit the HSE web page on health and safety priorities for body shops.

Storing hazardous waste

Sometimes it may be necessary for you to store hazardous waste on the premises before it can be disposed of or recycled. This waste may include liquids such as used engine oil and components such as asbestos and tyres.

Guidelines for storing hazardous waste include:

- Store the waste securely.
- Store different types of hazardous waste separately.
- Regularly check for leaks, deteriorating containers or other potential risks.
- Display written instructions for storing each type of hazardous waste.
- Keep an inventory of any hazardous waste in the workplace and where it is stored.
- All staff must be properly trained to deal with hazardous waste.

Safe disposal of special and hazardous waste

Hazardous waste is waste that may be harmful to human health or the environment. It can include materials that are **flammable**, **corrosive** or **ecotoxic**. The warning symbols for flammable and corrosive substances are shown in Figure 1.11; the symbol for ecotoxic substances is shown in Figure 1.12.

Examples of hazardous waste in a workshop include:

- car lead acid batteries (also known as wet cell batteries)
- contaminated rags
- used oil/fuel filters
- aerosols
- antifreeze
- brake fluids
- waste vehicle components
- fluorescent tubes
- energy-saving lightbulbs
- sodium lamps
- toner and ink jet cartridges
- old computer monitors
- asbestos
- solvents.

> **Safe working**
>
> Contact with chemicals can lead to dermatitis (skin soreness, itching, rashes and blistering). Some chemicals can also damage the eyes. Always wear the correct PPE for the job.

> **Safe working**
>
> Always wash your hands after using chemicals and hazardous substances, and before and after eating, drinking, smoking and using the lavatory. Never clean your hands with concentrated cleaning products or solvents.

> **Key terms**
>
> **Isocyanate-based paints** – in these paints, isocyanate hardeners or activators added to liquid resin react to produce a polyurethane film.
>
> **WELs** – acronym for work exposure limits. This is the maximum amount of time you can be safely exposed to the harmful activity.
>
> **Flammable** – catches fire and burns easily.
>
> **Corrosive** – causes damage to any part of the body it contacts.
>
> **Ecotoxic** – damaging to the environment.

Figure 1.12 Symbol for ecotoxic substances

Level 2 Light Vehicle Maintenance & Repair

> **Key terms**
>
> **Controlled waste** – any waste which cannot be disposed of to landfill, including liquids, asbestos, tyres and waste that has been decontaminated. There are three types of controlled waste listed under the Environmental Protection (Controlled Waste) Regulations 2004: household, industrial and commercial waste.
>
> **Contractor** – an individual or company that provides a service for an agreed fee, for example to remove and dispose of waste products.
>
> **Carcinogen** – a substance that causes cancer.
>
> **Halogens** – a group of five chemically related non-metallic elements.

> **Did you know?**
>
> Any premises that handles over 500 kg of hazardous waste in a 12-month period must be registered as a producer of hazardous waste.

The Environmental Protection Act 1990 (see pages 9–10) imposes a 'duty of care' on all those who import, produce, carry, keep, treat or dispose of **controlled waste**.

Disposal of hazardous waste should be done with minimal environmental impact. It is your employer's responsibility to register your workshop with the Environment Agency (EA) and keep records of all hazardous waste disposed of. The actual disposal of waste will often be done by a specialist **contractor**.

Recycling in the automotive industry

Once a light vehicle is no longer fit for purpose, it is likely to be sold on to a vehicle dismantler, where the following will be removed:

- parts which can be sold for reuse
- materials which may be potentially dangerous to the environment, such as fluids and batteries.

Many of the materials that are retrieved from a vehicle can be recycled.

Metals

After the shredding process, any metals in the tyres are removed and sold on to the steel industry.

Tyres

Tyres are classed as a controlled waste under the Environmental Protection Act 1990. Any waste producer, such as your workplace, must ensure that tyres are disposed of via a registered waste carrier or authority.

The European Landfill Directive was introduced in July 2006, banning all used whole and shredded tyres from being sent to landfill for disposal.

Tyres can be recycled in the following ways:

- **Reuse** – If a vehicle is not roadworthy but the tyres are only part-worn, then it may be possible to reuse them. The remaining tyre tread must be at least 2 mm and the tyre must be marked as part-worn on both sides when sold.
- **Reuse through landfill engineering** – Whole tyres can be used on landfill sites in the construction and preparation of draining systems. Tyres used for this purpose are exempt from landfill tax.
- **Retreading** – This is done by replacing the tread section of a tyre or resurfacing the outer surface of the tyre with rubber. If a light vehicle tyre has excessive wear and is below the minimum (UK) 1.6 mm tread depth limit, it will be unsuitable for retreading.
- **Energy recovery** – The burning of tyres under controlled conditions is an alternative method of producing energy. The amount of energy released from tyres when burnt is greater than that of coal.

- **Reforming** – Tyres can be reformed into new materials such as shoe soles and road asphalt.

Other uses for old tyres include:

- boat and dock fenders
- crash barriers at motor racing circuits
- roof tiles
- noise control products
- structural support for earth walls
- motorway embankments.

Plastics

Many of the plastics used in car manufacturing are recyclable.

Waste oil

Waste oil contains heavy metals, **carcinogens** and **halogens**. If it is not disposed of correctly it may cause harm to the environment and human health. Significant amounts of oil can also be found in used oil filters. This can be removed and recycled.

There are several methods of recycling waste oils, including:

- **Processing** – This removes excess water and filters out particles. The processed material is used as fuel for power stations.
- **Refining** – The used oil is refined and then reused as a lubricating oil.

Some of the consequences of pouring oil down the drain are shown in Figure 1.13.

> **Did you know?**
> The UK produces a billion litres of waste oil per year. As little as one litre of this could pollute a million gallons of drinking water. It is therefore extremely important to make sure that you dispose of all waste oil correctly.

> **Did you know?**
> 'Waste oil from nearly 3 million car oil changes in Britain is not collected. If it is collected properly, this could meet the annual energy needs of 1.5 million people'.
> (Source: Scottish Oil Care Campaign)

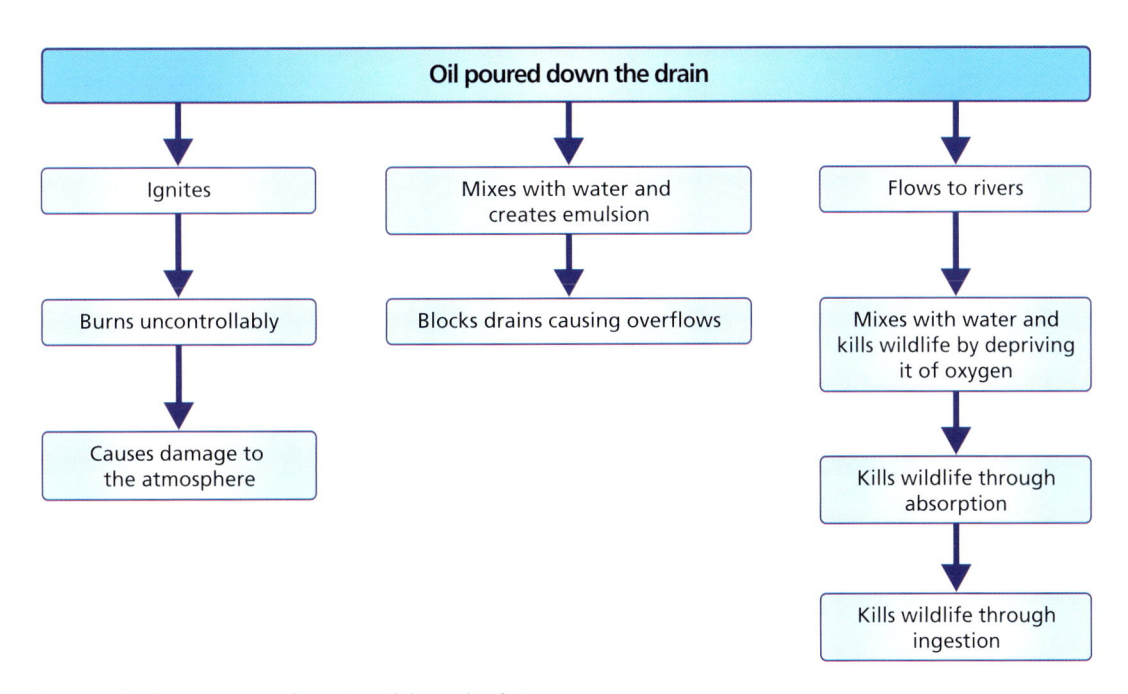

Figure 1.13 Consequences of pouring oil down the drain

Batteries

A light vehicle battery is made up of lead, electrolyte (battery acid) and plastic. Almost all of the lead in a light vehicle battery can be recycled. All car batteries must be recycled because they contain hazardous substances.

Table 1.25 gives some examples of reuse for battery components.

Table 1.25 How battery components are recycled

Battery component	Reuse
Lead	Manufacture of new LV batteries Manufacture of new plates used in LV batteries
Hard rubber	Used as a carbon additive in the lead recycling process
Plastic	Polypropylene used in the manufacture of new battery covers and cases
Electrolyte solution/battery acid	Sodium sulphate crystals treated and reused in battery manufacture, glass, textiles and laundry detergents

> **Did you know?**
> If not disposed of correctly, the lead from the battery may seep from the battery and contaminate the environment through ground water and surface water.

Legislation

Legislation covering the disposal of hazardous waste includes:

- Control of Substances Hazardous to Health Regulations 2002 (COSHH)
- Environmental Protection Act 1990
- Environmental Protection (Controlled Waste) Regulations 2004
- Environmental Protection (Duty of Care) Regulations 1991
- Hazardous Waste Regulations 2005
- Pre-treatment Regulations
- Waste Electronic and Electrical Equipment (WEEE) Directive
- Batteries Directive
- The European Union End-of-Life Vehicles (ELV) Directive
- Control of Asbestos at Work Regulations 2002
- The European Union Landfill Directive
- Waste Incineration Directive.

CHECK YOUR PROGRESS

1. What is the difference between a hazard and a risk?
2. How should you dispose of waste oil and why?
3. Give examples of three different toxic substances used in a workshop.
4. List four different uses for recycled tyres.

Accident prevention

This section covers **accident** prevention, including hazard areas in a workshop, first aid and reducing the risks from electricity. It also covers safe manual handling techniques and how to lift heavy objects using lifting gear. Finally this section looks at occupational health and fire prevention.

> **Key term**
>
> **Accident** – an unplanned, unexpected occurrence that may result in injury or even death.

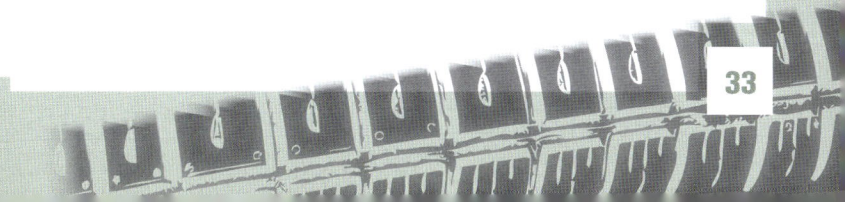

Eight fatal accidents in the motor vehicle repair (MVR) sector were reported to the HSE during the period 2003–2004:

- Three employees were hit by a moving vehicle due to the handbrake being left off.
- One roadside mechanic died when the vehicle he was working on was hit by another vehicle.
- Three employees were crushed to death when working under vehicles. One of these was at the roadside and another was when a vehicle fell off a lift.
- One apprentice was killed when mishandling petrol.

John Powell, from the HSE's Manufacturing Sector and Chair of the MVR Health and Safety Forum said, 'All of these deaths were avoidable. Simple things such as leaving parked vehicles with their handbrakes on or ensuring that vehicles were properly supported before going underneath them could be enough to save a life. Precautions to prevent almost all types of accidents in MVR are often simple and inexpensive.'

Accidents are often caused by a mixture of factors, including:

- **Occupational factors** – These are directly connected to specific tasks or occupations, for example a back injury from heavy lifting or repetitive strain injury from using a tool repeatedly.
- **Environmental factors** – These are created by the conditions in which people work, for example deafness caused by a noisy workplace. Other environmental factors might be inadequate heating, lighting, ventilation or space.
- **Human factors** – These include poor behaviour, carelessness, lack of attention, poor concentration, inexperience, poor training, haste, ignorance or being under the influence of drugs or alcohol.

In order to help prevent accidents, all three of these factors need to be controlled.

Hazard areas in a workshop

Some examples of hazards within the workshop environment which can cause accidents are:

- moving vehicle or vehicle's engine left running
- working with flammable liquids
- working with toxic and other harmful substances
- using machinery/equipment (e.g. welding, compressed air, drilling)
- storing and stacking spare parts

- dispensing petrol
- disposing of waste
- working on vehicles (e.g. jacking, working on electrical systems, fuel systems, tyres, wheels).

It is impossible to eliminate risks in the workplace entirely, but they can be controlled if you follow high standards of health and safety in your workplace.

✓ Keep activities under control by following management systems and your workplace health and safety policy.
✓ Pay attention to PPE and safety instructions.
✓ Follow risk assessment procedures.
✓ Observe good housekeeping (see pages 61–67).
✓ Always work safely.
✓ Report any machinery faults, accidents and dangerous occurrences.

> **Safe working**
>
> Regular inspections, supervision, maintenance and the identification of changes in the workplace or work tasks must also be carried out to ensure that control measures continue to reduce hazards and risks to an acceptable level.

Table 1.26 Potential causes of accidents in the workplace

Accident	Possible causes
Employee cut by broken glass in bin	• Broken glass not properly wrapped. • Employee pushing glass deep into bin. • Employee pressing down the rubbish in the bin without wearing PPE.
Employee's arm trapped in a machine with hot parts, causing severe burning	• Employee had removed the guards, allowing access to the dangerous parts of the machine. • Employee was wearing loose clothing. • Employee was distracted, e.g. chatting to their workmates.
Visitor falls down vehicle inspection pit	• Bad design where visitor has access to workshop. • Door to workshop which has an inspection pit left unlocked. • Poor signage so visitor went through the wrong door. • No warning of steps. • No fence around inspection pit.

Reporting accidents

Accident reports must be completed within 24 hours of the event and should contain:

- the date and time of the accident, near miss or incident of ill health
- the name of the person involved
- a description of the accident, near miss or incident of ill health
- a description of any injuries incurred
- what action was taken and by whom
- the final outcome, e.g. person involved sent home, hospitalised, etc.

1 General workshop operation principles for light vehicles

Accidents which require immediate reporting include those which result in:

- death
- major injuries
- any type of injury, dangerous occurrence or disease that is specified by law
- an injury resulting in absence from work for more than three days (including weekends and holidays)
- a member of the public needing to go to hospital immediately.

You must make your employers aware of accidents including near misses and all incidents of ill health that have occurred. They will then investigate the cause and take corrective action. As an employee you have a duty to report all occurrences in the accident book provided by your employer. (See the section on the Reporting of Injuries, Diseases and Dangerous Occurrences Regulations 1995 (RIDDOR) on page 9.)

Did you know?

Once completed, an accident report is protected under the Data Protection Act 1998 and must be kept confidential. It is recommended that the completed form should be kept for three years.

Working life

John was working under a customer's vehicle when the jack was released. The jack had been left with the handle halfway down and Bill got caught up on it as he walked past. Bill received a minor cut on his knee and John had damaged ribs. John's manager wrapped a bandage around John's ribcage and told him to make sure to leave the jack handle up in future.

- What is wrong with what the manager did?
- What should he have done that he didn't do?

Did you know?

Accidents where people escape injury are commonly called 'near misses'. These should always be investigated and recorded, as this will help prevent possible future injuries.

Figure 1.14 An accident record sheet

35

> **Key term**
>
> **First aid** – emergency treatment given to an injured or sick person to prevent their condition deteriorating before professional medical care is available.

Basic first aid

The advice that follows is not a substitute for a **first aid** course, and will only give you an outline of the steps you need to take. Reading this part of the unit will not qualify you to deal with these emergencies. Unless you have been on a first aid course, you should be careful about what you do, because the wrong action can cause more harm to the casualty. It may be better to get help.

There must always be at least one first aider available on site to deal with emergencies.

First aid box

The minimum level of first aid equipment in a suitably stocked first aid box should include:

- a guidance leaflet
- 2 sterile eye pads
- 6 triangular bandages
- 6 safety pins
- 3 extra large, 2 large and 6 medium-sized sterile unmedicated wound dressings
- 20 sterile adhesive dressings (assorted sizes)
- 1 pair of disposable gloves (as required under HSE guidance).

It is your employer's or the designated first aider's responsibility to ensure that the contents of the first aid box are in date and are sufficient, based on their assessment of your workplace's first aid needs. The law does not state how often the contents of a first aid box should be replaced, but most items, in particular sterile ones, will be marked with expiry dates.

Other equipment such as eye wash stations must also be available if the work being carried out requires it.

Calling the emergency services

999 is the official emergency telephone number. You can use it to call help for:

- Police
- Fire and Rescue
- Ambulance.

Calls to the 999 service are free from a landline.

You can call the emergency services free of charge from your mobile phone. Your phone will automatically unlock.

Figure 1.15 Every workplace needs to have a well-stocked first aid box

> **Did you know?**
>
> There should be NO pills, antiseptic creams, lotions or medicines in the first aid box.

1 General workshop operation principles for light vehicles

Making a 999 call

When you call the emergency services, ask for an ambulance and give the following information:

- your telephone number
- the location of the incident
- the type of incident
- the gender and age of the casualty
- details of any injuries you have observed
- any information you have observed about hazards, for example power cables, fog, ice, gas leaks you could smell.

> **Did you know?**
> Many lives could be saved each year if more people knew how to place a casualty correctly in the recovery position.

The recovery position

Many of the actions you need to take to deal with health emergencies will involve you placing someone in the recovery position. In this position a casualty has the best chance of keeping a clear airway, not inhaling vomit and remaining as safe as possible until help arrives. This position should not be attempted if you think someone has back or neck injuries, and it may not be possible if there are fractures of limbs.

Practise with a colleague.

Putting a casualty in the recovery position

Figure 1.16 The recovery position

| 1. Kneel at one side of the casualty, at about waist level. | 2. Tilt the head back – this opens the airway. With the casualty on their back, make sure that their limbs are straight. | 3. Bend the casualty's near arm so that it is at right angles to the body. Pull the arm on the far side over the chest and place the back of the hand against the opposite cheek. | 4. Use your other hand to roll the casualty towards you by pulling gently on the far leg, just above the knee. This will bring the casualty onto their side | 5. Once the casualty is rolled over, bend the leg at right angles to the body. Make sure the head is tilted well back to keep the airway open. |

Action in a health emergency

This section gives guidance on recognising and taking initial action in a number of health emergencies, including:

- electrical injuries
- severe bleeding
- shock
- loss of consciousness
- burns and scalds
- objects in the eye.

Remember that you will need to seek professional help in all emergencies.

Figure 1.17 Electrical hazard symbol

Electrical injuries

If someone receives an electric shock the electricity passes straight through them because of the large water content within the body. This may cause the skin to burn or look pale or bluish, and the person may not have a pulse. In this instance, the person may not be breathing and the heart may have stopped.

In the event of an electrical shock:

- Seek help.
- DO NOT touch the casualty.
- Switch off the current source.
- Stand on dry insulating material (wood, rubber or lino). Isolate the casualty by using material that does not conduct electricity, such as wood or plastic (for example, a wooden broom handle).
- If you are a qualified first aider, follow your training for dealing with electric shock.
- Obtain emergency medical assistance for the casualty.

Bleeding and severe bleeding

If you or a colleague has a cut, the qualified first aider will be able to issue plasters from the first aid box. Always make sure you wash the cut carefully with soap and running hot water then dry with a clean paper towel. (You will need to take care to protect yourself when dealing with casualties who are bleeding. See the section *Protect yourself* below.)

Severe bleeding in the workshop will most likely come from a cut caused by tools or glass. It could also be the result of a fall.

In the event of severe bleeding, the aim is to stem the flow of blood:

- Apply pressure to the wound that is bleeding, using a sterile dressing if possible or any absorbent material.
- Do not forget the precautions (see *Protect yourself* below).
- Apply direct pressure over the wound for 10 minutes to allow the blood to clot.
- Do not try to take any object from the wound, but apply pressure to the sides of the wound.
- Lay the casualty down and raise the affected part if possible. Make sure the casualty is warm, comfortable and safe.
- Obtain emergency medical assistance for the casualty.

Protect yourself

You should take steps to protect yourself when dealing with casualties who are bleeding. Skin provides an excellent barrier to infections, but you must take care if you have any broken skin such as a cut, graze or sore. Seek medical advice if blood comes into contact with your mouth or nose, or gets into your eyes. Blood-borne viruses (such as HIV or hepatitis) can be passed only if the blood of someone who is already infected comes into contact with your broken skin.

To avoid contamination:
- If possible, wear disposable gloves.
- If this is not possible, cover any areas of broken skin with a waterproof dressing.
- Wash your hands thoroughly using soap and water before and after treatment.
- Take care with any needles or broken glass in the area.
- Use a mask for mouth-to-mouth resuscitation if the casualty's nose or mouth is bleeding.

Shock

Shock occurs because blood is not being pumped around the body efficiently. It can be the result of severe bleeding or burns or a heart attack. The casualty:
- will look pale and grey
- will be very sweaty, with cold clammy skin
- will have a very fast pulse and may be breathing very fast
- may feel sick or vomit.

In the event of shock:
- Keep the casualty warm but not with direct heat – use a blanket or extra clothing.
- Lay the person down on the floor and raise their feet off the ground. This will help maintain the blood supply to the important organs.
- Loosen any tight clothing.
- Watch the person carefully. Check their pulse and breathing regularly.
- Obtain emergency medical assistance for the casualty.

Do not:
- ✗ allow the casualty to eat or drink
- ✗ leave the casualty alone, unless it is essential to do so briefly in order to summon help.

Loss of consciousness

Loss of consciousness can happen from fainting or as the result of injury, a fall or illness. The person will lack awareness and be unresponsive.

In the event of a person losing consciousness:
- Make sure that the person is breathing and has a clear airway.
- Maintain the airway by lifting the chin and tilting the head backwards.
- Place the casualty in the recovery position.
- Obtain emergency medical assistance for the casualty.

Do not:
- ✗ attempt to give anything by mouth
- ✗ attempt to make the casualty sit or stand
- ✗ leave the casualty alone, unless it is essential to do so briefly in order to summon help.

Find out if the casualty has an existing condition like diabetes or epilepsy (they may have an ID band with this information) which could help you give relevant information to the emergency services.

Minor burns or chemical injuries

Burns may occur in the workshop from contact with chemicals or hot components or burns from welding equipment.

In the event of a burn:

- Cool down the burn by flooding it with cold water for 10–20 minutes.
- If it is a chemical burn, flooding with cold water needs to be done for 20 minutes. Make sure that the contaminated water used to cool a chemical burn is disposed of safely.
- Cover the burn if possible with a clean dressing or even cling film. If there are no clean dressings available leave it uncovered.
- Obtain emergency medical assistance for the casualty if it is a significant burn.

Objects in the eye

Even though wearing goggles will prevent most eye injuries, there may be incidents where an object becomes lodged in the eye. In the event of an object entering the eye, it is helpful to have a colleague assist with flushing it out. You will need:

- good lighting
- a bottle of eyewash solution or clean tepid water
- some absorbent towel or a handkerchief.

When responding to an incident which results in an object entering a colleague's eye:

1. Ask the casualty to lie down if possible, otherwise stand beside or behind them.
2. Wash your hands with soap and water and rinse thoroughly.
3. Locate the object by gently lifting the eye lid and asking the casualty to look up and down and to the side until you see it.
4. Tilt the head to the side that is being flushed and place a towel or cloth against the ear to prevent the eyewash from entering the ear.
5. Using prepared eyewash or, if not available, tepid water, pour the liquid from the inner corner so that it drains onto the towel.
6. It may be possible to dislodge grit or dust by gently using a moist clean swab or tissue.
7. If after two flushings the object is not dislodged, stop and seek medical help.

> **Safe working**
> Never touch anything that is stuck in the eye but seek medical help.

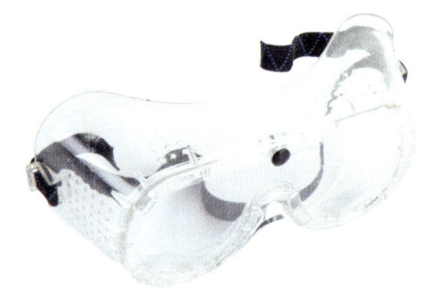

Figure 1.18 Wearing safety goggles may help prevent eye injuries

Accidents involving electricity

Electricity can cause electric shock, burns, fires and death. One quarter of all electrical accidents at work are fatal. Possible electrical incidents include:

- **Burns** – Electricity invariably burns and serious injuries take a long time to heal.

- **Flash** – Electrical flashes are very bright and can burn or damage the eyes.
- **Shocks** – Shocks occur when electricity passes through the body. They can burn or kill.
- **Fires** – One-fifth (20 per cent) of accidental fires in the workplace are caused by electricity.

Reducing the risks from electricity

There are two main dangers from using electricity in the workshop.

1. **Fire** – The overheating of electrical circuits and/or lightbulbs could ignite fuel.
2. **Electrocution** – Coming into contact with a live circuit, could result in burns, severe injury or even death.

Prevention is better than cure, so it is important to find appropriate controls to improve electrical safety. These include:

- **Insulation** – This will protect the individual from direct contact with electricity.
- **Earthing** – Providing a contact with earth will reduce the risk of shock.
- **Fuses** – These are protective strips of metal that break and melt if overheating occurs, so stopping the supply of electricity. There are 3 amp, 5 amp and 13 amp fuses available.
- **Circuit breaker** – This is designed to detect excess flow of electricity and stop the supply to the circuit when this occurs.
- **Residual current device (RCD)** – This is an electronic device that will shut off power in the event of a fault.
- **Voltage reduction** – Ensure that the lowest possible voltage is used.

> **Safe working**
>
> Always follow the manufacturer's recommendations when working on the electrical system of a hybrid vehicle as they can reach voltages of 330v.

Electricity – safety first

Your workplace will have safe working procedures in place, such as those shown in Figure 1.19, and you should always follow these.

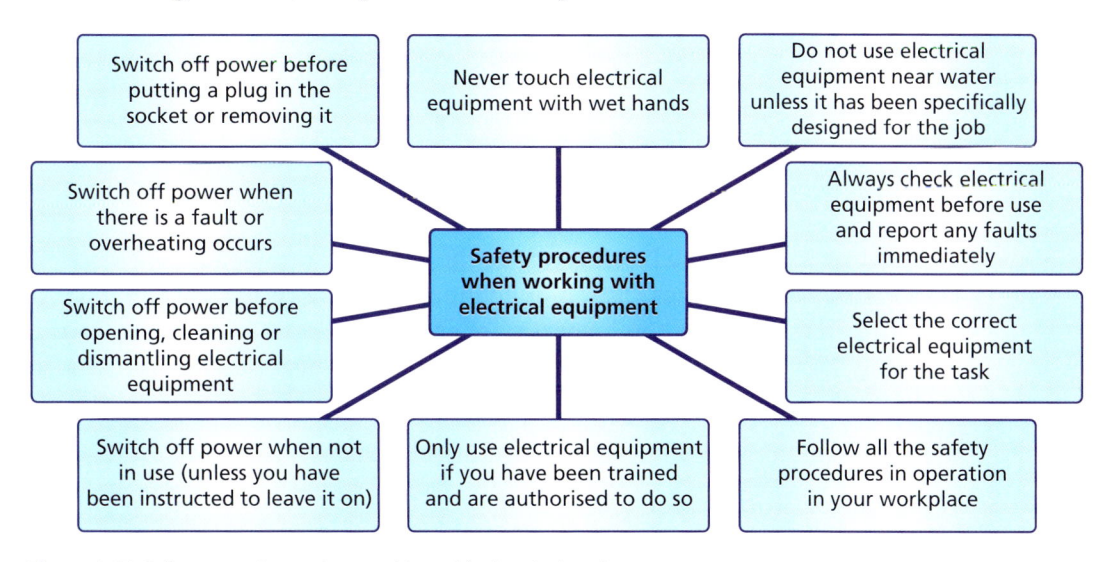

Safety procedures when working with electrical equipment:

- Switch off power before putting a plug in the socket or removing it
- Never touch electrical equipment with wet hands
- Do not use electrical equipment near water unless it has been specifically designed for the job
- Switch off power when there is a fault or overheating occurs
- Always check electrical equipment before use and report any faults immediately
- Switch off power before opening, cleaning or dismantling electrical equipment
- Select the correct electrical equipment for the task
- Switch off power when not in use (unless you have been instructed to leave it on)
- Only use electrical equipment if you have been trained and are authorised to do so
- Follow all the safety procedures in operation in your workplace

Figure 1.19 Safety procedures when working with electrical equipment

> **Did you know?**
>
> It is your employer's responsibility to make sure that all staff are fully informed, trained, supervised and receive correct information on using electrical equipment safely.

Everyone who works with electrical equipment must look out for faults through visual inspections. Any problems should be reported immediately. Signs of faults include:

- damaged sockets, plugs or cables
- burning smells or blackened sockets (this indicates overheating)
- frequent blown fuses.

Equipment and systems that use electricity must be checked regularly and maintained by trained staff.

Appliances that have a plug and are easily moved should be visually checked over every time they are used. Formal checks by competent people are required annually.

Table 1.27 lists the main hazards from electrical equipment in a workshop and the risks of these hazards.

Table 1.27 Main hazards from electrical equipment in a workshop

Electrical hazard	Risk
Using two or more plugs on a socket	Overheating sockets
Using incorrect extension	As the length of wire increases, so does the resistance of electricity. This makes the wire hotter and increases the risk of fire.
Poor or damaged insulation on cables	Electrocution to operator
Lack of adequate earthing for equipment	Electrocution to operator if appliance fails
Incorrect fuse used	Appliance becomes overheated and could set on fire

> **Did you know?**
>
> Hand-held electrically operated equipment, such as hand lamps and portable drills, work on reduced voltage compared to normal domestic use that works on 240 volts:
>
> Portable drills – 110 volts
>
> Hand lamps – 25 or 12 volts

Manual handling

The Manual Handling Operations Regulations 1992 were introduced to try to reduce the number of injuries caused by lifting and moving loads at work.

Manual handling means lifting and moving a piece of equipment or material from one place to another without using machinery. Lifting and moving loads by hand is one of the most common causes of injury at work. Most injuries caused by manual handling result from years of lifting items that are too heavy or are awkward shapes or sizes, or from using the wrong lifting technique. However, it is also possible to cause a lifetime of back pain with just one lift.

Poor manual handling can cause injuries such as:

- fractures
- trapped nerves
- damage to muscles, ligaments and tendons

- ruptured discs
- abrasions, cuts and crushing to parts of the body
- hernias.

The most common injury by far is spinal injury. Spinal injuries are very serious because there is little that doctors can do to correct them and, in extreme cases, workers have been left paralysed.

What you can do to avoid injury

The first and most important thing you can do to avoid injury from lifting is to receive proper manual handling training. **Kinetic lifting** is a way of lifting objects that reduces the chance of injury. It is covered in more detail in the next section.

Before you lift anything you should ask yourself the simple questions listed in Table 1.28.

> **Key term**
>
> **Kinetic lifting** – a way of lifting objects that reduces the risk of injury to the lifter.

Table 1.28 Questions to ask before lifting an object

Question	Further information
Does the object need to be moved?	
Can I use something to help me lift the object?	A mechanical aid, such as a scissor jack, or a hydraulic aid, such as an engine hoist, may be more appropriate than a person.
Can I reduce the weight by breaking down the load?	Breaking down a load into smaller and more manageable weights may mean that you need to make more journeys, but it will also reduce the risk of injury.
Do I need help?	Asking for help to lift a load is not a sign of weakness. Team lifting will greatly reduce the risk of injury.
How much can I lift safely?	The recommended maximum weight an adult male can lift is 25 kg, but this is only an average weight as each person is different. The amount that a person can lift will depend on their physique, age and experience.
Where is the object going?	Make sure that any obstacles in your path are out of the way before you lift. You also need to make sure there is somewhere to put the object when you get there.
Am I trained to lift?	Remember, the quickest way to receive a manual handling injury is to use the wrong lifting technique.

Level 2 Light Vehicle Maintenance & Repair

Lifting correctly (kinetic lifting)

When lifting any load it is important to keep the correct posture and to use the correct technique.

Get into the correct posture before lifting, as follows:

1. Stand with feet shoulder width apart, with one foot slightly in front of the other.
2. Knees should be bent.
3. Back must be straight.
4. Arms should be as close to the body as possible.
5. Your grip must be firm, using the whole hand and not just the fingers.

The correct technique when lifting is:

1. Approach the load squarely, facing the direction of travel.
2. Adopt the correct posture (as above).
3. Place hands under the load and pull the load close to your body.
4. Lift the load using your legs and not your back.

When lowering a load you must also adopt the correct posture and technique:

1. Bend at the knees, not the back.
2. Adjust the load to avoid trapping your fingers.
3. Release the load.

Preventing injuries

As with all areas of health and safety, the most effective method of prevention is elimination of the hazard. The hazard of manual handling is lifting and carrying. Ways you could reduce the risks include the use of trolleys, re-designing the workflow so that items do not need to be moved from one area to another or **automating** the process with the use of an overhead crane.

Before lifting a load, assess the load using the LITE assessment shown in Table 1.29. Avoid lifting a load by yourself if possible.

Table 1.29 The LITE assessment

L	Load	Can the load be broken down? Is it balanced? Where is the best place to hold the load for lifting?
I	Individual	Can someone else help you? Are they capable? Are they healthy? Have they had training?
T	Task	Do you need to stretch or twist? Can lifting devices be used to carry out the task?
E	Environment	Are there objects preventing you getting to where you need to go with the load? Is the floor slippery?

Safe working

Where manual handling tasks cannot be avoided, they must be carefully assessed to establish the risks associated with them. Control measures can then be put in place.

Safe working

Even light loads can cause back problems, so when lifting anything always take care to avoid twisting or stretching.

Key term

Automating – converting a process to an automatic operation.

Safe working

Report any problems such as sprains or strains immediately.

Where there are any changes to the activity or the load then the task must be re-assessed.

1 General workshop operation principles for light vehicles

1. Approach the loan squarely, facing the direction of travel.

2. Adopt the correct posture before lifting.

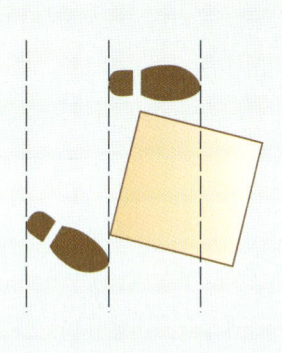

3. Place hands under the load and pull the load close to your body.

4. Adopt the correct posture when lifting. Lift the load using your legs and not your back.

5. Move smoothly with the load.

6. Adopt the correct posture and technique when lowering.

45

Figure 1.20 Hydraulic crane

Moving loads using lifting equipment

A load is any object that requires moving manually or by mechanical lifting equipment. Examples of items that may need to be moved using mechanical methods include:

- differential
- door or bonnet
- suspension assemblies
- gearbox.

Table 1.30 gives some examples of lifting devices used in an automotive workshop.

Table 1.30 Lifting devices used in a workshop

Type	Source of power	Used for:
Scissor jack	Manual/mechanical	Lifting the vehicle for repairing roadside punctures
Crane	Manual/hydraulic	Lifting engines, gearboxes and subframe assemblies
Hydraulic jack	Manual/hydraulic	Lifting vehicles distances above the ground
Hoist	Manual/electric motor	Lifting vehicles to a suitable working height

Rules when using lifting devices in a workshop

1. The maximum safe working load (SWL) must never be exceeded (see also page 15). This will be indicated on the equipment and the accessories used for lifting.
2. Never shock-load the lifting equipment.
3. Always maintain an even balance with the load.
4. Do not push or pull the load to adjust the balance.
5. Never transport loads over the heads of people and never walk under a load.
6. Do not leave a load hanging without support.

Lifting gear accessories

When using a crane to move heavy vehicle objects, the accessories required to support the lifting frame are hooks, chains or slings and eye bolts or shackles.

Lifting hooks

Lifting hooks are used to link the crane to the sling, chain or shackle. They have a safety lever to prevent the sling or chain from falling out of the hook. The lifting hook should always be painted yellow (see Figure 1.21).

Position of chains or slings

Chains and slings are used between the lifting device and lifting hook that connects to the load being lifted. If the angle between the slings

Figure 1.21 A lifting hook

is greater than 90 degrees, each sling will carry a weight equal to the weight of the load.

> **Example: Engine's weight 500 kg, therefore 90 degrees = sling on left 250 kg – sling on right 250 kg**

Rules when using slings

1. Do not bend the sling around corners or sharp edges, unless protection is used.
2. Do not twist or kink the sling to adjust the height of one side during lifting.
3. Do not use worn or damaged slings.
4. Never exceed the safe working load (SWL).

Eye bolts and shackles

The eye bolt shown in Figure 1.22 is called a D shackle because it is shaped like the letter D. The eye bolt must always be tight before lifting or the weight of the load may damage the threads of the bolt.

Transportation of loads

In the garage or parts department, heavy loads may be transported by:

- forklift
- flat trailer
- hand truck.

Table 1.31 shows a risk assessment for using a lifting device when changing a wheel.

Figure 1.22 D shackle

Table 1.31 Risk assessment for using a lifting device when changing a wheel

Steps	Risks and precautions
Step 1 Assess the risks.	• Lifting the vehicle • Keeping the vehicle up in the air • Vehicle moving
Step 2 Decide what precautions are needed.	• PPE and workshop equipment
Step 3 Prevent or adequately control exposure.	• Axle stands to hold the vehicle up • Chocks to stop the wheels moving
Step 4 Make sure that control measures are used and maintained.	• Observe the employees carrying out the task
Step 5 Monitor the exposure to the danger of the risk.	• Keep the wheel off for as little time as possible • Keep the vehicle in the air for as little time as possible
Step 6 Carry out appropriate surveillance to observe any risk.	• Observe the employees carrying out the task
Step 7 Prepare plans and procedures to deal with accidents, incidents and emergencies.	• First aid box available • Health and safety poster showing information in the event of an emergency
Step 8 Make sure employees are properly informed, trained and supervised.	• Train employees to safely lift the vehicle and use the correct equipment, including PPE

> **Key terms**
>
> **Chronic** – an effect on the body that happens after long or repeated exposure to a health hazard.
>
> **Acute** – an effect on the body that happens rapidly after a short exposure to a health hazard.
>
> **Control measure** – actions or procedures put in place to reduce a risk to an acceptable level.

Occupational health

Health problems directly relating to an individual's job are defined as occupational illnesses and injuries. For example, staff may breathe in dangerous substances from machinery and over time this may cause respiratory problems. A **chronic** illness or injury is one that develops over a period of time, for example straining an arm or back from working continuously in a poorly designed workstation. An **acute** illness or injury comes on quickly as the immediate result of a work hazard, for example cutting your hand on a circular saw.

Health hazards

There are various types of health hazard that could cause occupational illness or injury, as shown in Table 1.32.

Table 1.32 Examples of health hazards that could cause occupational illness or injury

Type of hazard	Examples
Chemical	Harmful dust, liquids and/or fumes
Biological	Infectious diseases and agents
Physical	Noise, heat and radiation
Ergonomic	Poorly designed tasks, work areas and equipment

Table 1.33 gives some examples of occupational illnesses and injuries.

Table 1.33 Occupational illnesses and injuries and their causes

Occupational illness/disease	Cause
Anxiety, stress, heart disease	Work-related stress
Cancer	Contact with carcinogenic substances such as asbestos
Heat stroke	Working in high temperature environments
Work-related upper limb disorder	Repetitive movements
Silicosis (a respiratory disease)	Inhaling harmful substances
Noise-related illness resulting in hearing loss	Long-term exposure to loud noise
Dermatitis (scaling, cracking and crusting of the skin)	Oil and grease getting into your skin

Control measures

Where health hazards cannot be avoided, **control measures** must be applied to minimise any possible harmful effects and their consequences.

The different types of control measures are described below. They are listed in order of priority. This is sometimes called the 'hierachy of controls' because the most important controls are at the top and the less important controls are at the bottom.

1. **Elimination** – complete removal of the hazard.
2. **Substitution** – providing a safer alternative in procedures or materials, for example replacing a highly toxic cleaning chemical with a less hazardous one that will still achieve the same results.
3. **Isolation** – moving the process to another area.
4. **Enclosure** – using physical barriers to separate the process.
5. **Local exhaust ventilation (LEV)** – trapping the contaminant close to its source and removing it directly by ventilation (hoods, ducting, fan) before it enters the work atmosphere.
6. **General ventilation** – using natural air movement through open doors and windows.
7. **Good housekeeping** – reducing risks from spillages, dust and debris.
8. **Reduced exposure time** – reducing the amount of time that people spend in contact with the hazard.
9. **Training** – making people aware of the control measures and their use in order to reduce risks.
10. **PPE** – using PPE if it has not been possible to reduce the risks using the above control methods.
11. **Welfare facilities** – minimising the effects of exposure to hazards by, for example, washing and showering facilities and the use of adequate first aid and emergency facilities.
12. **Medical surveillance** – this is used to detect early signs of ill health and identify anyone who is particularly susceptible to a hazard.

Welfare facilities

It is your employer's responsibility to provide adequate welfare facilities for your use, to help promote your well-being at work. They include:

- washing facilities with hot and cold running water, soap and hand drying facilities
- sufficient toilets that are clean, well lit and ventilated
- changing areas (where necessary) and facilities for storing clothes
- a supply of drinking water
- facilities for work breaks and to eat meals
- suitable rest facilities for pregnant women and nursing mothers.

Alcohol, drugs, stress and violence

Common issues that can affect health and safety in the workplace include alcohol and drug abuse, stress and the threat of, or actual, violence.

Alcohol

Drinking alcohol can put you, your colleagues and the general public at risk. It impairs and dulls the senses, slows down reaction times and reduces an individual's ability to carry out even simple tasks. Being

Figure 1.23 Being under the influence of alcohol at work is a serious health and safety risk

Level 2 Light Vehicle Maintenance & Repair

under the influence of alcohol at work is extremely dangerous and illegal if the job involves driving or operating dangerous machinery.

If you are found to be under the influence of alcohol while at work your employer may take disciplinary action, possibly even resulting in your dismissal.

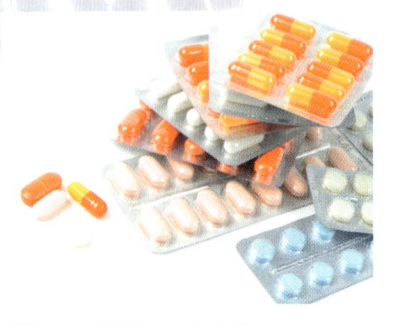

Figure 1.24 Drugs, including prescription medicines, can affect your performance at work

Drugs

It is your responsibility to check that any drugs prescribed to you will not affect your performance at work. If there is any risk that they will affect you, then you should report this to your supervisor or manager so they can make suitable arrangements to protect everyone's safety. It is your employer's responsibility to take action to protect all concerned against drug abuse in the workplace.

Stress

Millions of working days are lost each year due to stress, depression and anxiety. Health problems linked to stress include heart problems and skin and stomach conditions. Many factors can increase stress levels, such as working in cramped conditions, overworking, concerns over risk of injury or illness, poor communication with managers and lack of job security. Your employer can help to reduce stress levels by having procedures in place to address the causes of stress.

Violence

People react to stress in various ways – one reaction may be violence at work. You must report any violence to your manager immediately. Violence may include physical or verbal threats or abuse, or harassment and intimidation. It is your employer's responsibility to investigate incidents of violence and report them to the relevant enforcement authority if necessary.

Figure 1.25 Stress can cause health and safety problems at work

Figure 1.26 Never resort to violence at work – refer any problems or issues to your manager

Did you know?

It is illegal under the Health and Safety at Work Act 1974 for one employee to bully or harass another.

Harassment and bullying

In line with the Health and Safety at Work Act 1974 your employer must have a policy in place which clearly states that harassment and bullying are unacceptable and that this type of behaviour will lead to disciplinary action being taken. Your employer must ensure that there is a responsible person for you to go to if you are being harassed or bullied.

Figure 1.27 Harassment and bullying can seriously affect your performance at work and should be reported to your manager

1 General workshop operation principles for light vehicles

Fire prevention

Each year fire and its effects are responsible for a substantial loss of life. Fire prevention and control depend on managing the three elements that a fire needs to burn: fuel, oxygen and heat. These three elements are referred to as the 'fire triangle' as shown in Figure 1.28. Table 1.34 explains each of the elements. It also gives the controls needed to help prevent fire.

Table 1.34 The fire triangle elements and how to control them

Fire triangle element	What is it?	Controls
Fuel	Paper, wood, waste, rubbish and flammable substances	• Remove waste and rubbish regularly. • Avoid using flammable substances or keep to a minimum. • Store flammable substances away from sources of ignition (e.g. hot machine parts) and preferably in fireproof stores.
Oxygen	Approximately 20 per cent of the atmosphere is made up of oxygen.	• It is not normally possible to control the oxygen in the air. But fires can be put out by depriving them of oxygen – by smothering the flames.
Heat	There has to be a source of ignition to create a fire. This can be produced by friction in machines, naked flames, hot surfaces and smoking.	• Machines should be checked and serviced regularly. • Smoking should be banned anywhere near flammable substances. • Open flames should be avoided.

Figure 1.28 Fire triangle – the three elements involved in fire

Did you know?

The fire triangle includes oxygen as one of the main ingredients for fire. If more oxygen is added to the other two ingredients, heat and fuel, this increases the risk of fire.

Did you know?

Vapours given off from liquids are heavier than air. They can carry across a workshop floor or settle in the bottom of an inspection pit.

Flammable liquids

Types of flammable liquids and gases found in a workshop include:

- **petrol** – used as fuel for vehicles
- **diesel** – used as fuel for vehicles
- **thinners** – used to clean spray paint guns
- **paraffin** – used for cleaning oil off vehicle components
- **paint** – used for painting vehicles
- **brake cleaner** – used to damp down brake dust and clean brake linings.

A **volatile** liquid will quickly **vaporise** at room temperature. A substance such as petrol has a low **flash point**, which means it will vaporise at a low temperature. This makes petrol more dangerous than a substance such as oil, which has a high flash point.

Key terms

Volatile – a liquid that readily turns to vapour at room temperature.

Vaporise – to turn into a gas, for example water vaporises when it turns into steam.

Flash point – the lowest temperature at which a liquid will vaporise to form an ignitable mixture in air.

51

Working life

A customer has filled their tank with petrol instead of diesel. Sid is now emptying the tank before removing the fuel tank to clean it out. He pumps out the petrol into an open-topped container he has found in the back of the workshop. As he does this his supervisor enters the workshop and notices a strong smell of petrol. His supervisor frantically gestures for Sid to stop what he is doing.

1. Why is Sid's manager so concerned?

Causes of accidents involving flammable liquids

Table 1.35 shows some of the accidents that might happen involving flammable liquids.

Find out

Look at the 'Possible cause' column in Table 1.35.

- Try to decide how each of these accidents could be avoided.

Table 1.35 Accidents involving flammable liquids and their possible causes

Accident	Possible cause
Fuel tank explosion	• Welding near to the fuel tank • Smoking when filling a fuel tank • Repairing a fuel tank using heat
Battery explosion	• Disconnecting leads during charging • Smoking near charging battery
Fire in inspection pit	• Allowing vapours from fuel to spread • Using flammable substances to clean pit
Fire under bonnet	• Sparks from ignition system • Heat from the exhaust • Electrical short circuit
Fire at or near welding bench	• Not disposing of **combustible** waste • Excess oxygen on oily rags

Key term

Combustible – any material which is capable of igniting and burning.

Emergency

Flashing lights or similar alarm systems may be used where ear protectors are worn or where there are known to be individuals with hearing problems.

Fire detection

Your workplace will have systems in place for detecting and giving warnings of fire. These systems may detect high temperatures, smoke, radiation or certain gases produced by a fire. Automatic or manual alarms will be used to warn of danger. Fire alarms need to be tested weekly to make sure they are working, that everyone is familiar with the sound of the alarm and that it can be heard.

In the event of a fire – evacuation

It is your employer's responsibility to make sure that you know what to do in the event of a fire. At your induction your employer will make sure that you receive appropriate training and are given relevant information on safe evacuation. To make sure you are fully aware of the fire evacuation procedure there will be rehearsals at regular intervals.

Routes of evacuation

Your workplace must have adequate safe routes of escape. Routes of escape may include fire doors, emergency exits and fire-resistant staircases. Internal fire doors must be kept closed at all times. Leaving a door propped open would allow a fire to spread more quickly.

- **Emergency exit doors** – These must open outwards and they must not be locked unless strictly necessary. If they are locked, there must be a safe emergency opening system that is clearly labelled and explained. Fire exits and doors must never be obstructed.
- **Sign posts** – These must be clear and inform people of the location of emergency exits, escape routes, firefighting equipment and fire doors (see Figure 1.29).
- **Emergency lighting system** – This must highlight the location of fire exits and walkways. It must be checked regularly.
- **Record of all visitors and staff** – A daily record of all the people in the workplace should be kept, with details of when people arrive and when they leave. If fire does break out, the designated fire officer will have a list of everyone who was in the building and will be able to check whether everyone has got out of the building to the fire assembly point.
- **Fire assembly point** – Everyone should remain at the assembly point until they are given permission from someone in authority, such as a fire officer or senior manager, to re-enter the building.

Dealing with a fire

The main consideration in the event of fire is to evacuate the building. The building can be rebuilt after a fire, but people cannot. The only reason for having firefighting equipment is to allow people to escape from a burning building. However, if you do think it is safe to put out a fire, make sure that you are close to a fire exit if the fire gets worse.

The purpose of firefighting is to remove elements of the fire triangle (see page 51) by:

- **cooling** – this removes or reduces heat
- **smothering** – to remove air (oxygen)
- **starving** – this removes the fuel.

Firefighting equipment

There are several types of firefighting equipment found in the workplace, as shown in Figures 1.30 to 1.32 on pages 54–55.

Fire blankets

Staff should be trained in the correct, safe use of fire blankets. These can be used to smother small oil fires.

Red label extinguisher – water

- This is used for fires involving solid materials such as wood, cloth, paper, plastics and coal.

Emergency

In the event of a fire, keep internal fire doors closed. This cuts off the oxygen supply to the fire and helps prevent smoke and flames from spreading.

Figure 1.29 Fire exit sign

Emergency

In the event of a fire, lifts must not be used. They could fail if the electricity supply was disconnected, leaving the occupants trapped within.

Emergency

It is important that any fire extinguisher used is the correct type. You should only use an extinguisher if you have been trained to do so. You must never put yourself or others at risk.

Level 2 Light Vehicle Maintenance & Repair

- Do not use on petrol, oil or on electrical appliances.
- Use it by pointing the jet at the base of the flames and keeping it moving across the area of the fire.
- Make sure that all areas of the fire are out.
- It works mainly by cooling the burning material.

Black label extinguisher– CO_2 (carbon dioxide)

- This is used on liquids such as grease, oil, paint and petrol. It can also be used on electrical fires.
- Do not use on molten metal fires.
- This type of extinguisher does not cool the fire very well and you need to watch that the fire does not start up again.
- The fumes can be harmful if used in confined spaces – ventilate the area as soon as the fire has been controlled.
- Use it by pointing the jet at the base of the flames. Keep the jet moving across the area of the fire.
- It works because carbon dioxide gas smothers the flames by replacing oxygen in the air.

Cream label extinguisher– foam

- This is used on fires involving solids such as wood, cloth, paper, plastics and coal.
- It is also used for liquids such as grease, oil, paint and petrol.
- Do not use on electrical fires.
- For fire involving liquids, do not aim the jet straight into the liquid.
- Where the liquid on fire is in a container, point the jet at the inside edge of the container or on a nearby surface above the burning liquid.
- Allow the foam to build up and flow across the liquid. It works by forming a fire-extinguishing film on the surface of a burning liquid.

Figure 1.30 Fire blanket and four types of fire extinguisher

Blue label extinguisher– powder (dry)

- This is for fires involving solids such as wood, cloth, paper, plastics and coal.
- It can also be used on liquids such as grease, paints and petrol.
- Do not use on molten metal fires.
- Use by pointing the jet or discharge horn at the base of the flames. With a rapid sweeping motion, drive the fire towards the far edge until all the flames are out.
- If the extinguisher has a shut-off control wait until the air clears and, if you can still see the flames, attack the fire again.

Emergency

Do not use water (red label) or foam (cream label) extinguishers to put out liquid or electrical fires, as these extinguishers are water based. The water will spread the flames and intensify the fire. Water conducts electricity and may lead to electrocution if used on an electrical fire.

- It works by melting to form a skin smothering the fire and provides a cooling effect.

Sprinkler systems

These are automatic. They detect and control a fire in the early stages.

Hose reels

These are provided for the use of the Fire Service. They must be easily accessible.

Fire safety checks

Fire safety checks must be carried out regularly in the workplace. Table 1.36 shows what must be checked, how often and who is responsible for carrying out the checks.

Figure 1.31 Sprinkler system

Table 1.36 Fire safety checks

How often?	What is checked?	Who is responsible?
Daily	Fire exits	All staff
Weekly	Fire alarms in rooms	Health and safety officer/ maintenance department
Monthly	Electrical appliances	Maintenance department
Half yearly	External fire stairs	Manager Health and safety officer
Yearly	Firefighting equipment	Fire service representative

Figure 1.32 Fire hose reel

CHECK YOUR PROGRESS

1. Accidents are often caused by a mixture of factors. List three factors.
2. What type of fire extinguisher has a black label?
3. What is the meaning of flash point?

Risk assessment

Risk assessment is a technique for preventing accidents and ill health by helping everyone in the workplace to think about what could go wrong and ways to prevent problems.

Risk assessment is good practice and a legal requirement. It enables your workplace to reduce the costs associated with accidents and ill health. It will also help your employer to decide on their priorities, highlight training needs and assist with quality assurance programmes.

Did you know?

A bottle of bleach contains a hazard. But when the bleach is locked in a cupboard the hazard is reduced. The risk increases when the bleach is used.

Risk assessments help identify what could go wrong and how serious the results could be. The steps of a risk assessment are shown in Figure 1.33.

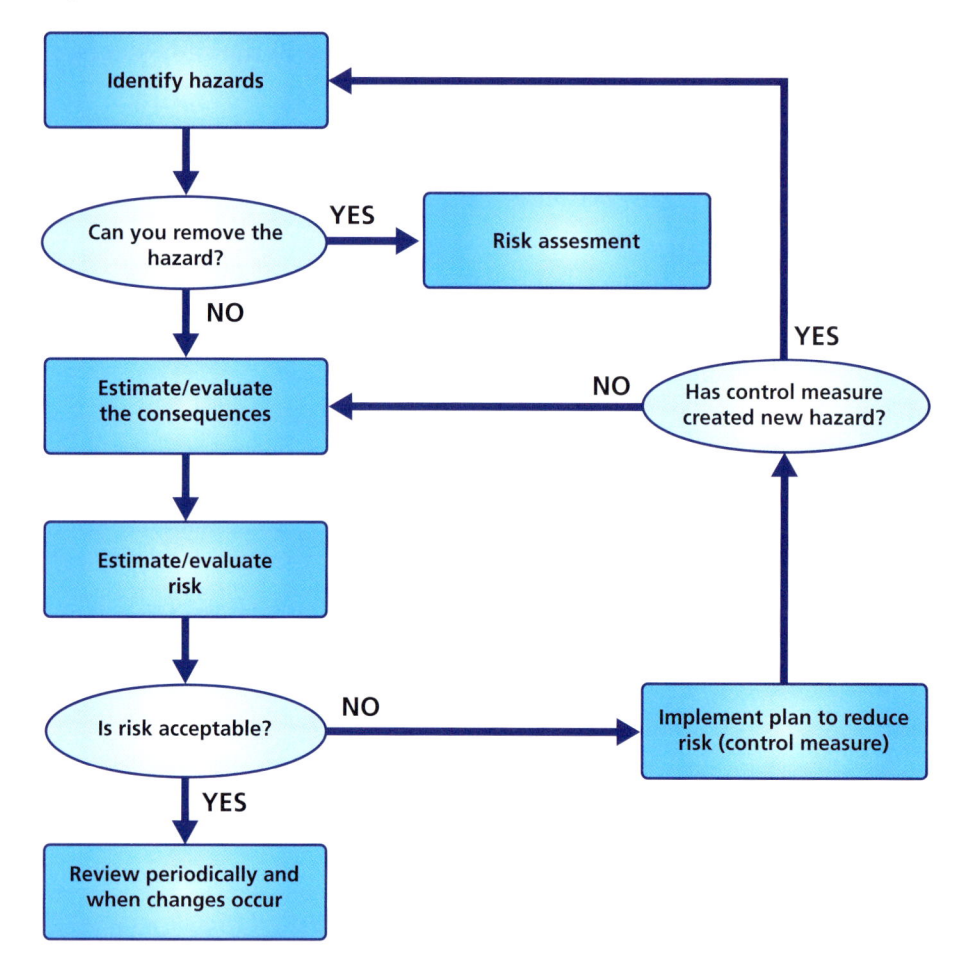

Figure 1.33 Flow chart showing the steps involved in risk assessment

The 1, 2, 3 system

Evaluating risk is an area that can be daunting. There are many different systems for achieving an evaluation. One of the most common is the 1, 2, 3 system where the assessor rates the severity of a possible injury and the likelihood of that injury occurring on a scale of 1 to 3 (see Table 1.37).

Table 1.37 The 1, 2, 3 system of risk evaluation

Severity of injury		Likelihood	
Minor	1	Seldom	1
Serious	2	Occasional	2
Major or death	3	Imminent	3

The person assessing the risk then multiplies the two numbers together to give a **risk rating**. The possible risk ratings are shown in Table 1.38.

1 General workshop operation principles for light vehicles

Table 1.38 Risk level ratings

Rating	Level of risk
1	Trivial risk
2	Low risk
3	Low to medium risk
4	Medium risk
6	Medium to high risk
9	High risk

Table 1.39 Risk severity ratings

Likelihood	Severity: Major or death	Serious	Minor
Imminent	9	6	3
Occasional	6	4	2
Seldom	3	2	1

A rating of 9 will be regarded as an unacceptable risk and work practices or equipment will need to be changed. Ratings of 3 to 6 require that control measures are introduced to ensure the safety of staff and customers. See Table 1.38.

> ### Risk rating example
>
> Consider the risk of being hit by a car that a pedestrian who is crossing a busy road faces.
>
> **Severity value:** Being hit by a car will usually result in serious injury or even death. So using Table 1.37, this has a severity value of 3.
>
> **Likelihood value:** The likelihood of the pedestrian being hit by a car and injured or killed depends on many factors, such as road conditions, traffic volume and the ability of the pedestrian. On a dangerous stretch of road, there might be a risk of occasional injury, giving a likelihood value of 2.
>
> **Risk rating:** Multiply together the severity and livelihood values:
>
> Severity × Likelihood
>
> = 3 × 2
>
> = 6
>
> **Control measures:** A rating of 6 represents a medium to high risk. Control measures need to be put in place to bring the risk down to an acceptable level. One possible control measure would be to install a pedestrian crossing to make crossing the road safer.

Risk assessment control measures

A control measure (extra measure) is introduced when a hazard or risk is not already adequately controlled.

It is not always possible to remove a hazard, but it may be possible to separate people from it. For example, a guard can separate a machine operator from a sharp or hot piece of equipment. Sometimes it is possible to make a substitution, such as replacing a dangerous chemical with a less hazardous one.

Level 2 Light Vehicle Maintenance & Repair

> ⚠️ **Safe working**
> Always think about what could go wrong so that the situation can be controlled before accidents or ill health occur.

As a last resort, staff may be provided with personal protective equipment (PPE). For example, chain mail gauntlets could be provided for someone whose work involves cutting, and earplugs could be issued for someone working in a noisy area. However, it is important to remember that the hazard is not sufficiently controlled, even though people are protected.

What other action is needed?

- Information and training on all hazards and control measures must be provided.
- Records of the assessments should be kept.
- Risk assessments must be reviewed from time to time to ensure that the control measures continue to be appropriate.

A review should always take place when changes are made such as the introduction of new equipment. Managers and staff who carry out risk assessments need full training in the technique and the legal requirements.

All risk assessments should be recorded and stored for inspection by internal and external authorities if required.

Recording a risk assessment

You may need to assist your employer in carrying out a risk assessment for your workplace. You should therefore be aware of the five steps of recording a risk assessment. These are shown in Figure 1.38 on page 60.

Safety signs

You will come across many safety signs in your workplace. They are there to communicate information, such as:

- warning of a hazard
- showing emergency exits
- to instruct you to wear personal protective equipment.

The type of warning is indicated by the colour of the sign.

Green signs

Green signs indicate a SAFE condition (useful information). These are often used for signs indicating where to go in emergency situations.

Look for boxes with this symbol throughout the book for SAFE conditions you should be aware of:

> **+ Emergency**
> If the low battery symbol appears on a digital multimeter screen, replace the battery straight away. Otherwise, you might get inaccurate readings that could lead to electric shock or personal injury.

Figure 1.34 Green safe condition signs often give information that you will need in an emergency

Red signs

Red signs indicate a PROHIBITION (something you **must not** do). They often have a red diagonal line over the symbol on the sign.

Do not oil or clean this machine whilst in motion
Warning to protect against damage to machine and operator

Do not drink the water

No smoking

Figure 1.35 Red prohibition signs tell you what you must not do

> Look for boxes with this symbol throughout the book for PROHIBITIONS you should be aware of:
>
> **Safe working**
> Do not measure the voltages of a hybrid drive system unless you have been specifically trained. Hybrid drives operate with high voltages that can cause electric shock and death.

Blue signs

Blue signs indicate a MANDATORY instruction (something that you must do).

Wear protective gloves

Guards must be in position before starting

Ear protection zone

Figure 1.36 Blue mandatory signs tell you what you must do

> Look for boxes with this symbol throughout the book for MANDATORY instructions you should be aware of:
>
> **Safe working**
> Following the replacement of any engine components, the car should be road tested to ensure correct function and operation.

Yellow signs

Yellow signs indicate a WARNING (hazard or danger).

Danger of fire

Tripping hazards

Electrical hazard

Figure 1.37 Yellow warning signs tell you about dangers that are nearby

> Look for boxes with this symbol throughout the book for WARNINGS you should be aware of:
>
> **Safe working**
> When checking for an electrical short circuit, only bypass the fuse with an electrical consumer like a bulb. Using other electrical components could cause a sudden discharge of electricity that may burn you.

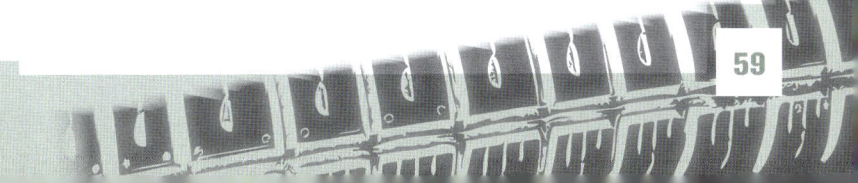

STEP 1

Hazard

Look only for hazards which you could reasonably expect to result in significant harm under the conditions in your workplace. Use the following examples as a guide.

- ☐ slipping/tripping hazards (e.g. poorly maintained floors or stairs)
- ☐ fire (e.g. from flammable materials)
- ☐ chemicals (e.g. battery acid)
- ☐ moving parts of machinery (e.g. blades)
- ☐ work at height (e.g. from mezzanine floors)
- ☐ ejection of material (e.g. from plastic moulding)
- ☐ pressure systems (e.g. steam boilers)
- ☐ vehicles (e.g. fork-lift trucks)
- ☐ electricity (poor wiring)
- ☐ dust (e.g. from grinding)
- ☐ fumes (e.g. welding)
- ☐ manual handling
- ☐ noise
- ☐ poor lighting
- ☐ low temperature

STEP 2

Who might be harmed?

There is no need to list individuals by name – just think about groups of people doing similar work or who may be affected, for example:

- ☐ office staff
- ☐ maintenance personnel
- ☐ contractors
- ☐ people sharing your workplace
- ☐ operators
- ☐ cleaners
- ☐ members of the public

Pay particular attention to:

- ☐ staff with disabilities
- ☐ visitors
- ☐ inexperienced staff
- ☐ lone workers

They may be more vulnerable.

STEP 3

Is more needed to control the risk?

For the hazards listed, do the precautions already taken:

- ☐ meet the standards set by a legal requirement?
- ☐ comply with a recognised industry standard?
- ☐ represent good practice?
- ☐ reduce risk as far as reasonably practicable?

Have you provided:

- ☐ adequate information instruction or training?
- ☐ adequate systems or procedures?

If so, then the risks are adequately controlled, but you need to indicate the precautions you have in place. (You may refer to procedures, company rules, etc.) Where the risk is not adequately controlled, indicate what more you need to do (the 'action' list).

STEPS 4 and 5

Review and revision

Set a date for review of the assessment.

On review check that the precautions for each hazard still adequately control the risk. If not indicate the action needed. Note the outcome. If necessary complete a new page for your risk assessment.

Making changes in your workplace, for example when bringing in new machines, substances, or procedures may introduce significant new hazards. Look for them and follow the 5 steps.

Figure 1.38 The five steps involved in carrying out a risk assessment

First aid

It is your employer's responsibility to provide first aid facilities. These will be based on the risk of accident or injury occurring. Your employer will have carried out a risk assessment to confirm the level of first aid provision required.

Points which will have been considered when assessing first aid requirements:

- the number of staff employed
- the history of accidents
- how far the workshop is from medical services
- working with hazardous substances or dangerous equipment.

There are two levels of first aid staff:

- **Trained first aider** – This is someone who has completed a recognised course in first aid and has achieved a first aid certificate approved by the Health and Safety Executive (HSE).
- **Appointed person** – This person may not be trained in first aid but has been appointed to take charge of the situation to ensure that the casualty receives the correct attention/treatment. An appointed person must be available during all times of the working day.

> **Did you know?**
>
> The Health and Safety (First Aid) Regulations 1981 state that employers must have emergency procedures in place in the event of an incident occurring.

CHECK YOUR PROGRESS

1. What is the 1, 2, 3 system of risk assessment?
2. What is a control measure?
3. What colour are mandatory warning signs?

Good housekeeping practices

Good housekeeping in the automotive environment means keeping the workshop clean and tidy at all times, so that it is a safe place to work. This includes making sure that all equipment is stored correctly after use and the floor is clear of obstructions.

Good housekeeping also involves looking after resources and using them economically.

Good housekeeping calls for a combination of:

- self-discipline
- organised storage
- effective supervision
- proper management
- staff training to develop the right attitude.

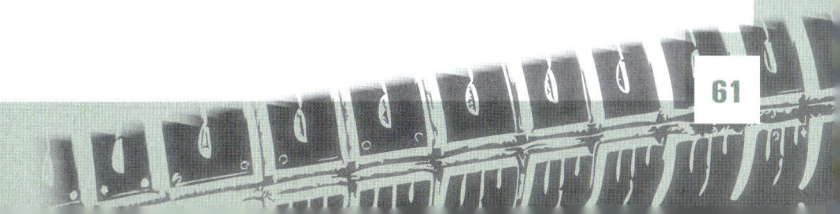

Level 2 Light Vehicle Maintenance & Repair

Figure 1.39 Can you spot all the examples of poor housekeeping in this workshop? (Answers can be found on the Level 2 Light Vehicle Maintenance & Repair Training Resource Disk)

1 General workshop operation principles for light vehicles

Housekeeping and cleaning are an essential part of preventing accidents and occupational ill health. Figure 1.40 provides a list of daily housekeeping activities.

Workshop housekeeping – a typical day

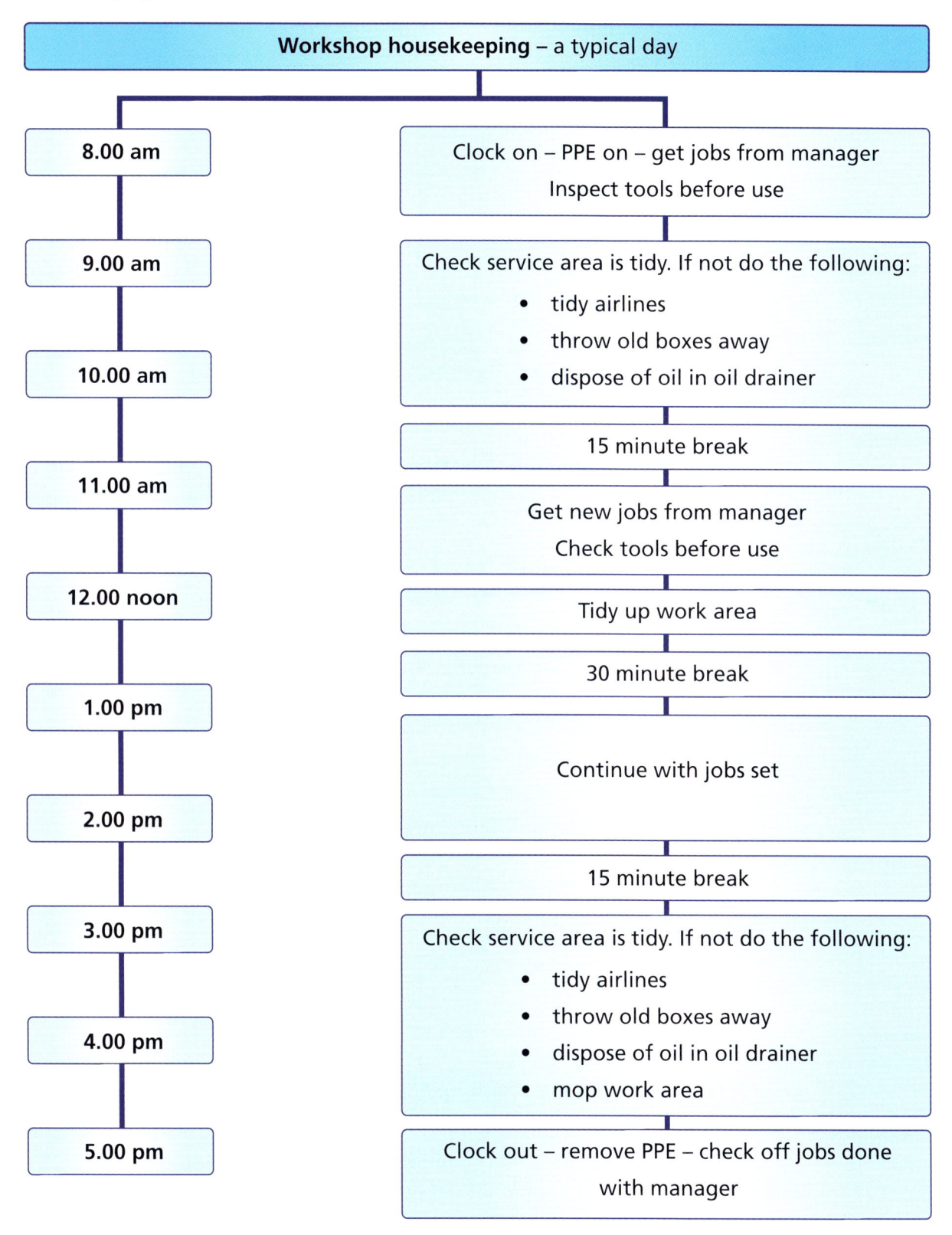

8.00 am	Clock on – PPE on – get jobs from manager Inspect tools before use
9.00 am	Check service area is tidy. If not do the following: • tidy airlines • throw old boxes away • dispose of oil in oil drainer
10.00 am	15 minute break
11.00 am	Get new jobs from manager Check tools before use
12.00 noon	Tidy up work area
	30 minute break
1.00 pm	
2.00 pm	Continue with jobs set
	15 minute break
3.00 pm	Check service area is tidy. If not do the following: • tidy airlines • throw old boxes away • dispose of oil in oil drainer • mop work area
4.00 pm	
5.00 pm	Clock out – remove PPE – check off jobs done with manager

Figure 1.40 Flow chart of daily housekeeping activities

Level 2 Light Vehicle Maintenance & Repair

Find out
- Look around your workshop and see if you can spot any examples of poor housekeeping.
- Report anything that you see to your manager.

Did you know?
'Falls are the most common cause of death or serious injury to people at work. In Motor Vehicle Repair (MVR) they account for nearly 10 per cent of injuries, mainly involving falls from ladders.'

(Source: HSE, *Reducing ill health and accidents in motor vehicle repair*)

Did you know?
The Health and Safety at Work Act 1974 requires employers to ensure the health and safety of all employees and anyone who may be affected by their work. Employers must also maintain the workplace and/or premises in a safe condition. This includes adequate cleaning and housekeeping arrangements. (For more on the Health and Safety at Work Act 1974 see page 2).

Poor housekeeping

Trips, falls and other accidents can easily occur if good housekeeping principles are not practised in the workplace. Table 1.40 gives some examples of poor housekeeping and how they can be avoided.

Table 1.40 Examples of poor housekeeping that may cause falls

Example of poor housekeeping	Control measure
Tools left on the floor	Keep a tidy workshop. Be aware of any potential trip hazards you may cause.
Trolleys obstructing exits	Never obstruct exits.
Broken floor tiles	All floor surfaces should be kept in good repair and maintained regularly.
Trailing wires	Keep all wires covered and do not leave trailing across walkways.
Standing on a box to reach an item	Use a stepladder or step stool to reach the item.
Wet floors	Clear away spillages immediately and use the appropriate signage to notify others that the floor is wet.

Good housekeeping rules

A checklist of good housekeeping rules might include:

✓ After using tools, clean them and return them to their correct storage place.
✓ Pick up or move any obstructions on the workshop floor.
✓ Store all chemicals in their correct location.
✓ Put all debris in the bins provided.
✓ Empty the bins when they get full.
✓ Clean up any spillages of oil or dirt from the floor using the correct materials and equipment.

Cleaning

The terms good housekeeping and cleaning are closely linked but have different meanings:

- Housekeeping means having a place for everything and keeping everything in its place.
- Cleaning means removing dirt and debris and disposing of it correctly.

You should immediately clean away all spillages using appropriate methods. Table 1.41 lists some common workshop spillages and the items you need to use to clean up the spillage.

1 General workshop operation principles for light vehicles

Table 1.41 Dealing with spillages and leaks

	PPE*	Absorbent granules	Brush and shovel	Mop and bucket	Water solvent	Extra precautions
Oil spillage	✓	✓	✓	✓	✓	Use a sign to warn people of the spillage
Fuel	✓	✓	✓	✓	✓	Ventilate area well
Coolant	✓	✓	✓	✓	✓	Use a sign to warn people of the spillage
Nuts and bolts	✓	–	✓	–	–	–

Note: PPE will be latex gloves, overalls and steel toe-capped boots

Risks involved when using solvents and detergents

You need to be aware of the risks involved when using **solvents** and **detergents** to clean up spillages and to clean tools after use. Risks include irritation to the skin which could result in dermatitis or burning of the skin.

Key terms

Solvent – a substance, usually a liquid, that is capable of dissolving another substance.

Detergent – a cleaning agent that increases the ability of water to penetrate fabric and break down grease and dirt. The molecules in detergents surround particles of grease and dirt, allowing them to be carried away.

Working life

William has been asked to remove brake dust from the front brake discs. His service manager informs him of the solvent to use. Eager to get the job finished, William uses the solvent without hand protection. He drops oil on the vehicle's seat and grabs a bottle of detergent to rub the seat clean before reading the label.

1. What could be the long-term consequences of not using hand protection?
2. Was William correct to use detergent without reading the instructions?

Good housekeeping of hazardous substances

Hazardous substances are any liquids, solid materials or gases which could cause you harm. They are described more fully in the section on Control of Substances Hazardous to Health Regulations 2002 (COSHH) on pages 10–11.

Before you use or dispose of any hazardous substance you must receive effective COSHH induction training. It is essential that hazardous substances are stored and used correctly:

- ✓ The storage of hazardous substances must be properly planned.
- ✓ All containers must be labelled accurately.
- ✓ Always keep hazardous substances out of reach of food and drink.
- ✓ Follow procedures for using and disposing of hazardous substances.
- ✗ Never smoke around hazardous substances.

Safe working

Don't forget: Always clean your tools and work areas after using or disposing of hazardous substances to avoid **cross-contamination**.

Key term

Cross-contamination – where one substance causes damage to another by coming into contact with it for example where an aluminium wheel sticks to a cast iron hub during removal.

Safe disposal of hazardous waste

Good housekeeping includes the safe disposal of hazardous waste. For more information on this, see the sections *Safe disposal of special and hazardous waste* and *Recycling in the automotive industry* on pages 29–30.

Using resources economically

Resources used in your workplace may include **consumables**, electricity, water and heating. Using these resources economically will help keep your employer's bills low (making the business more successful) and reduce harmful effects on the environment.

Consumables such as grease, freezing agents and sealants are used daily in motor vehicle workshops. It is important that you are aware that unnecessary use of these items impacts on dealer costs and the environment. Some examples of how to use consumables economically are given in Table 1.42.

> **Did you know?**
> In July 2010 legislation was passed so that everyone who works on light vehicle air conditioning systems must have a recognised qualification for refrigerant handling.

Table 1.42 Consumables in a workshop and how to use them economically

Consumable	Economic use
Grease	Only use enough grease as is necessary on the wheel bearings.
Sealant	Always seal the end after use.
Freezing agents	Ensure the air-conditioning system has no leaks before recharging the system.
Brake cleaner	Always replace the lid to prevent unnecessary spraying.
Epoxy resins	Only use enough body filler for the size of dent.
Engine oil	Never overfill an engine.

Using electricity economically

Electricity is used for lighting, heating and to power the equipment you need to work within the motor vehicle workshop.

It is important not to use electricity excessively for two reasons:

- **Economic reasons** – the more electricity you use, the higher your employer's bills will be and the less profit the business will make.
- **Environmental reasons** – creating electricity has an effect on the environment, as explained below.

Electricity can be produced in a variety of ways. Each technique involves the use of a turbine, which spins and converts **kinetic energy** into electricity. The most likely source of electricity to your workplace and home is a coal power plant. Burning the fuel to produce electricity creates the greenhouse gases which cause global warming. Also, coal and oil are not limitless resources. By using electricity economically you will reduce the strain on these resources.

> **Key terms**
> **Consumables** – goods intended to be bought and used.
> **Kinetic energy** – the energy of motion.

1 General workshop operation principles for light vehicles

You can save electricity by:

- turning off equipment when not in use
- switching off lights in rooms and corridors that aren't being used
- making the most of natural light
- using a task light instead of the overhead lighting (where sensible)
- not charging appliances or batteries needlessly.

Preventing pointless heat loss

You need be comfortable in your workplace, but you must make sure that if the heating is on you do not leave windows and doors open. This will reduce the impact on the environment, help save your employer money and also keep your workplace warm. If you are too hot, turn the heating down first before opening a window or door.

Using water economically

By reducing the amount of water wasted you will help:

- protect and preserve the Earth's most valuable resource
- save energy (used for treatment and pumping of water and waste water)
- save money (by limiting the amount of hot water and metered water used).

Find out

Take a look around your workshop and note down all the different uses of water within the workshop. See if you can make any suggestions for how to reduce water use and waste.

Did you know?

Water is essential for all life. Your body can survive for a month or so without food but, depending on the situation you are in, less than a week without water.

CHECK YOUR PROGRESS

1 Why do you need to turn off the light in a store room when it is not being used?
2 What effect does the wasting of resources have on a business?
3 Name three things that can be done to save electricity.
4 What should you use to scrub your hands with when using a bench drill? (Bear in mind that hands and forearms may come into contact with cutting oil.)

2. Job roles and communication

Organisational structures, functions and roles

A typical dealership will consist of the following departments, all of which work alongside each other in partnership:

- Reception – vehicle sales
- Service department
- Parts department/body shop/workshop
- Main office.

A smaller workshop may consist of a reception/service desk and vehicle repair workshop. In a large dealership you will be a specialist in your own department. In a smaller workshop you will need to have a diverse array of skills, including the ability to communicate effectively with others.

Typical job roles in a dealership or workshop are:

- **Manager** – This person has overall responsibility.
- **Supervisor** – This person ensures the members of the team do their jobs correctly.
- **Skilled technician** – He or she has in-depth knowledge in their role.
- **Trainee/apprentice** – This person is constantly learning from the skilled technician, supervisor and manager.

Figure 1.41 shows how all the departments are involved in the vehicle repair process.

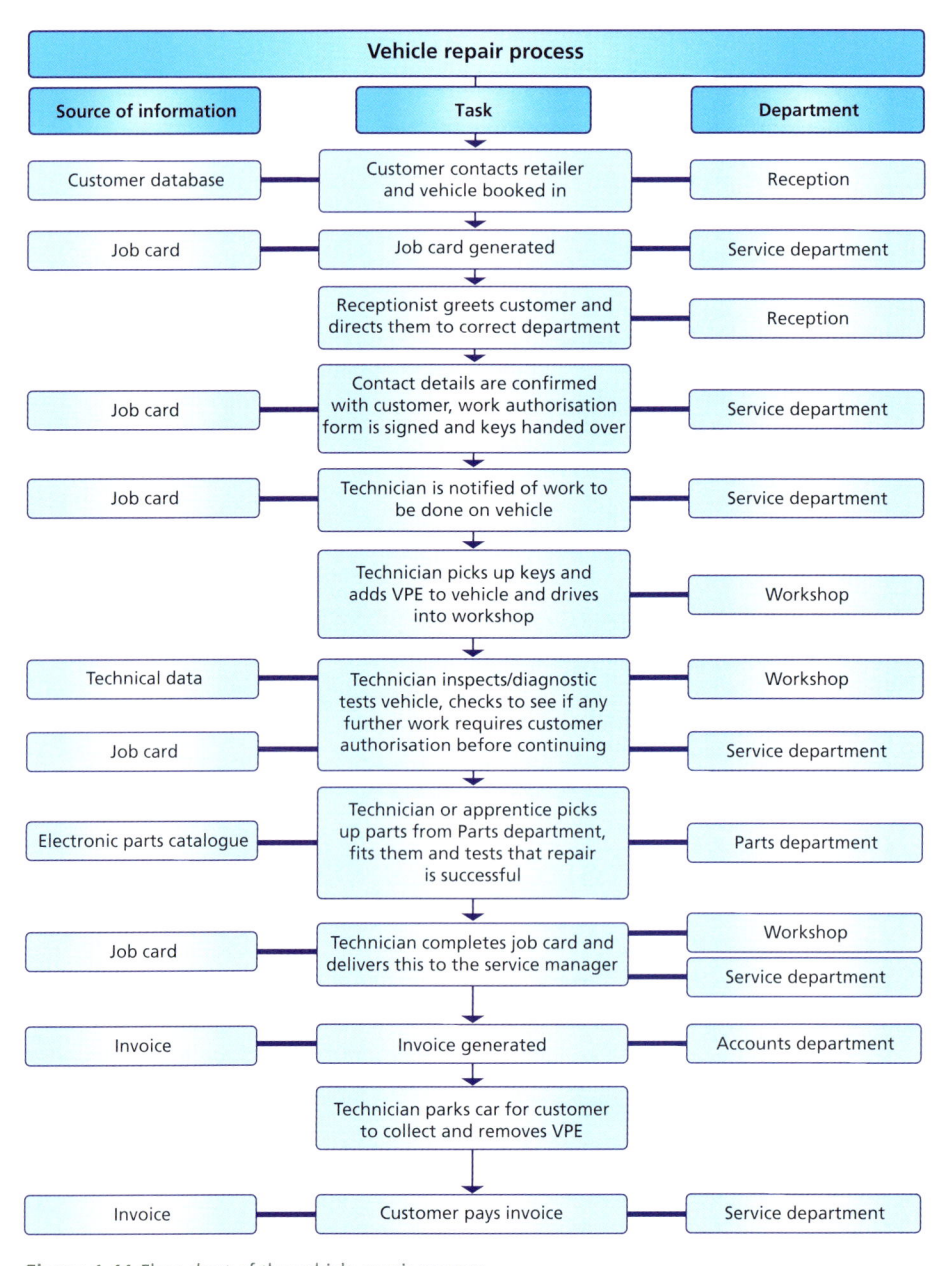

Source of information	Task	Department
Customer database	Customer contacts retailer and vehicle booked in	Reception
Job card	Job card generated	Service department
	Receptionist greets customer and directs them to correct department	Reception
Job card	Contact details are confirmed with customer, work authorisation form is signed and keys handed over	Service department
Job card	Technician is notified of work to be done on vehicle	Service department
	Technician picks up keys and adds VPE to vehicle and drives into workshop	Workshop
Technical data	Technician inspects/diagnostic tests vehicle, checks to see if any further work requires customer authorisation before continuing	Workshop
Job card		Service department
Electronic parts catalogue	Technician or apprentice picks up parts from Parts department, fits them and tests that repair is successful	Parts department
Job card	Technician completes job card and delivers this to the service manager	Workshop / Service department
Invoice	Invoice generated	Accounts department
	Technician parks car for customer to collect and removes VPE	
Invoice	Customer pays invoice	Service department

Figure 1.41 Flow chart of the vehicle repair process

Reception

In large dealerships the reception may be covered by a full-time receptionist. In smaller establishments it is likely that the role of receptionist will be taken on by any one of the workshop team.

When you are covering reception you must focus on the customer's needs. You will need to be able to confidently and accurately advise customers about minor technical and motor-related issues. As with all roles you will also need to achieve company objectives and targets set by your employer.

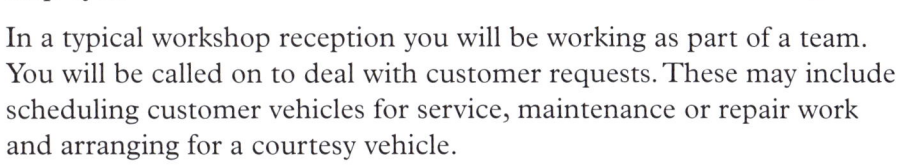

Figure 1.42 The reception in a large dealership

In a typical workshop reception you will be working as part of a team. You will be called on to deal with customer requests. These may include scheduling customer vehicles for service, maintenance or repair work and arranging for a courtesy vehicle.

You will greet customers in the service department, liaise with technicians in the workshop to chase work in progress, keep customers informed on progress, and gain authorisation for any additional work to be carried out.

Customer queries and problems may arise at any time. You should aim to resolve all issues to your customer's satisfaction.

> **Did you know?**
>
> A great deal of money is spent by larger dealerships to analyse customer satisfaction. Surveys are generally given out to customers for their feedback so improvements can be made.

> **Working life**
>
> Alex has just started as an apprentice vehicle technician at a small family-run workshop. His manager has asked him to cover the reception. A customer arrives to collect her car. Alex locates the customer's invoice and presents this to the customer. The customer states that she has agreed with the manager that she will pay next week. Alex gives the customer her car keys and the customer leaves. When the manager returns to reception, he notices that the invoice hasn't been paid. He asks Alex about this and is shocked to hear that the customer has taken her car without paying.
>
> 1   How could this situation have been avoided?

Vehicle sales

The sales department is usually located at the front of a dealership. The area for sales may include a showroom and outside sales forecourt. The manager responsible for this department is the sales manager.

In the role of vehicle sales you will approach and assist potential new customers who come to look at cars in the showroom. A great deal of your time will be taken getting to know your customers, asking them questions to determine their price range and preferences. Once you have this knowledge you will be able to sell the features and benefits of the vehicles you feel are most suited to them. As buying a new vehicle is an important investment, you must be aware of all finance options available to your customer.

Level 2 Light Vehicle Maintenance & Repair

Service department/workshop

This is normally the main hub of a dealership or workshop. It is responsible for all mechanical repairs, routine servicing, manufacturer warranty work and preparation of new and used vehicles. The service manager is responsible for this department.

Parts department

Parts and accessories are sold by this department to trade, external customers and internal customers. Potential sales areas include:

- customers whose vehicles are in the workshop (including warranty repairs)
- additional accessories sold alongside new and used car sales
- private retail customers via the retail counter
- other businesses (trade customers)
- in-house 'sister' companies (businesses within the same group).

Figure 1.43 The service department

Figure 1.44 The parts department

An after sales department may be combined with the parts department.

Body shop

Cosmetic and structural repairs are completed in a body shop. These will include insurance, retail and warranty body repairs and paint refinishing work. Your role within this department would be as an apprentice technician assisting the **body technician**.

The Kitemark pledge

PAS 125 is the recognised standard for the body repair industry. The standard exists to improve the quality of vehicle body repair. The following five elements of the repair process are audited to achieve the standard:

- **Equipment** – Suitable equipment is used to undertake repairs.
- **Methods** – Correct repair methods must be used.
- **Staff training** – Technicians must be properly trained and currently competent.
- **Process management (inspection)** – Checks are carried out to ensure that the job is to the correct standard.
- **Materials (quality)** – Checks are carried out to ensure that all materials meet the required quality standards.

Figure 1.45 The body shop

> **Key term**
>
> **Body technician** – qualified body repairer who replaces and repairs body panels and trim.

A body shop that has the Kitemark (see Figure 1.46) for vehicle body repair will have been inspected and shown to have complied with the Kitemark Vehicle Body Repair standards. For more information, go to hotlinks to visit the Kitemark website.

Main office

This is the back office of the dealership or workshop. It is responsible for all financial aspects of the dealership, including payment of suppliers, payment of wages, banking, VAT and other tax returns, the production of monthly accounts and other management reports.

Not all vehicle repair organisations are as large as the example mentioned in this section. There are many vehicle repairers operating with just a handful of staff or less. These independent outlets make up the majority of the automotive sector and work on all types and ages of vehicles.

Figure 1.46 British Standards Institution (BSI) Kitemark® Certification

> **Working life**
>
> Aston is thinking of opening his own garage.
>
> 1   Where would be a good place for him to have the reception?

Processes and procedures

Any job done on a vehicle needs to follow a logical process to make the work as efficient as possible for both the garage and the customer. One example of this is using a checklist when servicing a vehicle. A service checklist is carefully designed to provide a systematic way of working on systems, so it reduces unnecessary time spent within a small area. It also reduces lifting and lowering times for working underneath a vehicle. (For an example of a service checklist, see Figure 1.49 on page 75.)

A similar process is needed for locating faults on vehicles, particularly electrical systems. The process will set out a logical approach for testing each component in turn to narrow down the fault. When working on vehicles it is very important that you follow the set procedures and do not guess a fault, as this will take extra time. Figure 1.48 on page 72 gives an example of a process for finding the fault in a lamp that doesn't work.

Figure 1.47 The main office of a dealership

> **Working life**
>
> While Donny is servicing a vehicle for a customer he finds that a rear light is not working. He goes to get a new bulb from the parts store and fits it. The light still does not work.
>
> 1   What should Donny have done first?

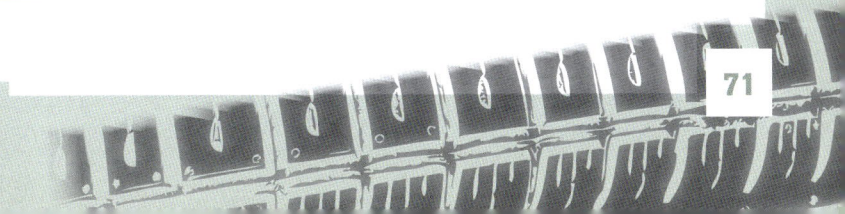

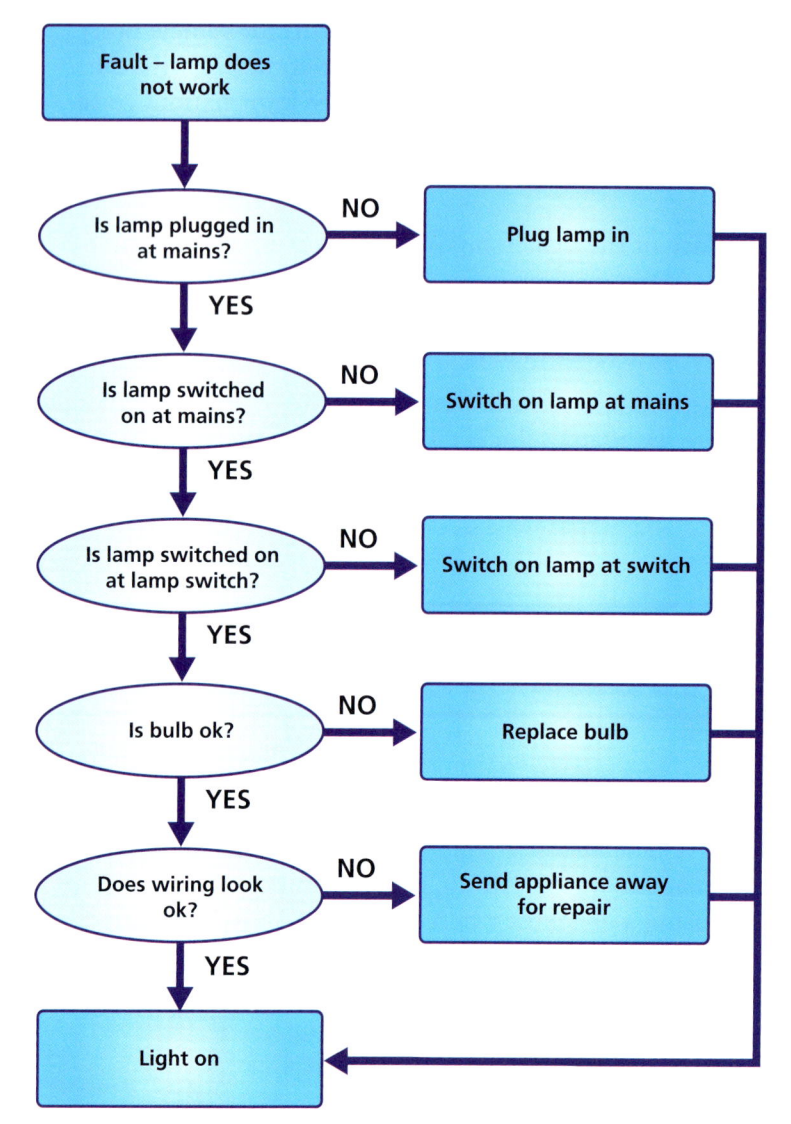

Figure 1.48 Process for finding the fault in a lamp that doesn't work

Sources of information

There are many sources of information to assist you in your job role.

Other staff

Your workshop team will have a vast array of knowledge and skills. While you are training, there will be many occasions when you will be asked to complete a task which you have never done before. By asking other members of your team for help and advice, you are demonstrating that you want to do the job right and that you appreciate their knowledge. You are also helping your own development by learning from others.

Service manual

An essential source of information for vehicle repair and testing is the vehicle's service manual. The service manual is generally available in paper form, from the Internet or on a CD.

Information that can be found in the service manual includes:

- repair procedures
- information for vehicle specifications
- vehicle identification codes
- vehicle service schedules
- wiring diagrams
- filling capacities.

Vehicle handbook

Vehicle handbooks are often supplied to the technician and kept in the toolbox for reference. They contain vehicle and component specifications. Specifications are the dimensions, measurements and identification codes a vehicle has, so that servicing and replacing parts can be standardised. Examples of these are how much oil the engine needs, tyre pressures and bulb sizes.

Internet

Vehicle manufacturer websites are an excellent information resource. They provide up-to-date information, with online workshop manuals, wiring diagrams, testing procedures and service bulletins. Information found on these sites can be printed out and kept as a reference for future repairs.

Tester's handbook

Information on MOT testing can be found in the MOT tester's handbook. The handbook states the items that must be tested and the limits allowed. Other information relating to MOTs can be found on the Internet. One useful site is the Directgov MOT page – go to hotlinks to visit this site.

The workshop manual

Information on vehicles is normally covered in the workshop manual. Examples of information include:

- general information
- technical specifications
- health and safety guidelines
- maintenance guide
- contact for further assistance or information
- fault diagnosis information
- removal and replacement procedures.

The equipment information given in the workshop manual will give you the appropriate instructions to understand how the equipment works, how to maintain it and how to take care of it.

EPC computer software

An electronic parts catalogue (EPC) is a catalogue of vehicle information held on CD-ROM or the Internet. The benefits of EPC are:

- All vehicle catalogues are stored in one place.
- Filter function – inputting the VIN or registration number brings up the information you need for that particular vehicle.
- Monthly updates are available on CD-ROM or to download.

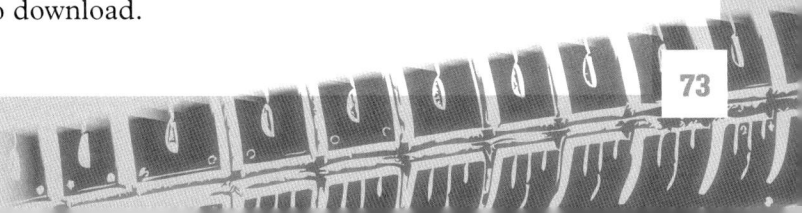

- It is quick and easy to use.
- It minimises the amount of incorrect part numbers.

Quick reference guides

Most vehicle component manufacturers publish quick reference guides. For example, a brake component manufacturer will supply a quick reference manual which will list all of the brake components made for every make and model of vehicle, all in one catalogue. The vehicle details are usually listed on the left of the page and the part numbers opposite on the right.

Many manufacturers publish quick reference guides for accident repairs. These usually come in the form of a pocket-sized book for each vehicle model, which lists the part numbers for commonly damaged panels.

Identification codes

A vehicle is made up of many different components. Each one has a different part number. These numbers mean that the part can be directly replaced with one from the vehicle manufacturer. Alternatively, an **aftermarket** part that is manufactured by a different parts company but is of a similar quality can be used. You can locate an aftermarket part by using the cross-referencing guide at the back of a parts book.

> **Key term**
>
> **Aftermarket** – a part which is manufactured by a company that is not the original vehicle manufacturer.

Documenting information

Documents are very important for keeping track of a vehicle's history. When a vehicle arrives for a service or maintenance you can use the recorded information to confirm what services and other work have been carried out on the vehicle. This information is also invaluable if a vehicle is sold, as the new owner will be able to trace the history of the vehicle.

The vehicle's documents must be completed accurately with the exact mileage, dates and any other information required.

Job cards

During the repair and maintenance of a light vehicle there is a flow of information. First the customer gives you or the service adviser their contact details and basic information about their vehicle. From this information a job card is created. This contains details of all the information the technician needs to carry out a repair or service.

Manufacturer's service schedules

Every manufacturer creates service schedules for each of their vehicles. When a vehicle is booked in for a service, the correct service schedule will be established from the vehicle's mileage and history. The corresponding service sheet is then printed off for the technician to carry out the service. The service is not only determined by the mileage, but also the age of the vehicle.

The technician will also need to use service information, which is found in the service manual or handbook.

1 General workshop operation principles for light vehicles

Inspection sheets

Some companies like to use inspection sheets alongside the service check sheet. Inspections sheets can also be used independently to carry out a vehicle inspection.

Service Inspection Checklist

Repair Order Number:	12345	Service Date:	22/05/2011	
Customer Name:	Mr Bloggs	Vehicle Make:	Sangsong	
Telephone Number:	07978 564564	Vehicle Model:	Coupe	
Vehicle Registration:	SJ58 SWJ	VIN:	SCBGA1114XHC37189	
Current Mileage:	40012	Engine Size:	1600cc	

FLUID LEVELS & REPLENISHMENTS	A	B	C	Recommendation
1. Engine Oil/Filter Change		✓		
2. Coolant, Washer Brake Fluid Level		✓		
3. Power steering/Auto Box Level		✓		
4. Air filter/Pollen Filter Change		✓		
5. Coolant Change				
6. Brake Fluid Change				
7. Timing Belt/Auxiliary Belt Change				

LIGHTS

8. Indicators / hazard warning lights		✓		
9. Side lights / number plate		✓		
10. Dipped / main beam headlights		✓		
11. Reversing Lights / Stop lights		✓		
12. Front / rear fog lights		✓		
13. Wiper blade wear and operation of front / rear wash / wipe		✓		

STEERING, BRAKES & SUSPENSION

14. Tyre wear / condition	✓	N/S/F 3 mm	O/S/F 3 mm	N/S/R 4 mm	O/S/R 6 mm	Spare 7 mm
15. Adjusted tyre pressures	✓	Front 1.8 bar		Rear 1.8 bar		Spare 1.8 bar

16. Exhausts (leaks, mounting)	✓	
17. Condition of front /rear dampers	✓	
18. Condition of brake system	✓	
19. Leaks in brake system (external)	✓	
20. Wheel Bearing/Steering Joints	✓	
21. Condition of drive shaft bellows	✓	

BODY & ROAD TEST

22. Horn	✓	
23. Mounting of bumpers	✓	
24. Windscreen	✓	
25. Rear view mirrors	✓	
26. Operation of doors and windows – Lubricate Locks/Hinges	✓	
27. Road Test	✓	

A – 10,000 Miles/1 Year Interval ✓ Inspected – Good Condition
B – 20,000 Miles/2 Years Interval R Inspected – Adjustment Made
C – 60,000 Miles/6 Years Interval X Inspected – Fault Found

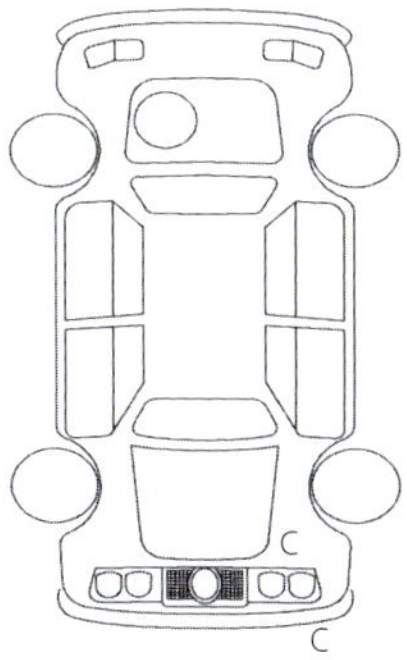

* Consumables in addition, a full estimate may be requested
** Price estimate only provided without dismantling work. This price can only illustrate approximate cost. You may request a full estimate detailing the exact nature and estimated cost of repairs.

Note Body Damage	Additional Comments
D = Dents ☐ S = Scratches ☐	
C = Chips ☒ R = Rust ☐	
O = Other ☐	
Estimate Given Y N	

Inspected by: JBarnes   Customer authorisation: CBloggs

Figure 1.49 An example of a service inspection sheet

In-vehicle service record

Every new vehicle is supplied with a service book which is kept by the owner. You should stamp the service book with the vehicle's mileage, the date and the type of service every time that the vehicle is serviced.

Customer requirements

There will be occasions where a customer requests additional work to be carried out. This information should be recorded on the job sheet as additional work for the technician to carry out.

Time, cost and profit

Most **retail repairs** you carry out will have a specific labour time allocated for completion. This will normally be written on the job sheet. It is very important to complete jobs within the time allocated (labour time). The same applies to **warranty repairs**, as the manufacturer will only pay the set labour time. In most cases the hourly rate for warranty is lower than that of a retail repair.

Any time you spend working on a vehicle that goes over the allocated labour time will have a direct impact on your company's profits. If all the technicians in the workshop exceed the allocated labour time, it will not be too long before the company suffers financially. This is why it is important to you and your company that you become efficient and productive as soon as possible. If you are inefficient and your productivity is poor, you could eventually lose your job.

Profit does not just depend on the quantity of work completed. It is equally important to complete a job to a high standard. If you are not 100 per cent happy with the work you have just carried out on a vehicle, make sure you inform your supervisor or mentor so that they can take a look and advise you on the best way to complete the job successfully.

Getting jobs done on time is not just about the individual technician. It requires the combination of all staff involved, from the servicing staff being given the right information to having the correct tools and the necessary parts in stock.

Service and repair contract

Any work on a vehicle needs customer **authorisation** before starting the job. This is because they and you are entering into a **contract**. If you are completing service and repair work and find the extra work is needed, the customer has to be informed. They must then give you authorisation before you carry out the additional work.

Delays and additional work

If during a repair job you realise that you are not going to be able to complete the repair in the customer's specified time, you should feed back to your line manager immediately, or at least by the end of business. This is so your manager can contact the customer to explain any problems. You should do this in the same way as you report vehicle findings.

Key terms

Retail repair – a repair that the customer pays for.

Warranty repair – a repair that is covered by the vehicle's warranty as part of a guarantee.

Authorisation – the customer's permission to carry out the work and their agreement that they will pay the invoice when the work is complete. The customer usually gives their authorisation by signing the contract.

Contract – a legally binding agreement that both parties must approve before work is done.

During a repair you may notice that there is additional work required. You may also encounter unforeseen delays such as those caused by seized components. It is important that you report any possible additional work to the relevant members of staff, who can promptly act on what you have found and agree a solution with the customer.

If the customer gives their authorisation to go ahead with the repair the service adviser will give the workshop controller the repair order. This will now show that the customer has given their authorisation to complete the additional work. The workshop controller will then give you the repair order.

SYKES MOTORS

SERVICE & REPAIR INVOICE

Job Number	12345	Date	22/05/2011
Customer Name	Mr Bloggs	Vehicle Make	Sangsong
Customer Address	1 Long Drive	Vehicle Model	Coupe
		Engine Size	1600cc
	Cardiff	Reg/Chassis No.	SJ58 SW
Post Code	GT1 1HP	Mileage	40012
Telephone	07978 564564	Reg. Date	09/09/2008

Customers Instructions

Carry out B Service on Vehicle

Description of Work carried out

B Service as per checklist

Supply Touch Up Stick

Pre-Work Inspection

Note Body Damage

D = Dents ☐   S = Scratches ☐
C = Chips ☒   R = Rust ☐
O = Other ☐

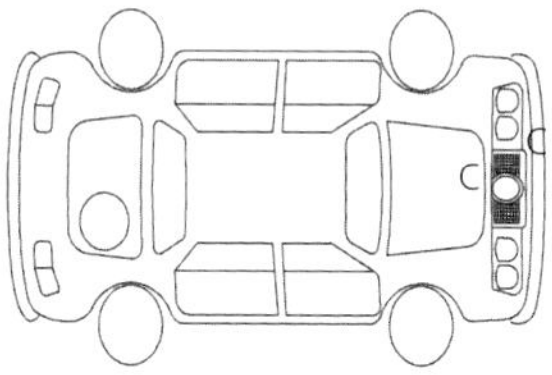

Parts	Cost of Parts £	VAT (20%) £	Total £
Oil Filter	6.00	1.20	7.20
Air Filter	14.00	2.80	16.80
Oil	30.00	6.00	36.00
Pollen Filter	15.00	3.00	18.00
Touch Up Stick	18.00	3.60	21.60
Labour Time (Hours)	Cost of Labour £	VAT (20%) £	Total £
1 Hour	35.00	7.00	42.00
Consumables	Cost of Materials £	VAT (20%) £	Total £
Screen Wash	3.00	0.60	3.60
		Total £	145.20

Customer Signature	C Bloggs	Date	22/05/2011
Technician Name	John Barnes	Signature	JBarnes

Figure 1.50 An example of an invoice for work carried out

Level 2 Light Vehicle Maintenance & Repair

CHECK YOUR PROGRESS

1 Name three sources of technical information used in the workshop.
2 Name three documents a technician needs to use and/or fill out in the workshop.
3 Why should you get customer authorisation for any additional work carried out?

Communicating with colleagues and customers

In your role you will deal with many different kinds of people and you will need to communicate with them in different ways. You will communicate with your workmates using informal **communication**, which means you will show each other a level of respect and a business-like attitude, but on equal terms. When communicating with your manager, you will use more formal language, showing respect for their position.

Methods of communication in the workplace include those shown in Figure 1.51.

Key term

Communication – the sending and/or receiving of information.

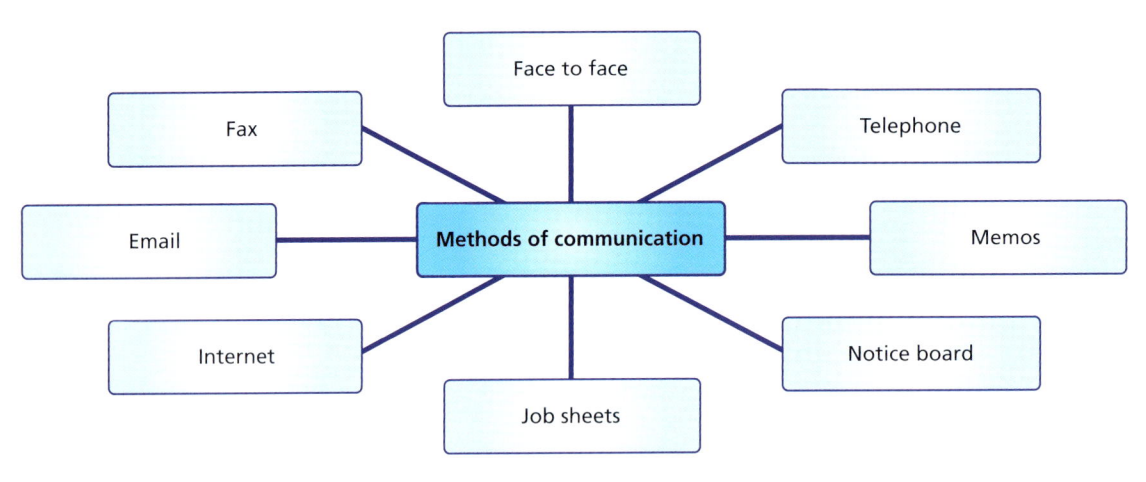

Figure 1.51 Methods of communication in the workplace

Find out

Write a list of all the methods of communication used in your workplace. Are there any not shown in Figure 1.51? Show your list to your manager and ask if there are any methods that you haven't included.

Effective communication

How well people communicate with others depends on the skills they have. When you communicate well you will use a range of skills. You will use body language and verbal skills for face-to-face contact, writing skills for emails, memos, messages, letters or reports, and a variety of techniques on the telephone. It is essential to use the right method of communication at all times so that the message you want to give to others is always clearly understood.

Communication takes place in three ways:

- **verbal** – speaking
- **non-verbal** (body language) – this includes gestures, facial expressions and body position
- **written** – for example documentation.

> **Find out**
>
> Imagine you need to contact a customer urgently to inform them of additional work required on a vehicle.
>
> - Which would be the best communication method for doing this?

Verbal communication

Verbal communication uses words to present ideas, thoughts and feelings. Effective verbal communication is the ability to explain yourself and present your ideas clearly so that the person or people you are speaking to have no doubts about what you are saying. It also involves the ability to listen and respond to what other people are saying.

Verbal communication is made up of three parts, as shown in Figure 1.52.

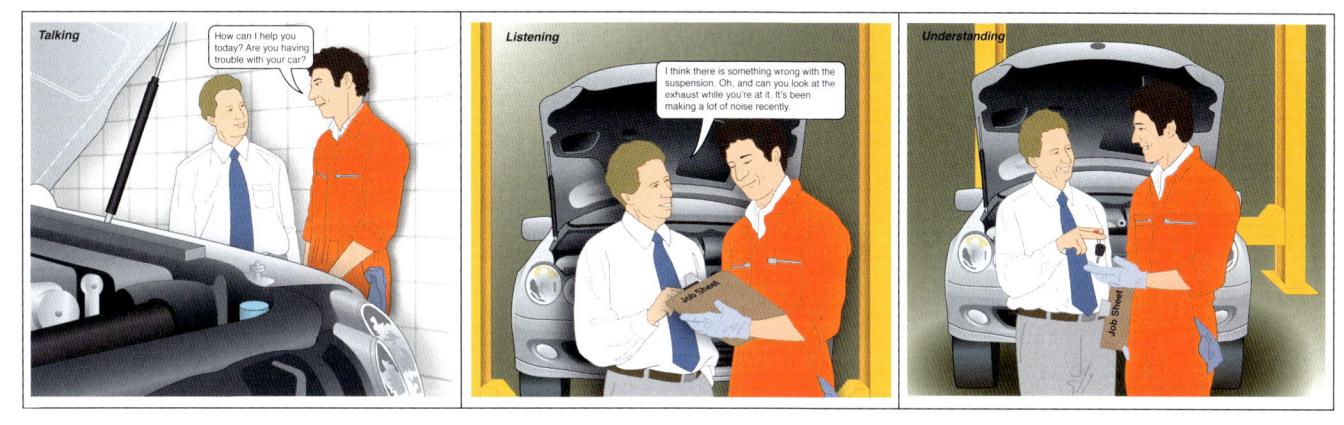

Figure 1.52 The three parts involved in verbal communication

Non-verbal communication

This refers to the messages we send out to people without talking, as shown in Figure 1.53.

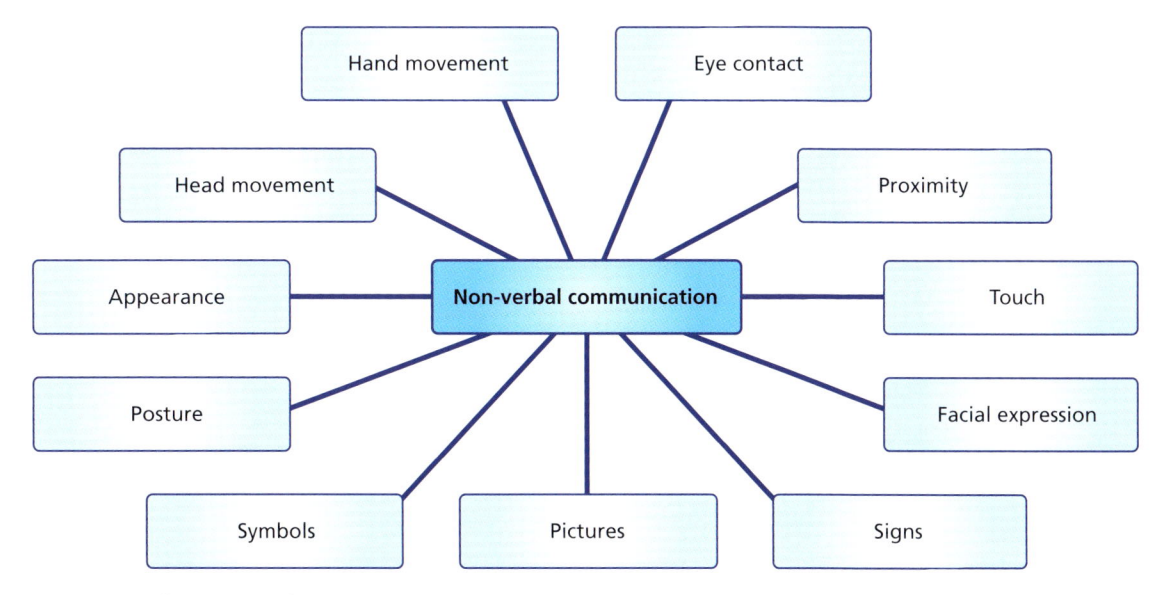

Figure 1.53 Different types of non-verbal communication

Level 2 Light Vehicle Maintenance & Repair

- **Eye contact** – This is a direct and powerful form of non-verbal communication. Eye contact with your customer will show openness and trust. Make sure you don't hold eye contact too long, as this could make your customer or workmates uneasy.
- **Proximity** – This means how close you are to someone else. You don't need a whole workshop to yourself but at the same time you would find it hard to change a wheel with someone under your feet. It's the same when you are with a customer. You may have a friendly personality but remember to respect their personal space, as invading it can be interpreted as intimidating and aggressive.
- **Touch** – You may need to assist a customer, for example to help them into their vehicle. First ask if they need assistance and check they are happy to receive it before rushing in.
- **Facial expressions** – A smile is contagious. Greeting a customer with a genuine smile shows you are welcoming them and in turn putting them at ease. Smiling too long could give the impression that you are laughing at them, which could make them feel uncomfortable. It's all about balance and being genuine.
- **Signs, pictures and symbols** – These might be used, for example, on a step-by-step poster on the store room door that shows how to clear up oil spills, what materials to use and which sign to erect. Signs, pictures and symbols are all around your workshop to give instructions, warn of hazards and remind you what to do. For more information on the different types of signs see pages 58 and 60.
- **Posture** – The way you hold your body will demonstrate if you are paying attention, how interested you are and whether you can be bothered. Slouching your shoulders when a customer asks you a question will not give them confidence in your abilities.
- **Appearance** – One of the first things your employer, workmates and customers will notice about you is your appearance. If you are smart and appropriately dressed in clean overalls, your customers will believe that you will take good care of their vehicle. If you turn up to work in torn, dirty overalls, your customers and employer will not be so positively impressed.
- **Head movements** – Nodding or shaking your head in response to a question shows you are responding, but this could be misinterpreted. So it is best to back up head movements with verbal communication.
- **Hand movements** – Gesturing with your hands can also easily be misread. Always make sure to use the most appropriate communication method.

Figure 1.54 Safety signs remind you of hazards and what to do

Find out

Select two types of non-verbal communication from Figure 1.51. In small groups or pairs, discuss how these can be important.

Create a short scenario showing these non-verbal techniques being used in a workshop situation. Then discuss your scenario with your colleagues.

Written communication

This is central to your role, as you will be keeping records and writing reports. You won't always have a computer to carry out a spell check, as you will have to do some of your record keeping by hand.

In this book many of the key words you need are spelt out for you in the *Key terms* boxes. Try to learn these, as you will be using them a lot in your professional life and you need to be able to write them correctly. **Legible** handwriting is also important.

Communicating using technology

Methods of communication using technology include fax, telephone, texting and email.

- **Fax** – This uses the public telephone system as a way of delivering paper documents from one place to another. The fax machine scans the paper document and sends the information electronically down the phone line to the receiver, where the data is converted back to a paper copy.
- **Telephone** – Talking on the phone is an immediate way of communicating information to people. Voice mail and mobile phones now make it possible to reach people at any time and almost anywhere.
- **Text** – Mobile phones also have a text facility to send short messages. This way of communicating is particularly helpful for sending quick reminders, for example to remind a customer that their vehicle is due for servicing.
- **Email** – This has become the primary method of delivering messages within organisations. Email is increasingly taking the place of fax messages, particularly as whole documents can be sent as email attachments. Email also provides many cost advantages when compared with using paper or the telephone.

Key term

Legible – clear enough to read.

Barriers to communication

Sometimes we don't manage to get our message across to other people, or we don't understand what other people are trying to tell us. Figure 1.55 gives some of the reasons why communications go wrong.

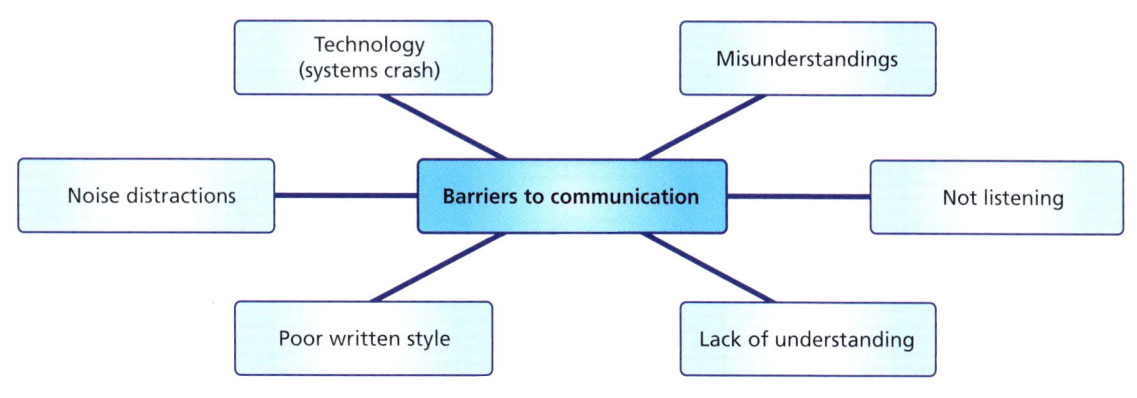

Figure 1.55 Some common barriers to communication

The longer and more complicated the line of communication is, the higher the risk of the communication going wrong. So always try to keep your communications as simple as possible. You should also make sure that you are communicating with the most appropriate person to receive the information – try to avoid information passing between lots of different people before it reaches the intended recipient.

Giving information and instructions

When you are giving specific instructions to others it is often useful to organise the message into a logical sequence. Here is a model sequence for you to follow.

1. Start by describing what the purpose of the message is. If you go straight into an instruction or explanation you may confuse the recipient.
2. Give the instruction or information. If the instruction or information is long or complicated, break it down into shorter steps so that it makes sense and there is no room for misinterpretation. To make sure your recipient understands what you are saying:
   - use plain language
   - don't use abbreviations or jargon
   - give enough information but not too much – overloading the recipient with too much information will make it harder for them to understand.
3. Get feedback. Check that the recipient has understood what you have communicated. There are three ways in which you can get feedback:
   - Ask the recipient if they have understood you. This is a simple but effective way of getting immediate feedback.
   - Get the recipient to summarise the instruction or information.
   - Check if the activity you have explained has been completed successfully.

Teamwork

A group of individuals becomes a team when they share the same goals and objectives.

Each individual in a company can be compared to a cog in a gear train – without the smallest cog the machine fails to operate. When an individual fails to perform their given activity in a satisfactory manner, although the company will not grind to a halt, the effect on the smooth running of the business is soon noticed.

Figure 1.56 Working together as a team ensures a positive approach to work

Effects of good teamwork

Working together as a team will ensure a strong and positive approach to work and an environment in which people feel valued, motivated and have respect for each other. Team members regularly communicate with each other to avoid any misunderstandings and to deal with any potential issues before they become bigger problems. Effective teams always celebrate success.

When the individual skills of all the team members are brought together as a whole, this is called team **synergy**.

Assisting others

It is important to remember that if a work colleague asks you for assistance, you should acknowledge the request. If you are busy with

> **Key term**
>
> **Synergy** – when people work together so that their combined effect is more than it would be if everyone was just working for themselves.

something, you should inform them of your current working situation. Although you may not be able to leave the task you are doing at that time, it is important to explain why to your work colleague in a polite and constructive manner and to offer to help later if possible.

If it is possible for you to help your work colleague, then it is only fair and polite to come to their assistance. You should be able to expect the same help when you need assistance. You must not be afraid to ask your work colleagues to help you when you need it, as it will make your job within the company a lot easier and you will learn faster too.

If you do not offer help when your colleagues ask you, they may not want to help you when you need help. Remember, nobody likes to deal with an unhelpful or impolite person.

During the whole of your working day, it is important to develop and maintain effective interpersonal relationships with the people that you come into contact with.

> **Find out**
>
> Think about your work today.
> - Who have you worked with?
> - Have you worked as part of a team?
> - How have you contributed to the work in your team?

What prevents a team working well together?

Some of the things that can have a negative effect on teamwork and team spirit are shown in Figure 1.57.

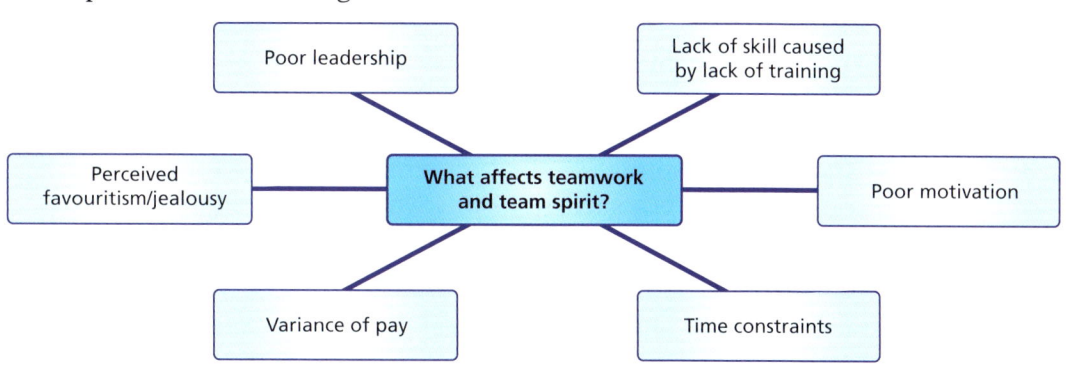

Figure 1.57 Factors that can stop a team from working well together

Factors affecting "What affects teamwork and team spirit?":
- Poor leadership
- Lack of skill caused by lack of training
- Perceived favouritism/jealousy
- Poor motivation
- Variance of pay
- Time constraints

What are your responsibilities?

To summarise, here are the things you need to do to help communication in your workplace. It is your responsibility to:
- make sure you are able to communicate effectively
- be helpful, responsive and courteous at all times
- be approachable at all times
- offer assistance to colleagues
- be a team player.

CHECK YOUR PROGRESS

1. Give four examples of non-verbal communication.
2. Give three methods of electronic communication.
3. State two barriers to communication.

3. Materials and tools

Hand tools and measuring devices

Measuring is a major part of your work as a technician. From time to time you may need to **fabricate** components and to do this you must be able to measure accurately. Accurate measuring skills are also vital when checking vehicle systems before making decisions to replace components. This will prevent unnecessary cost for the customer and will demonstrate your skills as a technician.

> **Key term**
>
> **Fabrication** – the process of making or modifying a component.

Standards

To make sure that manufacturing can be standardised across all countries, there are special organisations which agree on the standards for measuring.

- The British Standards Institution (BSI) is the recognised authority for the measuring standards in the UK. Go to hotlinks to visit the BSI website.
- Control of measuring systems internationally is carried out by the International Organization for Standardization (ISO). Go to hotlinks to visit the ISO website.

There need to be international standards for quantities and units so that the same units are used across different countries. The International System of Units (known as the SI system) exists so that all countries use the same measuring standard. This means that each unit of measurement will be exactly the same whatever the country you are in. For example, a metre is the same whether you are in the UK or any other part of the world.

Table 1.43 gives some examples of standard units of measurement. Go to hotlinks to visit the website of the Bureau International des Poids et Mesures (BIPM).

Table 1.43 Some examples of worldwide standard units of measurement

Quantity	Standard	Abbreviation
Length	Metre	m
Mass	Gram	g
Force	Newton	N
Velocity	Metre per second	m/s
Area	Square metre	m^2
Pressure	Pascal	Pa
Temperature	Celsius	°C
Torque	Newton metre	Nm
Time	Second	s
Volume	Litre	l

Measuring is the comparison between a component measured and the agreed standard. The metric system is now the adopted industry standard, although the older imperial system may still be used by senior technicians.

The advantages of a set standard of measurements are:

- Component parts can be changed easily since all parts are made to same size by all manufacturers.
- There are lower manufacturing costs. This is because manufacturers can share the production costs for different components to make one unit.
- There are lower costs to the consumer. For example, parts can be bought from other manufacturers rather than just one, which means prices will have to be competitive.
- It leads to improved quality since all manufacturers will need to stick to specific sizes.

Original equipment replacements and manufacturer parts

Replacement parts can be very costly when purchased from the vehicle manufacturer. For most parts, substitute components manufactured to similar standards are available from other manufacturers. These 'non-genuine' parts are available at a lower cost and are manufactured to copy the originals, using the same specifications such as measurements and fittings. Every part has a unique number that can be cross-referenced to other manufacturers. These part numbers are known as identification codes.

Measuring equipment

To make sure that the measurements given are accurate, it is very important to store and use measuring equipment correctly. Follow these guidelines to take care of measuring equipment:

1. Use it with care and keep it clean.
2. Where appropriate, lubricate the equipment to prevent rust.
3. Only use the equipment for its intended purpose.
4. When each measuring task is completed, return the instrument to its protective case.
5. If the instrument is faulty, either replace it or get it repaired professionally.
6. Keep the instrument in a stable temperature environment.

Find out

- Find out the part number of an oil filter for a popular vehicle model.
- Look in the back of a parts book and see how many different manufacturers there are for the same part.

Did you know?

Metal expands with an increase in temperature. This can reduce the accuracy of measuring equipment. Working in a room with a constant temperature of 20°C will ensure consistency when measuring components.

Working life

Johnny is late completing a task in the workshop and needs to get off home early. He leaves out a measuring instrument on the edge of the bench which is covered in water.

1 What could be the consequences of Johnny's actions?

Level 2 Light Vehicle Maintenance & Repair

Key terms

Calibration – making sure that the measuring tool is standardised before using it to measure components.

Parallax error – where a wrong measurement is taken because the measuring equipment is positioned incorrectly or is viewed from the wrong angle.

Calibration

Manufacturers of measuring equipment need to make sure that the measurements are standardised. For this reason, all equipment is **calibrated** at a temperature of 20°C in what is called a standards room. Whenever measurements are carried out, it is essential that the workshop is very close to this temperature to ensure absolute accuracy.

Parallax error

Figure 1.58 shows how to take measurements correctly to avoid making the **parallax error**. You should always make sure you position the measuring equipment correctly and view it from the correct angle.

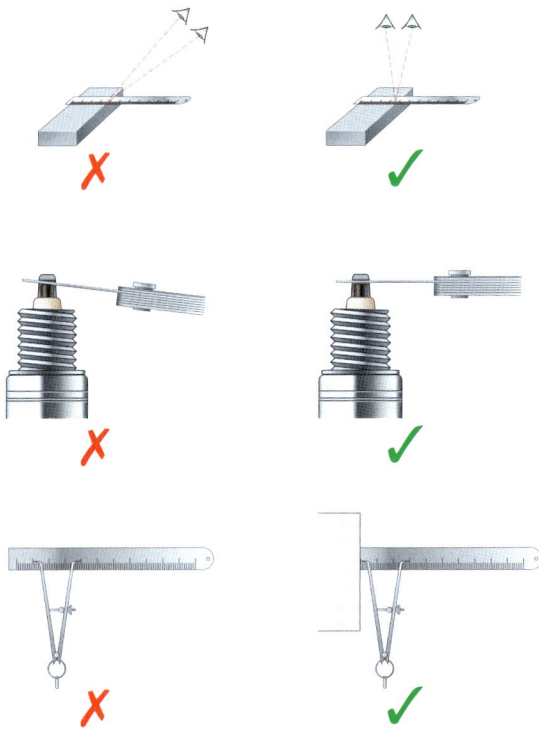

Figure 1.58 How measurements should be taken to avoid the parallax error

Figure 1.59 Tape measure

Figure 1.60 Steel rule

Key term

Cylinder head – the top part of the engine.

Types of measuring equipment

Tape measure

A tape measure is used to measure long distances in length, generally up to about 10m. It is accurate to 1mm.

Steel rule

A steel rule has an accuracy of 0.5mm depending on the scale used. It is generally used for marking out and measuring where you need to check the size of non-critical components.

The steel rule can also be used as a straight edge to measure the flatness of a component, such as a **cylinder head**, in conjunction with a feeler gauge (see page 93).

1 General workshop operation principles for light vehicles

> **Working life**
>
> Tony needs to measure the wiper blades on a vehicle.
>
> 1 Which measuring device should Tony use to make sure that replacements of the correct size are purchased – a tape measure or a steel rule?

Micrometer

A micrometer is used for measuring the thickness of a component to an accuracy of 0.01 mm. It is built up of many parts and should be calibrated to zero before use.

> **Did you know?**
>
> A metric micrometer has a screw pitch of 0.5 mm. This means that the spindle moves 0.5 mm for every turn of the thimble. The bevelled edge of the thimble is divided into 50 parts. This means each **graduation** is equal to 0.01 mm.

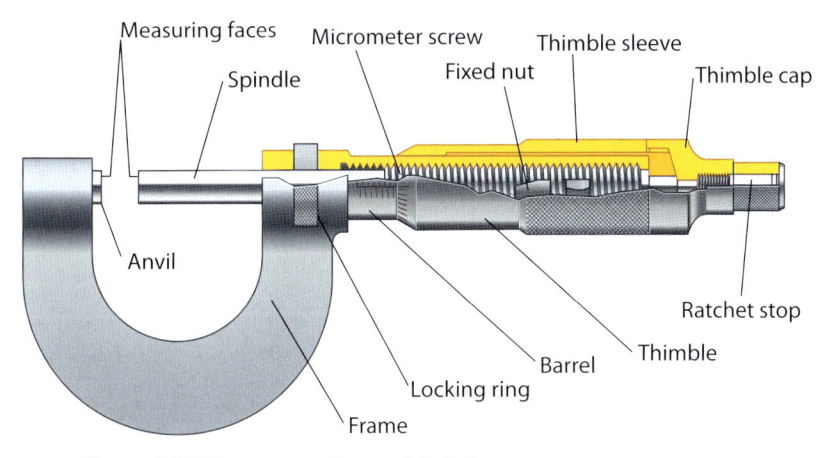

Figure 1.61 Micrometer, with parts labelled

Using a micrometer

You should hold the micrometer correctly when measuring components (as shown in Figure 1.62). This allows you a free hand to hold the component to be measured between the measuring faces.

When measuring a component **tension** is vital. Normally the micrometer will have a **ratchet** – when this is turned, it prevents overtightening of the measuring faces against the component.

> **Key terms**
>
> **Graduation** – the smallest unit that can be measured using a measuring instrument.
>
> **Tension** – how tight the grip of the measuring instrument is on the component to be measured.
>
> **Ratchet** – a mechanism which stops the micrometer from turning any more when the micrometer reaches a certain tightness.

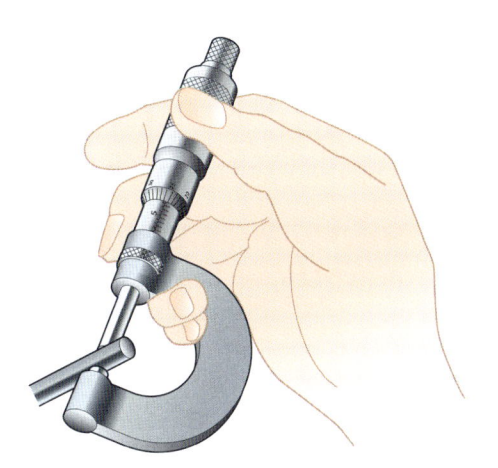

Figure 1.62 Measuring using a micrometer

87

Level 2 Light Vehicle Maintenance & Repair

Where there is no ratchet, you need to take care to obtain the correct tension on the component for an accurate reading. This is very important when measuring round objects. With round objects, you need to move the component across the measuring faces and record the largest **diameter**.

Calibrating a micrometer

Make sure that the micrometer is set to zero when fully closed before use. If the micrometer is not set to zero when the faces are together, you should use the G spanner (see Figure 1.63) to adjust the spindle to zero.

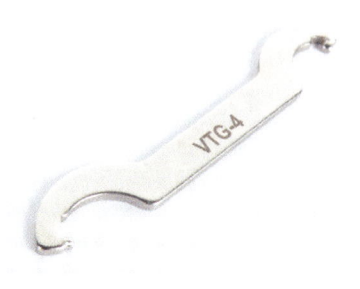

Figure 1.63 G spanner (which gets its name from its shape)

Reading a micrometer

A simplified micrometer reading is shown in Figure 1.64. You read the graduations in the following order:

1 = 1 mm graduations

2 = 0.5 mm graduations if shown after the 1 mm graduations

3 = 0.01 mm graduations.

The example shown in Figure 1.60 works out as:

1	=	7.00 mm
2	=	0.00 mm
3	=	0.25 mm
Total		7.25 mm

Please note: Some micrometers have their main scale below the line and some above the line on the spindle.

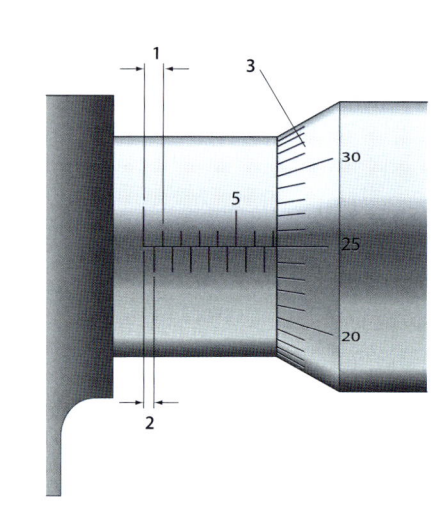

Figure 1.64 A simplified micrometer reading

Digital micrometer

Modern micrometers have a digital readout to prevent reading error. However, you still need to follow the correct process:

- Set the micrometer to zero before use.
- Make sure you apply the correct tension on the component to be measured.

Internal micrometers

You will need to use an internal micrometer to check the diameter of internal surfaces such as cylinder bores. When measuring wider diameters of 50 mm and above, you will need to use a set of **spacers** with the internal micrometer to ensure accurate measurements.

Figure 1.65 Digital micrometer

Find out

Create readings on a micrometer for the following measurements:

- 12.69 mm
- 1.89 mm
- 0.50 mm
- 19.05 mm

Check your readings with a colleague.

Key terms

Diameter – the width of a circle, passing through the centre.

Spacers – attachments to the micrometer set at different lengths. They are very closely machined, in sizes of 25 mm, 50 mm, 75 mm and 100 mm.

Vernier caliper

You can use a Vernier caliper to measure width, diameter and depth. You must take care with the caliper and check that the jaws are not sprained or worn excessively, as this will affect accuracy. The Vernier caliper has an accuracy of 0.02 mm.

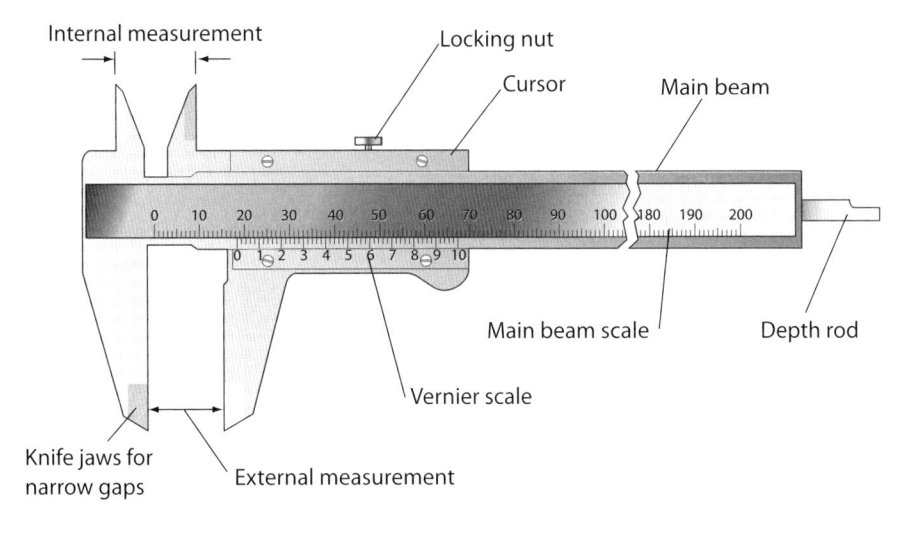

Figure 1.66 Vernier caliper with parts labelled

Part	Description
Locking nut	This will lock the cursor to the main beam to take and retain a measurement
Cursor	This is the sliding part of the vernier caliper
Main beam	This is the sliding part of the vernier caliper
Depth rod	This is used to measure the depth
Main beam scale	This is the main measuring scale
Vernier scale	This is the secondary measuring scale which provides greater accuracy
External measurement	This is for measuring external components
Knife jaws for narrow gaps	These surfaces are machined thin to provide greater accuracy of measurement
Internal measurement	This is for measuring internal components

For correct calibration of the caliper, close the jaws and check that the measurement is zero before use.

The top scale is the main beam. This is the fixed jaw. The Vernier scale is the sliding jaw.

Measuring with a Vernier caliper

1. Release the locking nut.

2. Hold the caliper as shown in Figure 1.67.

3. Slide the moving jaw towards the component with your thumb.

4. Lock the clamping screw.

5. Read the measurement in good light and without parallax error (see page 86).

6. Take a reading from the main beam scale where the Vernier scale shows zero.

7. Take the second reading where both scales line up exactly and add the two together.

Level 2 Light Vehicle Maintenance & Repair

Find out

Practise using a Vernier caliper to measure a range of items in your workshop. For example, you could measure:

- the gaps in open-ended spanners
- the depth of a threaded hole on an engine
- the diameter of a socket.

Write down the measurements you obtain and ask a colleague to check them.

The reading from the main beam scale will give you the number of whole millimetres. The second reading will give you the decimal part of the measurement. An example of a Vernier caliper reading is:

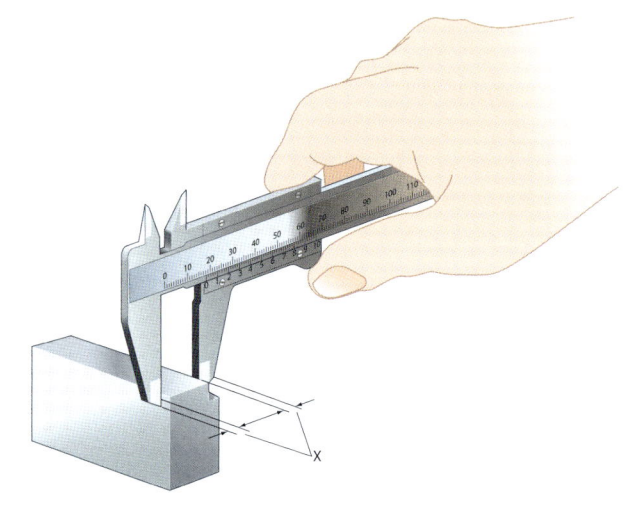

```
1  =  31.00 mm
2  =  00.66 mm
       ───────
       31.66 mm
```

Figure 1.67 Correct measuring with Vernier caliper

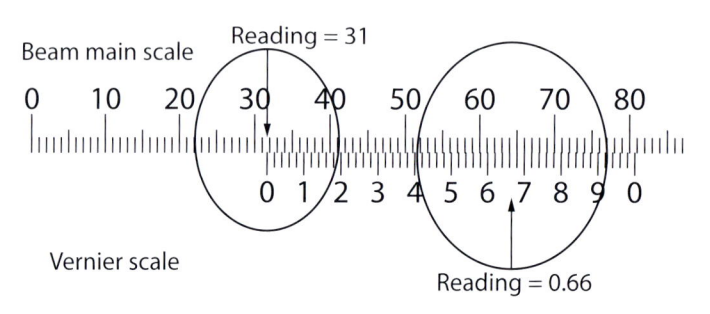

Figure 1.68 Example of a Vernier caliper reading of 31.66

Digital Vernier caliper

Modern Vernier calipers have a digital readout so there is no need to calculate the final total reading. A digital Vernier caliper can be used to measure width, inside diameter and depth, as with an analogue Vernier caliper. Examples of use include:

- width – shaft thickness
- inside diameter – cylinder bore
- depth – brake pad wear.

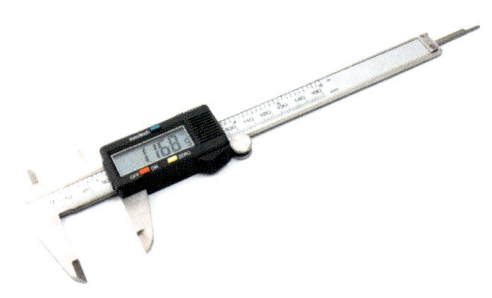

Figure 1.69 Digital Vernier caliper

Dial gauge

The dial gauge compares height or **run-out** between one component and another. It is sometimes called a dial test indicator (DTI) and has a measuring accuracy of 0.01mm. You can use a dial gauge to check the run-out of **brake discs** or the height of a **cam**.

Various attachments (as shown in Figure 1.71) ensure that the dial gauge can be positioned correctly, so that maximum movement of the plunger can highlight the **tolerance** of the component.

Figure 1.70 Dial gauge

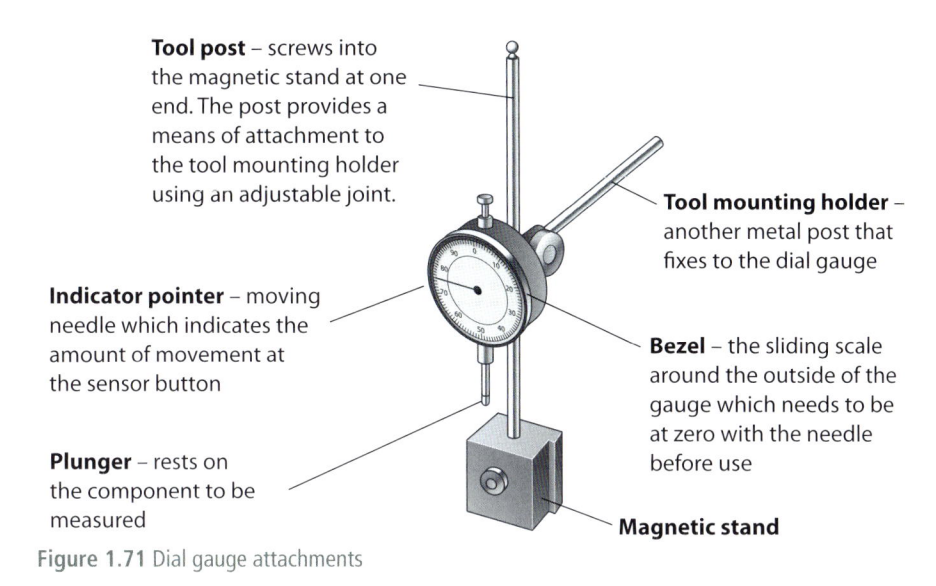

1 General workshop operation principles for light vehicles

Tool post – screws into the magnetic stand at one end. The post provides a means of attachment to the tool mounting holder using an adjustable joint.

Indicator pointer – moving needle which indicates the amount of movement at the sensor button

Plunger – rests on the component to be measured

Tool mounting holder – another metal post that fixes to the dial gauge

Bezel – the sliding scale around the outside of the gauge which needs to be at zero with the needle before use

Magnetic stand

Figure 1.71 Dial gauge attachments

> **Key terms**
>
> **Run-out** – the amount of variation measured between rotating and fixed parts.
>
> **Brake disc** – a rotating brake surface in the shape of a round flat disc.
>
> **Cam** – an oval shape which is fixed to a round shaft called the cam shaft. It is used to open and close valves in the engine.
>
> **Tolerance** – an allowable difference from the required measurement (to allow for slight imperfections in a manufactured object).

The magnetic stand is attached to the component. Magnetic force prevents it from moving from the component during measurement.

The plunger is attached to the sensor button. This rests on the component to be measured. The plunger transmits upward and downward movement to the gears inside the gauge that turn the indicator pointer so a reading can be taken.

Care of the dial gauge

The dial gauge is a delicate tool. You need to:

- Keep it in its case when it is not in use.
- Avoid sharp movement of the plunger to prevent overwork of the springs inside – this would greatly affect accuracy and could cause permanent damage.
- Prepare for measurement by using a magnetic stand, clamps and rods.
- Make sure that the components to be measured are clean and free from corrosion.

Figure 1.72 Preparation for using a dial gauge

Testing for roundness of a component

1. Use **Vee blocks** to support the component to be measured.	2. Adjust the height of the gauge so that the stylus just rests on the component under slight tension.	3. Zero the scale by turning the outer section of the gauge.	4. Rotate the component one full turn and record the tolerance below the zero line and above.	5. Add the two tolerances together to obtain the total run-out figure.

Digital dial gauge

Digital dial gauges are slowly replacing analogue gauges as they are easier to use and have a clearer tolerance reading.

Measuring electrical values

Electrical values are usually measured with a multimeter. Using a multimeter is covered in detail in Chapter 6, pages 385–390.

> **Key term**
>
> **Vee blocks** – blocks shaped like a letter 'V'. They are used to support and enable round components, like shafts, to rotate during measurement.

Level 2 Light Vehicle Maintenance & Repair

Did you know?

- Voltage is the electrical pressure on **electrons** through a **circuit**.
- Current is the rate of flow of electrons through a circuit.
- Resistance is the prevention of electrons flowing through a circuit.

For more information on electrical circuits, see Chapter 6.

Voltage: To measure electrical voltage, the multimeter selector needs to be set at V (volts) and the leads need to be put across the component to be measured, as shown in Figure 1.73.

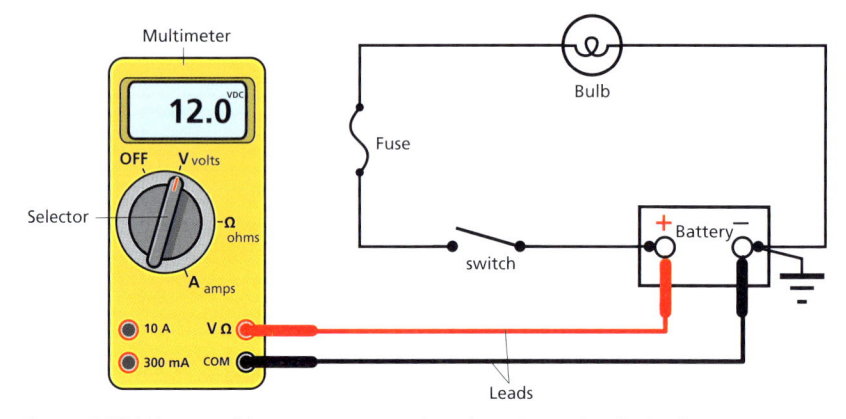

Figure 1.73 Using a multimeter to measure the voltage in an electric circuit

Current: To measure electrical current, the multimeter selector needs to be set at A (amps). The circuit needs to be opened (see Figure 1.74) and the leads need to be put across the open leads.

Key terms

Electron – a subatomic particle with a negative electric charge. It is the movement of electrons that creates electric current.

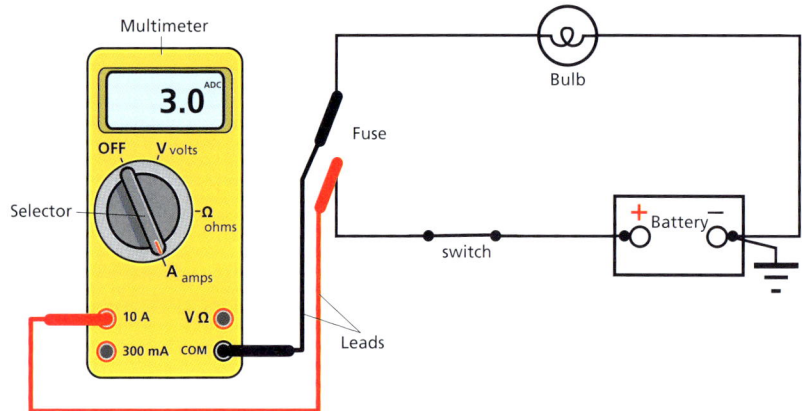

Circuit – an unbroken path that an electric current can flow around. The circuit consists of a power source, wires and components.

Figure 1.74 Using a multimeter to measure electrical current

Resistance: To measure electrical resistance the multimeter selector needs to be set at Ω (ohms), as shown in Figure 1.75. The component needs to be out of the circuit and the leads need to be put across the component to be measured.

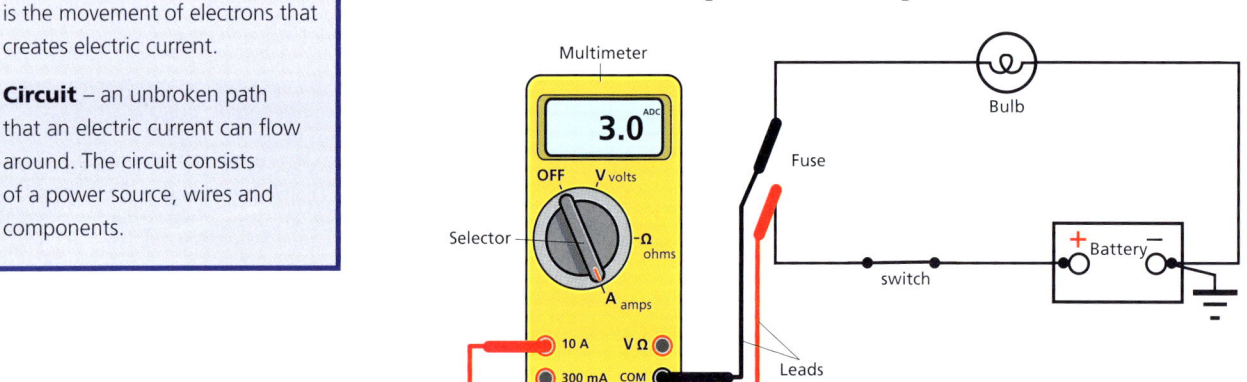

Working life

Brian is trying to measure the resistance of a component in a circuit.

1. What is he doing wrong?

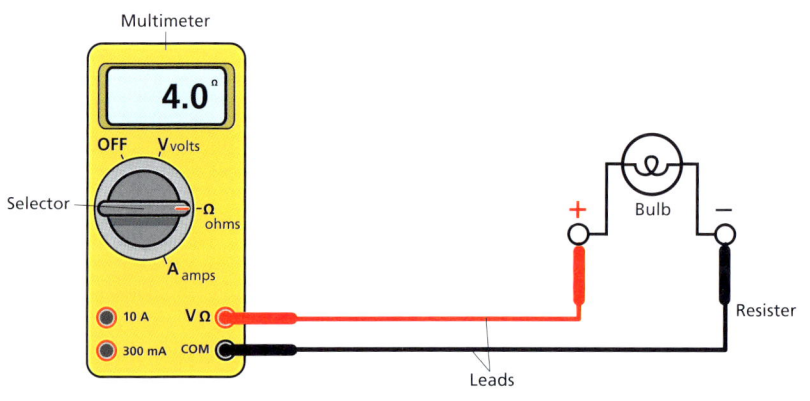

Figure 1.75 Using a multimeter to measure electrical resistance

Take care not to damage the multimeter, as it is a delicate measuring device. Also, make sure that you use the correct scale according to the size of the electrical property being measured.

Diagnostic equipment

Most vehicles on the road today have computers on board to make sure the vehicle runs efficiently and at maximum performance. These are wired up to a warning light on the dashboard that lights up if there is fault in the system. A diagnostic tester is a machine that can narrow down the possible number of faults the vehicle may have.

You should only carry out a diagnostic test after you have carried out some initial checks, for example checking battery voltage and wiring connections. The *Technical skills* feature below shows a typical general fault finding route.

Finding and repairing a fault

1. A fault is indicated when a light on the vehicle dashboard display is illuminated.
2. Make initial checks.
3. Plug in the tester.
4. Insert the vehicle details.
5. Request trouble code information.
6. Take and record the reading on the job card.
7. Use the manufacturer's repair guidelines to find the exact fault.
8. Repair or replace the faulty component.

Figure 1.76 Diagnostic tester

Trouble codes

These are codes that relate to different faults in electrical circuits. Earlier diagnostic testers had a small reading screen and could only show a code number. This had to be checked against the manufacturer's data to confirm which electrical circuit was faulty. Today, the screens give a full readout of the circuit fault. An example of a trouble code is:

P0118 – Engine Coolant Temperature Circuit High Input.

Feeler gauges

A feeler gauge is used to measure gap widths. It measures the **clearance** between two component parts. Feeler gauge blades range from 0.05 mm to 1 mm, as shown in Figure 1.77, and the measurement is stated on the blade.

Figure 1.77 Feeler gauges

Key term

Clearance – the gap between engine components.

Level 2 Light Vehicle Maintenance & Repair

Figure 1.78 Measuring jugs used to measure fluid volume

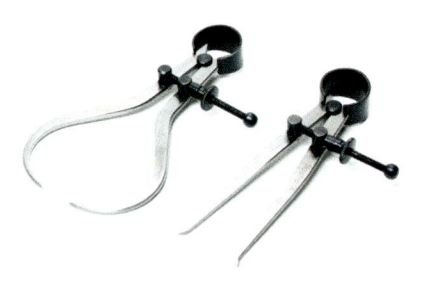

Figure 1.79 External (left) and internal calipers

Figure 1.80 Engineer's tri square

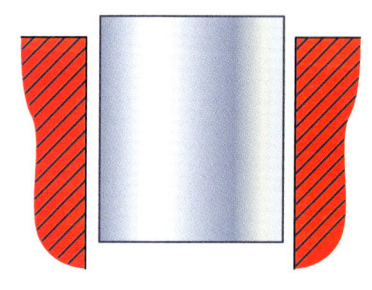

Figure 1.81 Clearance fit

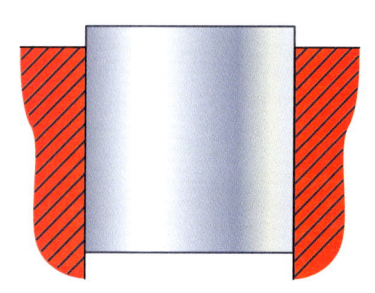

Figure 1.82 Transition fit

Measuring jugs

Measuring jugs are used to make sure that the correct amount of fluids can be transferred from one place to another. Graduations on the side of a measuring jug indicate the amount of fluid in the jug.

> **Working life**
>
> Aswad has 6 litres of oil in his jug and has just replaced the oil filter. The engine that he needs to pour oil into has a capacity of 4.5 litres and the oil filter takes 0.5 litres.
>
> 1 How much oil will be left in the jug when Aswad has filled the engine to capacity?

Internal/external calipers

Although these are not measuring devices, they are used to compare length and diameters against measuring devices where access is difficult. This can be achieved by opening the calipers against the component to be measured and then comparing the gap against a measuring scale to get the specific reading.

Dividers

These are used to draw circles and curves on metal surfaces. A thin coat of marking dye is applied to the area being marked. The point of the scriber scratches the surface to show boundary lines and centres.

Dividers can also be used as a measuring tool, when marking out repeat distances along a marked line.

Odd leg calipers

These are used to mark lines parallel to an edge. They are also useful for marking centre lines.

Engineer's tri square

This is used to check for squareness and to aid in scribing lines at a right angle to a machined edge.

Limits of fit

Motor vehicle components are joined together in various ways. A tolerance is an allowable difference in measurement from the required size. The difference between the limits of two mating component parts highlights the precision with which they must be manufactured.

Clearance fit

This is where the central component fits loosely inside the outer component (so there is clearance between them). This fit is used on shafts where movement is required and rotation needs to occur.

Transition fit

This type of fit is a cross between clearance fit and interference fit (see overleaf). Accuracy of manufacture will determine how close the fit is. This type of fit is used in engine construction for cylinder liners. In some cases, if the fit is too loose, a special adhesive may need to be used.

1 General workshop operation principles for light vehicles

Interference fit

This is where the inner component is forced into the outer component. An example of this is bearings pressed into hubs.

Examples of tolerance

Remember that tolerance is an allowable difference from the required measurement (to allow for slight imperfections in a manufactured object). For example, for a shaft which must have a diameter of 40 mm ± 0.2 mm:

- the minimum allowable diameter would be 39.8 mm (40.0 − 0.2)
- the maximum allowable diameter would be 40.2 mm (40 + 0.2).

Table 1.44 shows other examples of tolerances and the highest and lowest measurements possible within each tolerance.

Figure 1.83 Interference fit

Table 1.44 Examples of tolerances

Specification	Highest measurement	Lowest measurement	Tolerance
50 ± 0.4 mm	50.4 mm	49.6 mm	0.8 mm
80 ± 0.06 mm	80.06 mm	79.94 mm	0.12 mm
24 + 0.1 − 0.0 mm	24.1 mm	24 mm	0.1 mm

Hand tools

Screwdrivers

Loosening and tightening screws requires the use of a screwdriver. These are available in many different shapes and sizes and have different blades to accommodate various screw heads.

Figure 1.84 Screwdrivers

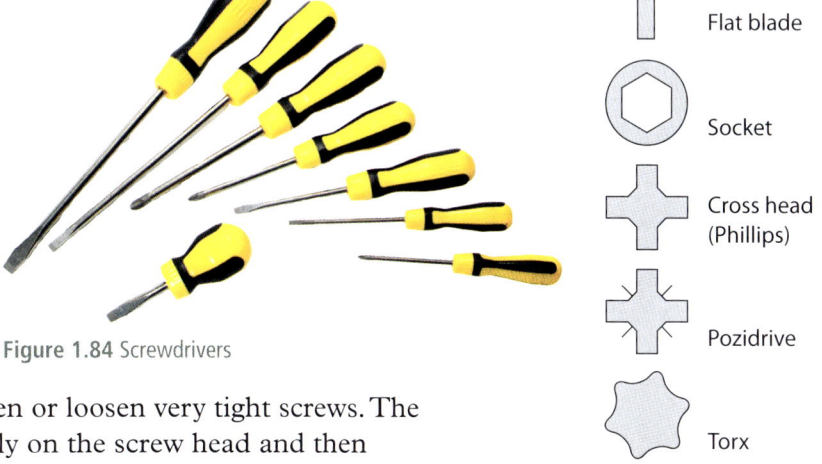

- Flat blade
- Socket
- Cross head (Phillips)
- Pozidrive
- Torx

Figure 1.85 Screwdriver heads – different cross-sections

Impact driver

An impact screwdriver is used to tighten or loosen very tight screws. The screwdriver blade is positioned carefully on the screw head and then struck firmly with a steel hammer. Assisted by the operator, an internal rotating system applies a turning force to the screw. The impact and the sudden turning force tightens or slackens the screw.

Safe working

Always use goggles when using an impact driver.

Did you know?

A torx is a hexagon-splined screw head or socket.

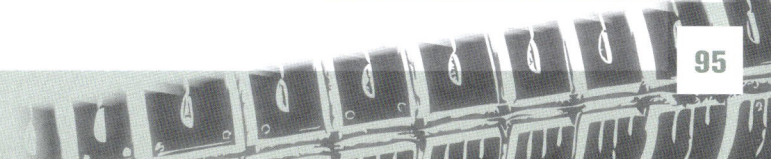

Figure 1.86 Impact driver

95

Pliers

Gripping, turning, crimping or cutting are all operations carried out by pliers.

- **Slip joint pliers** or water pump pliers allow you to adjust the jaws to various positions.
- **Needle nose pliers** or long nose pliers are used to handle pins in tight positions.
- **Side cutting pliers** are used to cut or trim wire or to remove split pins.
- **Crimping pliers** are used to strip and cut electrical wire. The top jaws are also used to crimp electrical connectors.
- **Vice grips** or **locking pliers** are used where an extremely strong grip is required. They can be particularly useful when welding parts and removing studs or rounded bolts.
- **Circlip pliers** are used for removing and replacing internal and external circlips from round shafts.

Figure 1.87 Slip joint pliers

Figure 1.88 Needle nose pliers or long nose pliers

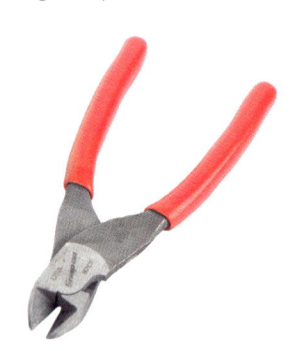

Figure 1.89 Side cutting pliers

Figure 1.90 Crimping pliers

Figure 1.91 Vice grips/locking pliers

Figure 1.92 Circlip pliers

Spanners

Because of the many different sizes of nuts and bolts, and the need to access these mechanical fixings in a number of different positions, various types of spanner and of spanner jaw gaps are produced.

- **Open-ended spanners** have jaws at each end that are of slightly different widths.
- **Ring spanners** have a ring at each end, usually of slightly different sizes.
- **Combination spanners** are a cross between an open-ended spanner and a ring spanner. They are the same size at both ends.
- **Pipe spanners** are used to release and tighten pipes and make sure that the union does not get damaged. They are similar to a ring spanner but with a slot so that the pipe can slip through the gap.

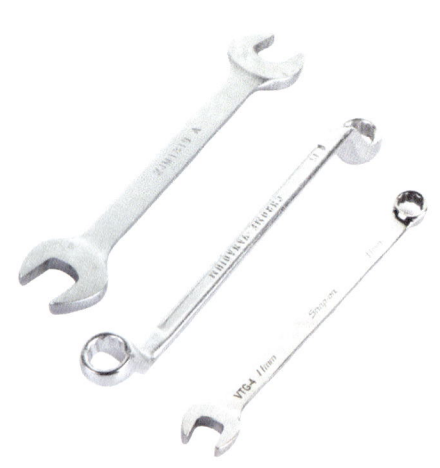

Figure 1.93 Different types of spanner: open-ended spanner, ring spanner, combination spanner

Figure 1.94 Pipe spanner

1 General workshop operation principles for light vehicles

- **Adjustable spanners** are a versatile tool that can be used for all nut and bolt sizes within a certain limit. Make sure you set the jaws correctly to prevent the spanner slipping off the fixing.

> **Working life**
>
> Jamela has the choice of using a ring spanner or an open-ended spanner to slacken a nut.
>
> 1 Which one is the safest to use?

Figure 1.95 Adjustable spanners

Socket set

You will need an arrangement of sockets, extensions and driving tools to slacken and tighten mechanical fixings. This is a called a socket set.

- **Breaker bar** – This is used to slacken and tighten nuts and bolts with the aid of sockets and extensions.
- **Extension bars** – These give extra reach to fixings in difficult areas.
- **Ratchet** – This is the most common tool for slackening and tightening nuts and bolts with speed.
- **T bar** – This is used to speed up the tightening and slackening of nuts in easily accessible places.

Figure 1.96 Breaker bar

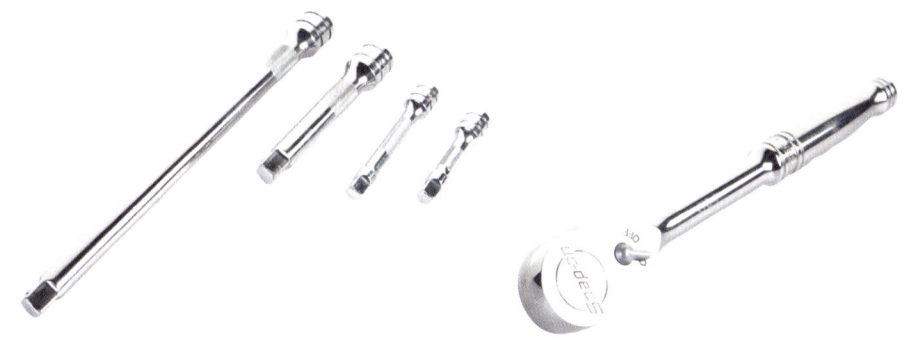

Figure 1.97 Extension bars

Figure 1.98 Ratchet

Figure 1.99 T bar

- **Universal joint** – This is an especially useful adapter and small extension. It is used where the normal extensions cannot reach because a fixing is slightly off centre and can be used through different angles.
- **Speed brace** – As the name suggests, the speed brace is used to slacken and tighten sockets where there is easy access at pace.

> **Safe working**
>
> When slackening and tightening nuts and bolts, you must always pull the spanner or wrench towards you. This prevents damage to your knuckles if the tool falls off the fixing.

Figure 1.100 Universal joint

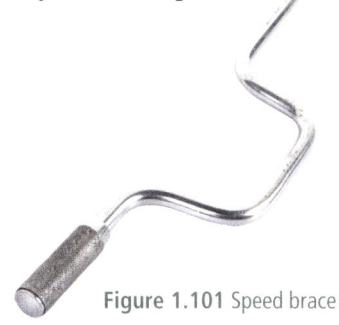

Figure 1.101 Speed brace

Level 2 Light Vehicle Maintenance & Repair

Figure 1.102 Air ratchet

> **Safe working**
>
> When using compressed air make sure that it is not pointed at any part of the body. Compressed air can penetrate the skin or cause blindness if pointed at the eyes.

Figure 1.103 Impact wrench

> **Did you know?**
>
> Torque is turning force. If a longer bar is used and the same force is applied, the torque will increase.
>
> Always reduce the torque setting to zero after use. This ensures that the torque wrench stays calibrated.

- **Air ratchet** – To speed up the removal and replacement of assemblies, an air ratchet offers an alternative to the normal hand ratchet. It reduces operator fatigue when lots of nuts and bolts need turning. Compressed air set at high pressure is used to operate the wrench.

- **Impact wrench** – This is another device used to speed up the assembly and dismantling of units. It is a heavy-duty wrench mainly used where tight fixings need to be removed. It can produce a high amount of torque and reduces the possibility of injury that might be incurred using other manual methods. If you use an impact wrench as a tightening method, you need to reduce the torque and use a torque wrench for the final tightening of the components.

- **Torque wrench** – This is the most important tool when tightening nuts and bolts. It measures the turning force applied and releases with a 'click' when a pre-determined value is reached. This means that fixings are not over- or under-tightened. In this way it prevents overstretching of the threads or premature loosening of the joined components. Torque wrenches are adjusted by twisting the bottom part of the handle until a setting is reached.

- **Socket adapters** – These are set at different sizes and work in conjunction with the other items in the socket set to remove the fixings. Sockets can be single hexagon, double hexagon, allen key or torx fixing. If using an air wrench, stronger 'impact' sockets must be used.

Figure 1.104 Torque wrench

Figure 1.105 Socket adapters

Angle gauge

This is used as a further tightening method when inserting stretch bolts, for example cylinder head bolts following the use of a torque wrench. Attached to a breaker bar, the angle gauge stay is fixed to the component and the pointer is turned through the angle specified by the manufacturer.

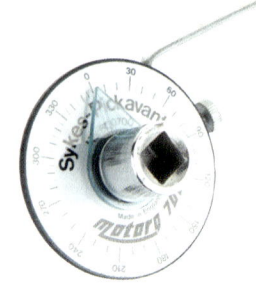

Figure 1.106 Angle gauge

1 General workshop operation principles for light vehicles

Hammers

Hammers are a means of providing a sharp shock or gentle persuasion to remove, replace or reshape components. Figures 1.107 to 1.112 show the different hammers and what they are used for.

Figure 1.107 Ball peen hammer – used for general engineering work

Figure 1.108 Copper and hide hammer – copper on one side and leather on the other to prevent damage to metal components, e.g. bearings on assembly

Figure 1.109 Lump hammer – used for heavy engineering work

Figure 1.110 Plastic hammer – used for light engineering work

Figure 1.111 Panel hammer – used for flattening vehicle body panels

Figure 1.112 Rubber hammer – used for inserting rubber seals

Chisel

A chisel is used for removing and cutting small components and removing waste or damaged materials, for example when removing seized nuts from bolt threads. The sharp end is held at an angle of around 45 degrees, while the other is struck with a ball peen hammer.

Figure 1.113 Chisels

Did you know?
Engineers mainly use 'cold' chisels. These are so called because they cut cold steel.

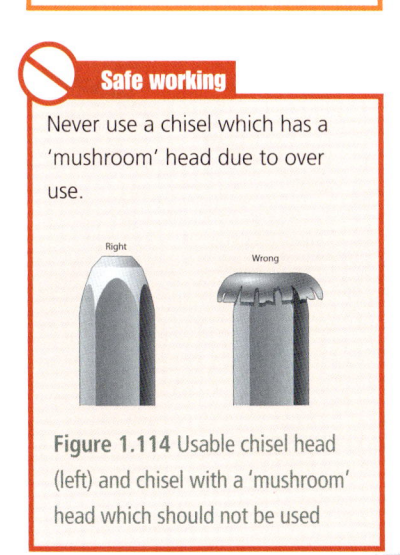

Safe working
Never use a chisel which has a 'mushroom' head due to over use.

Figure 1.114 Usable chisel head (left) and chisel with a 'mushroom' head which should not be used

Working life
Jimmy cannot find his chisel to remove a nut from a seized bolt, but he has a screwdriver nearby.

1 Why would it be a bad idea to use a screwdriver instead of a chisel?

99

Level 2 Light Vehicle Maintenance & Repair

Punches

Punches are used with a hammer to transmit a large pressure to a small diameter. These come in different shapes for a variety of specific uses.

- **Parallel punch** – This is used to remove **split pins** and **roll pins**.
- **Centre punch** – This is used with a hammer to create a guide when drilling components. It can also be used to mark components before dismantling so they can be reassembled correctly.
- **Taper punch** – This is used for aligning holes and removing **clevis pins**.

Figure 1.115 Punches – from left to right: parallel, centre and taper

> **Safe working**
> Always use goggles when using any punch or chisel to prevent **swarf** rebounding into your eyes.

> **Working life**
> Nairoshi cannot find her taper punch. She tries to align two holes in metal panels using her finger.
> 1. Why is it a bad idea for Nairoshi to use her finger?

Hacksaws

Hacksaws are used to cut various metallic and non-metallic materials.

- **Large hacksaw** – This has a replaceable blade made of heat-treated high carbon steel. The number of teeth on the blade determines what type of material it can cut. As the number of teeth per blade length increases, the harder the material it can be used for. A general **tooth pitch** for a hacksaw is 24 teeth per 25 mm.
- **Junior hacksaw** – This is a smaller hacksaw that can give better access when in tight spaces.

Figure 1.116 Large hacksaw

Figure 1.117 Junior hacksaw

With both types of hacksaw, the blade is inserted into the frame with the pointed part of each tooth facing forwards. When using the saw, it is important to keep the material to be cut as tight to the vice as possible. This will prevent vibration, noise and damage to the hacksaw blade.

> **Working life**
> Eric is sawing a piece of metal in a vice. This is making a lot of noise and he keeps breaking hacksaw blades.
> 1. What could be the fault?

1 General workshop operation principles for light vehicles

Files

Files are classified by shape, grade of cut and type of cut. Each one is designed for a different use. The shape used will be determined by the shape of the finish required. The grade and cut are determined by the softness of the material to be filed and the quality of finish required.

- **Flat file** – used for filing surfaces flat.
- **Round file** – used for enlarging holes.
- **Triangular file** – used for filing into corners.
- **Half-round file** – used for filing curved surfaces.
- **Needle files** – these come in the same shapes as the files listed above, but they are smaller and thinner so they can be used for finer working.

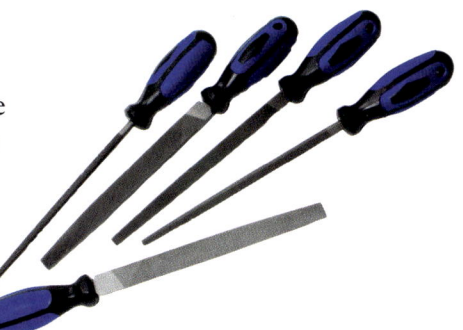

Figure 1.118 Files

File grade

Files are graded according to their smoothness or roughness. This is determined by the number of teeth per centimetre. The different grades of file available and their uses are shown in Table 1.45.

Table 1.45 File grades

Grade	No. of teeth	Use
Smooth	More	Draw filing, finish cutting
Second cut	↓	Rough cutting hard steel
Bastard	↓	Rough cutting steel, filing hard brass
Rough	Less	Filing soft materials such as plastics and aluminium

Type of cut

The file cut refers to the arrangement of the teeth on the blade. The type of cut will help you decide which file to use, as shown in Table 1.46.

Table 1.46 The different types of cut on files

Cut	Use
Curved	Soft materials, e.g. plastics
Single cut	Finishing soft materials and rough filing hard materials
Double cut	Fine working and finishing

Key terms

Split pin – a pin that is halved at one end and bent to prevent release (see page 108).

Roll pin – a pin that is shaped like a cylinder with a slight split down the length of it. It is made of sprung steel and is inserted into a hole under pressure.

Clevis pin – this is used in a joint where there needs to be some movement. A split pin is used to stop the clevis pin falling out of the joint.

Swarf – a discarded part of a larger metal component.

Tooth pitch – the distance between the point of one tooth and the point of the next.

Safe working

Never use a file without a handle. The file tang (the sharp part that inserts into the wooden or plastic handle) could pierce your wrist.

Level 2 Light Vehicle Maintenance & Repair

Care of your files

Take care of your files by following these guidelines:

- Clean the file with detergent after using.
- Remove ground-in swarf using a file brush.
- Place the tool down carefully as it is very brittle.
- Store the file in a rack or protective casing.

> **Working life**
>
> David can hear someone filing a piece of metal in another part of the workshop and it is very loud.
>
> 1 What could he suggest to his colleague to reduce the noise during the task?

Screw thread cutting set

A screw thread is produced using a tap or die. These normally come in a set with a range of different threads and sizes.

Taps

The tap is used to cut a thread in a hole. There are different types of tap and the one used will depend on the operation to be carried out:

- **Taper tap** – used to create a new thread. It allows a gradual cut into the metal.
- **Intermediate** or **second tap** – used to complete or finish the thread.
- **Plug tap** – used to repair threads in a **blind hole**.

Figure 1.119 shows a tap used for screw thread cutting.

Figure 1.119 Screw thread cutting set

> **Key term**
>
> **Blind hole** – a drilled hole in a component that does not go the whole way through.

Using a tap

1. Ensure the tap is clean and firmly fixed in the T wrench.
2. Apply cutting oil to the cutting edges.
3. Keep the tap at a right angle and turn clockwise through 90 degrees for one turn.
4. Turn half a turn anticlockwise.
5. Repeat steps 3 and 4 until the component reaches the tap shoulder.

Dies

Dies include:

- **Split die** – used for creating threads on a stud or bolt.
- **Solid die** – used for cleaning up old threads (see Figure 1.120).

As the thread develops, tension is put on either side of the split to cut the threads deeper.

A die nut (solid die) is used for repairing or 'chasing' damaged threads.

Figure 1.120 Dies

Thread restorer

Taps and dies can be used to repair damaged threads. Alternatively, a thread restorer can be used. This is used like a file across the threads to remove burrs and rust.

Spark plug thread chaser

This is used to clean the threads in the cylinder head so spark plugs can be inserted and removed more easily.

Figure 1.121 Spark plug thread chaser

> **Did you know?**
>
> It is good practice to smear grease between the gaps of the spark plug thread chaser teeth to catch any loose swarf so that it does not enter the engine.

Stud extraction

The easiest way to remove a stud is to screw two nuts on to the stud. Tighten them together and then turn the bottom nut anticlockwise.

If the stud has broken in the component, it may be difficult to use the above method. In this case, first cut the stud flat. Use a centre punch to guide the drill, then drill a hole and use a stud extractor.

Drills

The most common type of drill bit is the twist drill. It comes in three different lengths:

- **long** (long series)
- **short** (stub).
- **normal** (jobber's series)

and several diameters.

Figure 1.123 shows the different parts of a drill.

Figure 1.122 Stud extractor set

Figure 1.123 Parts of a drill

The drill bit is made from high carbon steel. When using a drill, you need to take care not to put too much pressure on it when forcing the drill through the hole, as this will make the drill bit blunt or will break it. Another consideration is the drill speed. If the drill speed is too fast, this could also make the drill bit blunt.

Drilling machines

Portable drills are the most popular type of drill, but in some large workshops pillar drills are used. A drilling machine can be used for a range of jobs in the workshop. The most commonly used types of drills are:

- air drill
- rechargeable drill
- electric pillar drill
- electric drill.

Figure 1.124 Electric drill

> **Did you know?**
>
> A drill can also be used for other operations such as sanding, grinding and, if it is fitted with a clutch, as an electric screwdriver.

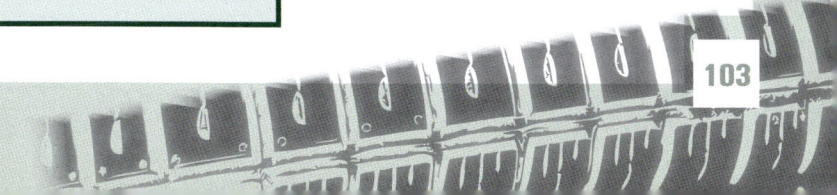

Working life

Ahmed is drilling a piece of metal and is pressing down hard on the drill. The drill bit starts to smoke.

1. What two faults could be the result?

Level 2 Light Vehicle Maintenance & Repair

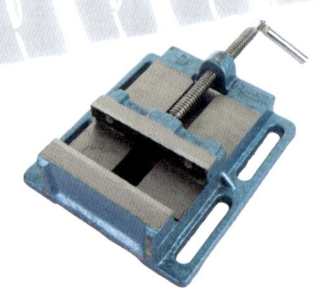

Figure 1.125 Hand vice

Figure 1.126 Bench vice

Vices

Wherever possible, you should use a vice to hold components and materials safely. With the pillar drill, you should use a hand vice (see Figure 1.125). If you are using a portable drill, where possible, you should use a bench vice (see Figure 1.126). This is also used for many other holding activities, for example filing, sawing and unit overhaul.

Care of tools

It is important to take good care of tools and equipment, so that they work well, last a long time and cause no health and safety hazards to the user.

Hand tools

Whenever you are going to use a tool, check it for visible damage, especially damage to areas where contact is made with the components to be worked on. Also check any piece of equipment used to move the tool.

Misuse of tools may also cause unnecessary damage. Table 1.47 gives some examples of damage to tools and the causes. To prevent damage to tools, make sure you use them correctly.

Table 1.47 Examples of damage to tools and the possible causes

Tool	Damage	Possible cause of damage
Screwdriver	• Blunt/broken tip	• Excessive or incorrect use
	• Damaged handle	• Using incorrectly, e.g. striking with a hammer
Socket	• Rounded	• Using wrong size
	• Cracked	• Using with air wrench
Hacksaw blade	• Broken teeth	• Holding the component incorrectly
	• Blunt teeth	• Blade inserted incorrectly
Files	• Broken	• Using as a lever
	• Worn	• Teeth not cleaned regularly
Chisel	• Blunt tip	• Chiselling hard materials
	• Mushroomed head	• Excessive use
Drill bit	• Blunt	• Using too much speed or force
	• Broken	• Drilling at incorrect angle

Pneumatic tools

Air leaks are the main causes of pneumatic tool faults. To repair joints, first disconnect the tool from the power. Then seal the joint using special thread tape made of **polytetrafluoroethylene** (**PTFE**). Lubricating air tools daily will extend the life of the equipment.

> **Key term**
>
> **Polytetrafluoroethylene (PTFE)** – a synthetic substance used to make a thin white tape which can be wrapped around a thread to prevent fluid and gas leaking past it.

1 General workshop operation principles for light vehicles

Electrical tools

Most electrical equipment in the workshop must operate at a reduced voltage, for example hand lamps at 24 volts and drills at 115 volts. Always check wires for damage before use and operate the equipment in dry conditions.

Hoists and lifts

Lifting and supporting any vehicle or heavy component needs to be done safely to prevent injury. Hydraulic jacks, whether a trolley jack or bottle jack, and hydraulic cranes need to have some sort of permanent support. This is because the hydraulic seals can weaken and leak when constantly under pressure.

Axle stands and component stands are used to provide extra mechanical support to prevent hydraulic failure.

Other larger equipment, such as a vehicle hoist, can be electric, pneumatic or hydraulic. Whatever the type of lift, it will have a mechanical locking device to prevent injury.

> **Find out**
>
> Check three tools in your workshop for damage and report any problems to your line manager.

Figure 1.127 Trolley jack

Figure 1.128 Bottle jack

Figure 1.129 Hydraulic crane

Figure 1.130 Axle stands are used to safely support a vehicle

Figure 1.131 Two-post hoist or ramp – used for lifting vehicles off the ground

> **Did you know?**
>
> Safe working load (SWL) is the maximum load the lifting or supporting device can be used for. This is indicated on the side of the equipment and must never be exceeded.

105

> **Working life**
>
> Ollie has just raised a vehicle on a hoist to work on the steering. He suddenly realises that the steering lock is on and the keys for the vehicle are in his overall pocket.
>
> 1   What would happen if Ollie decided to climb up the ramp to put the keys back in the ignition of the vehicle?

Checks to be made on hydraulic equipment should centre around the hydraulic ram of the lifting device. If there are any visible oil leaks, the piece of equipment needs to be sent away for repair.

> **Working life**
>
> A jack is found to be leaking. Its safe working load is 3 tonnes and the vehicle to be lifted is 1.5 tonnes.
>
> 1   Is it safe for Diane to use the jack given that the load is less than the SWL?

Extraction equipment

Vehicle exhausts emit poisonous gases and so they should not be left to run in the workshop without exhaust extraction. A pipe is fixed to the exhaust (see Figure 1.132) and the fumes are sucked from the running vehicle by a pump, which directs the gases to the atmosphere outside the workshop.

Welding is another operation that requires the use of extraction. Harmful fumes are cleansed using a series of filters in a portable extraction unit, as shown in Figure 1.133.

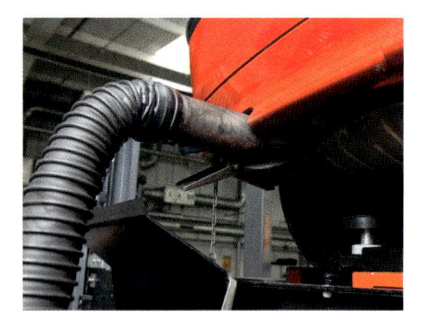

Figure 1.132 Workshop extraction equipment for exhaust fumes

Equipment instructions and use

When working with all types of tools, you must follow the manufacturer's instructions. These can be found on the manufacturer's website or in the original packaging of the equipment when new. The instructions state:

- any PPE you need to wear
- how you can avoid injury
- how to use the equipment correctly and safely
- how you must look after the equipment
- how to store the equipment safely to prolong its service life.

If you do not know how to use a piece of equipment or if it is damaged in any way, you must report this to your workshop supervisor or mentor.

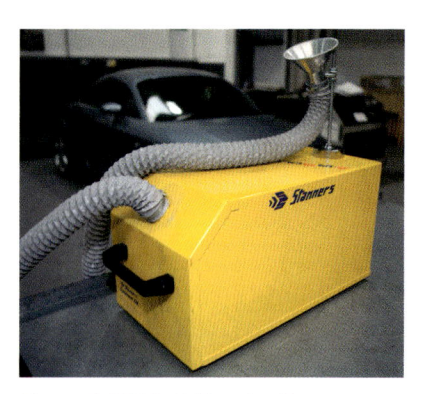

Figure 1.133 Portable extraction equipment for welding

1 General workshop operation principles for light vehicles

> **Working life**
>
> Steven has just bought a new **variable speed** drill which can also double up as a screwdriver. He takes it out of the box and inserts a screwdriver attachment.
>
> 1 What could be the consequences of Steven not reading the instructions before use?

> **Key term**
>
> **Variable speed** – the speed can be adjusted from slow to fast depending on the work to be done.

Using tools and equipment safely

Table 1.48 gives some of the precautions you need to take when using different types of tools and equipment.

Table 1.48 Precautions to take when using tools and equipment

Tool/equipment	Precautions
Jack	• Make sure the vehicle is on level ground. • The vehicle must be supported by axle stands. • The wheels on the ground should be chocked. • The jack handle should be upright.
Compressed air	• Never allow a jet of air to be directed at any part of the body. Small particles of dirt are held in the air and can penetrate the skin. • Never use compressed air for clearing away swarf or filings from benches and machinery. The swarf could be blown into someone's eyes. • Make sure the quick release couplings are fully engaged before use or the coupling may detach and be sent with force across the workshop.
Hoist	• Do not exceed safe working loads. • Position the vehicle centrally on the lift and chock the wheels. • Before lowering the lift, remove all equipment, warn other people and make sure your feet are clear.
Bench drill	• Keep clothing away from the spindle. • Securely clamp the work. • Use the drill guard.
Grinding wheel	• Ensure that the distance between the work rest and the grinding wheel is less than 3 mm. • Use the machine guards. • Wear goggles.

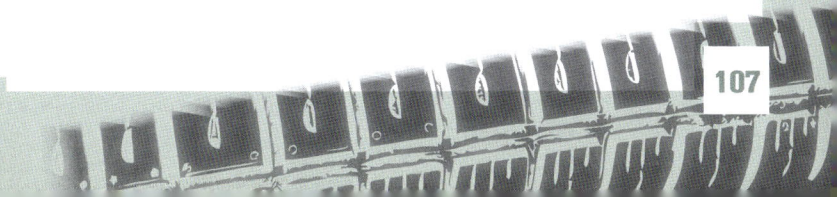

Fixing devices

The fixing devices shown in Table 1.49 prevent components loosening as a result of strain and vibration. There are two types of fixing devices: positive and frictional.

- Positive means that another consumable is used with the device.
- Frictional is where the force of the touching surfaces prevents the fixing from coming undone.

Table 1.49 shows the different types of fixing device and gives the use and the type for each one.

Table 1.49 Fixing devices, their uses and type

Fixing device	Use	Type of fixing
Slotted nut/split pin	Wheel hub nut	Positive
Spring washer	Gearbox casing	Frictional
Shake-proof washer	Body panel fixing	Frictional
Lock tab washer	Wheel hub nut	Positive
Castellated nut/split pin	Wheel hub nut	Positive
Self-locking (nyloc) nut	Suspension joints	Frictional
Locking plate	Heavy vehicle wheel nuts	Positive
Locking plate/locking wire	Differential bolts	Positive
Locking nut	Removing studs from components	Frictional

Did you know?

Nyloc nuts are fitted with a nylon insert in order to keep the tightened nut in place.

Other fixing devices may be used to prevent components rotating. Thread lock, applied to the threads during assembly, sets before service like a metal glue and ensures that the bolt does not loosen.

Thread designation

The most popular assembly method on a motor vehicle is to use nuts and bolts. Most vehicles now use metric fixings and threads with a right-hand thread. This means that the thread goes clockwise to tighten and anticlockwise to loosen the assembly. The fixings come in standard measurements, for example M6 × 1.0 × 50.

M = metric
6 = bolt diameter in millimetres (mm)
1.0 = thread **pitch** in millimetres (mm)
50 = length in millimetres (mm)

Metric threads with a fine pitch are less likely to work loose when subject to vibration. However, this isn't the case when soft materials such as plastic are used. With soft materials, a coarse thread pitch is preferred to prevent thread damage. This is because the **thread** will have a greater **depth** and so each thread will have more strength.

Table 1.50 lists other thread types.

Table 1.50 Other thread types

Abbreviation	Meaning
ANC	American National Coarse
ANF	American National Fine
BA	British Association
BSF	British Standard Fine
BSP	British Standard Pipe
BSW	British Standard Whitworth
UNF	Unified National Fine
UNC	Unified National Coarse

> **Did you know?**
>
> A coarse thread has a greater pitch and **thread depth** than a fine thread.

> **Key terms**
>
> **Pitch** – the distance a nut moves along a screw thread during one rotation.
>
> **Thread depth** – the distance between the bottom (root) and the top (crest) measured vertically.

Left-hand threads

Left-hand threads are used where rotation may affect the tightness of the fixing or where extension of a shaft is required. Some hub nuts and track rod ends will be subject to these forces.

Fastening devices

Keys

Keys are a way of locating components to shafts. They allow the component to rotate with the shaft. An example of this is the key used to connect the timing gear to the crankshaft in an engine.

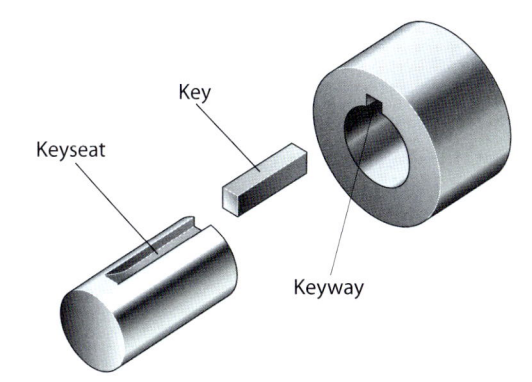

Figure 1.134 Flat key – flat-shaped locating key

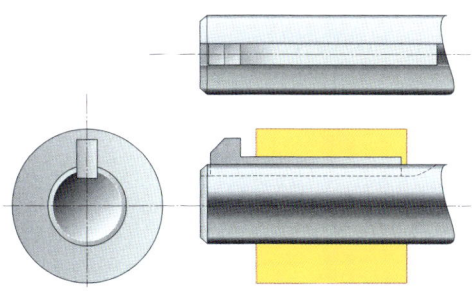

Figure 1.135 Gib head key – flat key with a raised end portion

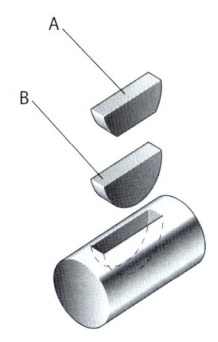

Figure 1.136 Woodruff key – semi-circular locating key

Level 2 Light Vehicle Maintenance & Repair

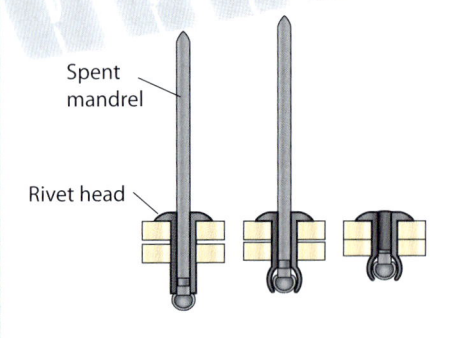

Spent mandrel

Rivet head

Figure 1.137 Pop rivets

Did you know?

As more and more vehicles use aluminium for their body panels, self-piercing riveting (SPR) has become an important alternative joining technique for the automotive applications of bonnets, tailgates and doors. Most existing SPR machines use electrical motors to drive a rivet into the sheets, however newer systems use gunpowder to drive the riveting process.

Did you know?

For general sealing, a gasket is shaped to the joint that needs to be sealed and can be made of resin-impregnated paper. Other materials are used, including rubber, silicone, metal, cork, felt, neoprene, nitrile rubber, fibreglass or a plastic polymer (such as polychlorotrifluoroethylene) where high pressure joints are required, for example the cylinder head gasket. (For more on the cylinder head gasket, see Chapter 3, page 251.)

Rivets

These are a method of joining together body panels. The most common rivet is the blind rivet. It is called a blind rivet because in most cases you cannot see the other side once a panel is joined together.

A riveting gun pulls the central rivet pin upwards and the outer section squeezes outwards at the same time as the body panels are brought together.

Other fixings

Other fixings are used in a variety of locations throughout the motor vehicle. They include:

- stamped wing nuts
- cold-formed wing nuts
- hex head cap screws
- wood drive screws
- machine screws
- recessed head and slotted machine screws
- type 'U' drive screws
- thumb screws
- elevator bolts
- step bolts
- hanger bolts
- square head machine bolts
- cotter pins
- tubular rivets
- flat washers
- pipe plugs.

Trim clips

Body panels and trim need to be fixed to the vehicle body to prevent unnecessary rattles and poor appearance. This is done using the methods shown in Table 1.51.

Table 1.51 Types of trim clip and their uses

Trim clip		Uses
	Hole plug	Used to fix thin trim under dashboard
	Panel clip	Used to fix the trim panel to the door
	Screw rivet	Used to fix thin trim under dashboard
	Push rivet	Used for carpets in the luggage compartment of vehicles
	Trim button	Used for carpets in the luggage compartment of vehicles

A captive nut is used for fixing bumpers and some metal trim panels. Where you cannot reach the nut when screwing in a bolt or screw, you can use a captive nut.

Gaskets and seals

Gaskets

Wherever two components that come together need to retain fluids, a gasket or seal should be used. For fixed components, this is because it is not possible to machine each part exactly to prevent leakage. A gasket and/or sealing compound will be used for this purpose.

Sealant

Sealant is used between component faces and is applied to gaskets. Sealants are mainly flexible rubber-based products. They may need to be applied with a sealant gun.

Seals

Where there is one fixed and one rotating part that need to be fluid-tight, the most common seal is a lip seal.

Adhesives

Adhesive, or glue, is a mixture in a semi-liquid state that bonds items together. Types of adhesive include:

- **Contact** – Apply a thin coating of the adhesive to both surfaces of the mating parts and allow it to go 'tacky' (almost dry). Then place the surfaces together, apply pressure and leave for up to 24 hours.
- **Pressure-sensitive adhesive** – This requires pressure to make sure the bonding process takes place.
- **Synthetic adhesives** – These are mainly epoxy-resin-based and normally come in two parts. The filler and hardener are mixed together to give a paste to apply to the components to be bonded.
- **Instant** – 'Super glue' is an example of an instant adhesive.
- **Hot adhesives** – These are applied in molten form from a heating device, such as that shown in Figure 1.138. They can give a very strong metal-to-metal bond.

More and more vehicle manufacturers use adhesives to bond panels and chassis sections together as this reduces the heat distortion caused by welding without affecting the strength of the joint.

Failure of adhesive joints

A joint may split at any time, usually as a result of failure of the adhesive used. Failure of joints can occur in the situations shown in Figure 1.139.

> **Safe working**
> When using sealants, always take care to make sure that you keep skin contact to a minimum.

Figure 1.138 Hot adhesive gun

> **Safe working**
> Take care that instant adhesives such as 'super glue' do not touch the skin, as they can stick your fingers together.

> **Key term**
> **Mating faces/parts** – the areas which need to be joined together.

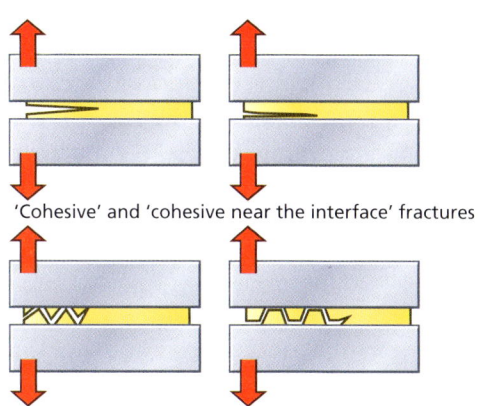

'Cohesive' and 'cohesive near the interface' fractures   'Adhesive' or 'interfacial' fracture

Fractures jumping from one interface to the other   Fractures in the adherent

Figure 1.139 Failure of adhesive joints

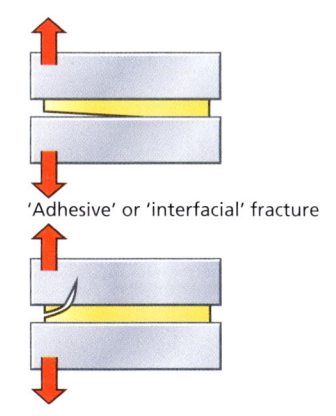

CHECK YOUR PROGRESS

1 Name a measuring tool that can measure inside diameter, outside diameter and depth.
2 What is the standard unit of measurement for temperature?
3 What is meant by the pitch of a hacksaw blade?
4 What does the M signify on a M12 x 1.5 x 60 bolt?

Materials used in fabricating, modifying and repairing vehicles and fitting components

All materials exhibit properties naturally. Properties can also be given to materials by using different heating and cooling processes. The different properties used to describe materials are shown in Table 1.52.

Table 1.52 Properties of materials

Property	Description
Hardness	The ability to resist indentation, abrasion and wear
Toughness	The ability to withstand shock loading without breaking
Strength	The amount of pull a material will stand before it breaks
Conductivity	The ease with which a material allows the passage of electricity or heat
Softness	The ability to be easily shaped or cut
Plasticity	The ability to be moulded and forged to shape when hot
Ductility	The ability of a material to stretch before fracturing
Elasticity	The ability of a material to return to its original shape after being deformed
Malleability	The ability to be worked into different shapes without breaking
Brittleness	A material that allows very little change in shape before fracturing

When you are working on components, it is important to know about the properties of the materials you are working with. For example, if you are working with softer materials, you will need to take greater care to prevent damaging the material.

Ferrous and non-ferrous metals

Ferrous metals are metals which contain iron. The different types of ferrous metals (iron and steel) are defined by the amount of carbon contained in the material. As the percentage of carbon increases, the metal becomes harder and more brittle. Table 1.53 lists some different types of ferrous metal and the percentage of carbon in each.

Key term

Ferrous metals – metals which contain iron, for example different types of steel.

Table 1.53 The percentage of carbon in different ferrous metals

Ferrous metal	% carbon
Cast iron	3.00
High carbon steel	1.20
Medium carbon steel	0.50
Mild steel	0.25
Wrought iron	0.01

> **Find out**
>
> Write the headings 'Ferrous' and 'Non-ferrous' on a sheet of paper. Use a magnet to help you find 10 ferrous and 10 non-ferrous vehicle components in your workshop. Make a note of them under the correct headings.

Some facts about ferrous metals:

- Ferrous metals are magnetic.
- If ferrous metals are exposed to the atmosphere and are unprotected for long periods, oxidising (rust) will occur.

Uses of ferrous metals

Table 1.54 shows some common uses of ferrous metals in car components.

Table 1.54 Some uses of ferrous metals in car components

Typical motor vehicle components	Typical alloy steel material
Gudgeon pins	Low carbon, nickel steel
Valves	Nickel-chromium steel
Crankshaft	Nickel-chromium-molybdenum steel
Road springs	Silicon manganese steel

Effects of hardening and tempering steel

Various materials are changed through the application of heat and cooling. For example, metal expands when the temperature increases (you will use this principle in the workshop when you use heat to loosen tight components). However, if materials are heated or cooled incorrectly, they will change their state and the component will fail. For example, you should never use heat on steering joints as they are made of hardened steel, which will soften through incorrect heating application.

During the manufacture of steel, different processes are used to produce different properties in the metal:

- **Hardening** – Hardening steel gives it the ability to resist scratching, wear, abrasion or indentation by harder objects.
  To harden steel, it needs to be heated to cherry red colour (880°C) and quenched in water.
- **Tempering** – Hardening can make steel too brittle. This destroys its resistance to impact and shock. Tempering steel removes some of the hardness.
  To temper steel, it is first hardened and then heated to tempering temperature for the properties required. Finally it is quenched.

- **Annealing** – This is done to soften the material, increase its ductility and relieve some of the internal stresses.
  To anneal a metal, it is heated to a cherry red colour and then allowed to cool slowly in ashes or hot sand.
- **Normalising** – This restores the grain structure of steel after it has been subjected to cold or hot working. If steel is kept cold when bending without heating this is known as cold working. The internal structure of the grains in the steel become deformed and stressed. If steel is kept hot for a long period of time, such as when welding, the grain in the steel becomes large and coarse.
  To normalise steel, it is heated to a cherry red colour and then allowed to cool naturally in the air.

Non-ferrous metals

These tend to be soft and do not withstand large stresses. Table 1.55 gives some examples of non-ferrous metals and their properties.

Table 1.55 Some non-ferrous metals and their properties

Metal	Appearance	Properties
Aluminium	White	Soft, ductile, good conductor of electricity and heat, very light
Copper	Reddish brown	Soft, ductile, good conductor of electricity and heat
Tin	Silver white	Malleable and ductile
Lead	Silver white	Soft, malleable, unaffected by acid

Copper scratches and dents easily. It is reddish in appearance and acquires a green oxide if left unprotected in the atmosphere. If the oxide is heated and quenched, the copper returns to its normal state.

If non-ferrous metals are mixed to form **alloys**, they take on whole new properties. For example, bronze (copper + tin) and brass (copper + zinc) scratch easily and vary in yellowness according to the amount of copper content.

Aluminium and its alloys

Pure aluminium is not commonly used in vehicle manufacture because it is too soft and ductile. Alloys can be produced by mixing aluminium with other materials that are strong, hard, retain strength at high temperatures and are resistant to corrosion. However, if unprotected and exposed to the atmosphere for long periods of time, a powdery oxide forms on the surface.

Table 1.56 gives two types of aluminium alloys and their uses in vehicle manufacture.

> **Key terms**
>
> **Alloy** – a mixture of two chemical elements, at least one of which is a metal.
>
> **Extruded** – when hot, the metal is forced under pressure through a die to provide a specific shape.

Table 1.56 Aluminium alloys: properties and uses

Aluminium alloy	Property	Use
Cast alloys	Easy to cast	• Pistons and cylinder heads
Wrought alloys	Can be cold worked and **extruded**	• Frames and body panels • Wheels and body mouldings

Plastics

Plastics are man-made materials. They can be formed into a variety of shapes with the use of heat and pressure.

The two groups of plastics are:

- **Thermosetting** – When this type of plastic has set it cannot be reshaped after reheating. These plastics are produced from a binder, consisting of filler for strength and colouring, a hardening agent and an accelerant.
- **Thermoplastic** – If reheated, this type of plastic can be reshaped. Thermoplastic items are produced from a binder, consisting of a filler for strength and a colouring material.

Three common plastics used in vehicle manufacture are nylon, PTA and PVC. Table 1.57 shows their properties and uses in the automotive industry.

Table 1.57 Plastics used in vehicles: properties and applications

Materials	Properties	Vehicle application
Nylon	• Strong heat resistance • Low coefficient of friction	• Bushes and bearings • Speedometer drive gear • Steering and suspension • Electrical switches
PTA	• Self-lubricating	• Light-duty bearings in small electric motor drive gears
PVC	• Good insulator • Can be coloured • Chemically resistant • Reasonably heat resistant	• Electrical cable insulation • Upholstery • Fuel pipelines

Forces generated in vehicle construction

Vehicles are made up of many different types of materials. The different materials are chosen because they are subject to different types of forces. The different forces are:

- **Tensile stress** – This is created by stretching a material by applying opposite pulling forces, as shown in Figure 1.140. This stress will cause the material to elongate and become weak (as when you pull an elastic band).

- **Shear stress** – This is caused when opposing forces act on the face of a material, as shown in Figure 1.141. An example of shear stress is cutting metal with snips.
- **Compressive stress** – This is the load on the material which squashes it together by direct pushing forces (see Figure 1.142). It is the opposite of tensile stress. A hand vice uses compression to hold a component in place.
- **Yield stress** – This happens when a material is bent and cannot go back to its original shape. For example, when a ring pull on a can is lifted slightly, it will return, but when it is lifted higher, it stays in the same position. The yield point is the position the ring pull will get to after which it will not return to its original shape. The graph in Figure 1.143 shows the amount of stress put on a material until the strain becomes too much.

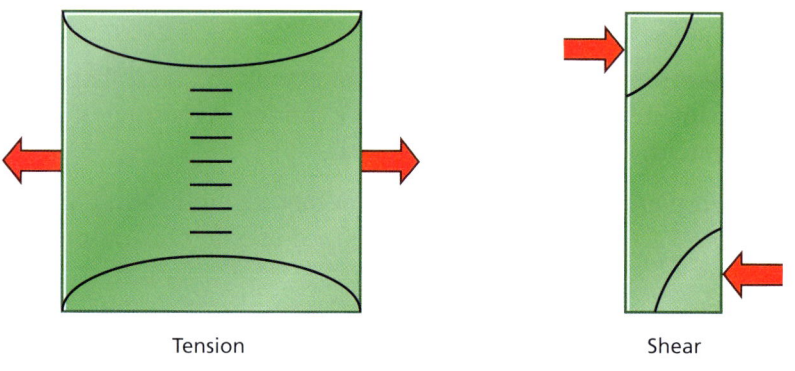

Figure 1.140 Tensile stress

Figure 1.141 Shear stress

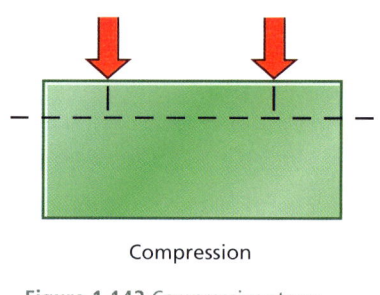

Figure 1.142 Compressive stress

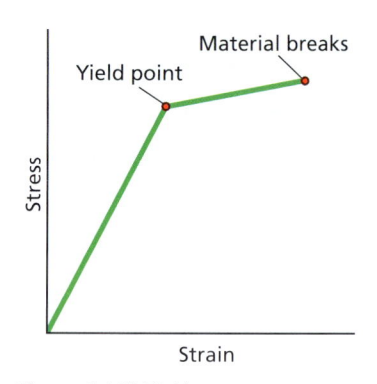

Figure 1.143 Yield stress

CHECK YOUR PROGRESS

1 Which hammer should you use to avoid damaging components when dismantling?
2 What should you do to ensure correct reassembly of a component, before you dismantle it?
3 What type of force is used to hold a component in a hand vice?

1 General workshop operation principles for light vehicles

FINAL CHECK

1 Under the Provision and Use of Work Equipment Regulations (PUWER), who has the main responsibility for ensuring that all equipment supplied for use in the workshop is in safe working order?

   a your employer
   b trade union officials
   c you
   d the Health and Safety Executive inspector

2 What is the correct PPE to wear when charging a battery?

   a helmet, heat-resistant gloves, apron and overalls, safety boots
   b goggles, Kevlar gloves, overalls, safety boots
   c helmet, full face mask, ear defenders, latex gloves, apron, safety boots
   d goggles, chemical-resistant gloves, mask, overalls, apron, safety boots

3 Which of the following is the correct definition of a hazard?

   a something that has the potential to cause harm or damage
   b the likelihood of an accident happening
   c a serious accident resulting in an injury
   d an unexpected and unplanned event in the workplace

4 Which of the following types of fire extinguisher should not be used on fires involving flammable liquids?

   a cream label – foam
   b black label – carbon dioxide
   c blue label – powder
   d red label – water

5 The chemical name for poisonous vehicle exhaust fumes is:

   a carbon dioxide
   b carbon monoxide
   c epoxy resin
   d trichloroethylene

6 If there is a serious accident and the person is bleeding severely, the first action to take would be:

   a make them a cup of tea
   b apply pressure to the wound
   c put a plaster on the wound
   d apply a tourniquet

7 Which one of the following is the correct way of lifting a weight from the ground?

   a back bent, legs straight
   b back bent, knees on the ground
   c back straight, knees bent
   d back straight, legs straight

8 Some nuts are fitted with a nylon insert in order to:

   a allow the nut to be loosened easily
   b lubricate the thread as the nut is fitted
   c keep the tightened nut in place
   d protect the thread on the bolt

9 The most effective equipment for raising a vehicle a short distance from the ground is:

   a a pulley block
   b a winch
   c a hydraulic jack
   d a lifting chain

10 When a metal has become hard due to cold working, it can be changed to its softest state by:

   a annealing
   b tempering
   c nitriding
   d normalising

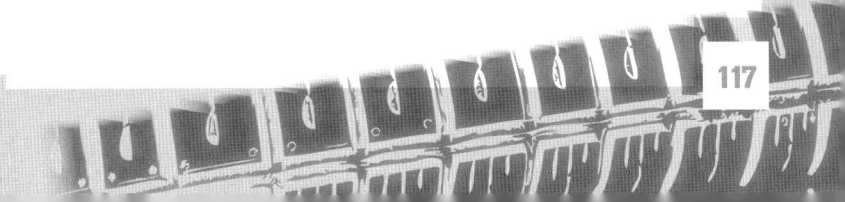

Level 2 Light Vehicle Maintenance & Repair

GETTING READY FOR ASSESSMENT

By reading and completing this chapter you have gained the knowledge and skills you need to work safely within the workplace. You have learned the importance of using and caring for personal protection equipment, as well as the skills required when using tools and measuring devices and when fabricating materials in an automotive environment. Throughout your working you will have gained understanding of the associated legislation, the hazards and risks involved in your working environment and the importance of effective communication.

You will be assessed on the following topics:

- Health and safety legislation
- Appropriate personal and vehicle protective equipment
- Hazards and risks
- Accident prevention
- Risk assessment
- Effective housekeeping practices
- Organisational structures, functions and roles
- Communicating with colleagues and customers
- Hand tools and measuring devices
- Materials used in fabricating, modifying and repairing vehicles and fitting components
- Applying automotive engineering, fabrication and fitting principles when modifying and repairing vehicles and components

You now need to apply the knowledge you have gained in this chapter in your day-to-day working activities. For example, you need to be able to work efficiently while observing effective housekeeping practices. You can do this by making sure all equipment is stored correctly after use and keeping the floor clear from obstructions. You also need to show awareness of the hazards and potential risks associated with poor housekeeping and the regulations surrounding general health and safety: how to report accidents, clear up spillages, use warning signs and be aware of the safe disposal and storage of hazardous substances. You can use this knowledge in your workplace by working safely, observing set down processes and regulations and communicating effectively.

This chapter has provided the basic knowledge that will help you with both theory and practical assessments.

Before you try a theory end-of-unit test or multiple-choice test, make sure you have reviewed and revised any key terms and read all the questions carefully. Take time to digest the information so that you are confident about what the question is asking you. With multiple-choice tests, it is very important that you read all of the answers carefully, as it is common for two of the answers to be very similar, which may lead to confusion.

For practical assessment tasks, it is important that you have had enough practice and that you feel that you are capable of passing. Before you begin a task make sure you have the correct PPE, VPE, tools and equipment to hand and that you have a plan to follow.

When you are doing any practical assessment, always make sure that you are working safely throughout the task. Take care to observe all health and safety requirements and use the recommended personal protective equipment (PPE) and vehicle protection equipment (VPE). When using tools, make sure you are using them correctly and safely.

Good luck!

2 Light vehicle chassis units & components

This chapter is an introduction to light vehicle chassis systems. You will learn how a light vehicle body is constructed and how all the systems relating to the chassis unit operate, including steering, suspension, braking, wheels and tyres. You will learn about safe working practices when you work on chassis systems. By the end of this chapter, you will know how to identify the main components used in light vehicle steering, suspension, braking, wheels and tyres. This chapter also covers some of the skills you will need to check, remove, replace and test light vehicle chassis units and components. These skills are covered in the sections on the different systems involved in the vehicle chassis.

This chapter covers:

- Safe working on light vehicle chassis units
- Construction of light vehicle bodies
- Light vehicle steering systems
- Light vehicle suspension systems
- Light vehicle braking systems
- Light vehicle wheel and tyre systems

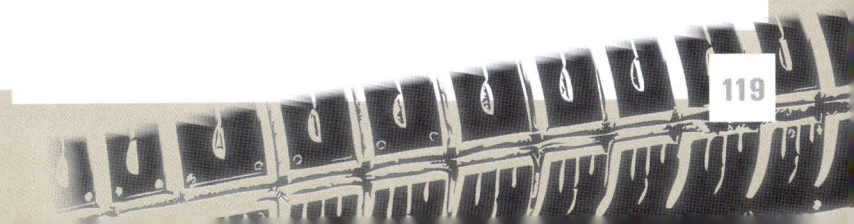

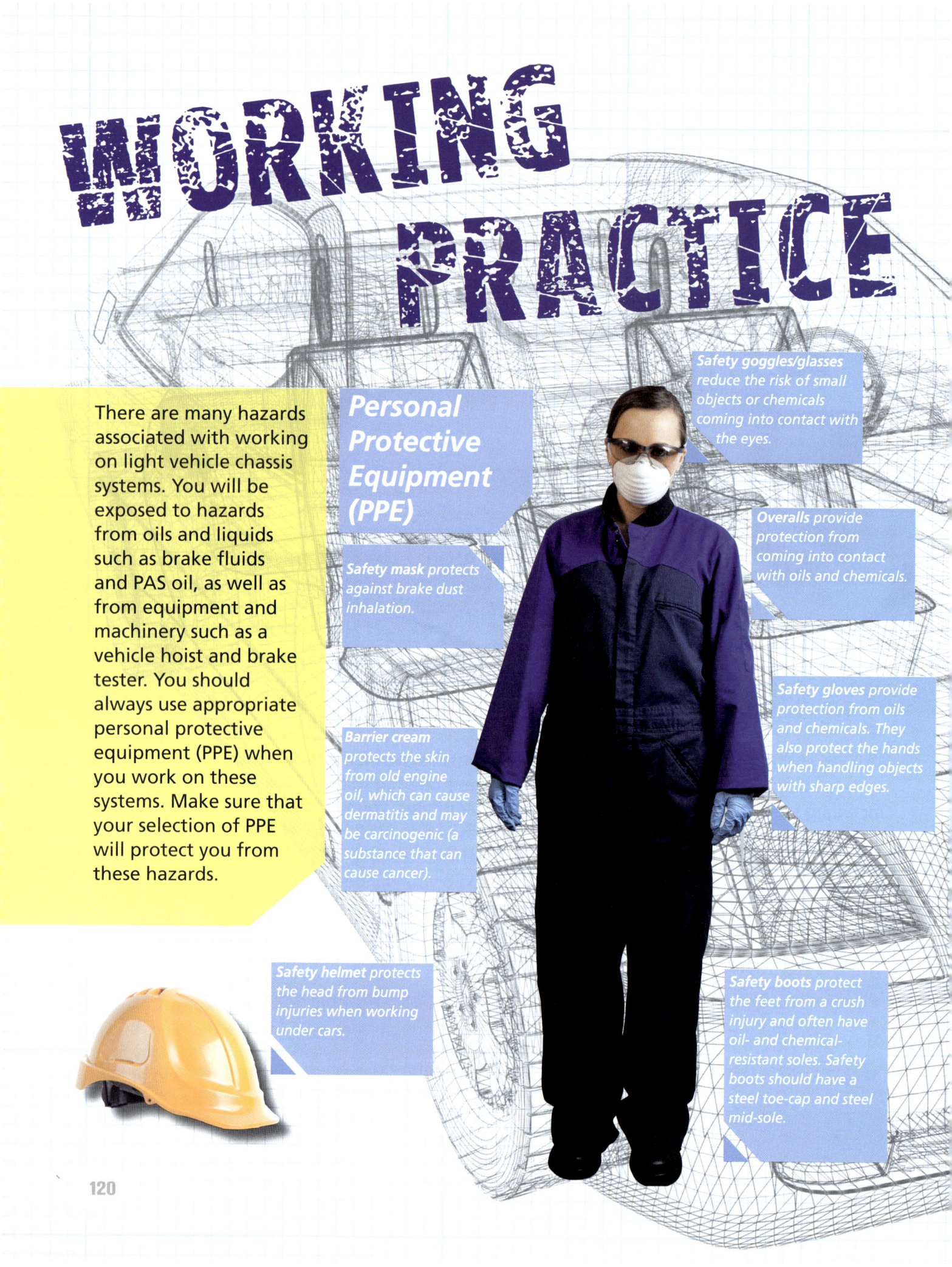

WORKING PRACTICE

There are many hazards associated with working on light vehicle chassis systems. You will be exposed to hazards from oils and liquids such as brake fluids and PAS oil, as well as from equipment and machinery such as a vehicle hoist and brake tester. You should always use appropriate personal protective equipment (PPE) when you work on these systems. Make sure that your selection of PPE will protect you from these hazards.

Personal Protective Equipment (PPE)

Safety mask protects against brake dust inhalation.

Barrier cream protects the skin from old engine oil, which can cause dermatitis and may be carcinogenic (a substance that can cause cancer).

Safety helmet protects the head from bump injuries when working under cars.

Safety goggles/glasses reduce the risk of small objects or chemicals coming into contact with the eyes.

Overalls provide protection from coming into contact with oils and chemicals.

Safety gloves provide protection from oils and chemicals. They also protect the hands when handling objects with sharp edges.

Safety boots protect the feet from a crush injury and often have oil- and chemical-resistant soles. Safety boots should have a steel toe-cap and steel mid-sole.

To reduce the possibility of damage to the car, always use the appropriate vehicle protection equipment (VPE):

Wing covers

Steering wheel covers

Seat covers

Floor mats

If appropriate, safely remove and store the owner's property before you work on the vehicle. Before returning the vehicle to the customer, reinstate the vehicle owner's property. Always check the interior and exterior to make sure that it hasn't become dirty or damaged during the repair operations. This will help promote good customer relations and maintain a professional company image.

Vehicle Protective Equipment (VPE)

Safe Environment

During the repair of chassis systems you may be required to dispose of waste such as tyres, brake fluid and used or damaged PPE, for example gloves. Under the Environmental Protection Act 1990 (EPA), you must dispose of all waste in the correct manner. You should store all waste safely in a clearly marked container until it is collected by a licensed recycling company. This company should give you a waste transfer note as the receipt of collection.

To further reduce the risks involved with hazards always use safe working practices, including:

1. Immobilise the vehicle by removing the ignition key.

2. Prevent the vehicle moving during maintenance by applying the handbrake or chocking the wheels.

3. Follow a logical sequence when working. This reduces the possibility of missing things out and of accidents occurring. Work safely at all times.

4. Always use the correct tools and equipment. To avoid damage to components, tools or personal injury, check tools and equipment before each use.
   - Inspect any mechanical lifting equipment for correct operation, damage and hydraulic leaks.
   - Never exceed safe working loads (SWL).
   - Check that measuring equipment is accurate and calibrated before you take any readings.

5. If chassis components need replacing, always check that the quality meets the original equipment manufacturer (OEM) specifications. (If the vehicle is under warranty, inferior parts or deliberate modification might make the warranty invalid. Also, if parts of an inferior quality are fitted, it might affect vehicle performance and safety.)

6. Following the replacement of any vehicle steering, suspension, braking, wheel or tyre components, thoroughly road test the vehicle to ensure safe and correct operation. Make sure that all work is correctly recorded on the job card and vehicle's service history, to ensure that any maintenance work can be tracked.

Preparing the car

Tools

Hand tools

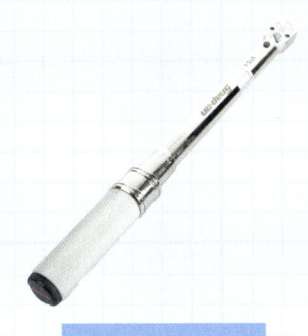

Torque wrench

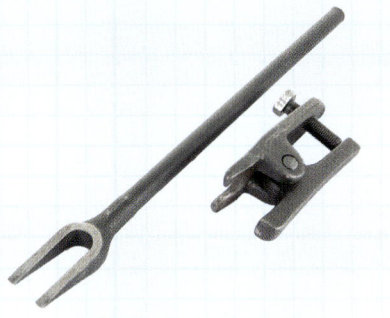

Ball joint splitter

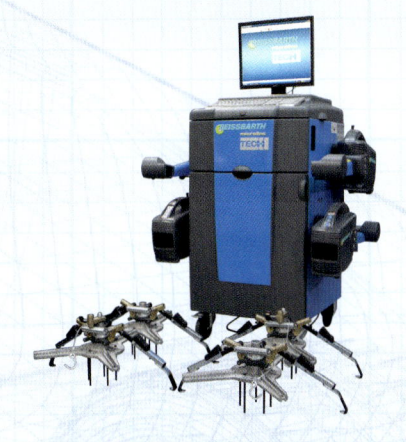

Wheel alignment gauges

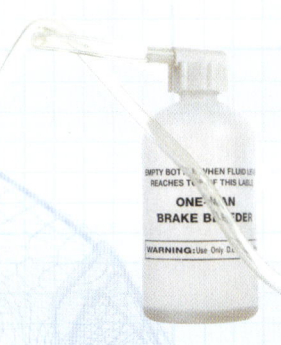

Brake bleeding equipment

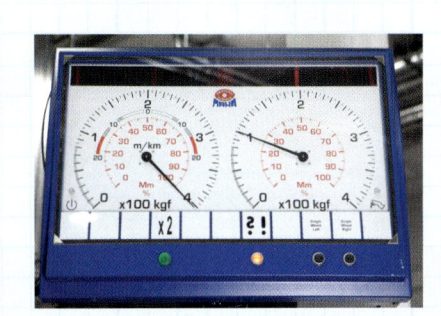

Brake tester

Compressor

Tyre pressure gauge

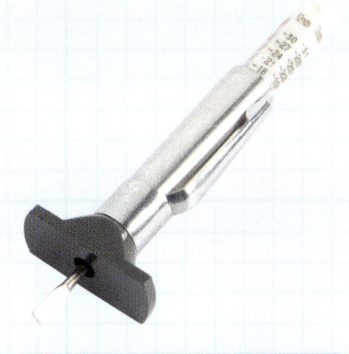

Tread depth gauge

Safe Working

- Always clean up any fluid spills immediately to avoid slips, trips and falls.
- Always work in a well-ventilated area.
- Always use a mask when you are working with brake linings.

123

Key terms

Power train – the combination of the engine, gearbox and **final drive**.

Final drive – the unit which transmits the drive from the gearbox to the wheels through drive shafts at a reduced gear ratio.

Figure 2.1 Coupé

Figure 2.2 Convertible

Figure 2.3 Saloon

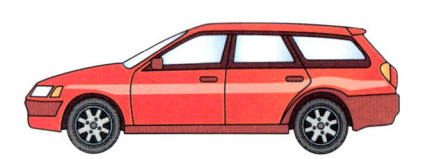

Figure 2.4 Estate

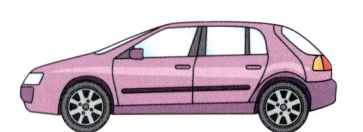

Figure 2.5 Hatchback

Construction of light vehicle bodies

You are probably familiar with different shapes of light vehicle. If you look around any car showroom you will find a whole range of car body shapes.

Vehicle body types

The purpose of the vehicle's body is to hold the driver and passengers. It must also look appealing and allow adequate space for luggage. The vehicle body must provide fixing points and access for the **power train** and protect the occupants from all weather conditions and accidental damage.

Coupé

This solid-roofed, two-door vehicle normally seats only two people: the driver and the passenger. Some manufacturers make 2 + 2 coupés, which have rear seats. These rear seats are often cramped and are only suitable for occasional use. Coupés are often referred to as two-door sports cars or just sports cars.

Convertible

Convertibles are also sometimes called cabriolets. They have a drop top or rag top with either a metal or cloth roof. The vehicle can be converted into an open air car by removing the metal roof or folding the cloth down into a compartment that is normally behind the seats or in the boot. On most modern cloth top vehicles, the top folds electrically at the touch of a button.

Saloon

This is a solid-roofed, two- or four-door vehicle which has ample room for four or more passengers. It also has a substantial area for luggage which is covered by a separate door called a boot lid. This type of vehicle is a favourite with families and is often referred to as a family car.

Estate

The roof of this type of vehicle is fully extended to the back, which increases the amount of space inside the car. The folding rear seats provide a vast storage/luggage area which is covered by a separate door (tailgate). The tailgate allows large items to be easily loaded into the vehicle.

Hatchback

A hatchback is in between a saloon and an estate car. It has a tailgate which is classed as an additional door, and it comes in either three-door or five-door versions. The three-door hatchback is also referred to as a 'hot hatch' and is thought of as being small and quick.

Multi-purpose vehicle (MPV)

This type of vehicle is often called a people carrier. This vehicle is often built on a car chassis, but it is much taller than a standard car and provides more passenger and luggage room. Many multi-purpose vehicles have seven seats.

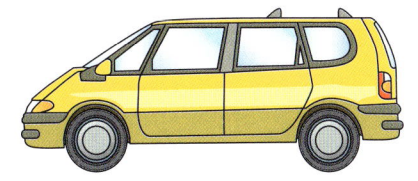

Figure 2.6 Multi-purpose vehicle (MPV), also called a people carrier

All terrain (4 x 4)

This type of vehicle is built to be driven on- or off-road. It is much taller than the average people carrier and more robust. There are three- and five-door versions with tailgates, as with the hatchback.

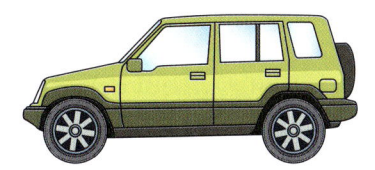

Figure 2.7 All terrain vehicle (4 x 4)

Did you know?

A car-derived van is a car that has had its rear seats removed and turned into a box-shaped load compartment.

Van

This is a type of light vehicle used for carrying loads. It normally has seating for three at the front. The rest of the large box-shaped area at the rear is for the load.

Figure 2.8 Van

Find out

Make a list of a vehicle manufacturer and model for each one of the body types listed on pages 124–125.

Types of vehicle layout

You will notice that the position of the engine and transmission can differ in vehicles. The location depends on the vehicle's purpose and body design. Other factors which influence the design choice include cost, complexity, reliability, packaging (location and size of the passenger compartment and boot), weight distribution and the vehicle's intended handling characteristics.

Vehicle layout can be divided into two categories: front-wheel or rear-wheel drive. Four-wheel drive vehicles may take on the characteristics of either, depending on how power is distributed to the wheels.

To move the vehicle forwards or backwards, **drive** is directed to the wheels from the engine and transmission in various ways (shown in Figures 2.9 to 2.11).

Key term

Drive – the mechanism by which force or power is transmitted in a machine.

- **Rear-wheel drive** (RWD) – the back wheels push the vehicle forwards.

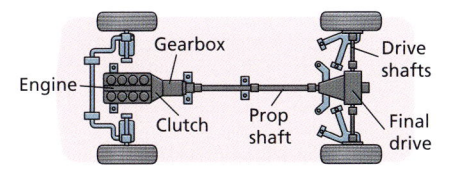

Front engine RWD (Longitudinal)

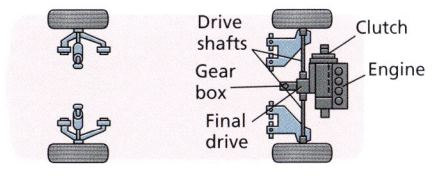

Rear engine RWD (Transverse)

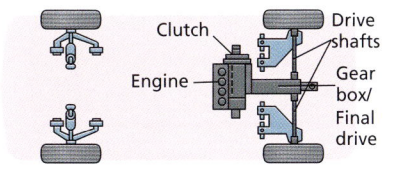

Mid engine RWD (Transverse)

Figure 2.9 Different drive component layouts for rear-wheel drive (RWD)

Find out

Make a list of the first ten vehicles that come into your workshop. Note down whether they are front-wheel drive, rear-wheel drive or four-wheel drive.

- **Front-wheel drive** (FWD) – the front wheels pull the vehicle.

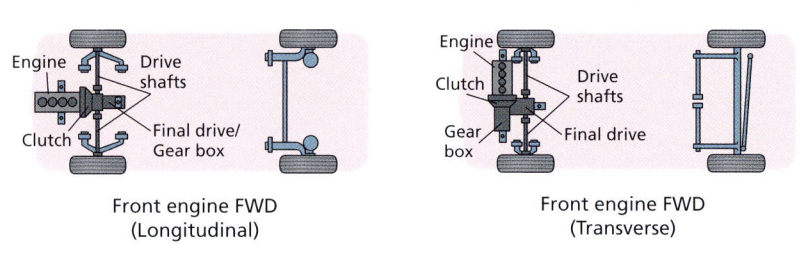

Front engine FWD (Longitudinal)

Front engine FWD (Transverse)

Figure 2.10 Different drive component layouts for front-wheel drive (FWD)

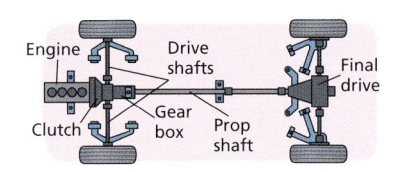

- **Four-wheel drive** (4WD) – all wheels drive the vehicle.

Different types of power train layout include:

- **transverse** – when the engine and **transmission** run the width of the vehicle
- **longitudinal** – when the engine and transmission run the full length of the vehicle.

Front engine 4WD (Longitudinal)

Figure 2.11 Different drive component layout for four-wheel drive (4WD)

Vehicle body construction

Vehicle bodies are made from various materials. Steel is the most popular but aluminium, glass and glass-reinforced plastic are also used on lightweight vehicles such as sports cars. On some vehicles a combination of materials is used, taking advantage of lightweight materials for doors and bonnets and the more cost-effective steel for the chassis.

Key terms

Transmission (system) – consists of the clutch, gearbox, **driveshafts** and final drive.

Driveshaft – the means of transmitting drive from the final drive to the hub.

Live axle – the axle where the drive is transmitted from the power train to the road wheels through a final drive unit.

Monocoque – French for 'single shell'. This chassis type may also be referred to as structural skin, stressed skin, unit body, unibody, unitary construction or body frame integral (BFI).

Spot welding – a process in which contacting metal surfaces are joined by the heat obtained from resistance to electric current flow.

Chassis construction

A chassis is a skeletal frame on which various mechanical parts are fixed, such as engine, suspension, **axle** assemblies, brakes and steering. These are fixed either directly to the chassis or through rubber by way of rubber mountings. To provide a rigid structure to the vehicle, the chassis is made up of fabricated sections.

The chassis is considered to be the most significant component of a vehicle. It gives strength and stability to the vehicle under different conditions.

Monocoque chassis

Most modern light vehicles have a **monocoque** construction (see Figure 2.12). This means that the chassis and vehicle body are made in one unit by **spot welding**. This steel-plated unit supports the structural load by using the exterior instead of an internal frame. It is suitable for robotised production and is therefore cost-effective to produce.

Ladder chassis

One of the oldest types of chassis is the ladder chassis. It is still used today on sport utility vehicles (SUVs) and heavy commercial vehicles. A ladder chassis resembles the shape of a ladder, as it has two longitudinal rails interlinked by several lateral and cross braces. The body is bolted to the chassis with all the other parts of the power train.

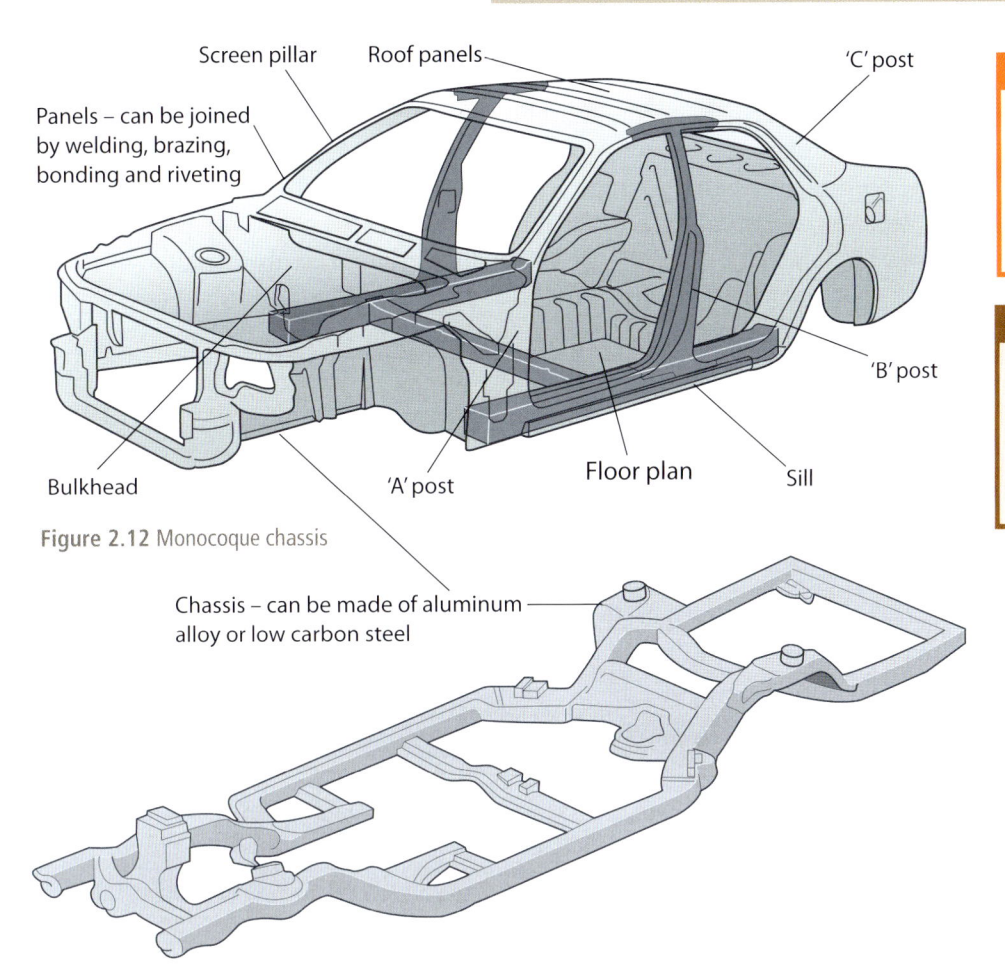

2 Light vehicle chassis units & components

Figure 2.12 Monocoque chassis

Labels: Screen pillar, Roof panels, 'C' post, Panels – can be joined by welding, brazing, bonding and riveting, 'B' post, Bulkhead, 'A' post, Floor plan, Sill

Figure 2.13 Ladder chassis

Label: Chassis – can be made of aluminum alloy or low carbon steel

> **Did you know?**
> Monocoque construction was first widely used in aircraft in the 1930s. It was first applied to vehicle construction in the late 20th century.

> **Find out**
> To see the differences between chassis types, take a look around your workshop. Compare the chassis of an older vehicle to a newly manufactured one.

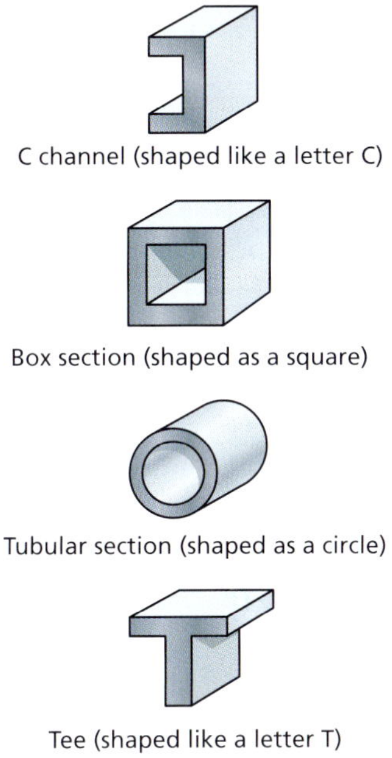

Figure 2.14 Different types of chassis section
- C channel (shaped like a letter C)
- Box section (shaped as a square)
- Tubular section (shaped as a circle)
- Tee (shaped like a letter T)

Backbone chassis

The backbone chassis has a rectangular tube like a backbone. This is usually made of metal and attached to a glass fibre body. This type of chassis is strong and powerful enough to provide support in smaller sports cars. The backbone chassis is easy to make and is cost-effective. Some race cars may have a carbon fibre backbone but this would make them too expensive for most road cars.

Fabricated sections of a chassis

To provide a rigid structure to the vehicle, the chassis is made up of fabricated sections. The different types are shown in Figure 2.14.

Figure 2.15 gives the names of the different body panels on a light vehicle.

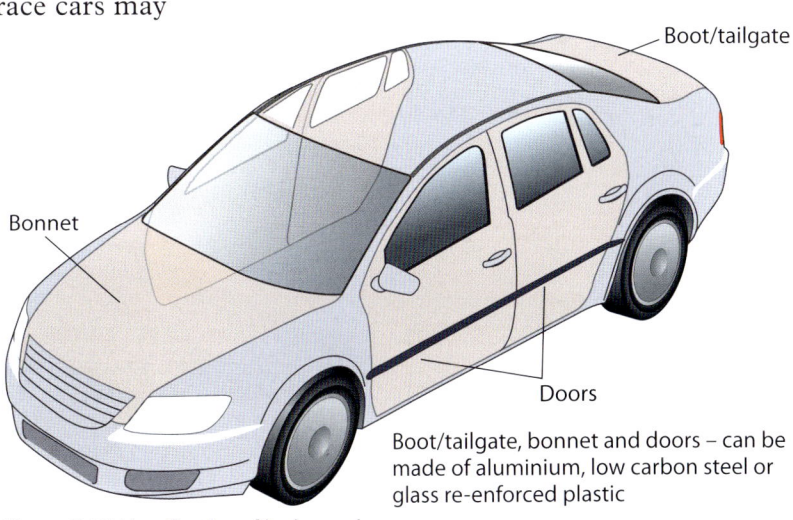

Figure 2.15 Identification of body panels

Labels: Bonnet, Boot/tailgate, Doors, Boot/tailgate, bonnet and doors – can be made of aluminium, low carbon steel or glass re-enforced plastic

127

Level 2 Light Vehicle Maintenance & Repair

Collision safety

Every motor vehicle on the road has been tested to ensure the safety of its occupants in the event of a collision. You will notice that around the seating area extra strength is given to the body construction – this is called a safety cage. To the front and rear of this cage are the crumple zones (shown in Figure 2.16). These are designed to absorb most of the impact during a collision so that the safety cage remains intact and some of the impact energy is taken away from the passengers.

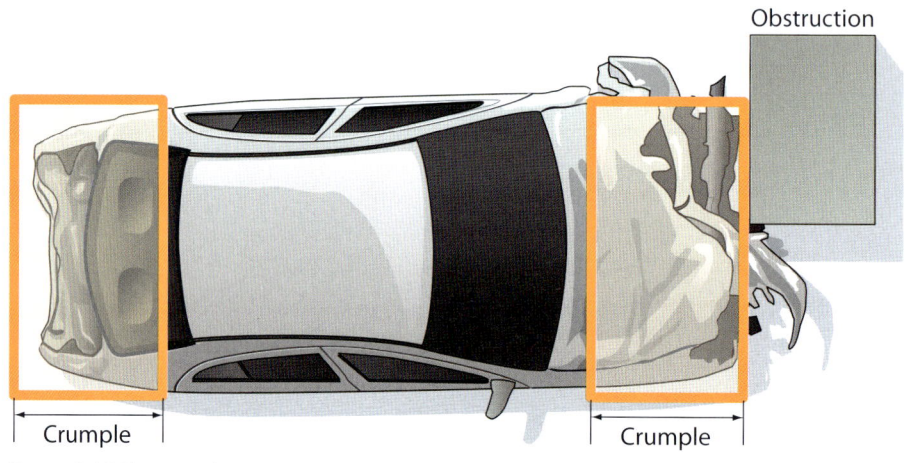

Figure 2.16 The crumple zones

Aerodynamics

A motor vehicle body must be designed to prevent resistance or **drag** to air while it is in **motion**. The term drag is a shortened version of the 'coefficient of drag (CD)' value. This is a number between 0 and 1. When designing the vehicle body, manufacturers aim to get this value as low as possible.

If the value is incorrect and too much air gets under the vehicle while in motion it could act like the wing of an aeroplane and take off. Careful testing of the vehicle shape is carried out in a wind tunnel to make sure the aerodynamics are correct before final production goes ahead.

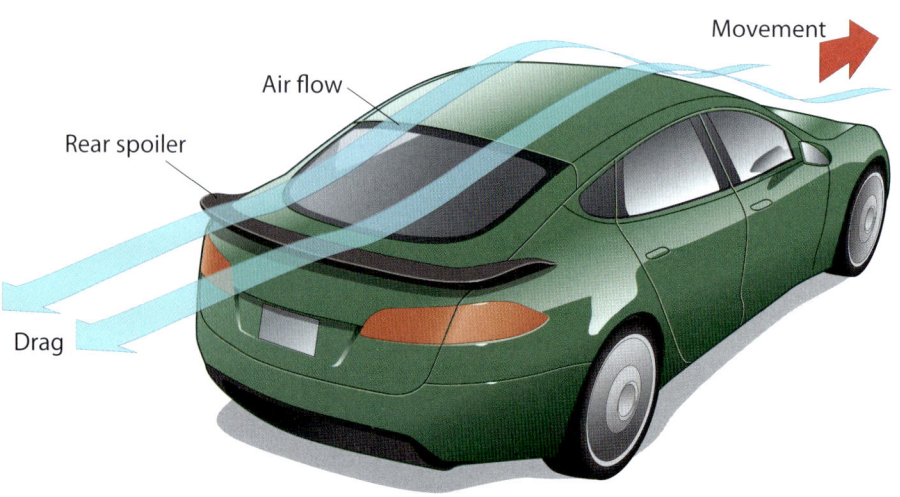

Figure 2.17 Airflow over the vehicle and rear spoiler

Key terms

Drag – an object's resistance to airflow while it is in motion.

Motion – any movement or change in position or place.

Spoilers – moulded plastic sections that direct airflow to force the vehicle body closer to the road surface.

Did you know?

The study of how air moves when it interacts with a moving object is called aerodynamics. Manufacturers design their vehicles so that there is minimal interruption to the flow of air over the vehicle body. This results in a speedier and more economical drive.

Find out

Find out the coefficient of drag (CD) values of different vehicle body types. Go to an Internet search engine and type 'automobile_drag_coefficient' to find results.

Spoilers may be used on vehicles to help reduce resistance. As the vehicle moves forward, the shape of the spoiler directs the airflow to force the vehicle downwards. This ensures that the wheels are in contact with the road surface for maximum traction.

CHECK YOUR PROGRESS

1 State three purposes of the vehicle body.
2 Name five different body styles and their uses.
3 What vehicle units go together to make the drive train?

Light vehicle steering systems

The steering column is a supported shaft that connects the driver's steering wheel to the steering box. The steering column in a modern vehicle is a complex mechanism. It is designed to collapse in a collision to protect the driver.

Purposes of the steering system

The steering system of a vehicle allows the driver to control the direction of the vehicle. This is done using a system of gears and linkages that connect the steering wheel to the front wheels. The steering system must perform the following:

- Turn the vehicle safely and under control.
- Give the driver a 'feel' of the road surface.
- Prevent the driver feeling excessive shock on an uneven road.
- Prevent excessive tyre wear.

Steering system types

Many types of steering system are used in modern vehicles, depending on the vehicle. An older vehicle is likely to use a manual steering system. Manual steering is a mechanical system where the force required to turn the steering is produced by the driver.

Rack and pinion

The rack and pinion is the most frequently used steering system. Look at Figure 2.18 when you are reading this description.

- The steering wheel is connected to a steering rack via a steering column.
- The rotational movement is controlled through any angle by the universal joint, which connects the column to the pinion housing.
- The pinion housing contains the pinion, which is connected to the steering rack (commonly known as the steering gear).
- The steering rack is enclosed in the rack housing.
- The steering gear is made up of two components: the pinion, which is the driving gear, and the rack, which is the driven gear.

Level 2 Light Vehicle Maintenance & Repair

Steering wheel
Steering column
Universal joint
Track rod ends
Pinion
Pinion housing
Rubber gaiters
Steering rack
Track rods
Ball joints

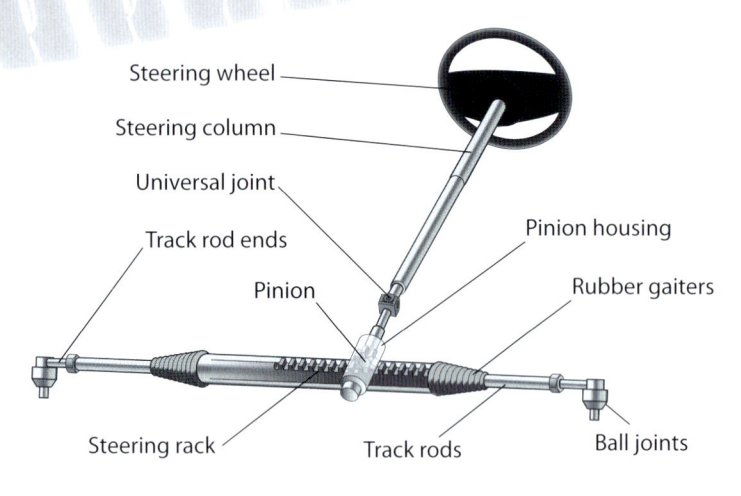

Figure 2.18 Rack and pinion

When you turn the steering wheel, the teeth on the pinion move the steering rack from side to side. This transmits the movement to the **track rods** (also known as tie rods), which are connected to the rack by ball joints on the track rod ends. The ball joints are connected to the road wheels by the **stub axle**. Rubber gaiters are fitted to each end of the rack housing to protect the track rod end from dust, dirt and moisture.

Steering gear ratio

The steering gear ratio is the number of teeth on the driven gear (rack) divided by the number of teeth on the driving gear (pinion).

$$\text{Steering gear ratio} = \frac{\text{Number of teeth on driven gear (rack)}}{\text{Number of teeth on driving gear (pinion)}}$$

Example:

Where there are 40 teeth on the rack and 10 teeth on the pinion, the steering gear ratio will be:

$$\frac{40 \text{ rack teeth}}{10 \text{ pinion teeth}} = 4:1$$

This means the steering wheel will turn four times from left-hand steering lock to right-hand steering lock.

Ball joints

Ball joints are ball-and-socket type joints used to connect components together where movement has to take place between the components. They generally consist of a tapered pin with a **case hardened** steel ball on the end that fits into a steel cup lined with nylon or plastic to provide lubrication.

Ball joints are fitted to each end of the track rod on a rack and pinion system. The outer joint is called the track rod end. This is more prone to wear because of the work it has to do. You can check for wear by turning the road wheel side to side when the wheel is safely off the ground. If you feel movement, the track rod end needs to be changed.

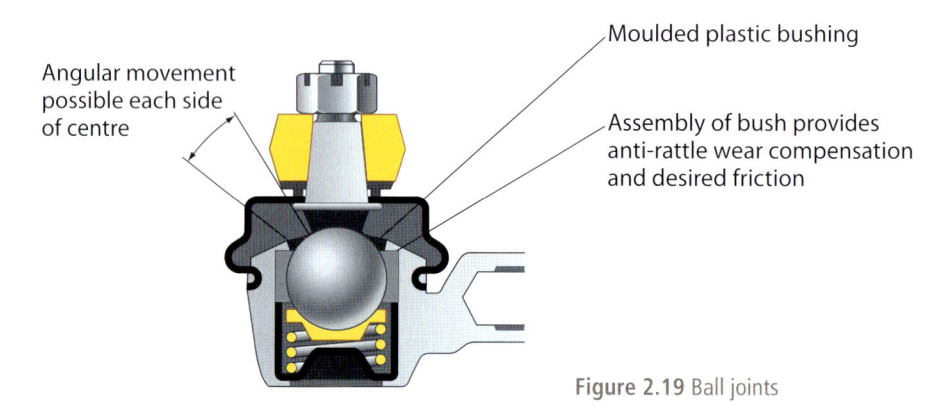

> **+ Emergency**
>
> Steering columns are built with a safety feature which allows them to collapse to protect the driver in the event of an impact.

> **Key terms**
>
> **Track rod** – the adjustable rod that determines what track the wheels take when moving forwards.
>
> **Stub axle** – a short metal shaft that carries the road wheel supported by bearings.
>
> **Case hardening** – a process where the outer surface of a metal is hardened by heating and rapid cooling, but the inside of the component remains in its original state.

Angular movement possible each side of centre
Moulded plastic bushing
Assembly of bush provides anti-rattle wear compensation and desired friction

Figure 2.19 Ball joints

2 Light vehicle chassis units & components

Replacing a track rod end

Checklist			
PPE	**VPE**	**Tools and equipment**	**Source information**
• Steel toe-capped boots • Overalls • Latex gloves	• Wing covers • Steering wheel cover • Seat covers • Floor mat covers	• Vehicle hoist • Wire brush • Torque wrench • Socket set • Ball joint splitter • Spanners • Wheel alignment gauges	• Tracking information • Track rod end • Wheel nut torque setting information • Job card

1. Support the vehicle safely and remove the road wheel.

2. Clean around the track rod thread, using a wire brush.

3. Slacken the track rod lock nut half a turn.

4. Remove the nut on the ball joint.

5. Use the ball joint splitter to release the ball joint taper from the hub.

6. Unscrew the joint from the track rod.

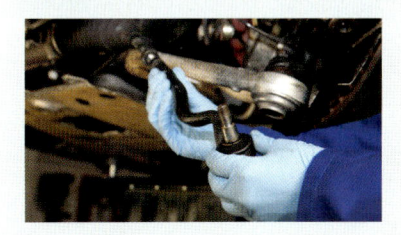

7. Screw the new track rod end on to the track rod up to the lock nut and back off half a turn.

8. Press the ball joint taper into the steering hub and fit the retaining nut, using a torque wrench.

9. Tighten the lock nut to the track rod end.

10. Replace the wheel and torque the wheel nuts.

11. Lower the vehicle and roll it backwards and forwards for the suspension to settle.

12. Check the wheel alignment (see Chapter 7, page 465).

Level 2 Light Vehicle Maintenance & Repair

> **Safe working**
> When replacing track rod ends, check that the old and new track rods are the same length before fitting. If you don't do this, the vehicle track will be severely affected.

> **Working life**
> Steve is working on a steering system and has just replaced a wheel.
> 1 What could happen if Steve does not use a torque wrench to tighten the wheel nuts?

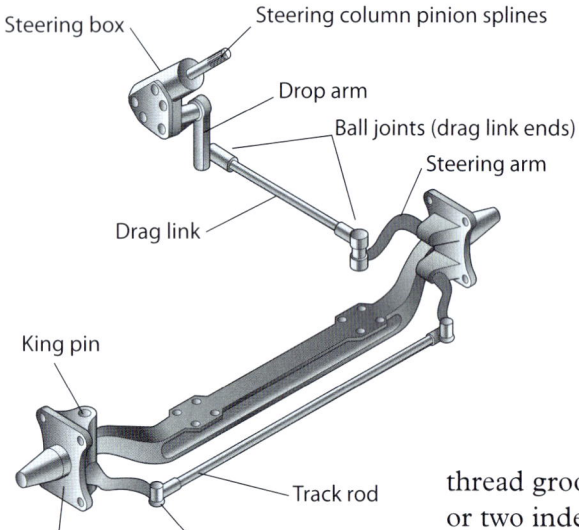

Steering boxes

A steering box (see Figure 2.20) is a means of transmitting circular motion from the steering wheel to side-to-side movement at the road wheels. It also needs to have a gear mechanism to assist the driver moving the wheels.

There are four main types of steering box: recirculating ball, worm and roller, cam and peg, and screw and nut.

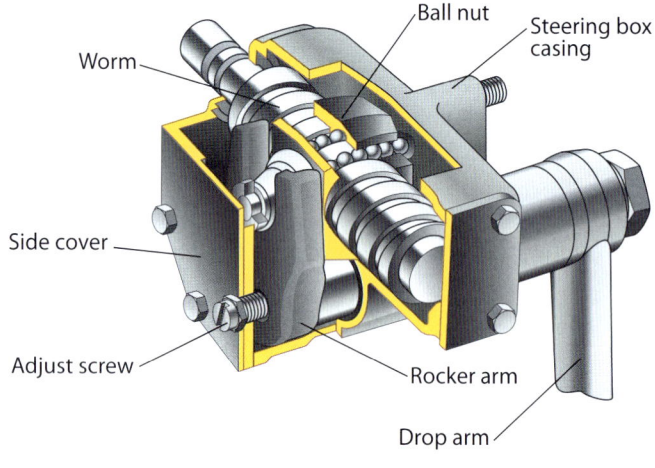

Recirculating ball steering box

The recirculating ball type of steering box, as shown in Figure 2.21, is used mainly for modern 4 × 4 vehicles, light commercial vehicles and trucks. Many older vehicles use worm and roller type arrangements. The recirculating ball system uses a chain of ball bearings installed between the thread grooves of the rotating worm and its translating nut. Either one or two independent ball bearing circuits are used (depending on how much work is required of the box).

When the steering wheel is turned, the column (worm) rotates inside the ball nut. As this happens, the ball nut moves up and down the worm and transmits movement on to the drop arm.

Figure 2.20 Steering box

Worm and roller steering box

This type of steering box is shown in Figure 2.22. It is likely to be found in older vehicles.

- The end of the shaft from the steering wheel has a worm gear attached to it.
- A roller follower is mounted on a roller bearing shaft that **meshes** with the worm gear.
- As the worm gear turns, the roller is forced to move along the roller bearing shaft, but because it is not allowed to move sideways, it twists the rocker shaft.

Figure 2.21 Recirculating ball steering box

- Typically in these designs, the worm gear is actually an hourglass shape so that it is wider at the ends. Without the hourglass shape, the roller might disengage from the worm gear at the extent of its travel.
- An adjusting screw controls the **backlash** and rocker shaft adjustment end float.

132

2 Light vehicle chassis units & components

Cam and peg steering box

The cam and peg steering box (see Figure 2.23) can be found in older vehicles. It is no longer used in the construction of modern vehicles. A tapered peg in the rocker arm engages with a special cam formed on the input shaft (cam shaft). The end float of the column is controlled by preload **shims**. An adjusting screw on the side cover controls the backlash and end float of the rocker shaft.

Figure 2.22 Worm and roller steering box

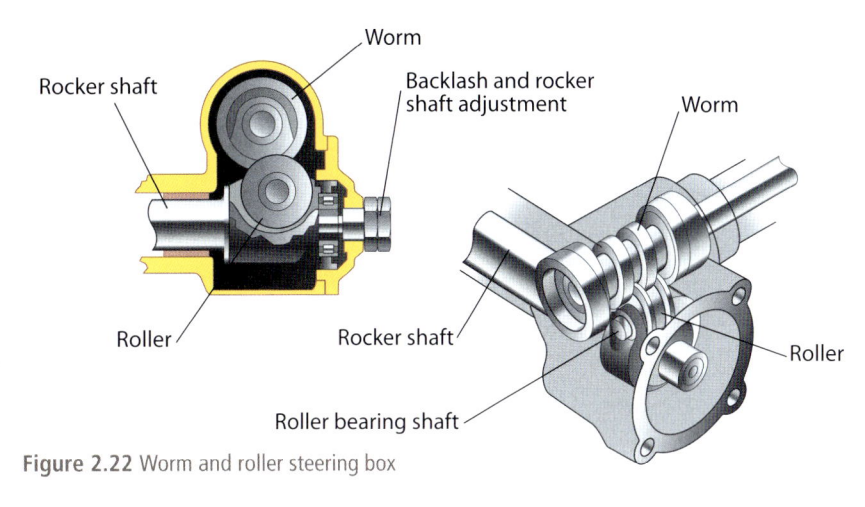

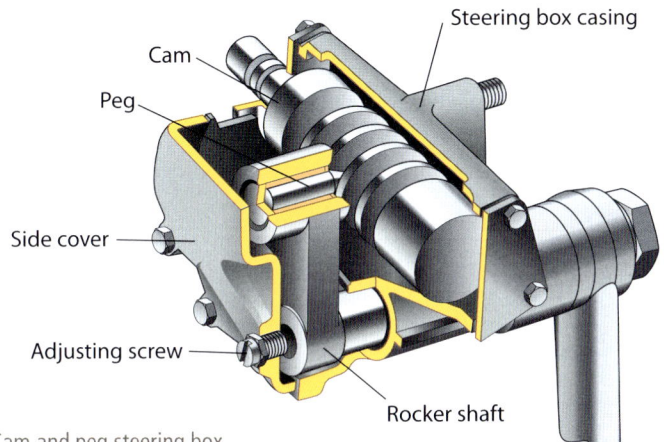

Figure 2.23 Cam and peg steering box

> **Key terms**
>
> **Mesh** – to interlink with each other (used to describe two gears interlinking).
>
> **Backlash** – a clearance between two meshing gears.
>
> **Shims** – thin metal plates that reduce any unrequired movement between two fixed components.
>
> **Splined** – grooves cut into a shaft that prevent rotational movement.

Screw and nut steering box

This type of steering gearbox can be found on vintage vehicles. It is shown in Figure 2.24.

- It consists of a phosphor-bronze or steel nut screwed on to a *screw* formed on the inner column.
- Rotation of the nut is prevented by a ball fitted on the rocker shaft.
- When the steering wheel is turned, the **splined** end rotates.
- This causes the threaded part to rotate, but the ball prevents the nut from turning.
- This results in the nut moving up and down the thread, transmitting the movement via the rocker shaft to the steering linkages.

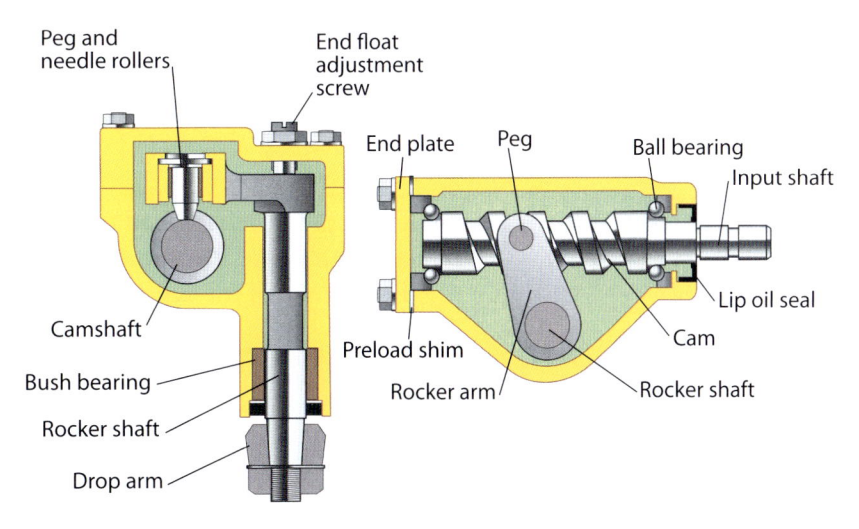

Figure 2.24 Screw and nut steering box components

133

> **Key term**
>
> **Parallelogram** – four-sided shape with opposite sides parallel.

Steering linkage

The steering linkage is a series of rods, arms and ball joints. Their job is to connect the steering mechanism to the stub axle and absorb most of the shock going to the steering wheel. This type of linkage is known as **parallelogram** steering linkage.

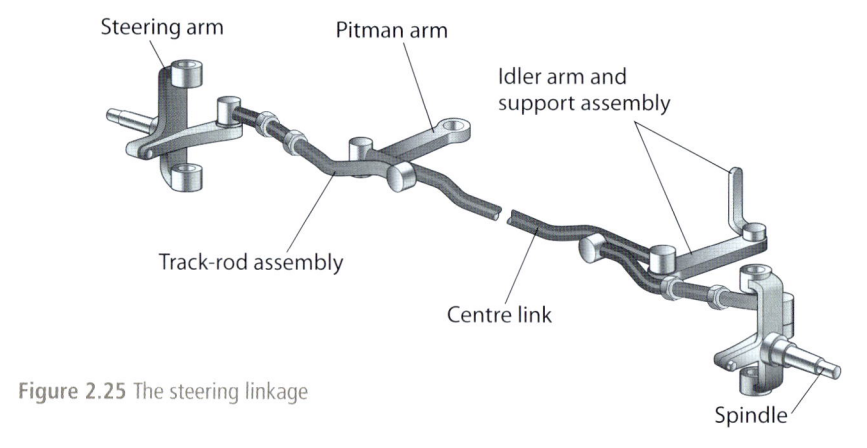

Figure 2.25 The steering linkage

The steering linkage used with most manual and power steering mechanisms typically includes the following:

- pitman arm
- centre link
- idler arm
- track rod assemblies.

Pitman arm

The turning action of the steering mechanism is transmitted to the centre link through the pitman arm. The pitman arm is splined to the steering mechanism's output shaft (pitman arm shaft). A nut and lock washer secure the pitman arm to the output shaft. The outer end of the pitman arm uses a ball-and-socket joint to connect to the centre link.

Centre link

A centre link is also known as an intermediate rod, track rod or relay rod. It is basically a steel bar that connects the steering arms (pitman arm, track rod ends and idler arm) together.

Idler arm

The centre arm is bolted to the chassis or subframe and is hinged on the opposite end of the pitman arm by the idler arm. The idler arm supports the free end of the centre link and allows it to move left and right with ease.

Track rod assemblies

The track rod assemblies link the steering arms to the centre link. They also provide a means of adjusting the vehicle tracking. (See the section on *Steering geometry* on pages 137–143.).

2 Light vehicle chassis units & components

Power-assisted steering

Most vehicles today have power-assisted steering (PAS). This is a steering system that provides mechanical steering assistance to the driver. A vehicle without PAS requires more **torque** when turning the wheel of a vehicle driven at low speeds.

The advantages of power-assisted steering (PAS) include:

✓ It lightens the steering and reduces driver effort.
✓ It reduces the required number of turns lock to lock (a higher gear ratio can be used with PAS).
✓ It counteracts road shock to reduce **kick back** at the steering wheel.
✓ It resists any sudden swerving of the vehicle in the event of a puncture.
✓ It allows the vehicle to be designed so that extra weight can be distributed to the steered wheels to give more passenger and cargo space.

Hydraulic PAS operation

A hydraulic **servo** mechanism is incorporated in most PAS systems. This operates whenever the resistance to turning the steering wheel goes above a set amount. The result is that the driver needs to apply less effort to steer, especially when steering the vehicle at low speeds.

An engine-driven pump circulates fluid around a closed circuit. This fluid pressure is limited by a pressure relief and flow-limiting valve to prevent excessive pressure around the system. As the driver turns the steering wheel, a hydraulic control valve opens and allows the hydraulic fluid under pressure into a power cylinder. The fluid then creates force that acts on one side or the other of a servo piston. The side that the force acts on depends on which way the steering wheel is turned.

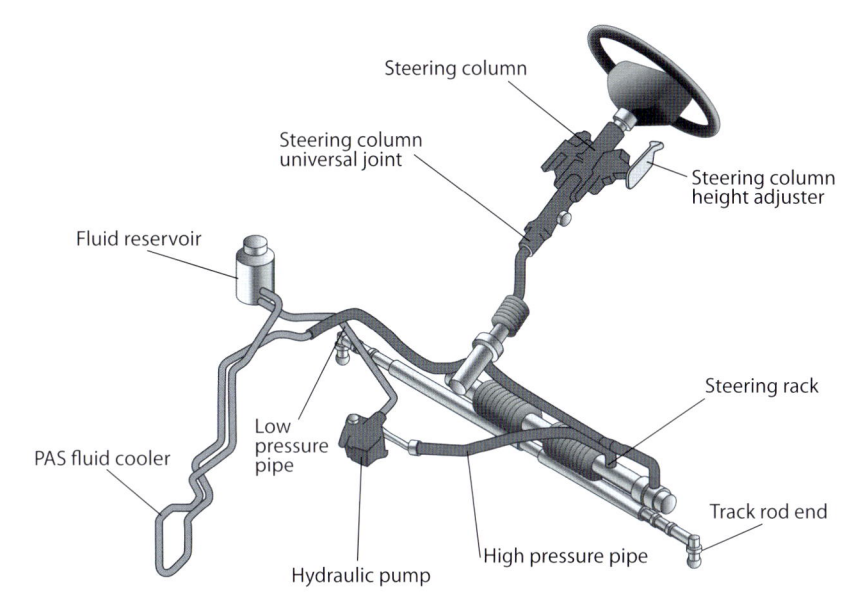

Figure 2.26 Power-assisted steering column

Key terms

Torque – turning effort. In this case, the turning effort made by the driver to turn the steering wheel.

Kick back – if a vehicle wheel hits a bump this force can be transmitted to the steering wheel if the linkages and steering gear do not allow a small amount of flexibility.

Servo – assistance given to a component by fluid or gas pressure.

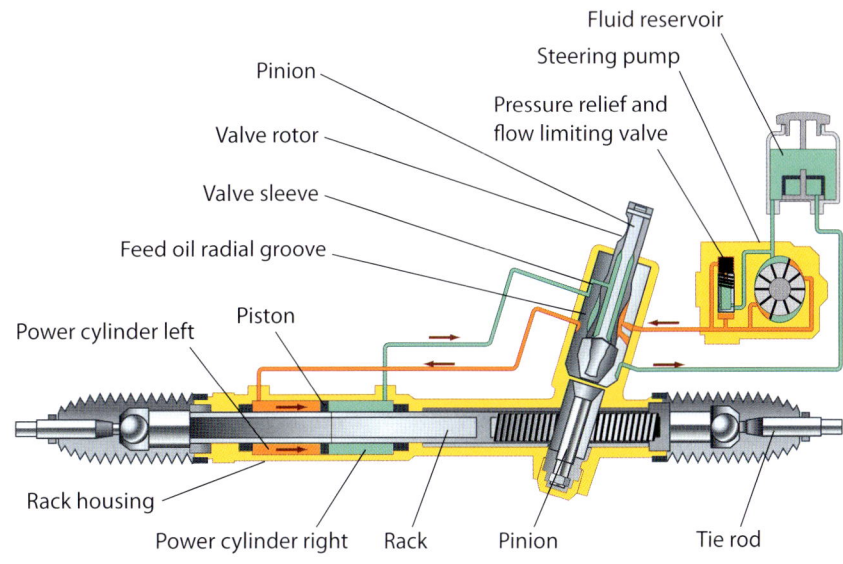

Figure 2.27 Hydraulic PAS operation

135

Level 2 Light Vehicle Maintenance & Repair

> **Key term**
>
> **Combined unit** – single unit which consists of a number of components put together.

> **Emergency**
>
> **Fail-safe** means that in the unlikely event of a component failing, the driver is still able to control the vehicle system safely through a series of mechanical linkages.

Other PAS systems

An electro-hydraulic system is now used on some modern vehicles. In this system, the power steering pump is driven by an electric motor in a **combined unit** and is used only when the driver requires assistance.

Some systems incorporate a switch so that the power steering assistance can be increased when parking or manoeuvring slowly. In fully electronic PAS systems an electric motor acts directly on the steering column to provide the power assistance.

The advantage of such systems is that the power steering force can be varied with road speed. Some cars also have a switch to give extremely light steering for parking. This means that maximum assistance is given at parking speeds and little or no assistance at high speeds.

Desirable features of PAS operation

✓ All PAS systems must be '**fail-safe**'.
✓ The steering system should give the driver the 'feel' of the road.
✓ In the event of tyre slippage, the driver must be able to feel the road and keep the vehicle under control.

Servicing

Servicing of the PAS system is minimal.

- Inspections of the fluid circuit must be done every service to check for any leaks.
- Fluid levels need to be checked and topped up and drive belts checked for wear and tension.
- Over-tightening of the drive belt must be avoided to prevent bearing damage.
- If a large amount of new oil is added to the PAS reservoir, the system may have to be bled. This is done by turning the steering from lock to lock when the engine is running.
- Regular topping up of the PAS reservoir is required to prevent more air entering the system.

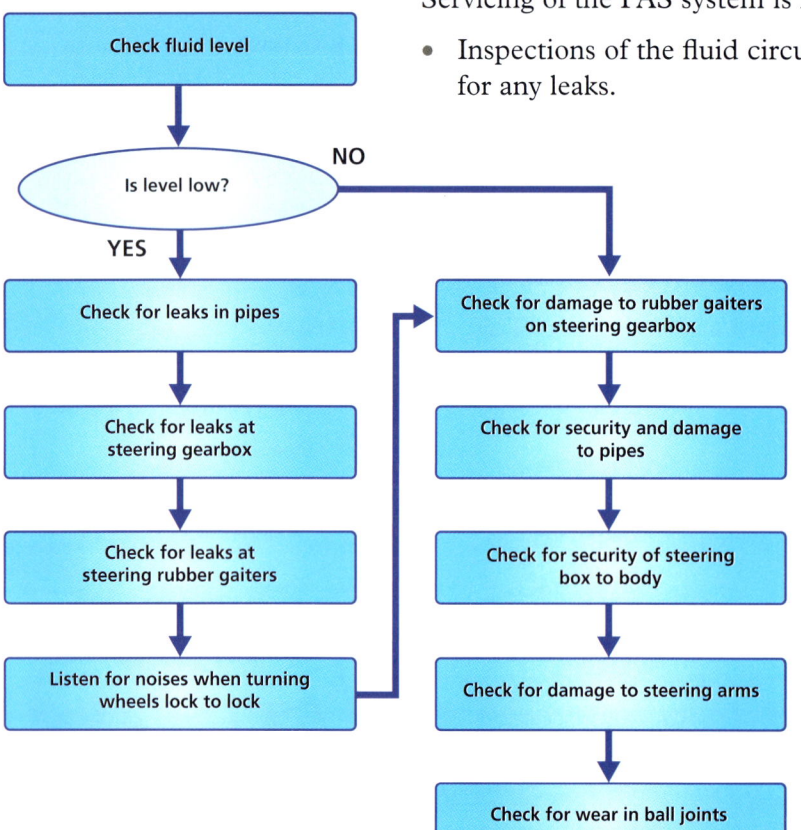

The PAS system service process is shown in Figure 2.28.

PAS system service process flowchart:

Check fluid level → Is level low?
- YES → Check for leaks in pipes → Check for leaks at steering gearbox → Check for leaks at steering rubber gaiters → Listen for noises when turning wheels lock to lock
- NO → Check for damage to rubber gaiters on steering gearbox → Check for security and damage to pipes → Check for security of steering box to body → Check for damage to steering arms → Check for wear in ball joints → Report faults and inform customer

Figure 2.28 PAS system service process

> **Working life**
>
> Ben has just topped up the PAS reservoir during a service. He found that it has taken quite a lot of fluid.
>
> 1. What may be the cause of this problem?

136

2 Light vehicle chassis units & components

Steering geometry

Suspension systems contribute to the vehicle's road handling and braking. They also help to keep passengers comfortable and reasonably well protected from road noise, bumps and vibrations.

It is important for the suspension to keep the road wheel in contact with the road surface as much as possible. This is because all the forces acting on the vehicle do so through a small surface area of the tyre where it meets the road. The suspension also protects the vehicle itself and any cargo or luggage from damage and wear. The design of the front and rear suspension of a car may be different.

It is common practice to associate the geometry of a vehicle only with the steering system. But in fact, it is the **set up** of both the suspension components and the steering components combined that creates the **steering geometry**.

To get the required features from a vehicle steering system, the components from both the steering and suspension systems must be set up in a way that suits the purpose and layout of the vehicle. For example, a commercial vehicle will have different requirements from a sports car. This is because the commercial vehicle will need to carry much heavier loads than the sports car, whereas the sports car will have a greater need to corner at speed.

The steering geometry angles found in light vehicle maintenance include: camber angle, king pin inclination and castor angle.

Camber angle

The camber angle is the angle that the wheel tilts from top to bottom when the wheel is viewed from the front of the vehicle. There are two main settings for camber, as shown in Figure 2.29:

- **Positive camber** – This is where the top of the wheel tilts or leans outwards in relation to the bottom of the wheel.
- **Negative camber** – Here, the top of the wheel tilts or leans inwards in relation to the bottom of the wheel.

Positive camber

Positive camber reduces the stress placed on the steering and suspension components and tends to give lighter steering. Tyre wear and vehicle handling are also reduced because of the additional positive camber created during cornering.

Positive camber angles are best suited to heavy goods vehicles and agricultural vehicles.

Negative camber

Negative camber results in the vehicle sitting lower and wider on the road. This is because of the **splayed** effect at the bottom of the wheels, which makes for better

> **Key terms**
>
> **Set up** – refers to the positions and angles of the steering and suspension components where they are fitted to the vehicle.
>
> **Steering geometry** – a combination of various angles of suspension components (castor, camber, toe-in). These are set to enable the vehicle wheels to point in the direction the driver intends them to go, whatever the road conditions.
>
> **Splayed** – spread out.

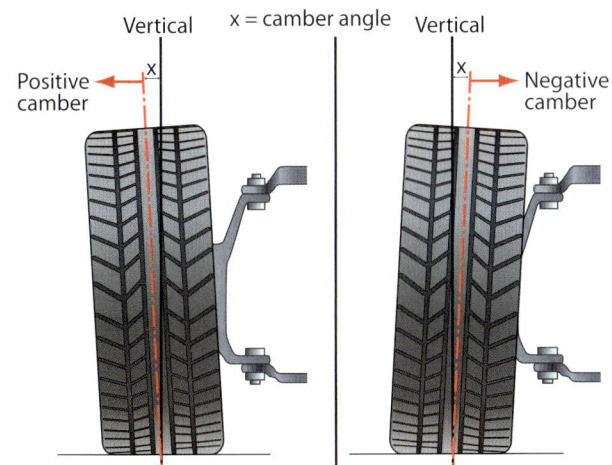

Figure 2.29 Positive and negative camber

137

Level 2 Light Vehicle Maintenance & Repair

Did you know?

- The camber angle is rarely more than two degrees, as the splaying-out effect can cause excessive tyre wear.
- Extreme tyre pressures can affect these angles, so it is really important that the tyre pressures are set correctly before continuing to check the angles.

Key terms

True vertical – an imaginary line that is at a right angle to a flat road surface.

Self-centering action – where the front wheels return to the straight ahead position after cornering.

Hub – the assembly that the wheel is attached to.

handling and cornering at speeds. However, this means that more stress is placed on system components, resulting in heavier steering.

Negative camber angles are best suited to light and performance vehicles.

King pin inclination (KPI)/Swivel axis inclination (SAI)

King pin inclination (KPI) is also known as swivel axis inclination (SAI). It is the tilt of the suspension leg (called the king pin or swivel axis) from the **true vertical** when viewed from the front or rear of the vehicle, as shown in Figure 2.30. It is used together with the camber angle to give some of the same effects.

KPI also provides a **self-centering action** of the wheels. This has an important effect on the steering, as it assists the steering in returning to the straight ahead or centre position after cornering. The KPI may be between 5 and 10 degrees, depending on the manufacturer's **hub** design.

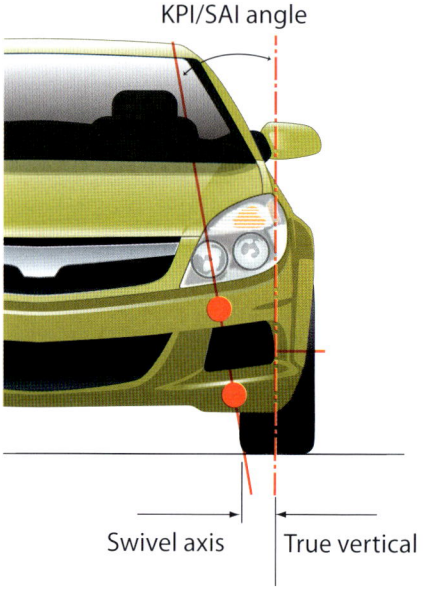

Figure 2.30 King pin inclination

Included angle

The included angle is the total of the KPI angle and the camber angle added together. This is useful when manufacturing the stub axle.

Caster angle

The caster angle is the forward or backward tilt of the king pin (or swivel axis) from the true vertical, when viewed from the side of the vehicle (see Figure 2.31). The caster angle also gives a self-centering action to the steering, so that the wheels point straight ahead during forward motion.

Too little caster angle creates wander and too much gives hard steering.

- **Negative caster** is where the swivel axis centre line strikes the road behind the wheels' centre line. It is generally used for a front-wheel drive vehicle.
- **Positive caster** is where the swivel axis centre line strikes the road in front of the wheels' centre line. This set up is common for a rear-wheel drive vehicle.

The choice of the caster set up used is determined by the forces acting on the front wheels from the forward motion of the vehicle.

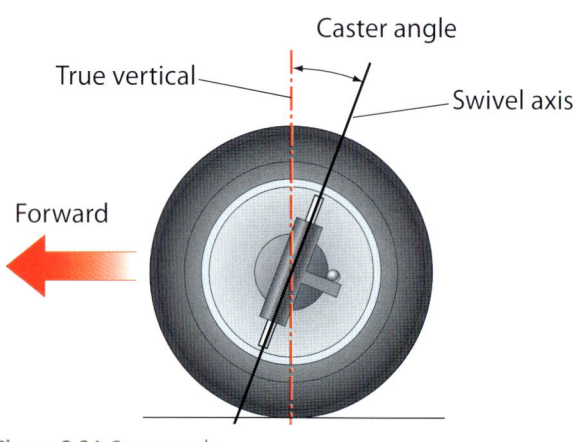

Figure 2.31 Caster angle

Positive offset (Positive scrub radius)

Where the centre line of the wheel meets the swivel axis at the point just below the road surface, this is called positive offset. It is measured at the road surface between the two centre lines (see Figure 2.32).

During normal driving conditions, the offset distance must be equal on both sides. If the vehicle you are driving has a puncture to one of the front tyres, the positive offset will increase. This causes the vehicle to pull violently to that side and makes the vehicle difficult to control.

Negative offset (Negative scrub radius)

Where the centre line of the wheel meets the swivel axis at the point just above the road surface this is called negative offset. It is measured at the road surface between the two centre lines (see Figure 2.33).

> **Emergency**
>
> If a puncture occurs, the point at which the wheel centre line and the swivel axis meet will be at the road surface. This ensures that the steering will remain controlled. If a front brake imbalance exists, negative offset also helps to give straight-line stability.

Figure 2.32 Positive offset (positive scrub radius)

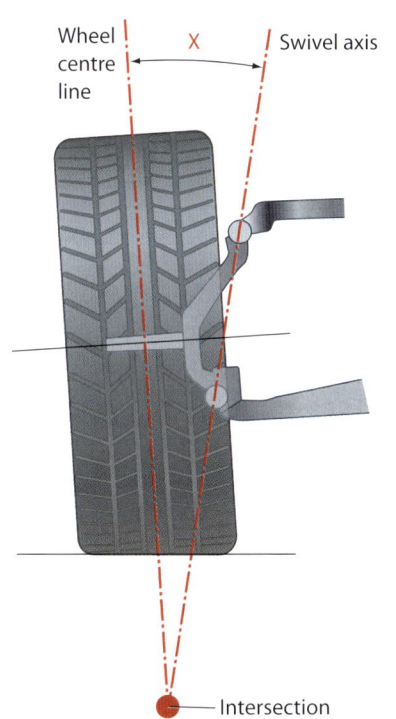

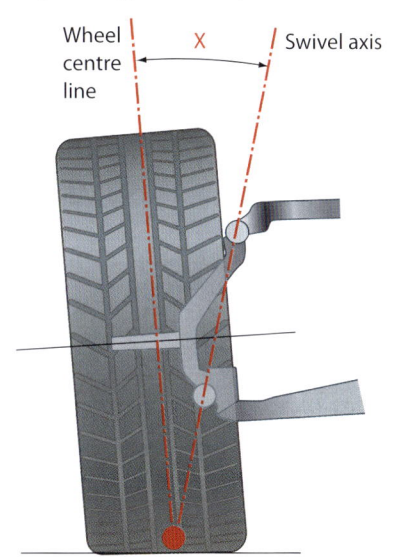

Figure 2.33 Negative offset (negative scrub radius)

> **Find out**
>
> Look on TV or online for clips of athletics events involving athletes running round a track. For example, in the 200m, the runners have to be staggered so each runner covers the same distance. If they all started along the same line, the runner on the outside would have to run further and would get worn out faster, just like tyres would get worn on cornering.

Scrub radius

Scrub radius is the distance between the points where the centre line of the wheel and tyre assembly and the centre line of the king pin or swivel axis intersect with the ground.

True rolling

To prevent tyre wear when cornering, all wheels must roll from the same radius point. This is called true rolling and is the basis of the Ackerman system.

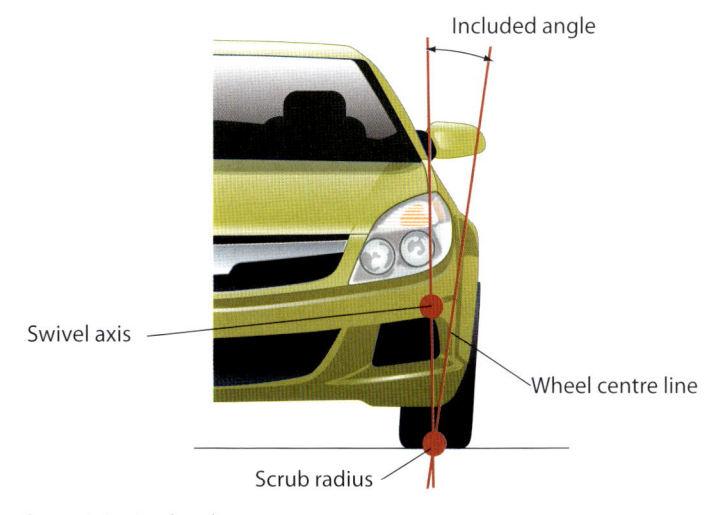

Figure 2.34 Scrub radius

Ackerman system

The Ackerman system (see Figure 2.35) is determined by positioning the track rod joint and the swivel axis on an imaginary line. This imaginary line crosses the centre line of the vehicle, usually just in front of the rear axle centre line. The angle between these two lines is called the Ackerman angle.

- If the steering rack is fitted to the front of the swivel axis, then the track rod is longer.
- If the steering rack is fitted to the rear of the swivel axis, then the track rod is shorter.

This will make sure the wheels move outwards at the front during cornering to give true rolling and prevent the tyres from premature wear.

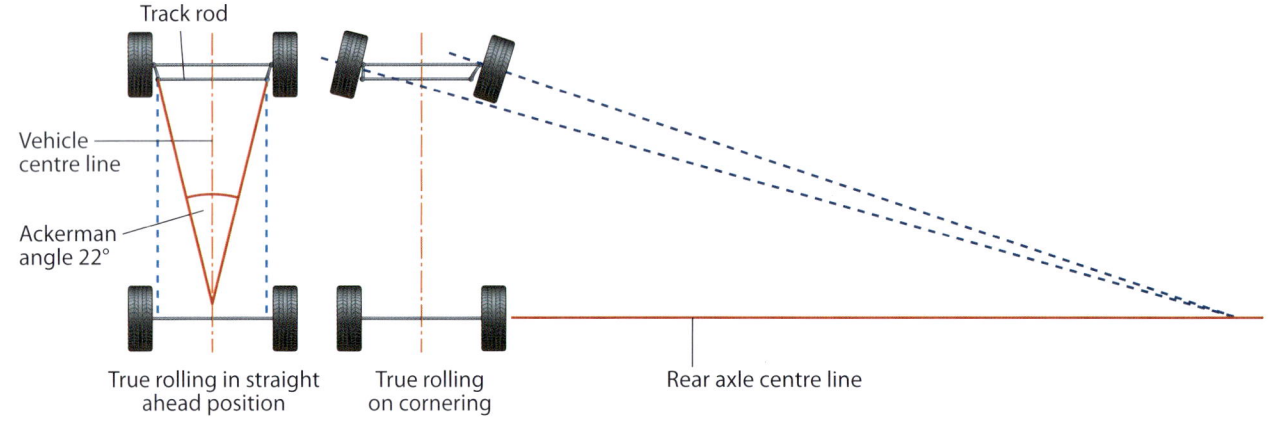

Figure 2.35 Ackerman system

Toe-in and toe-out

When the vehicle is driven forward in a straight line, both wheels must be parallel. Slight **free play** in the steering and suspension systems prevents this from happening. If the wheels were set parallel when the vehicle is at rest, the forces generated by forward motion of the vehicle in the direction of the **thrust line** may cause the wheels to splay in or out. To counter this tendency, the front wheel alignment is given a degree of toe. Toe is a difference in width between the extreme front and rear of the front wheels when measured at axle height.

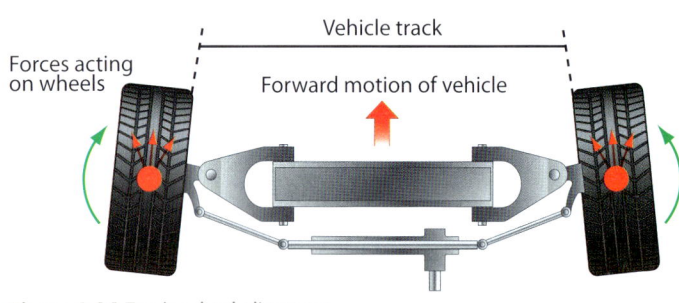

Figure 2.36 Toe-in wheel alignment

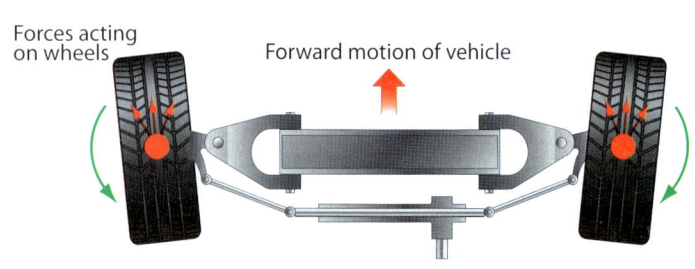

Figure 2.37 Toe-out wheel alignment

- **Toe-in** is when the width of the front of the wheel is less than the width of the rear of the wheel, as shown in Figure 2.36. This is usually used on RWD vehicles.
- **Toe-out** is when the width of the front of the wheel is greater than the width of the rear of the wheel, as shown in Figure 2.37. This is usually used on FWD vehicles.

2 Light vehicle chassis units & components

Toe-out on turns

In the Ackermann system and to ensure true rolling, the wheels have to rotate around a single radius point during cornering to prevent tyre wear. This means that the inner wheel will have to turn at a greater angle than the outer one just like the runners on a race track. This is known as toe-out on turns (TOOT). The amount of toe-out on turns is set by the track rod arms and will depend on the track length, its angular set and the track width.

You can check the angular set and track width as follows (see Figure 2.38):

1. Put the vehicle on graduated turntables.
2. Turn the steering to 20 degrees on the outer wheel.
3. Check the inner wheel reading. This should be a slightly larger reading, e.g. 22°.

The reading should be the same as the reading given when the steering is turned in the opposite lock. To ensure this happens, when you are setting the vehicle tracking, both track arms need to be moved an even amount.

Slip angle

When there is a side force acting on a wheel and tyre, the tyre will change shape and take a different path from the one the wheel is steered in.

The angle between the actual path of the tyre along the ground and the centre line on the planned path of the wheels is called the **slip angle** (see Figure 2.39). ('Creep angle' might be a more accurate term to use, as the wheel does not actually slip against the road surface.)

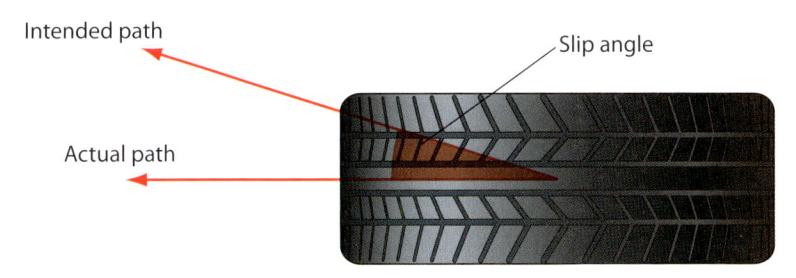

Figure 2.39 Slip angle

Oversteer

When the slip angle for the rear tyres is greater than for the front tyres, oversteer occurs (see Figure 2.40).

If this happens, the vehicle will turn more sharply into the curve than expected. Side force acting on the tyres will increase and this will increase the slip angle, creating even more oversteer. To correct oversteer, the steering wheel has to be turned through the opposite lock.

> **Did you know?**
>
> RWD vehicles tend to want to toe-out. When stationary, the wheels are set to toe-in. This enables the wheels to run parallel when the vehicle is moving. The opposite takes place on FWD drive vehicles.

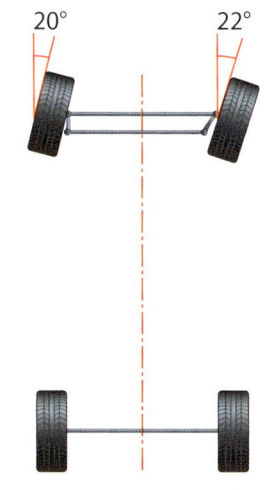

Figure 2.38 Toe-out on turns (TOOT)

> **Key terms**
>
> **Free play** – a small amount of movement in order for components to move slightly.
>
> **Thrust line** – an imaginary line at right angles to the rear axle centre line that does not point straight ahead due to movement in the rear wheel suspension joints during forward motion of the vehicle.
>
> **Slip angle** – the angle between a rolling wheel's actual direction of travel and the direction towards which it is pointing.

141

Level 2 Light Vehicle Maintenance & Repair

Figure 2.40 Overseer   Figure 2.41 Understeer

Understeer

When the slip angle for the front tyres is greater than for the rear tyres understeer occurs (Figure 2.41). This is when the steering wheel has been turned insufficiently for the curvature of the corner or if there is a lack of grip in the front tyres. This can be corrected by reducing speed and applying more rotation to the steering wheel.

> **Working life**
>
> A customer's vehicle has left the garage without having the rear tyres blown up correctly.
>
> 1 What steering problem could occur?

Self-aligning torque

Self-aligning torque occurs when a side force acts on a tyre wall and deflects it to give a slip angle. At the same time, an equal and opposite force is generated at ground level, at a point where the tyre leaves the road. This could happen during hard cornering where the tyre eventually slides due to excessive side force.

Self-aligning torque helps tyres to return to their original position after turning in the road. This assists the action of KPI and caster to return the steered wheels to the straight ahead position.

Adjusting and checking the front end geometry

Forms of adjustments used to set the camber and caster angles on independent front suspension systems are:

- slotted shims
- eccentric bushes
- cam-headed pivot bolts
- clearance holes
- screws or adjustable-length diagonal links.

Adjustment of the camber and caster angles is rare unless the vehicle has been involved in an accident or suspension components have been changed. For this reason, it is important that you mark all suspension bolts before removing them, so they can be returned to the same position.

> **Did you know?**
>
> Neutral steer is where the vehicle takes the intended path the driver wishes to go.

> **Did you know?**
>
> Wheel alignment is often confused with wheel balancing. When a wheel is out of balance, it will cause a vibration at high speed that can be felt in the steering wheel and/or the seat. If the alignment is out, it can cause excessive tyre wear and steering or tracking problems.

> **Working life**
>
> Over the last few months, Terry has noticed that his tyres are wearing on one edge.
>
> 1 What could be a cause of his problem?

2 Light vehicle chassis units & components

Steering system faults

Table 2.1 lists some common steering system faults and how to correct them.

> **Key term**
>
> **Feathering** – where a tyre is worn gradually across the width of the tread so that it feels rough to the touch in one direction and smooth in the other.

Table 2.1 Steering system faults and how to correct them

Symptom	Fault	Remedy
Steering vibrations	Wheels not balanced	Balance wheels
	Wear in linkage	Replace linkage bushes and joints
	Wear in joints	Check all tracking and geometry
Steering wheel not returning to centre Vehicle wanders	Damaged linkage	Replace linkage and check all tracking and geometry
Excessive steering wheel free play before road wheels turn	Worn linkages Worn joints	Replace linkage bushes and joints
	Worn column bearings	Replace bearings
	Worn steering gear	Adjust or replace steering gearbox
Feathering on one edge of the tyre Steering wheel not returning to centre Vehicle wanders	Incorrect wheel alignment	Set wheel alignment
Tyre wear on one edge Steering wheel not returning to centre Vehicle wanders	Incorrect steering geometry	Adjust steering geometry

Wheel bearings

Bearings reduce rolling friction. They consist of smooth metal balls or rollers, and a smooth outer and inner metal surface for the balls to roll against, as shown in Figure 2.42. The balls or rollers bear the load, and this ensures that the assembly spins smoothly.

Table 2.2 shows the two types of load acting on the bearings and the applications in which they would appear.

Table 2.2 Loads acting on the wheel bearings

Type of load	Application
Radial load – any load that is in the same direction as the axis of rotation	Load on the bearing during forward movement
Axial load – any side-to-side load on a bearing	Load on the bearing during cornering

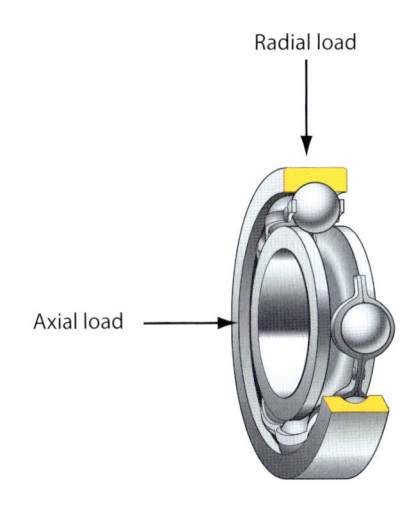

Figure 2.42 Wheel bearings

Level 2 Light Vehicle Maintenance & Repair

Key terms

Bearing race – the ring of metal that the balls or rollers are in contact with.

Sphere – round object in the shape of a ball.

Roller bearing

Roller bearings use solid metal cylinders with a length that is slightly longer than the diameter. The metal cylinders revolve around an inner and outer **bearing race** in a bearing cage. Unlike a plain roller, the roller bearing spreads the load-carrying capacity over a larger area. Spreading the load is necessary during cornering to counteract sideways movement.

Tapered roller bearing

Tapered roller bearings are usually mounted in pairs facing opposite directions. This ensures that they can handle axial loads in both directions. Large radial loads are also absorbed by the construction of this type of bearing, making them the most popular bearing for hub assemblies.

Needle roller bearing

Similar to a roller bearing, the needle roller bearing uses long, thin cylindrical needles to revolve around an inner and outer race. They are much more compact than the tapered roller bearing and can be used where there is limited space between an outer and inner revolving assembly. The disadvantage is that because of their size, only small loads can be imposed on them.

Ball and plain bearings

Ball bearings, sometimes called plain bearings, can support both radial and axial loads. Instead of cylinders, they use **spheres** that rotate around an outer and inner race. Contact of each sphere with the inner and outer race is at a very small point, which helps it spin very smoothly. However, the small contact area between the spheres and the race makes it prone to damage if high loads are placed on it.

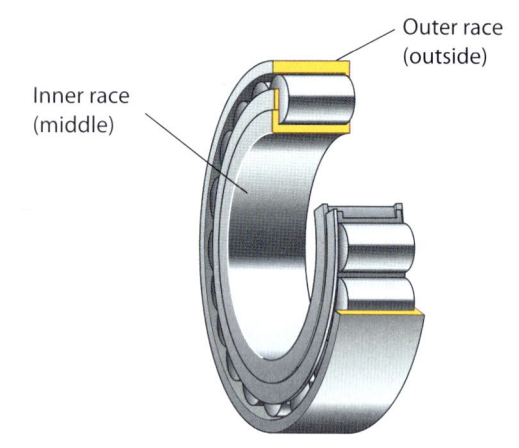

Figure 2.43 Roller bearing

Figure 2.44 Tapered roller bearing

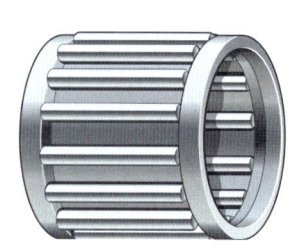

Figure 2.45 Needle roller bearing

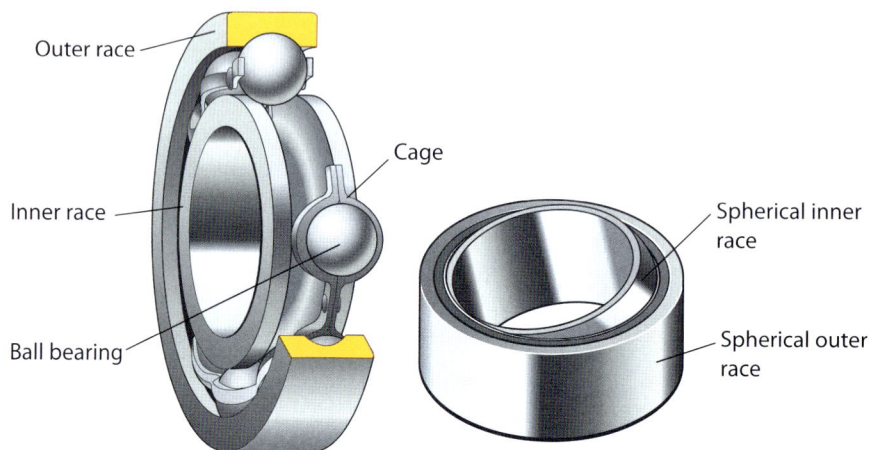

Figure 2.46 Ball bearing (left) and plain bearing

2 Light vehicle chassis units & components

Wheel bearing arrangements

Front hub

The construction of the hub and types of bearing used depend on whether the wheels are driving or non-driving. A driving front hub is used for a FWD vehicle or 4WD vehicle where it has to incorporate a **drive shaft**.

Non-driving front hub

The construction of a non-driving front hub is simpler as it does not need to accommodate the drive shaft. It is constructed of two adjustable tapered roller bearings fitted between the stub axle and the malleable iron or steel cast hub. The hub is lubricated by packing each bearing, and half the cavity between the bearings, with grease. A lip-type synthetic rubber seal prevents the grease escaping on to the brake linings.

The hub bearing in most cases requires a small clearance. This allows for heat expansion as a result of friction and heat from the brakes. The bearing is adjusted by tightening the hub nut until all clearance is eliminated and then slackening the nut so that the hub turns without any bearing resistance.

When adjusting any bearing assembly, you must use the manufacturer's specifications to make sure that you set the end float correctly.

The assembly shown in Figure 2.47 uses a lock nut and split pin as a locking device on to the stub axle. Where the assembly uses a **nyloc nut** to secure the hub to the stub axle, this must be renewed every time the hub is removed.

An alternative arrangement is to use two angular contact-type ball bearings held apart by a rigid spacer. In this type of assembly, the hub nut must be tightened fully to the correct torque loading.

Driving hub

The driving hub (Figure 2.48) consists of a stub axle housing containing two bearings. These support both the wheel hub and the drive shaft. The load carried by the road wheel determines which type of bearing is used. This could be a pair of ball-race bearings or a pair of tapered roller bearings.

> **Key terms**
>
> **Drive shaft** – round bar that transmits the force from the final drive to the road wheel.
>
> **Nyloc nut** – nut with a nylon insert that acts as a frictional locking device on the threads it is screwed to.

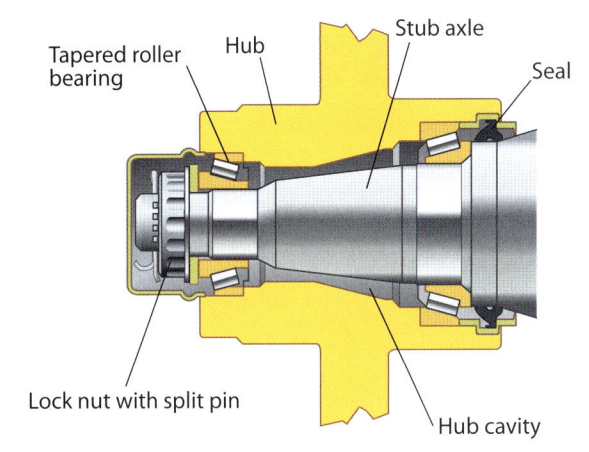

Figure 2.47 Non-driving hub

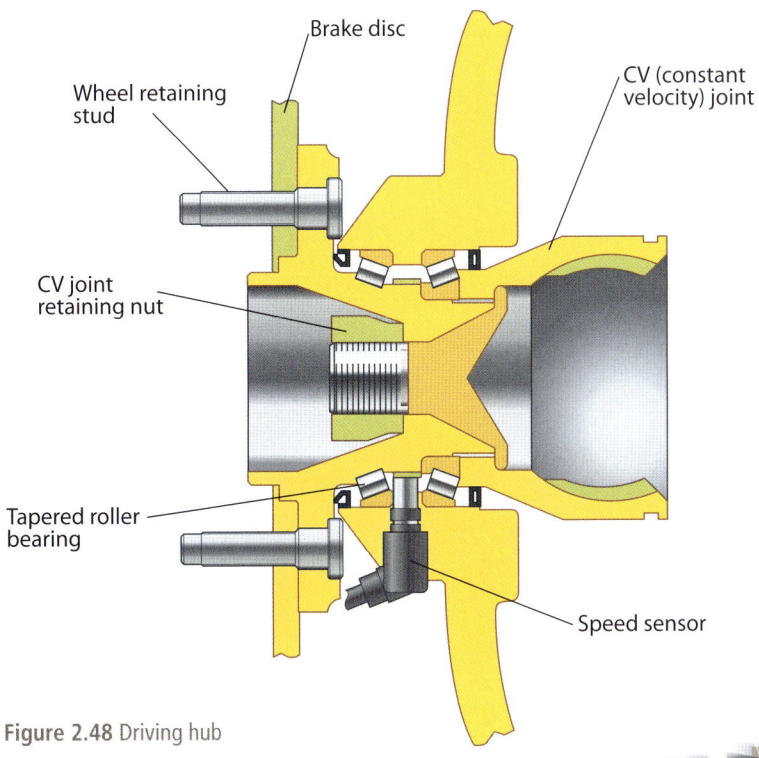

Figure 2.48 Driving hub

145

Level 2 Light Vehicle Maintenance & Repair

Checking wheel bearings

Checklist			
PPE	**VPE**	**Tools and equipment**	**Source information**
• Steel toe-capped boots • Overalls • Latex gloves	• Wing covers • Steering wheel cover • Seat covers • Floor mat covers	• Jack • Chocks • Axle stand	• Jacking points from handbook • Free play allowed in bearing from workshop manual

1. Chock the vehicle wheels.

2. Jack the vehicle wheels off the ground safely.

3. Use an axle stand to ensure the wheel stays up in the air if the jack fails.

4. Release the handbrake.

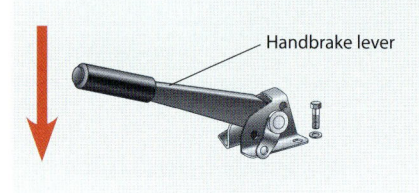

   Handbrake lever

5. Hold the top and bottom part of the wheel.

6. Move the wheel top forwards and bottom backwards.

7. Move the wheel top backwards and bottom forwards.

8. Any movement felt through the wheel can indicate bearing wear.

9. Roll the wheel to listen for scraping or whirling noises that indicate bearing damage.

10. Apply the handbrake.

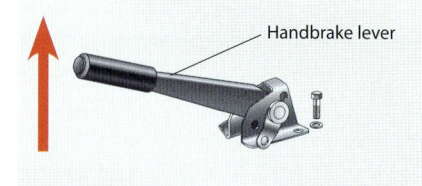

    Handbrake lever

11. Lower the vehicle with the jack.

12. Remove the chocks.

Axle shaft arrangements

The axle shaft transmits the drive from the final drive to the rear hub. It is exposed to various stresses, including:

- torsional (twisting) stress due to driving and braking torque
- shear stress or tearing due to the weight of the vehicle
- bending stress due to vehicle loading
- compressive and tensile stress due to cornering.

You can find out more about the stresses in Chapter 1 on pages 115–116.

The following three arrangements protect the vehicle from the various loads and stresses placed on the axle shaft. The arrangement used depends on the forces which are generated by that particular vehicle.

Semi-floating axle shaft

The semi-floating axle shaft is used for light vehicles. The axle rolls on a straight roller bearing against the outer part of the axle shaft casing. See illustration Figure 5.45 on page 350.

Three-quarter floating axle shaft

An alternative to the arrangement above is the three-quarter floating axle shaft. Shear stress on the shaft is eliminated, but the shaft still has to resist all of the other stresses. This type of axle shaft is used for light commercial vehicles and uses a single or double row ball bearing. See illustration Figure 5.46 on page 351.

Fully floating axle shaft

The fully floating axle shaft is more suitable for commercial vehicles, as most of the vehicle load is taken up by the axle housing. Two opposed tapered roller bearings are used with this type of arrangement. See illustration Figure 5.47 on page 351.

CHECK YOUR PROGRESS

1. State two purposes of the steering system.
2. Name the most popular light vehicle steering box.
3. What is meant by 'fail-safe'?
4. 'Toe-out' wheel alignment is set on which type of vehicle layout?
5. Why do manufacturers set wheel alignment to TOOT?
6. What is the difference between caster and camber?
7. What is a non-driving hub?
8. What is the difference between radial and axial load on a bearing?
9. State three safety precautions when checking wheel bearings.

Light vehicle suspension systems

The suspension system works closely with the steering system to ensure that the vehicle handles well under all conditions. If you have ever experienced a bumpy ride in a car, it was most likely down to a fault with the suspension. Figure 2.49 shows the purposes of the suspension system.

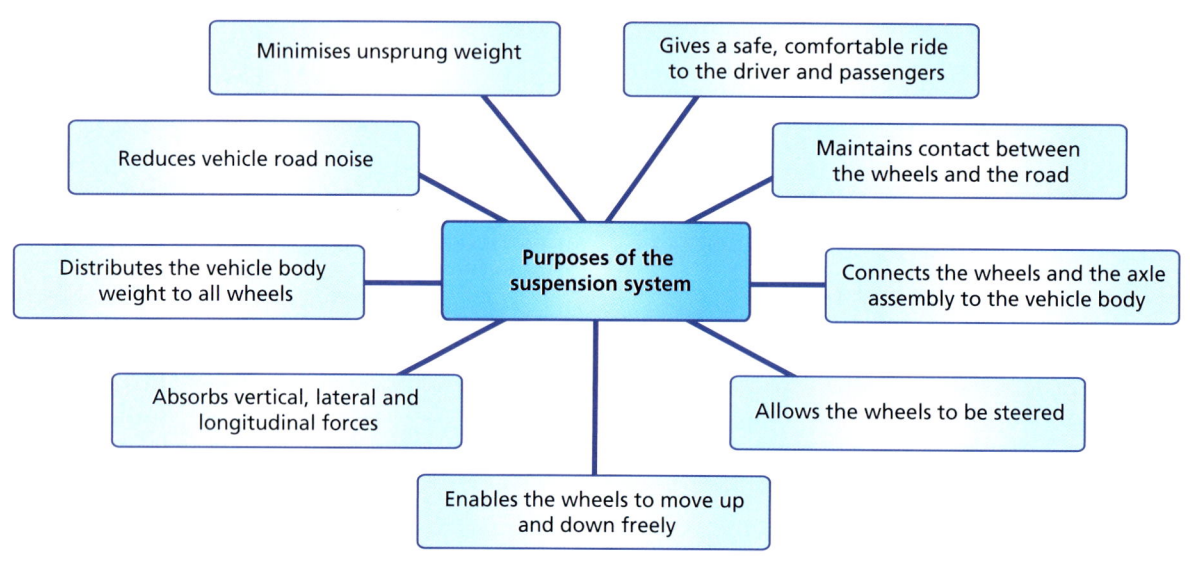

Purposes of the suspension system:
- Minimises unsprung weight
- Gives a safe, comfortable ride to the driver and passengers
- Reduces vehicle road noise
- Maintains contact between the wheels and the road
- Distributes the vehicle body weight to all wheels
- Connects the wheels and the axle assembly to the vehicle body
- Absorbs vertical, lateral and longitudinal forces
- Allows the wheels to be steered
- Enables the wheels to move up and down freely

Figure 2.49 The purposes of the suspension system

Unsprung weight (mass)

An **unsprung weight** (mass) on a vehicle will have a negative effect on the vehicle's handling and behaviour. Therefore, when a vehicle is designed, components that make up the unsprung weight need to be kept to a minimum. (All weight which is held on top of the springs, such as the vehicle body, is called the sprung weight.)

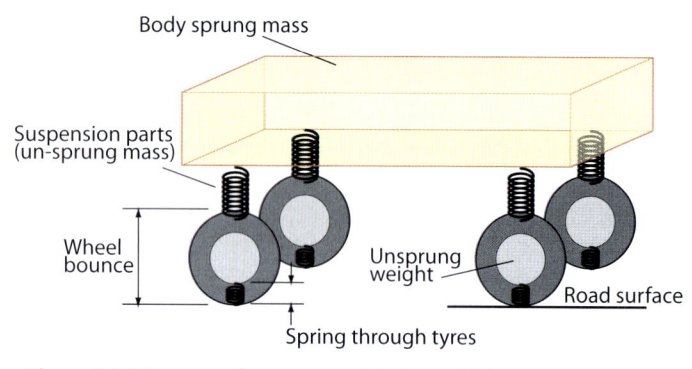

Figure 2.50 Sprung and unsprung weight in a vehicle

Maximum comfort to the driver and passengers is given by producing a suspension system that responds sensitively to road irregularities. This is achieved by reducing the number and the weight of unsprung components (components that are not supported by the suspension springs). Table 2.3 gives some examples of sprung and unsprung components.

> **Key term**
>
> **Unsprung weight** – the weight of the suspension, wheels and other components directly connected to them, that are not supported by the suspension.

Table 2.3 Sprung and unsprung weights in a vehicle

Sprung components	Unsprung components
• Body • Steering column • Interior trim • Steering rack • Engine • Gearbox	• Suspension parts • Wheels and tyres • Hubs and wheel bearings

Suspension terms

Table 2.4 shows some different suspension conditions, with a description and illustration of each condition. Some other suspension terms are listed below the table.

Table 2.4 Suspension conditions

Suspension condition	Description	Front suspension	Rear suspension
Pitching and bouncing	Similar to bump and rebound: the alternate rising and dipping of the front and rear of the vehicle body on uneven road surfaces		
Dive (braking)	Downward movement of the front and upward movement of the rear as the brakes are applied		
Drive (accelerating)	Upward movement of the front and downward movement of the rear during acceleration		
Squat	Downward movement of the chassis and upward movement of the wheels after travelling over a hump back bridge at speed		
Float	Where all vehicle wheels are level and subjected to the same forces acting on the suspension		
Bump	Where a vehicle wheel goes over a raised section in the road.		
Rebound	Where a wheel drops into a dip in the road surface		
Roll	Leaning movement of the vehicle body when turning		

Level 2 Light Vehicle Maintenance & Repair

Some other suspension terms are:

- **Compliance** – This is the amount of movement the wheel can make before the vehicle body height will be affected.
- **Vehicle attitude** – This describes the relative levelness or tilt of the vehicle body, both front-to-back and side-to-side.
- **Damping** – As the suspension rises and falls, compression and expansion of the **damper** occurs. Damping is where a resistance is created as a plunger moves through hydraulic fluid.
- **Vehicle ride height** – A characteristic of the suspension is to keep the vehicle body as level as possible during vehicle movement. The height of the vehicle body in relation to the centres of the front and rear wheels is known as the ride height. This is measured from the centre of the wheel to just under the wheel arch. Both measurements on the same axle must be the same.

The layout and components of suspension systems

Non-independent suspension

In a non-independent suspension (see Figure 2.52), the wheels at the front or rear of the vehicle are connected by a tubular single, solid axle. This is also known as a rigid or beam axle system. Non-independent suspensions are suitable for off-road vehicles.

The advantages of non-independent suspension are:

- ✓ Good wheel travel, axle position and ground clearance for off-road use.
- ✓ Simplicity: as there are very few components, manufacturing and assembly are less complicated with this suspension system.
- ✓ Strength: large loads are supported by the beam axle because of its strength.
- ✓ Camber control: camber will not be changed as the wheels are joined by the rigid axle assembly.

The disadvantages of non-independent suspension are:

- ✗ Movement of a wheel on one side of the axle affects the other, resulting in road handling difficulties.
- ✗ Unsprung weight: the axle is heavy, which is an undesirable feature as this is classed as unsprung weight.
- ✗ Wheel movement: **shimmy** occurs because the wheels are locked together on the rigid axle. This decreases the stability of the vehicle around corners.
- ✗ Bad bump steer: if beam axles are used on the front axle, steering will be difficult to control as wheel movement on one side affects wheel movement on the other side.
- ✗ Size: non-independent axles are large. This reduces space for passengers and luggage.
- ✗ Ride quality: if one wheel is moved the other is affected. The passengers and load are also subject to this movement.

> **Key term**
> **Damper** – a shock absorber.

Figure 2.51 Checking vehicle ride height

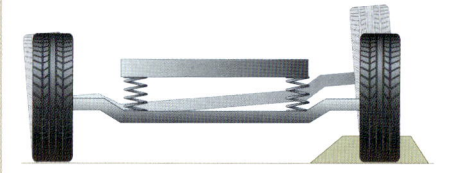

Figure 2.52 Non-independent suspension

> **Key term**
> **Shimmy** – where a wheel moves from side to side during rotation.

2 Light vehicle chassis units & components

Independent suspension

Independent suspension (see Figure 2.53) is more suitable for road vehicles. The wheels are able to move on their own to follow the road surface more closely. This gives improved handling and increased driver comfort.

Most vehicles have an independent rear suspension (IRS), also known as trailing link suspension. This suspension design uses a set of arms located ahead of the wheels to support the unsprung weight. The wheel 'trails' and the suspension 'links' the wheels to the chassis. Almost all vehicles today use this suspension set up.

Independent front suspensions (IFS) make it possible for each wheel on an axle to move in a vertical direction independent of the others. Except for commercial vehicles, vans and 4 × 4s, this suspension system is also used in most vehicles today.

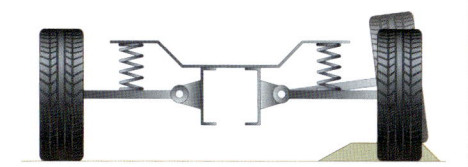

Figure 2.53 Independent suspension

The operation of suspension systems and components

Leaf spring

Before independent suspension was developed, the leaf spring was the most popular type of **semi-elliptic** spring used for all vehicles. The spring assembly is inexpensive compared with other suspension systems because it can also be used to connect the rear axle to the body.

- In a **laminated** spring assembly, there is a main leaf with a number of leaves of decreasing lengths fixed below, giving extra strength. This prevents the main leaf from breaking at a central point.
- At each end of the main leaf is a spring eye into which a rubber or phosphor-bronze bush is inserted. This bush prevents wear as the bolt moves slightly in operation when attached to the frame.
- When the vehicle is subjected to the various forces acting on it from the road or through driving and braking, the spring gets longer and shorter. A swinging shackle is fitted at the rear end of the spring to allow for this movement.
- A central retaining bolt passes through the leaves to prevent relative movement of the leaves.
- Two 'U' bolts clamp the spring to the axle.

> **Key terms**
>
> **Semi-elliptic** – in a half oval shape. (An 'ellipse' is an oval shape and 'semi' means half.)
>
> **Laminated** – describes a component made up of a number of layers.

> **Did you know?**
>
> - Upward movement (bump) is where the main leaf is deflected and the bending is controlled by all of the lower leaves.
> - Downward movement (rebound) is a deflection that tends to open up the leaves. This concentrates the load on the main leaf. Rebound clips are fitted to make the shorter leaves take some of this load.

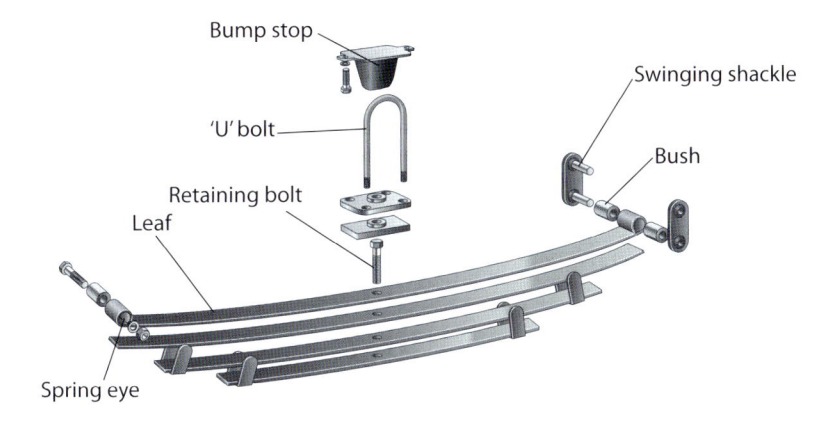

Figure 2.54 Leaf spring assembly

151

Level 2 Light Vehicle Maintenance & Repair

A major problem of the laminated spring is friction between the leaves. This causes noise, wear and a 'hard' ride for the vehicle occupants and load. Buttons at the end of each leaf made from low-friction material assist in solving this problem. A rubber bump-stop fitted between the spring and the vehicle body limits the movement of the spring in extreme bump conditions.

Single leaf spring

To reduce unsprung weight, single leaf springs were introduced. These are constructed using either spring steel or carbon fibre. The location of the spring is the same as the laminated spring.

Coil spring

Coil springs are the most commonly used springs in today's independent suspension arrangements. Steel rods formed into the shape of a coil create a **helical** coil spring design. This design gives a smoother action, as it removes any friction from the spring operation and it allows for a good range of movement.

When a load is placed on the coil spring, the entire rod is twisted as the spring is compressed. This absorbs the energy of the suspension forces and the road shock is cushioned.

One disadvantage of the coil spring is that it cannot resist forward, backward and sideways forces on its own. This means that extra suspension links need to be added to resist these forces, but this will increase the unsprung weight.

Progressive (variable rate) spring

If the width of a spring is the same from the top to the bottom, the amount of compression will increase at the same rate as the load placed on it. This means that for a small suspension movement, the spring will be stiff and will provide an uncomfortable ride for the passengers.

However, if the width of the spring is reduced at each end, the initial movement of the suspension will be taken up by the smaller width section of the spring. As the load increases, the larger central portion will resist the force placed on it. This is known as progressive suspension.

Taper coil spring

This type of spring has a thinner diameter spring steel coil at either end making it thicker in the centre section, as shown in Figure 2.55. This has the same effect as the variable rate spring described above.

Torsion bar suspension

A torsion bar is a spring steel rod that uses its torsional elasticity to resist twisting (see Figure 2.56).

> **Key terms**
>
> **Helical** – shaped in a round spiral.
>
> **Oscillation** – the amount of energy a spring retains after force is put on it. The spring will continue to expand and contract until movement energy is turned into heat energy.

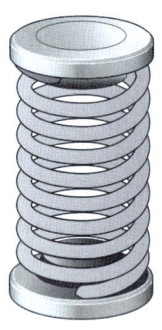

Constant rate coil spring   Variable rate coil spring

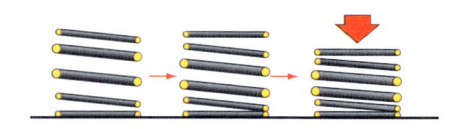

Taper coil spring

Figure 2.55 Coil spring and taper coil spring

2 Light vehicle chassis units & components

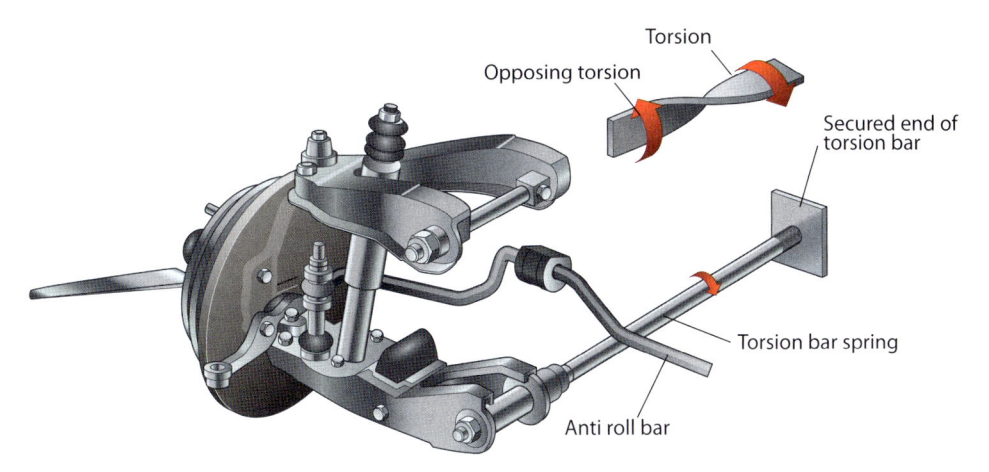

Figure 2.56 Torsion bar and torsional elasticity

One end of the torsion bar fits into the suspension components, while the other secures to the body/chassis frame of the vehicle. During bump and rebound forces, the torsion bar is subjected to a torsional load.

The spring strength is determined by the length and the diameter of the bar, as shown in Table 2.5.

Table 2.5 The spring strengths of different constructions of torsion bar

Construction of torsion bar	Spring strength
Long bar, narrow diameter	Soft spring
Long bar, wide diameter Short bar, narrow diameter	Medium spring
Short bar, wide diameter	Hard spring

> **Did you know?**
> Stabiliser bars can also provide a degree of torsional spring.

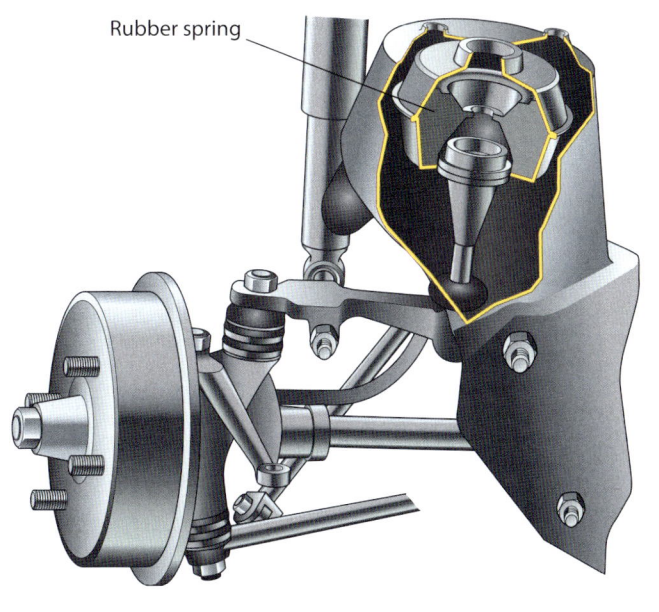

The advantages of torsion bar suspension are:

- ✓ simple suspension layout
- ✓ takes up little space underneath the vehicle
- ✓ inexpensive
- ✓ light in comparison with leaf springs.

The disadvantage of torsion bar suspension is:

- ✗ It does not control **oscillation**, so dampers have to be used.

Rubber components

Rubber is extremely light and compact and is a very effective spring material. It can absorb movement energy and has been used in suspension design to act as a spring to reduce the need for a damper (see Figure 2.57).

Rubber springs are naturally 'progressive'. This means that the more weight is put on the spring, the stiffer the suspension becomes.

Figure 2.57 Rubber springs

153

Level 2 Light Vehicle Maintenance & Repair

The advantages of rubber springs are:

✓ silent during use
✓ do not need to be lubricated
✓ can be made in any shape.

The disadvantages of rubber springs are:

✗ not effective at supporting heavy loads
✗ prone to splitting and deterioration.

Because of the disadvantages, rubber is mainly used to supplement other suspension systems. It is used to make:

- spacers
- bushes
- stoppers
- cushions
- supports for other suspension components.

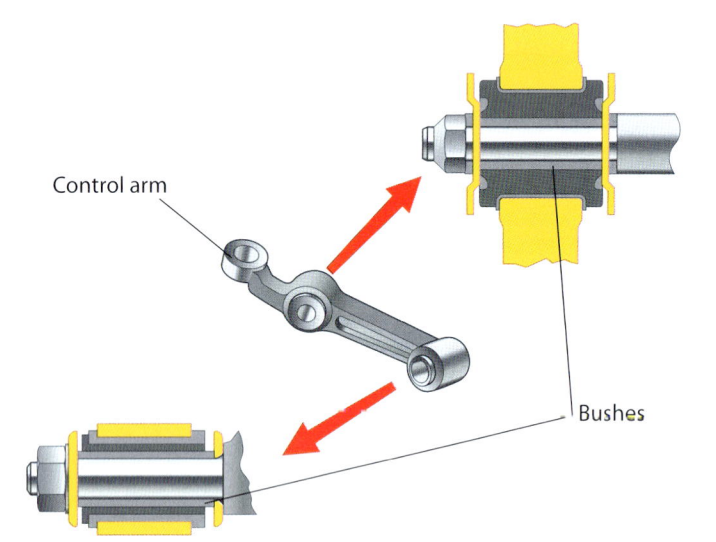

Control arm bushes

Figure 2.58 shows an example of rubber bushes. They are used to provide a degree of flexibility on the control arm shown. The bushes can also be aligned to provide a degree of steering assistance as the arm rises and falls due to cornering.

Rubber bushes

The bushes shown in Figure 2.59 are positioned at the top of a MacPherson strut. The bump stop provides a soft cushion if the suspension reaches its limit and the rubber cushion provides a flexible mounting between the strut top and vehicle body.

Figure 2.58 Control arm bushes

Figure 2.59 Rubber bushes

MacPherson strut

The MacPherson strut is a combined spring and damper assembly. It is located at the top of the inner wing of the vehicle by a bearing or bush that allows for steering movement. The lower part of the strut is located directly or indirectly by a ball joint to a track control arm. This allows the unit to swivel as well as giving sideways attachment of the assembly.

These characteristics contribute to vehicle control under driving and braking conditions. Anti-roll bars or tie rods may give additional support to the track control arm. The strut can be manufactured in two ways:

1. Fixed to a hub using bolts. This is known as a semi-strut and is the most commonly used front suspension assembly today.
2. As part of the hub carrier, where the strut is part of the hub assembly that holds the wheel.

A disadvantage of the MacPherson strut is that track alters as the wheel moves up and down. Premature wearing of the damper piston may occur due to increased side loading during cornering or excessive bump and rebound.

2 Light vehicle chassis units & components

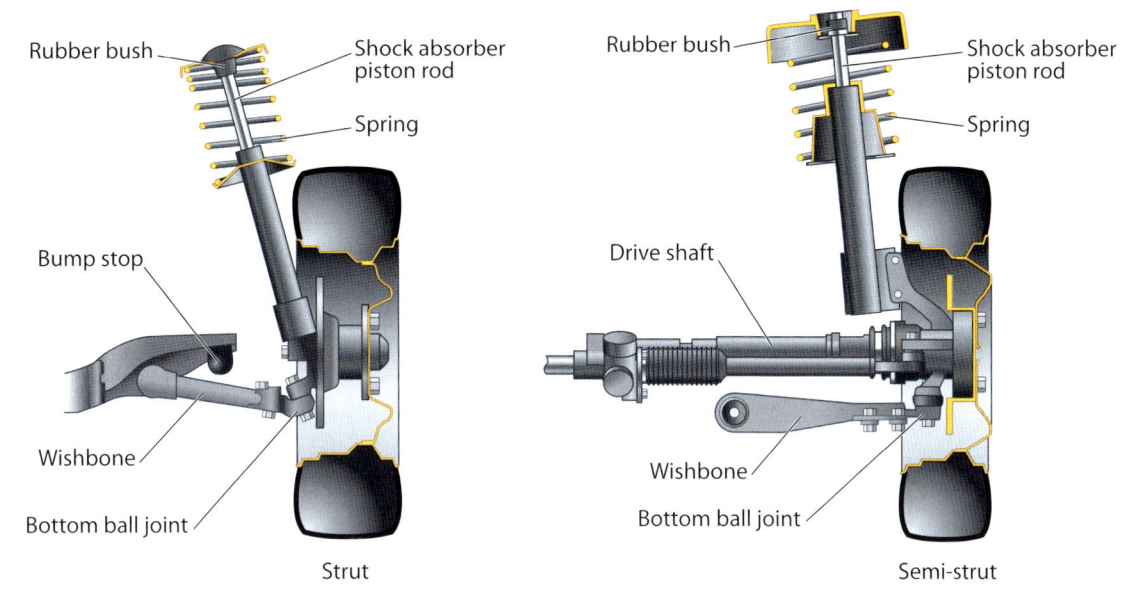

Figure 2.60 MacPherson strut suspension layout

Hydragas suspension

Hydragas suspension uses fluid and a compressed gas in sealed chambers at each wheel instead of the more usual coil springs and dampers. Although pipes are used to link the chambers front to rear, hydragas suspension is still referred to as an independent suspension.

The hydragas suspension unit is made up of three chambers. It uses fluids and nitrogen gas to absorb shocks from the road surface.

- **Chamber 1** – This contains pressurised nitrogen, sealed with a special rubber **diaphragm**. The nitrogen gives a 'progressive' spring rate.
- **Chamber 2** – The central chamber is filled with a special suspension fluid consisting of water, alcohol and a **corrosion inhibitor**. This is used to transmit the suspension forces to the nitrogen chamber (chamber 1).

Key terms

Diaphragm – a flat, flexible rubber plate that prevents leakage of fluids and gases.

Corrosion inhibitor – chemical that prevents the system corroding.

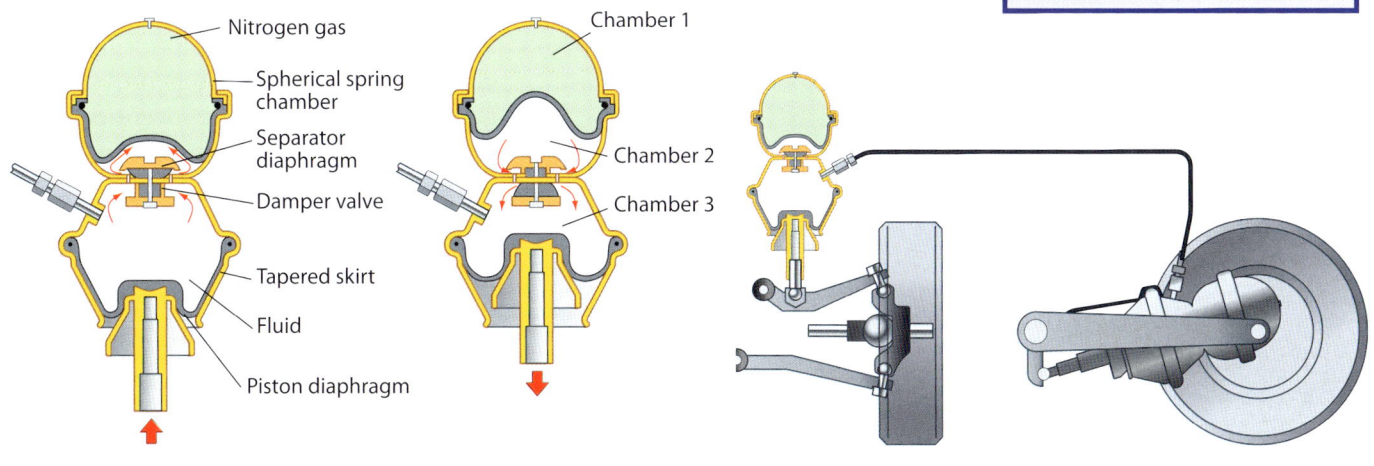

Figure 2.61 Hydragas unit

155

- **Chamber 3** – The lower chamber is filled with the same fluid as above in chamber 2. This is used to transmit the suspension forces to chamber 1.

An essential component in the hydragas unit is the damper valve. The valve reduces fluid flow from chamber 2 to chamber 3. This absorbs small suspension movement to prevent pitching, which would make the vehicle unstable during braking and accelerating.

Air suspension

In light vehicles, air suspension is used in the luxury car sector. This system consists of four air springs, a compressor and an electronic control unit (ECU).

- The ECU receives signals from four height sensors located on each corner of the vehicle.
- The compressor builds up air pressure and the flow of this air is controlled by valves at each of the air springs.
- The valves at the air springs are controlled by the ECU.
- The vehicle ride height and stiffness of the suspension can also be manually controlled within the driving compartment. For example, off-road vehicles will need a greater ride height, so more air can be added to each air spring in these conditions.

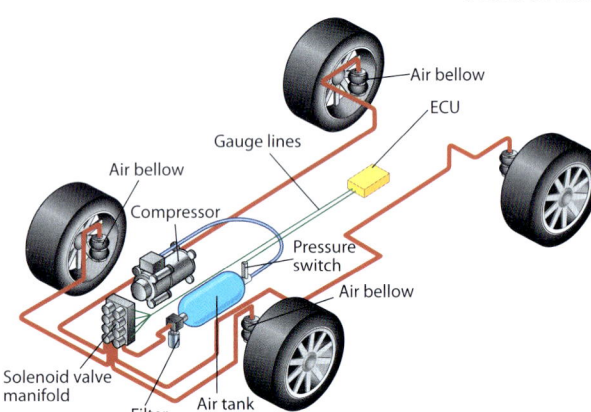

Figure 2.62 Air suspension

Key term

Actuator – a mechanism that puts something into action.

Setting ride (trim) height

- Before doing any height checks, you will need to make sure that the tyre pressures are set to the manufacturer's settings.
- Check the ride (trim) height by placing a tape measure from the hub centre to the base of the wheel arch.
- The measurements need to be the same on each axle and within the manufacturer's recommendations.

Electronic suspension systems

Electronic suspension is an adaptation of air suspension. Linear electromagnetic **actuators** provide all the functions of the conventional shock absorber and spring system as they respond quickly to eliminate movement and vibration of the car under all driving conditions. They can also prevent body roll when cornering by automatically stiffening the suspension. This gives the driver better control and improved safety.

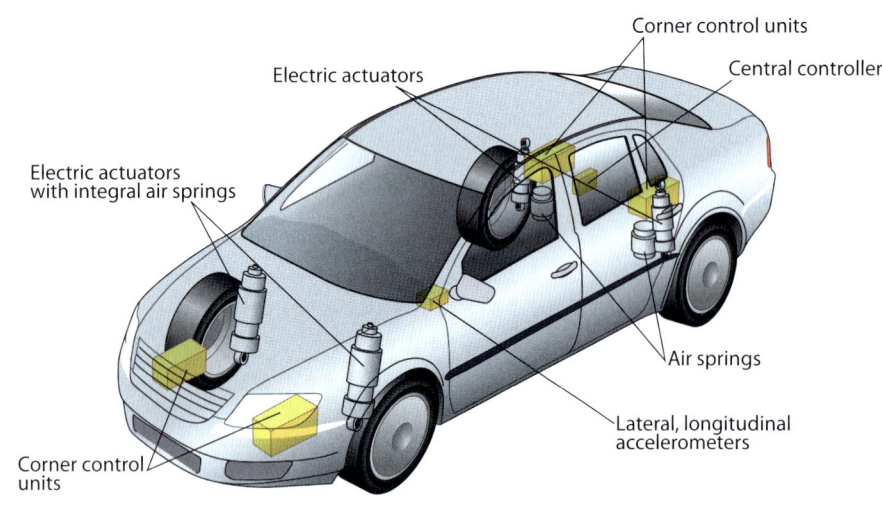

Figure 2.63 Electronic suspension system

2 Light vehicle chassis units & components

The damper

The damper absorbs shocks from uneven road surfaces and during various driving conditions. It is sometimes known as a shock absorber.

The damper is a hydraulic device consisting of a piston, valves and a confined cylinder. The piston is connected to the vehicle body. It slides up and down the cylinder, which is connected to the hub. The valve attached to the piston limits the motion of the suspension spring by controlling hydraulic fluid flowing through it.

Lever-type damper

Some older vehicles with torsion bar suspension, or where space was limited, used a lever-type damper (see Figure 2.64). The lever-type damper restricts the fluid transfer between two piston chambers at varying degrees, to compensate for bump and rebound conditions.

Consequences of worn dampers

Dampers in poor condition increase the wear on the mechanical parts of the vehicle, such as steering, suspension, transmission and tyres. Other consequences of worn dampers are:

- driver fatigue
- poor road holding
- increased braking distance
- increased risk of **aquaplaning** in wet weather.

Suspension layouts

Semi-trailing arm and coil spring suspension

The semi-trailing arm system is a fully independent unit used on rear suspension systems (see Figure 2.67). The hubs are attached to semi-trailing arms, which are fixed by bushes to the subframe. Also fitted to the subframe is the final drive assembly. The final drive assembly housing and trailing arm transmit the driving and braking forces. Therefore, softer springs can be used, as they only perform suspension duties. This makes for a more comfortable ride. Unsprung weight is reduced as the final drive is fixed to the body.

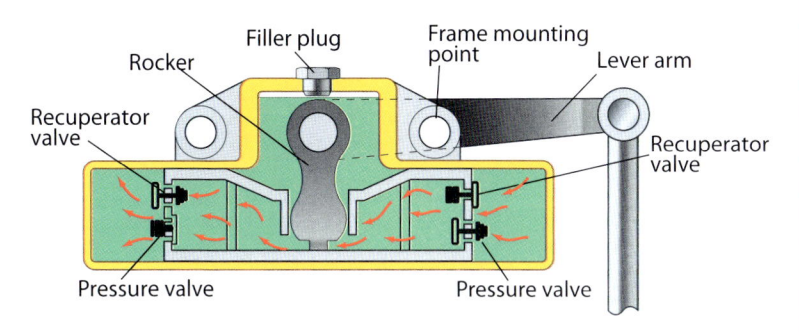

Figure 2.64 Lever-type damper

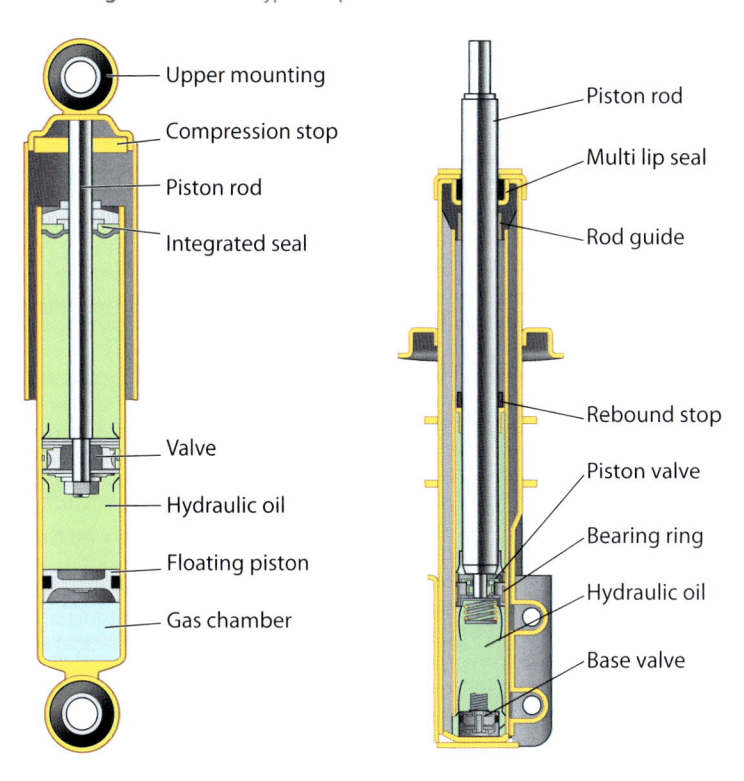

Figure 2.65 Bump — where the wheel rises when coming into contact with a raised section on the road surface

Figure 2.66 Rebound – where the wheel drops into a dip in the road surface

Key term

Aquaplaning – where a film of water forms between the tyre and road. The tyre can lose contact with the road and start to float on the water. This condition is so dangerous it can lead to total loss of control of the vehicle.

157

Level 2 Light Vehicle Maintenance & Repair

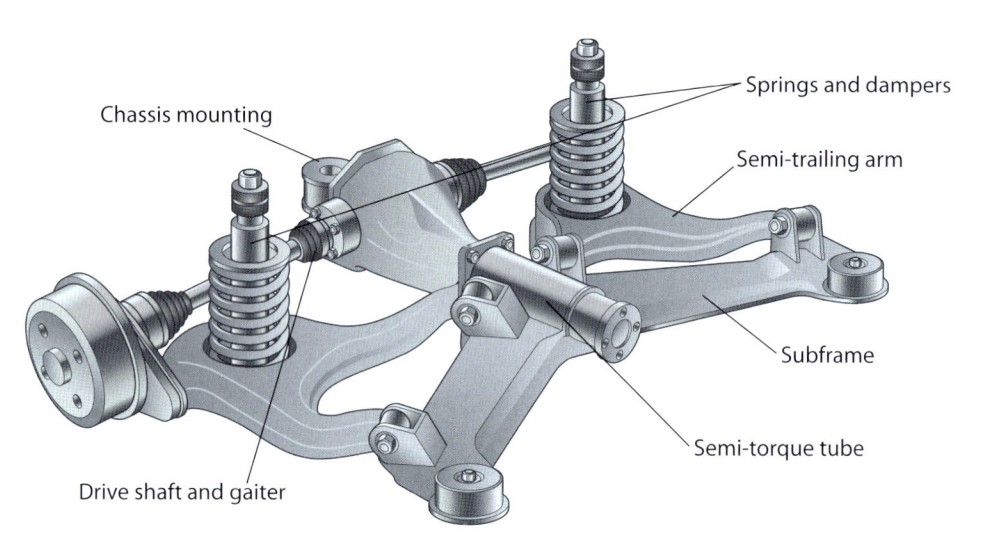

Figure 2.67 Semi-trailing arm rear suspension

The rear suspension system (see above) consists of:

- a rear subframe
- two semi-trailing arms
- two springs and two dampers.

The subframe is bolted to the body and provides the mounting points for the fixed and adjustable links. The semi-trailing links are attached to the hub and subframe, which will resist drive and braking forces. The springs and dampers absorb the suspension movement.

The suspension is designed to allow the wheel to move backwards and upwards in response to bumps and dips in the road surface. This **longitudinal** movement allows the springs and dampers time to react to surface changes. It also improves ride quality.

Trapezoidal links suspension

A **trapezoidal** link suspension assembly consists of a MacPherson strut, a subframe which is attached to the body, and parallel rods linking the hub to the subframe. A tie bar is also fitted to resist driving and braking forces, as shown in Figure 2.68.

Trailing arm suspension

The trailing arm suspension (see Figure 2.69) is similar to the semi-trailing arm suspension used on rear axles. It is connected at the front of the rear axle to the chassis by a pivoted mounting and at the rear by a damper. The linkages allow the rear to swing up and down. Side to side movement of the suspension is resisted by using a Panhard rod. This suspension is not often used because of the space it takes up. Its advantage is that it doesn't suffer from the side-to-side scrubbing problem of double wishbone systems.

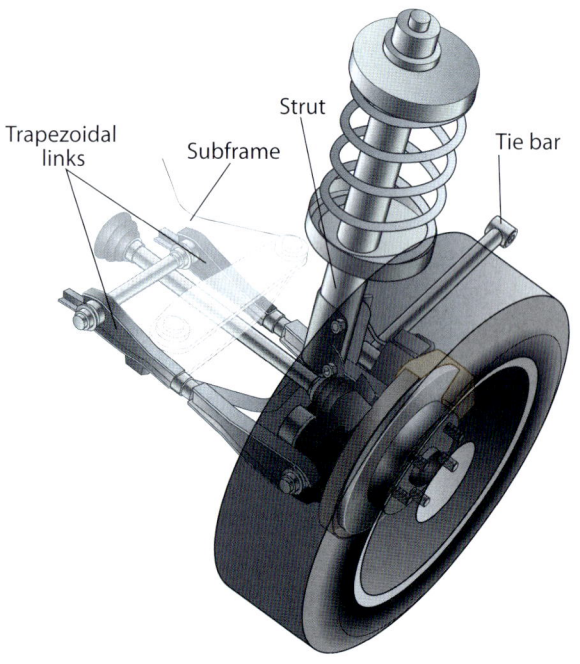

Figure 2.68 Trapezoidal links suspension

> **Key terms**
>
> **Longitudinal** – parallel with the centre line of the vehicle.
>
> **Trapezoidal** – describes a four-sided shape with one pair of parallel sides.
>
> **Vehicle track** – the distance from wheel to wheel on the same axle.

Coil spring/wishbone suspension

Coil spring suspension is also called wishbone suspension, as it is shaped like a chicken wishbone. It consists of two wishbone-shaped components, usually made from cast alloy or pressed mild steel. The components are fixed at the stub axle with ball joints and to the subframe or chassis with rubber bushes.

- The wishbones provide location for the stub axle assembly to the chassis and withstand side braking, acceleration and cornering forces.
- The ball joints allow movement when the wheel moves up and down, as well as allowing the stub axle to swivel for steering.

Parallel link suspension

The parallel link suspension linkage causes a change in **vehicle track** as it rises and falls due to road irregularities. This results in increased tyre wear. For this reason this type of suspension system is rarely used.

Unequal length wishbones

Unequal length wishbones are a development on parallel link systems. As the suspension rises and falls in operation, the vehicle track does not alter. This ensures vehicle stability during cornering. Extreme wishbone movement is limited by using bump and rebound rubbers.

This unequal length wishbone suspension system, as shown in Figure 2.71, uses a slightly more complicated arrangement than the parallel link type. It combines an independent double wishbone with coil springs and carefully selected dampers with an anti-roll bar at the front.

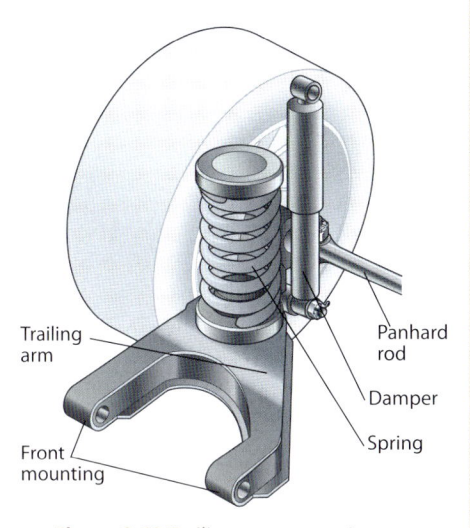

Figure 2.69 Trailing arm suspension

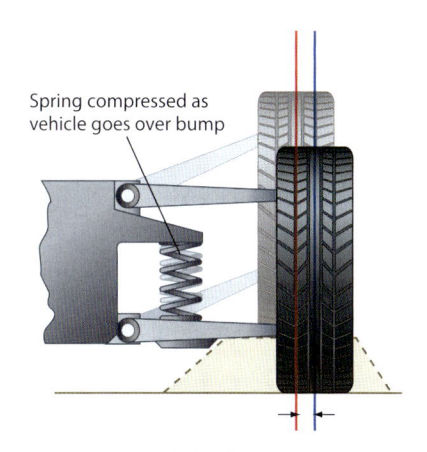

Figure 2.70 Parallel link suspension

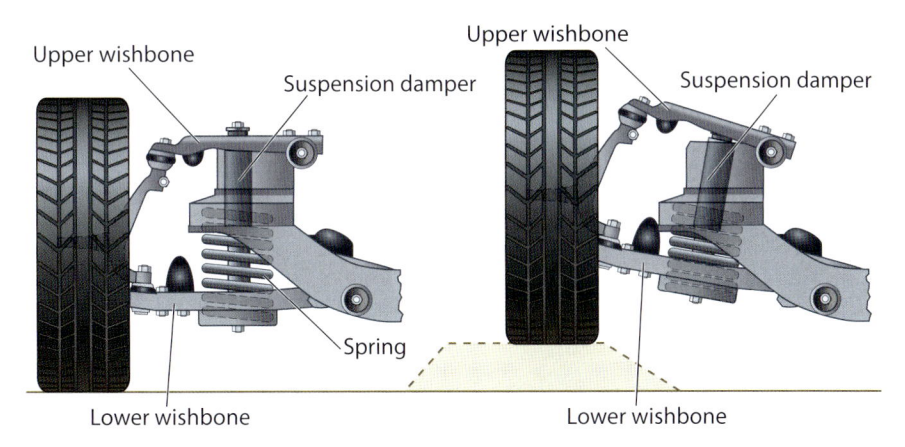

Figure 2.71 Unequal length wishbone suspension

Double wishbone front suspension

A popular choice of suspension on sports cars is the double wishbone front suspension (see Figure 2.72). This is because the vehicle body needs to be as low as possible to assist aerodynamics. To make this possible the MacPherson strut requires a high fixing in the inner wing.

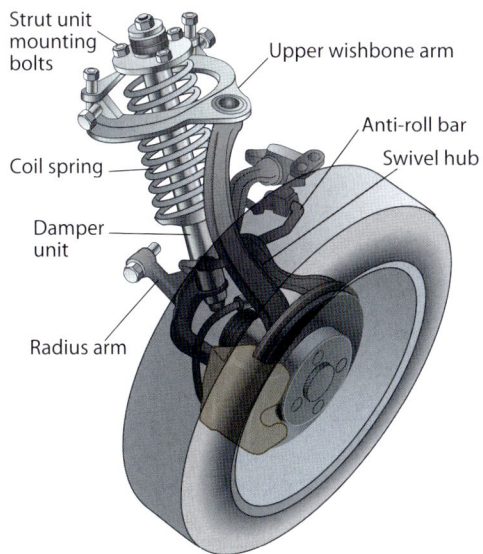

Figure 2.72 Double wishbone front suspension

159

Level 2 Light Vehicle Maintenance & Repair

Passive steer

Passive steer is where the rear suspension assembly is designed to steer slightly as the wheel moves up and down. This is done by pivoting the front suspension mounting at an angle (see Figure 2.73). This improves vehicle handling when cornering hard.

Transverse link and strut

The rear of a FWD vehicle may use the transverse link and strut assembly, as shown in Figure 2.74. This arrangement incorporates a separate damper and spring arrangement. Additional support is provided by the tie bar, which connects the wheel assembly to the body. The tie bar can sometimes be referred to as a tie rod or stabiliser bar.

The tie bar prevents forward movement of the wheel during acceleration and backward movement when the brakes are applied.

Off-road vehicle live axles

Panhard rods and radius arms provide linkage for the axle where coil springs are used. The advantage of using coil springs on off-road vehicles is the amount of axle movement the springs allow.

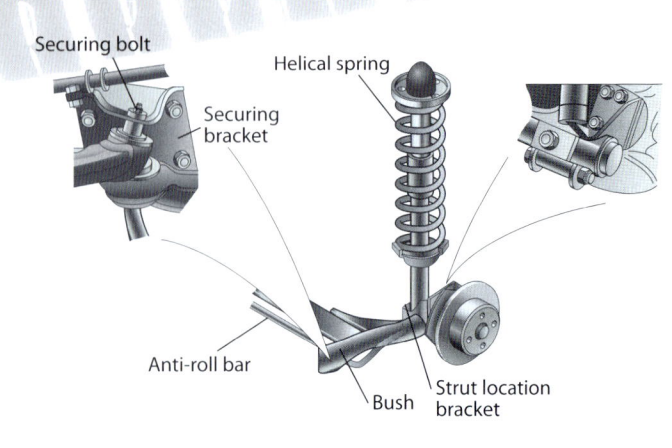

Figure 2.73 Passive steer

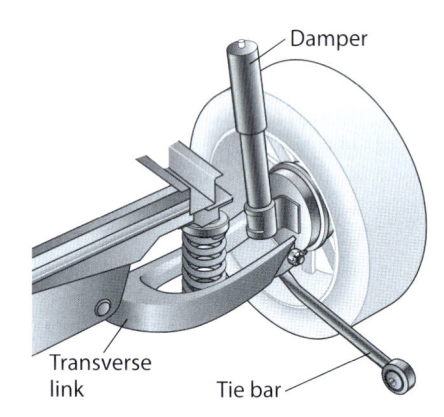

Figure 2.74 Transverse link and strut

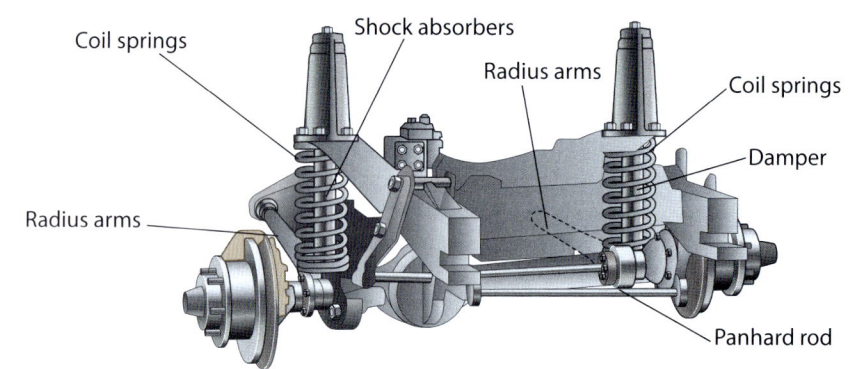

Figure 2.75 Off-road vehicle live axles

Layout of suspension components

Table 2.6 lists the layouts and uses of some key suspension components.

Table 2.6 Layouts and uses of suspension components

Suspension component	Layout	Use
Radius arm	From chassis to the suspension linkages in line with the vehicle centre line	Takes up drive during acceleration and brake torque and braking
Panhard rod	From chassis to the suspension linkages across the vehicle	Limits lateral axle movement during cornering
Track control arm	From the hub to the lower subframe, chassis or body	Controls the vehicle track as the front wheel moves up and down
Tie bar	From chassis to the suspension linkages	Takes up wheel movement during acceleration and braking

2 Light vehicle chassis units & components

Anti-roll bar

An anti-roll bar connects the vehicle body to the suspension arms to reduce roll during cornering. When an anti-roll bar is used in the suspension of a vehicle, it also enables low rate soft springs to be used. Under normal driving conditions, soft springs make the vehicle more comfortable.

The anti-roll bar does not contribute to the suspension spring stiffness. Yet it does become more effective if one wheel is raised higher than the other. This could be when the vehicle passes over a hump in the road or when the body starts to roll while cornering. When cornering, the stiffness of a spring increases in proportion to the movement of rebound for each wheel or the movement of body roll.

Vehicle stability

Suspension systems are designed to counteract the effect of forces acting on the vehicle, as these affect the handling and stability. Table 2.7 shows the forces that the suspension needs to absorb. Some of the forces are also illustrated in Figure 2.76.

Figure 2.76 Forces acting on suspension systems

Table 2.7 Forces acting on a vehicle and how they are absorbed

Forces	Method of absorbing force
Static and dynamic vertical loading of the vehicle	• Elastic compression, shear, bending or twisting action of the spring
Body roll due to cornering	• Anti-roll bar
Twisting reaction due to driving and braking torque	• Stiffness of the spring • Stabiliser arms • Triangular wishbone arms
Driving and braking thrusts between the sprung body and the road wheels	• The rigidity of the springs • Wishbone arms • Tie rods/tie bars
Side-thrust due to the centrifugal force, crosswinds, camber of the road, going over a bump or pot-hole	• Rigidity of the spring • Hinged linkage arms of the suspension • Attachments between the wheel stub axle and the chassis • Radius and Panhard rods

Inspecting the serviceability and condition of suspension systems

Bushes and bearings

Rubber has a tendency to age-harden. As it gets older, it gets harder and cracks. Visual checks around all rubber bushes must be made for signs of cracks and complete failure. Use a pry bar to move the joint so you can carry out a clearer visual inspection. Check the swivel bearings at the top of a MacPherson strut.

Figure 2.77 Visual inspection of bushes and bearings

161

Springs

Springs are made from brittle steel. They have a tendency to corrode and snap. Careful inspection is required, particularly in the suspension cups for coil types, to ensure a portion of the spring has not snapped and fallen off the vehicle.

Dampers

When testing a damper, you should push down on one corner of the vehicle. The vehicle should bounce one and a half oscillations.

If there is any difference in the amount of bounce the valve inside the damper may be damaged or the fluid may have leaked out. You can visually check for traces of leaking fluid from underneath the vehicle at the top of the lower part of the damper.

Failure of either the damper or the spring means that the MacPherson strut assembly on both sides needs to be removed (see opposite).

Linkages

Check for signs of corrosion and damage caused by accidental side forces such as hitting the curb. Make sure that any fixings are tight.

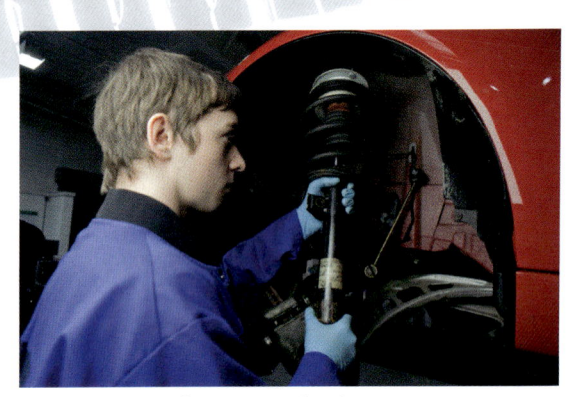

Figure 2.78 Visual inspection of springs

Figure 2.79 Visual inspection of dampers

Figure 2.80 Visual inspection of linkages

Common suspension system problems

Table 2.8 lists some common suspension system defects and how to correct them.

Table 2.8 Suspension system defects – causes and how to remedy them

Symptom	Cause	Remedy
Knocking noise over bumps	Damaged bushes	Replace bushes
	Damaged ball joints	Replace ball joints
	Loose fixings	Tighten all fixings
Spongy suspension	Worn dampers	Fit new dampers
	Leaking dampers	Fit new dampers
Hard suspension	Damaged dampers	Fit new dampers
Tyre wear (suspension fault)	Tyre pressures incorrect	Check tyre pressures
	Bent linkages	Fit new linkages
	Steering out of track	Set tracking
Noise in steering	Steering pivot bush damaged	Replace steering pivot bush
	Worn bushes	Replace bushes
Steering wander (suspension fault)	Damaged linkages	Replace linkages
	Damaged ball joints	Replace ball joints
	Loose fixings	Tighten all fixings

2 Light vehicle chassis units & components

MacPherson strut replacement

Checklist			
PPE	**VPE**	**Tools and equipment**	**Source information**
• Steel toe-capped boots • Overalls • Latex gloves	• Wing covers • Steering wheel cover • Seat covers • Floor mat covers	• Vehicle hoist • Wire brush • Torque wrench • Socket set • Ball joint splitter • Spanner • Wheel alignment gauges • Dot punch and hammer	• Steering geometry information • Torque setting information • Job card

1. Support the vehicle safely and remove the road wheel.

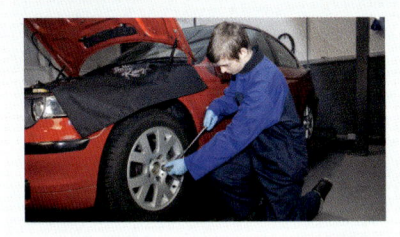

2. Use a wire brush to clean around the threads holding the strut assembly to the hub.

3. Mark the bolt head with a dot punch and hammer. You also need to make an aligning mark on the strut assembly using the dot punch and hammer.

4. Remove any tie bar linkages or brake pipe fixings that may prevent removal.

5. Slacken the hub retaining bolt but leave the bolt in position.

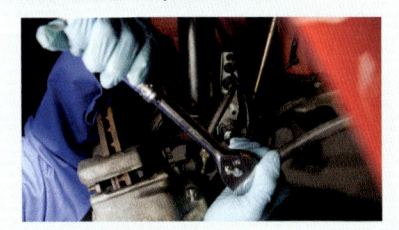

6. Lower the vehicle with the caliper just off the ground and lift the bonnet.

7. Remove the strut top nuts and top bearing plate.

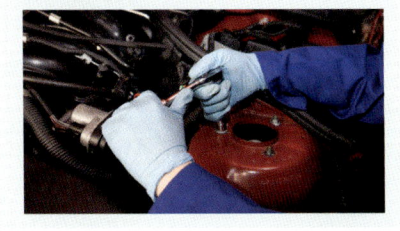

8. Lift the vehicle and fully remove the hub retaining bolt and MacPherson strut.

9. Replace the strut assembly by following the reverse procedure. Take care to set nuts to the correct torque and align the hub bolt(s) in the correct position(s).

10. Replace the wheel and torque the wheel nuts. Repeat for the other side of the vehicle.

11. Lower the vehicle and roll it backwards and forwards for the suspension to settle. Check the wheel alignment (see Chapter 7, page 465).

> **Safe working**
>
> If the spring needs to be removed from the MacPherson strut assembly, make sure you use the correct spring compressor. Do not attempt this without the correct equipment.

163

Level 2 Light Vehicle Maintenance & Repair

> **CHECK YOUR PROGRESS**
>
> 1. State two purposes of the suspension system.
> 2. What type of spring does a MacPherson strut use?
> 3. What is meant by 'pitching'?

Light vehicle braking systems

The purposes of a vehicle braking system are to:

- reduce the vehicle speed
- stop the vehicle
- hold the vehicle in a stationary position.

Kinetic energy

Kinetic energy is the energy of motion. As a vehicle is brought to rest, friction is applied by force on the **brake linings** and this converts the kinetic energy to heat energy. The speed of energy conversion controls the rate of deceleration.

Braking system fundamentals

The main requirements of a braking system are:

- The driver must be able to apply a suitable braking force without excessive effort.
- There must also be a back-up brake system in the event of a main system failure.
- There must be a mechanical means of keeping pressure on the linings to keep the vehicle stationary.

When you brake, the brake shoes come into contact with a rotating drum by the force from the wheel cylinders. The friction grip between the two surfaces will slow, stop or hold the vehicle. On light vehicles this is done using:

- a hydraulic circuit to apply the footbrake
- a mechanical linkage to apply the handbrake.

Drum brakes

On vintage vehicles, brake drum assemblies were used on both front and rear axles. Discs and pads replaced drum brakes on the front axle about 20 years ago. They are becoming more and more popular on the rear axles of vehicles as they improve the braking operation and efficiency.

The drum brake assembly consisted of two **brake shoes**:

- leading shoe
- trailing shoe.

> **Did you know?**
>
> Energy cannot be created or destroyed. However, it can be changed from one form to another. In a braking system, movement energy at the wheels is converted into heat energy at the brake linings.

> **Key term**
>
> **Brake linings** – these are made of non-asbestos material and are bonded to the brake pads and shoes. They contact the metal faces of the discs and drums to slow the vehicle down.

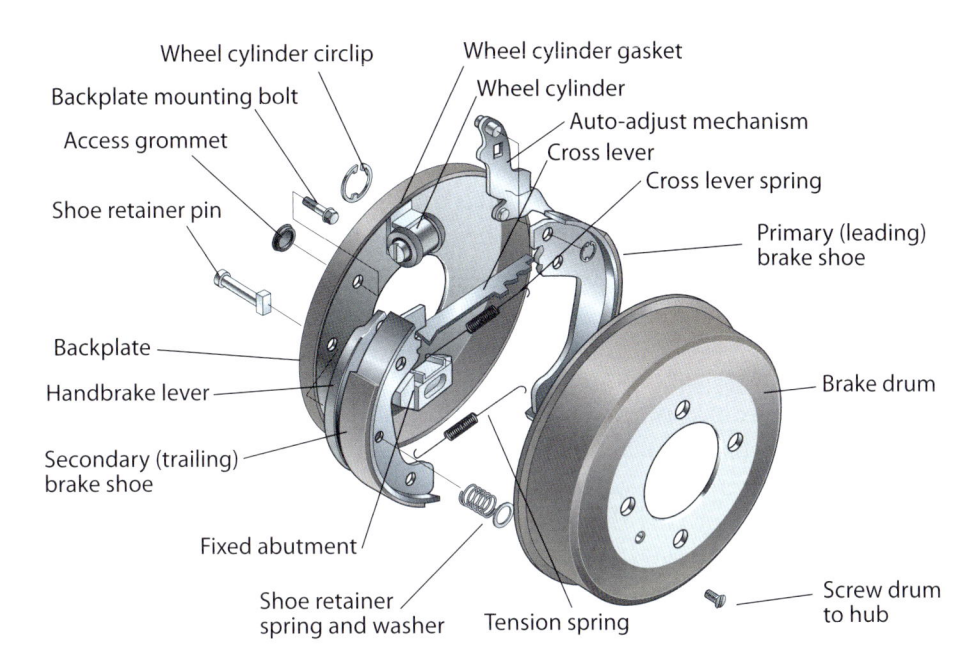

Figure 2.81 Drum brakes

Leading brake shoe

This is the first shoe after the wheel cylinder in the direction of drum rotation. Rotation of the drum drags the leading shoe into the friction surface, causing the brakes to bite harder, which increases the force holding them together – this is known as the self-servo action. It contributes three-quarters of the total braking effort of the assembly. Consequently, the leading shoe will have greater lining wear than the trailing shoe.

Trailing brake shoe

This is the shoe before the cylinder in the direction of drum rotation. Rotational force on the drum tries to push the shoe to the off position. For this reason, the shoe only contributes about one-quarter of the braking effort of the assembly.

Twin leading shoes

This type of construction (see Figure 2.82) was usually fitted to the front brakes of a vehicle that had drum brakes at every wheel. (The first Mini had front drum brakes.) This arrangement was ideal because of the self-servo action of both shoes, as larger braking effort is needed on the front axle due to weight distribution during braking.

Leading and trailing arrangement

This type of construction is usually fitted on the rear axle and will provide self-servo in both directions of rotation. This is particularly important when travelling in reverse or using the handbrake. It also gives a simple way to reduce the braking effort to the rear wheels, to prevent skidding under heavy braking.

> **Key term**
>
> **Brake shoe** – a half-moon-shaped metal component which has a brake lining bonded to the outside. This provides a surface for the metal drum to rub against to slow the vehicle down.

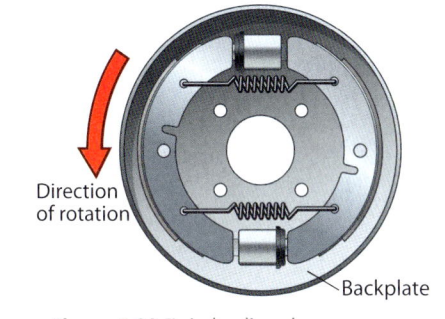

Figure 2.82 Twin leading shoes

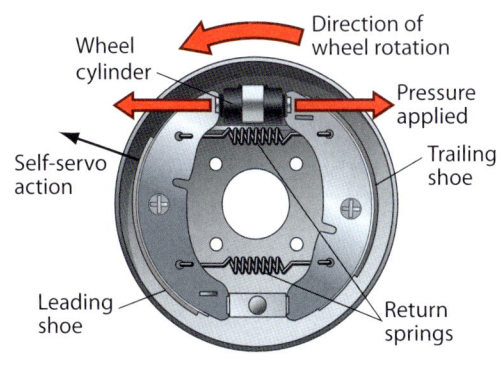

Figure 2.83 Leading and trailing arrangement

165

Level 2 Light Vehicle Maintenance & Repair

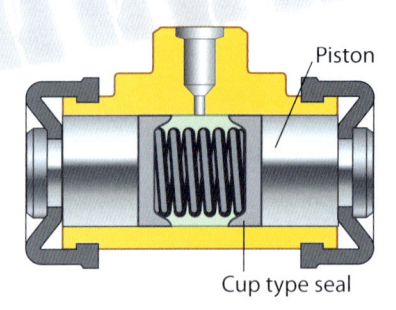

Figure 2.84 Wheel cylinders

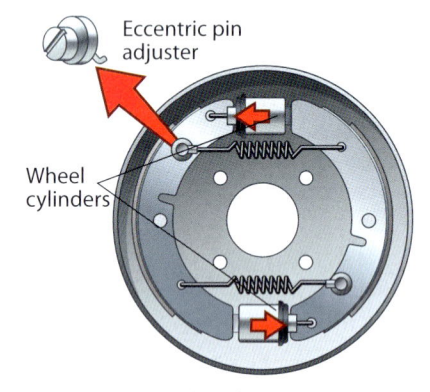

Figure 2.85 Brake adjusters

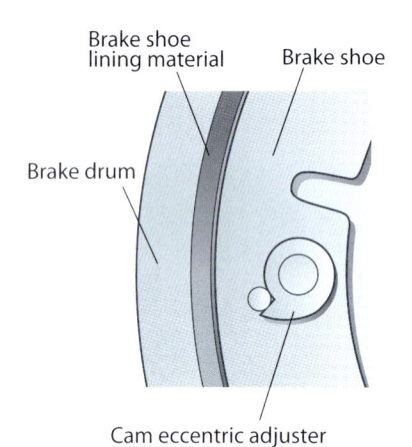

Wheel cylinders

Wheel cylinders are used to convert hydraulic pressure produced by the brake master cylinder back into mechanical force. This mechanical force then applies the brake shoes on to the drum. Wheel cylinders consist of either one or two pistons in a sealed tube which have a dust cover to prevent brake dust entering the cylinder.

Brake adjusters

Optimum performance of the braking system requires a small clearance between the brake linings and friction surface. As a friction lining wears down, the clearance increases. This in turn requires greater movement of the wheel cylinder and the master cylinder to apply the brakes. If adjustment is not made, the **pedal travel** distance will become excessive and eventually the system will not work. Brakes can be adjusted either manually or automatically.

Adjusting drum brakes manually

On certain older vehicles, such as the early Mini, the rear brakes need adjusting manually. This is performed by bringing the shoes into closer contact with the drums. There are various types of adjusters available, all with different adjustment procedures.

The three different adjusters are:

1. snail cam or eccentric (see Figure 2.86)
2. wedge type (see Figure 2.87)
3. star wheel (see Figure 2.88).

Star wheel automatic brake adjuster

As the brake is applied the shoes expand. A **cranked lever** applies force directly on one shoe and the reaction pushes the other side of the lever against the star wheel. When the brakes are released, the thread in the star wheel is turned and the clearance between the shoe and the drum is reduced.

Figure 2.86 Snail cam or eccentric brake adjuster

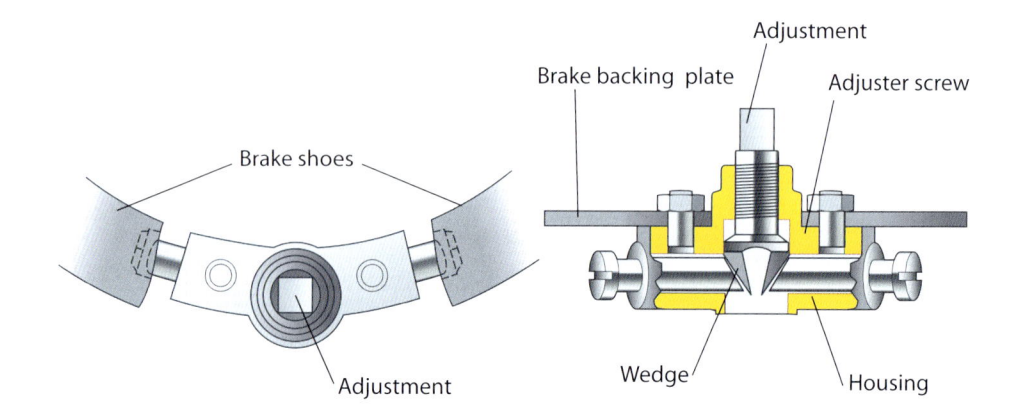

Figure 2.87 Wedge-type brake adjuster

166

2 Light vehicle chassis units & components

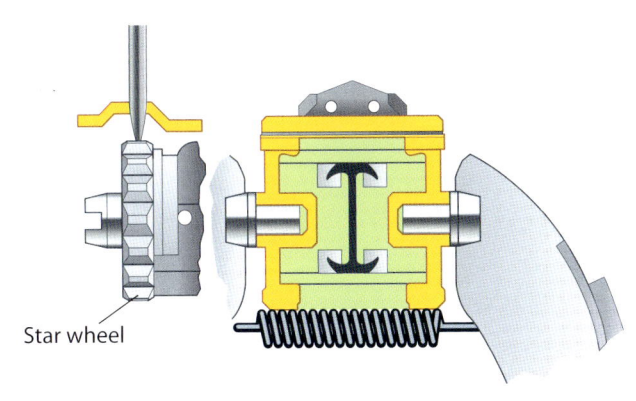

Figure 2.88 Star wheel automatic brake adjuster

Ratchet and pawl automatic brake adjuster

A threaded strut is placed between the shoes which adjusts the shoe-to-drum clearance by using the movement of the handbrake. As the handbrake is applied, if the shoe-to-drum clearance is over a certain limit, the **pawl** is able to jump a tooth on the ratchet and this turns the thread slightly. This movement lengthens the rod and so pushes the shoe nearer to the drum to take up the wear.

> **Key terms**
>
> **Pedal travel** – the amount the pedal moves when the driver presses it.
>
> **Cranked lever** – lever that is formed at a slight angle.
>
> **Pawl** – a circular nut with saw teeth shapes around the circumference.
>
> **Brake fade** – a condition that occurs when brake friction material can no longer absorb any more heat and can therefore no longer convert movement energy.

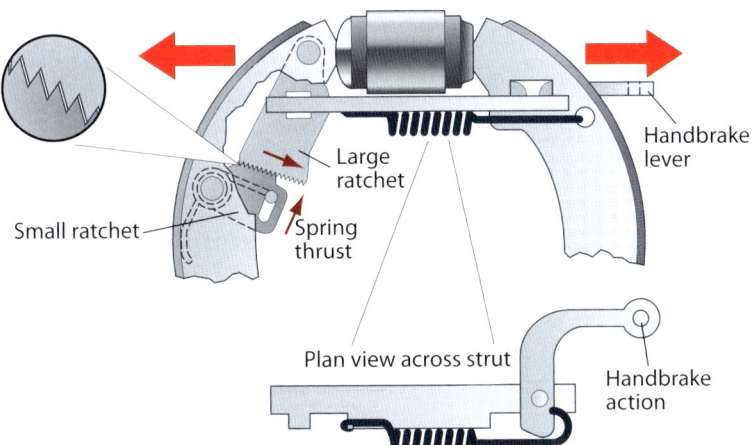

Figure 2.89 Automatic brake adjuster

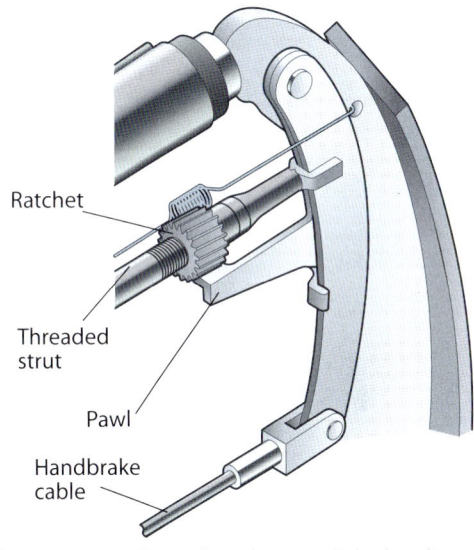

Figure 2.90 Ratchet and pawl automatic brake adjuster

Brake fade

One disadvantage of drum brake systems is that there is an increased possibility of **brake fade**.

Disc brakes

Disc brakes (see Figure 2.91) are based on the principle of friction pads being squeezed against a wheel-driven disc to slow it down.

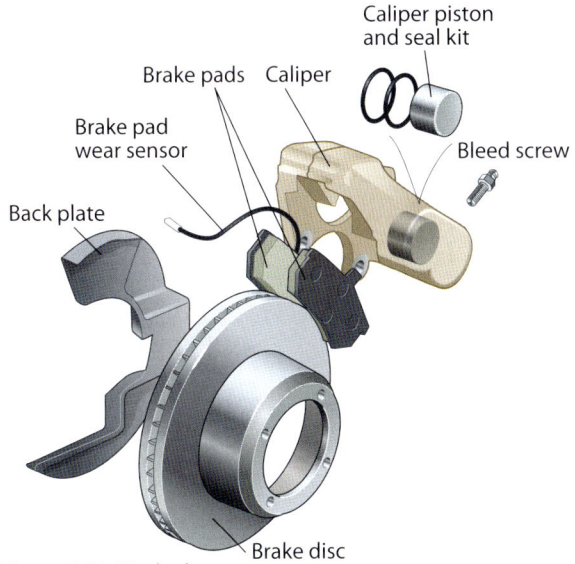

Figure 2.91 Disc brakes

167

The caliper is mounted on a torque arm which is secured to the suspension. The pads are carried in the caliper that straddles the disc.

The advantages of disc brakes are:

- ✓ heat dissipation is greater
- ✓ braking surfaces are cleaner
- ✓ they are self-adjusting
- ✓ construction is simpler
- ✓ servicing is easier
- ✓ there is equal braking during forward and backward motion of the vehicle
- ✓ they are lightweight.

The disadvantages of disc brakes are:

- ✗ there is no self-servo action
- ✗ if used on all wheels, the handbrake mechanism becomes complicated.

Types of disc

The brake disc is usually manufactured from grey cast iron, although some high-performance vehicles can have carbon-fibre-based or ceramic-based discs. The disc is bolted to the wheel hub and therefore rotates at the same speed as the road wheel. The brake disc friction surface receives a good flow of air as it is exposed to the atmosphere. This is important as the brakes need to be kept cool. The brake disc is either of a solid or ventilated type.

Solid discs

A solid disc (see Figure 2.92) has two frictional surfaces. When the brakes are applied excessively, the centre of the disc material can overheat and lead to brake fade and warping of the brake disc. Solid discs are used on the front of smaller vehicles. It is becoming more and more common to find discs fitted to the rear of light vehicles and these would normally be solid discs.

A small set of brake shoes, fitted to an internal drum, are used to provide the parking brake in some rear disc brake assemblies.

Figure 2.92 Solid discs

Ventilated discs

A ventilated disc (see Figure 2.93) has two frictional surfaces, which are separated by an air space. The space allows for a flow of cooling air to reduce the effects of overheating. Ventilated discs are used on medium and large cars, where more braking force is required because of the greater weight of the vehicle. Other means of removing the heat on high performance vehicles is by drilling small holes in the disc and/or cutting grooves in the disc surface. These are known as 'cross drilling' and 'cross cutting' respectively.

Brake pads

A bonded friction lining is fitted to the brake pad. When the brakes are applied, the lining is forced against the disc by the caliper. The friction

Figure 2.93 Ventilated discs

lining material has to be hard-wearing and heat-resistant. All brake pad friction material today is asbestos-free because of the health risks involved in using asbestos.

Asbestos-free materials are less able to disperse heat. The heat can raise the temperature of the brake fluid to dangerous levels. In some cases, the brake fluid can boil, resulting in loss of braking. Underlays fitted between the disc pad and the piston are used to act as a heat barrier from the caliper. Slots formed in the lining assist airflow and improve flexibility for **bedding in**, to further reduce this heat.

The **coefficient of friction** for non-asbestos materials is 0.43 compared to 0.3 for asbestos types. This higher coefficient means greater braking efficiency. A downside to this is that high friction results in brake squeal, which is caused by vibration between the pad and piston. To overcome this problem the back of the pad is coated with a copper compound or rubber seal.

Inspecting, removing and replacing brake pads

See Chapter 7, pages 456–457, for how to inspect, remove and replace brake pads.

Types of caliper

There are two main types of caliper assembly: fixed caliper, floating caliper and handbrake linkage layouts.

Fixed caliper

The caliper is bolted to the hub and it straddles the disc. In each side of the caliper there is either one piston or a pair of pistons (see Figure 2.95). Fluid pressure forces the pistons out to press the pads against the disc. When the pressure is released, the piston seals flex and return to their normal shape. This pulls the piston and pads clear of the disc, maintaining the correct pad to disc clearance.

The advantages of the fixed caliper are:

- ✓ most moving parts are lubricated by brake fluid
- ✓ pad location is simple
- ✓ four-piston dual-circuit assemblies are easier to arrange.

The disadvantages of the fixed caliper are:

- ✗ it requires more room inside the road wheel
- ✗ there is reduced airflow to the outboard cylinder
- ✗ complicated handbrake mechanism if used on rear wheels.

Figure 2.94 Brake pads

> **Safe working**
>
> Even though brake pads are no longer manufactured using asbestos, the brake assembly still has a lot of dust that you must not breathe in. Make sure you wear a dust mask when working on the braking assembly.

> **Key terms**
>
> **Bedding in** – the term given to wearing in new brake linings.
>
> **Coefficient of friction** – a value between 0 and 1. The higher the coefficient value the greater the amount of friction produced, resulting in more efficient braking.

> **Safe working**
>
> When new brake linings are fitted, the friction surfaces do not have full contact until the linings have worn slightly. For this reason, when fitting new linings, always advise the customer to drive steadily for the first few hundred miles to enable the new brake linings to 'bed in'.

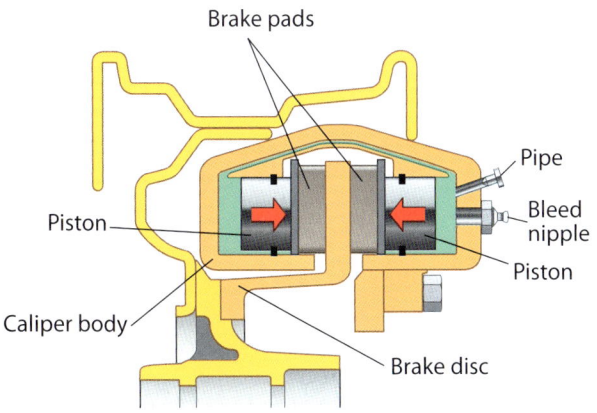

Figure 2.95 Fixed caliper

Level 2 Light Vehicle Maintenance & Repair

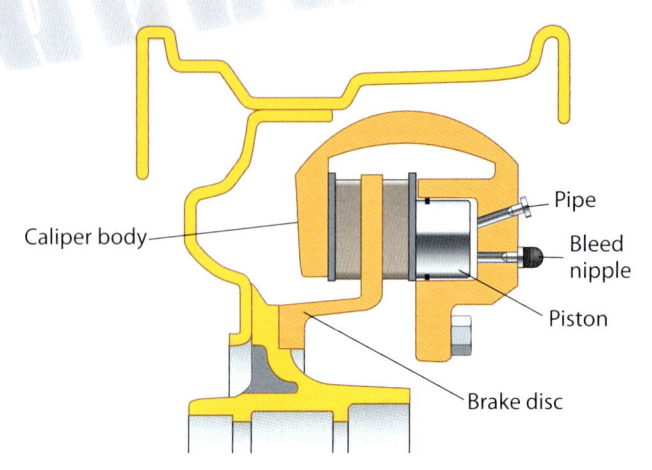

Figure 2.96 Floating caliper

Floating caliper

Fluid pressure forces an inner piston and pad towards the disc. As soon as the pad meets the disc, the build up of pressure inside the cylinder causes the caliper to react and move across the disc. This pulls the other pad into contact with the disc.

The advantages of the floating caliper are:

✓ good airflow to cool the disc and pad
✓ less space required so thinner wheels can be used
✓ easy to accommodate.

The disadvantages of the floating caliper are:

✗ seizure of caliper sliding mounting, resulting in loss of brake to one side or brake sticking on.

Handbrake linkage layouts

The handbrake usually operates on the rear wheels. The handbrake lever operates either a cable or rod linkage – this links together the various components which operate the brake shoes or pads. It is normal these days for light vehicles to use a cable linkage.

The handbrake cables are installed so that effort applied at the lever is distributed evenly to both wheels. A number of layouts have been designed, but the layout shown in Figure 2.97 is the most commonly used today.

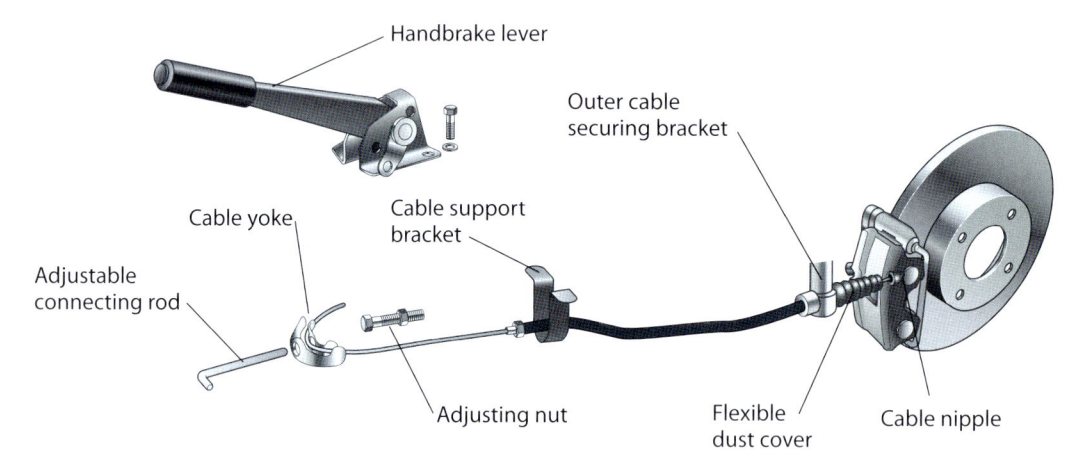

Figure 2.97 Handbrake linkage layout

Floating caliper/handbrake arrangement

This is similar in construction to the floating caliper. When the handbrake is applied, a cable pulls a lever attached to the caliper. This mechanically applies the pads.

You will have to use the manufacturer's technical data to replace the pads, as they require special tooling to wind the adjusting mechanism back as the pads wear.

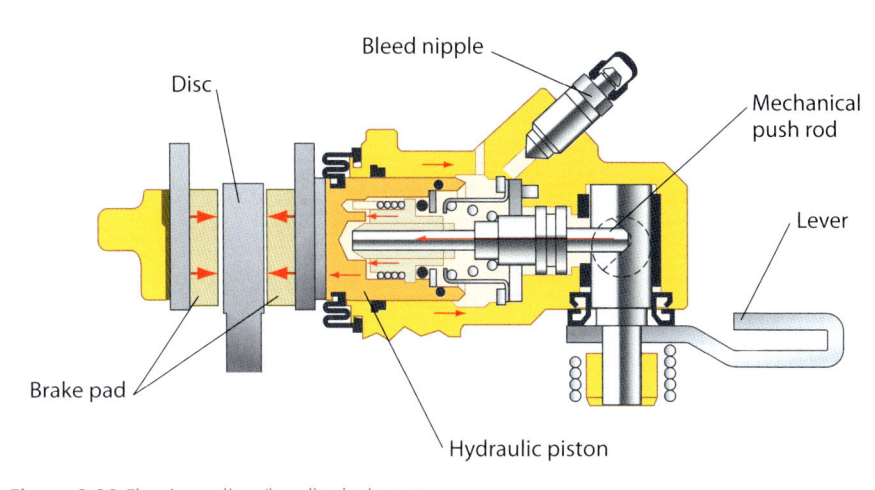

Figure 2.98 Floating caliper/handbrake layout

Electronic parking brakes (EPB)

The electronic parking brake (EPB) uses electrical components to apply and release the brakes when the vehicle is stationary. There are two main types:

1. The cable puller type uses an electrical actuator to pull on the brake cable.
2. The caliper type uses an electrical actuator situated in the rear brake calipers.

'Cable puller' type system

This type of system (see Figure 2.99) can be added to a conventional handbrake system. An electric motor is used to pull on the rear cables through a sleeve, which is shaped in the form of a screw thread. As the switch is pressed, the motor turns and the thread in effect shortens the rear cables, pulling them inwards and applying the brakes. An ECU is fitted inside the unit to sense the amount of movement and tension of the cable, to make sure the correct tension is applied.

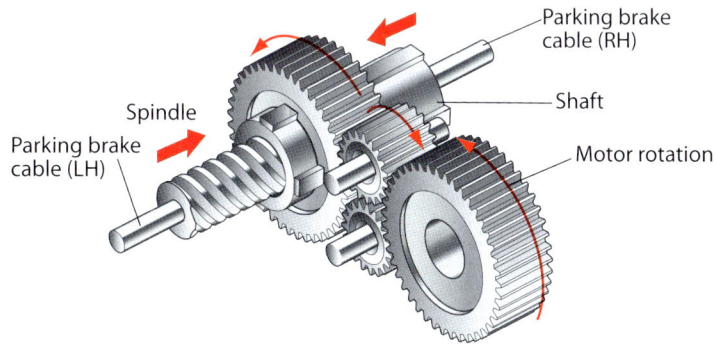

Figure 2.99 Cable puller type EPB system

Caliper type system

In each rear brake caliper is a piston that transmits hydraulic force from the brake fluid to mechanical force, pushing the piston on to the brake pads to apply the brakes. In the electronic handbrake caliper, a mechanical spindle within the piston pushes the piston against the pad. This is operated by an electrical switch when the handbrake is applied. This circuit also incorporates an ECU to prevent actuation when the vehicle is travelling at speed.

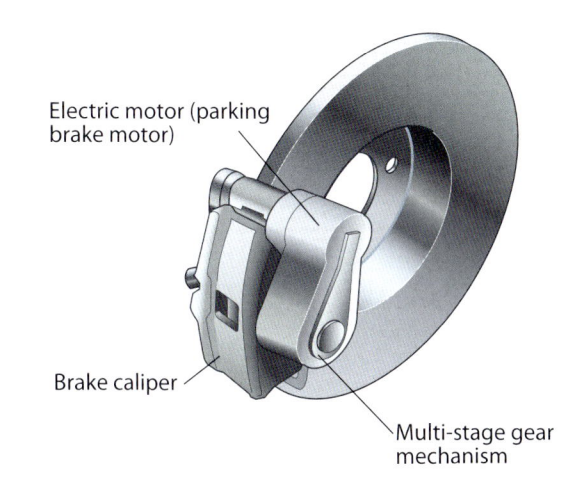

Figure 2.100 Caliper type EPB system

Level 2 Light Vehicle Maintenance & Repair

The construction and operation of the hydraulic braking system

The hydraulic braking system consists of:

- the driver's pedal
- linkages to a master cylinder
- fluid and hydraulic pipes that link the master cylinder to a hydraulic actuator at each wheel.

There are many different layouts of the pipes and this will determine how the master cylinder is used.

Single acting-type master cylinder

This type of cylinder is rarely used today for braking purposes. It is operated by the driver's pedal, pushing fluid by means of a sealed piston through a cylinder. The initial movement closes off the reservoir and the fluid is then forced under pressure through the pipes to the brake actuators. When the pedal is released, a spring inside the master cylinder returns the rod to its original position.

> **Did you know?**
> A hydraulic actuator can be a wheel cylinder or brake caliper. It uses fluid pressure to move the friction surfaces together.

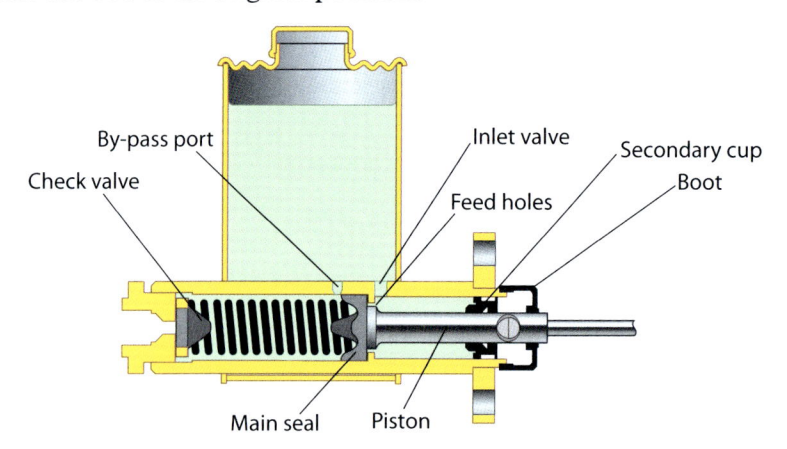

Figure 2.101 Single acting-type master cylinder

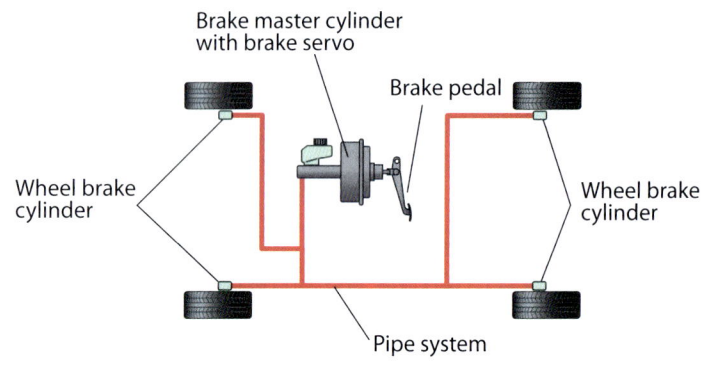

Figure 2.102 Single line braking system

Single line braking system

On actuation by the driver, a single acting-type master cylinder forces fluid through the pipes to each hydraulic actuator and applies the brakes at each wheel. As there is only one outlet from the master cylinder, there can only be one single line that supplies fluid pressure to all brake actuators. Flexible pipes are used in the system allowing for the chassis to suspension/steering movement.

Tandem master cylinders

The disadvantage with the single acting master cylinder is that if a pipeline leaks there will be a complete loss of braking pressure. For a long time, vehicles have been fitted with tandem master cylinders that use two pistons rather than one. One piston is connected to the

2 Light vehicle chassis units & components

footbrake pedal through a series of linkages and the other is moved by fluid pressure from the first piston. Each piston is fitted with a pair of seals and its own outlet. In the event of failure of one line, 50 per cent of the total braking pressure will still be available.

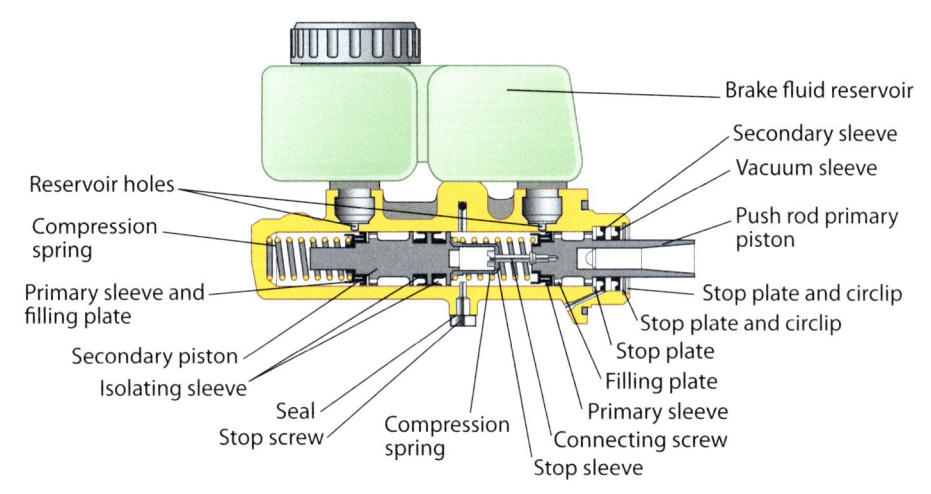

Figure 2.103 Tandem master cylinders

Other duties of the tandem master brake cylinder are:

- it produces brake circuit pressure
- it rapidly reduces the pressure when the brakes are released
- it ensures the correct amount of brake fluid is in the piston chambers if there are changes in temperature.

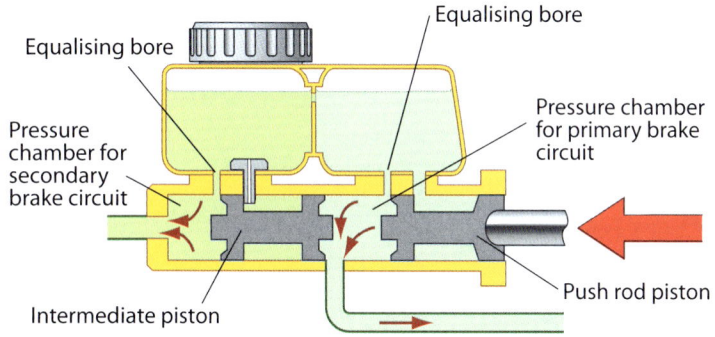

Figure 2.104 Actuated tandem master cylinder

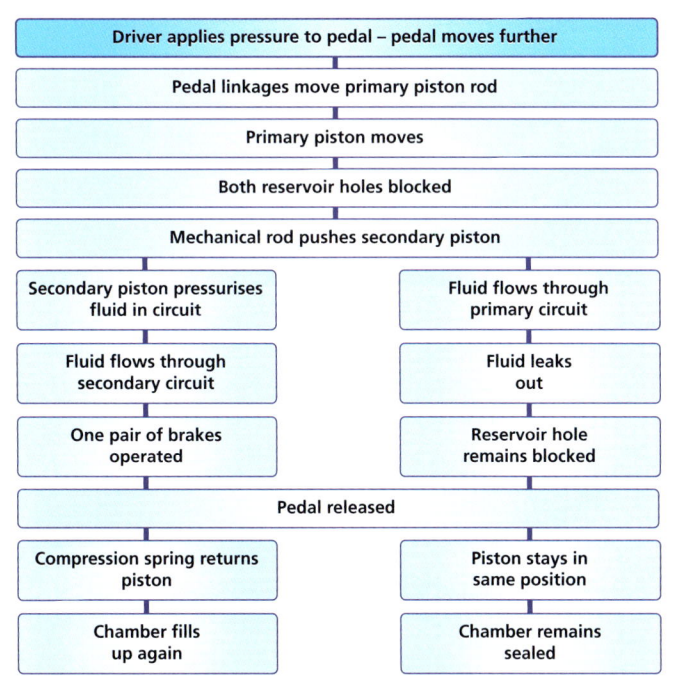

```
Driver applies pressure to pedal – pedal moves further
          ↓
Pedal linkages move primary piston rod
          ↓
Primary piston moves
          ↓
Both reservoir holes blocked
          ↓
Mechanical rod pushes secondary piston
       ↓                    ↓
Secondary piston        Fluid flows through
pressurises fluid       primary circuit
in circuit                  ↓
   ↓                    Fluid leaks out
Fluid flows through         ↓
secondary circuit       Reservoir hole
   ↓                    remains blocked
One pair of brakes
operated
          ↓
Pedal released
       ↓                    ↓
Compression spring      Piston stays in
returns piston          same position
   ↓                        ↓
Chamber fills           Chamber remains
up again                sealed
```

Figure 2.105 Tandem master cylinder components leak on primary chamber brake line

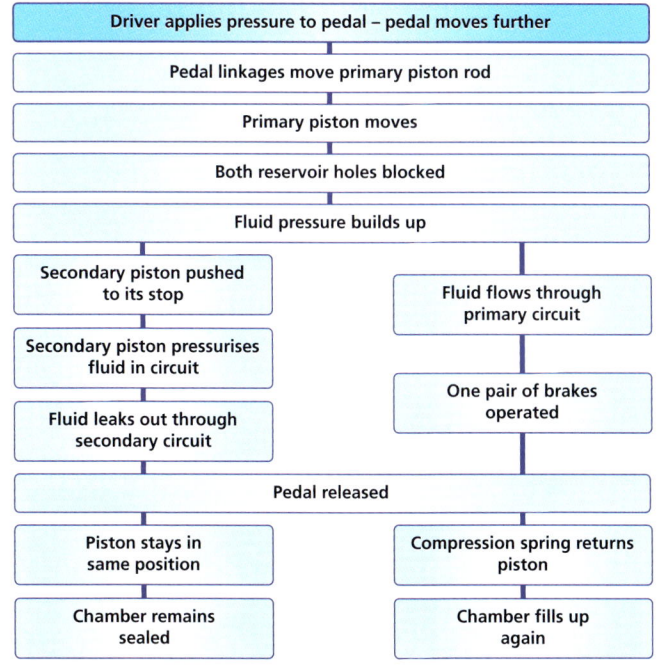

```
Driver applies pressure to pedal – pedal moves further
          ↓
Pedal linkages move primary piston rod
          ↓
Primary piston moves
          ↓
Both reservoir holes blocked
          ↓
Fluid pressure builds up
       ↓                    ↓
Secondary piston        Fluid flows through
pushed to its stop      primary circuit
   ↓                        ↓
Secondary piston        One pair of brakes
pressurises fluid       operated
in circuit
   ↓
Fluid leaks out through
secondary circuit
          ↓
Pedal released
       ↓                    ↓
Piston stays in         Compression spring
same position           returns piston
   ↓                        ↓
Chamber remains         Chamber fills
sealed                  up again
```

Figure 2.106 Tandem master cylinder components leak on secondary chamber brake line

173

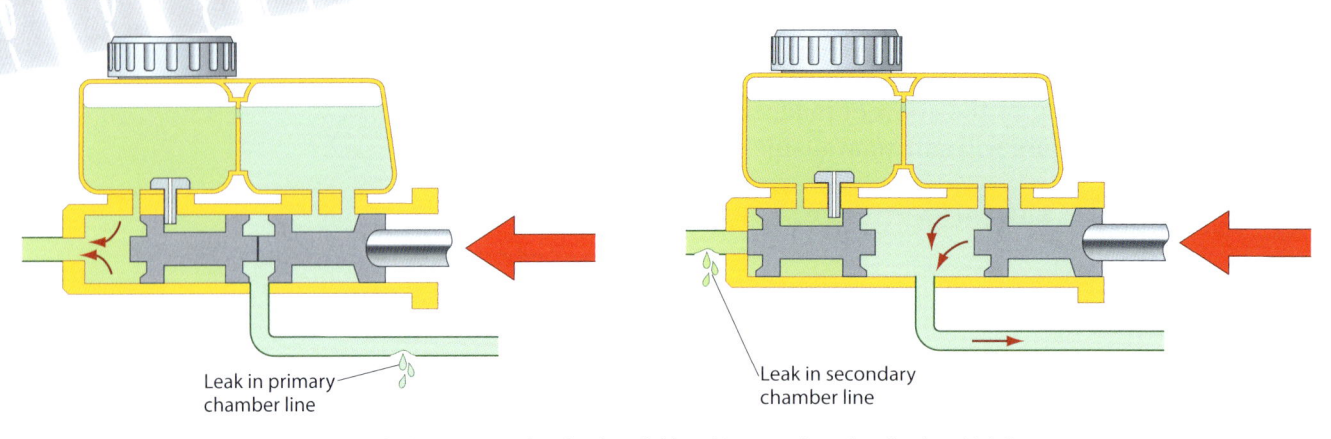

Level 2 Light Vehicle Maintenance & Repair

Leak in primary chamber line

Leak in secondary chamber line

Figure 2.107 Actuated master cylinder – leak in primary chamber line (left) and in secondary chamber line (right)

Replacement of master cylinder

Where there are problems with the master cylinder, it will need to be replaced.

Replacing the master cylinder

Checklist				
PPE	VPE	Tools and equipment	Consumables	Source information
• Steel toe-capped boots • Overalls • Latex gloves	• Wing covers • Steering wheel cover • Seat covers • Floor mat covers • Brake pipe plugs	• Socket set • Pipe spanner • Brake bleeding equipment • Fluid-proof vessel	• Brake fluid	• Brake bleeding procedure • Reservoir fluid capacity • Job card

1. Fit VPE and lift the bonnet.

2. Place a fluid-proof vessel under the master cylinder to catch any brake fluid.

3. Remove the brake fluid level indicator wire from the reservoir top.

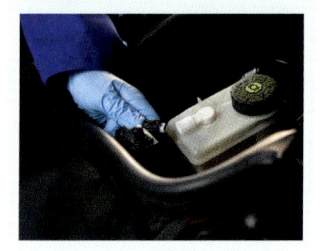

4. If applicable, remove the clutch fluid feed pipe and plug the pipe.

5. Remove each brake pipe with the pipe spanner and plug the pipe and master cylinder with the pipe plugs.

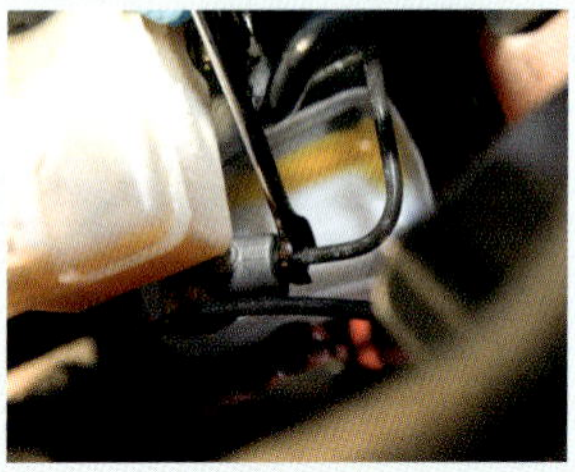

6. Remove the nuts holding the master cylinder to the servo and remove the master cylinder. Take care not to spill fluid on the vehicle paintwork.

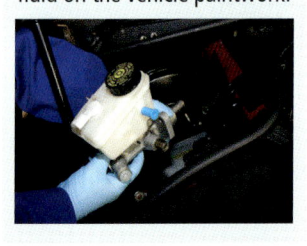

7. Replace with the new master cylinder, pipes and all fixings in the reverse procedure.

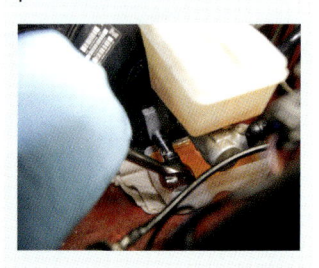

8. Fill the fluid reservoir up to the maximum and bleed the brakes as described on page 180.

2 Light vehicle chassis units & components

Split line systems

The single line system is unsafe in the event of a leak. Therefore, a split line system in conjunction with a tandem master cylinder provides safer braking in the event of leakage in one of the circuits. In a split line system, the brakes can be connected in various ways. Figure 2.108 shows front axle to rear axle split and diagonal split.

The diagonal system is safer than the front to rear system. This is because of the amount of work the front wheel brakes have to do due to weight transfer. A diagonal split system with twin piston calipers is one of the safest systems available on light vehicles. If there is leakage of one circuit, three out of the four brakes will still work.

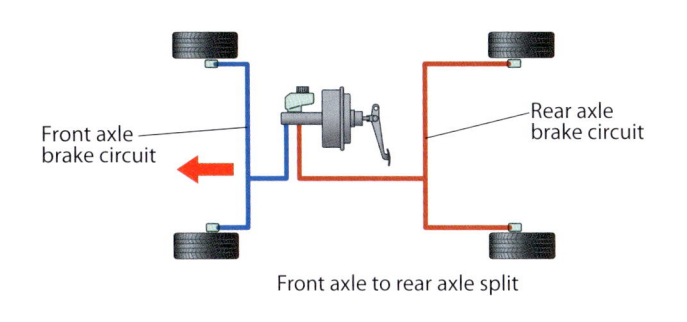

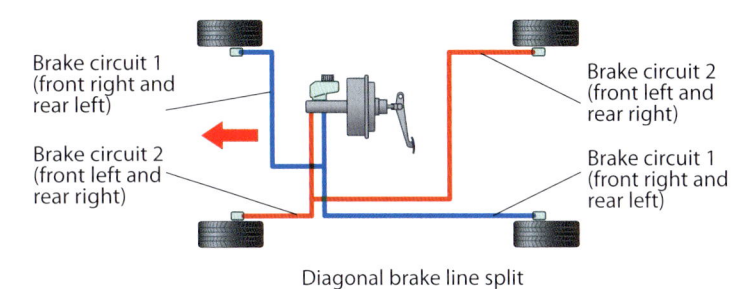

Figure 2.108 Split line system

The brake servo

The brake pedal effort applied by a driver is assisted by using a brake servo. This consists of two sealed chambers separated by a diaphragm. The diaphragm is connected to the rod which moves the master cylinder from the driver's pedal.

When the engine is running but the brakes are not being used, both chambers are at the same pressure, which is a vacuum. When the brakes are applied, atmospheric pressure enters the same side of the servo as where the brake pedal is situated. As this pressure is higher than the pressure of a vacuum, the diaphragm assists the pedal movement towards the master cylinder – this reduces the braking effort that the driver has to apply.

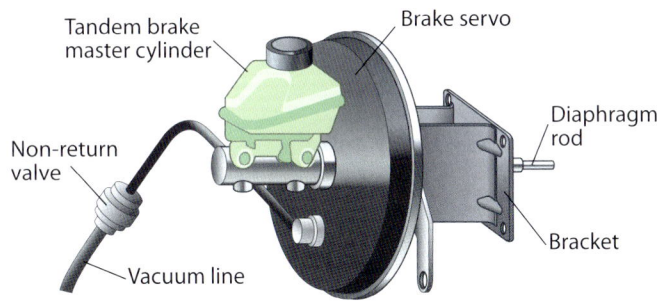

Figure 2.109 Brake servo

The brake servo must also be fail-safe in the event of the servo failure.

Vacuum assistance

For the brake servo to work correctly a vacuum needs to be generated. With a spark ignition type of engine, the 'vacuum' assistance is provided by a depression formed in the inlet manifold.

On a compression ignition type of engine, there is no depression formed in the inlet manifold and so an engine-driven 'vacuum' pump is used to provide the required assistance. This pump, or exhauster, produces a vacuum of about 500–600 mm of mercury.

Direct-acting type servo

With the direct-acting type of servo, the pedal linkage directly connects the master cylinder to the servo.

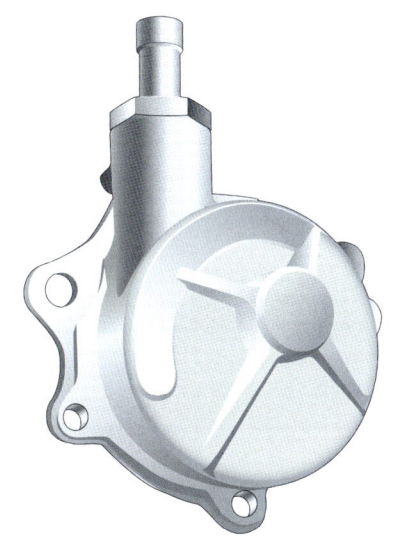

Figure 2.110 Vacuum assistance

175

Level 2 Light Vehicle Maintenance & Repair

Look at Figure 2.110 to understand how the direct-acting type of servo works:

- **Brake off** – Both sides of the diaphragm are exposed to the vacuum thorough the valve. The diaphragm is kept in its original place by the spring.
- **Brake on – pedal depressed** – The vacuum passage is closed and the exterior passage is opened. Atmospheric pressure enters one side of the diaphragm through this exterior passage. This creates a difference in pressure and this assists driver effort.

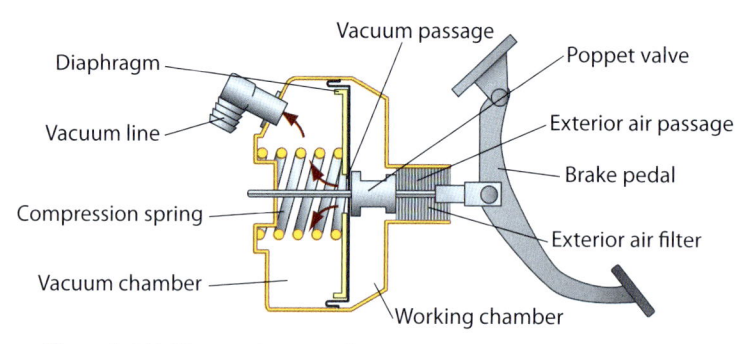

Figure 2.111 Direct-acting type of servo

Brake pressure regulating valves

The shortest emergency stopping distance can only be achieved when all four wheels of a vehicle are just at the point of locking. Weight transfer during braking will prevent this from happening as more braking force will be carried out by the front wheels and, therefore, the rear wheels will skid. This means the pressure to the rear wheels needs to be reduced progressively during heavy braking.

Under normal braking conditions, the pressure limiting valve allows the free passage of fluid through light pressure to the rear brakes. Under heavier braking conditions, if the fluid pressure exceeds a predetermined limit, it pushes back the return spring loading on the valve plunger. The plunger moves forwards and seals off the two outlets to the rear brakes. Any further increase in fluid pressure from the master cylinder is not transmitted to the rear brakes but to the front brakes where more braking force is required.

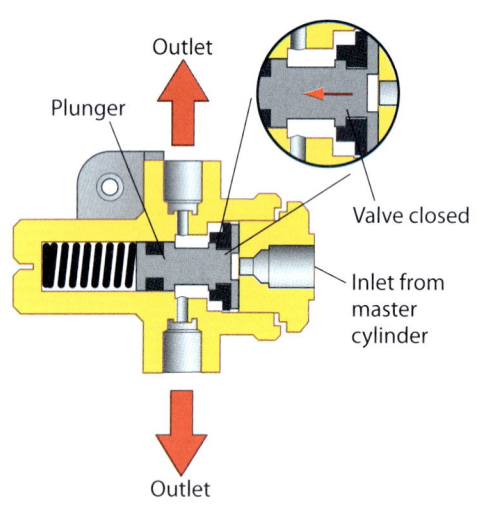

Figure 2.112 Pressure limiting valve

Inertia valve

An **inertia**-type valve is found in the rear brake lines. It senses the vehicle's deceleration under different braking forces and reduces the hydraulic fluid flowing through the rear brake lines in proportion to the amount of 'pitching' of the vehicle.

Under severe braking, a ball valve temporarily shuts off the rear line so that any extra force only increases the front brake pressure. If the pressure remains constant, the ball valve opens slightly to allow a reduced amount of fluid to the rear brakes.

> **Key term**
>
> **Inertia** – the resistance of an object to change its motion.

The unit is mounted on the vehicle body at a particular angle so the ball valve will be facing forwards. Under light braking, fluid passes around the ball and through a piston to the rear brakes. As the braking pressure increases, the force on the piston increases and this closes off fluid pressure to the rear brakes.

Load-proportioning valve

The load-proportioning valve supplies hydraulic pressure to the rear brakes in proportion to the loading on the rear wheels. This provides good braking when the rear wheels are heavily loaded and reduces the risk of a rear-wheel skid when the rear of the vehicle is lightly loaded.

The valve housing is bolted to the vehicle body. Either a tension or compression spring is used to sense loading on the rear wheels. The spring connects a lever that operates a ball valve to a part of the suspension system that moves in when load is applied to the vehicle. The spring also adds a delay, so the valve is only affected by load and not as the suspension moves up and down over bumps.

- **Brakes off** – When the brakes are off the piston is at the bottom of the bore and the ball valve is held open by a pushrod fixed to the valve body. Fluid passes freely between the inlet and outlet ports.
- **Brakes applied** – When hydraulic pressure is applied to the valve, the piston moves upwards. The amount of piston movement depends on how much load is on the vehicle. With a heavy load, the piston movement will be limited, and a higher brake fluid pressure is needed to lift the piston and allow fluid to pass to the rear brakes.

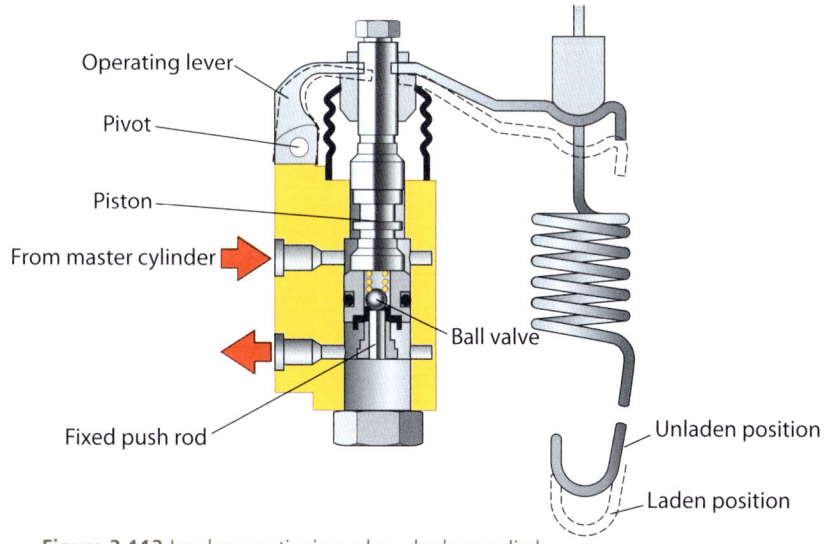

Figure 2.113 Load-proportioning valve – brakes applied

Braking efficiency

Vehicle braking efficiency is how well the brakes work, compared to the weight of the car, driver and load, shown as a percentage.

The Road Vehicle (Construction and Use) Regulations 1986 require that:

- The primary braking system must give a retarding (slowing down) force of more than 50 per cent of the vehicle weight.
- The secondary braking system must produce a retarding force of more than 25 per cent of the vehicle weight. This is also regarded as the emergency brake.
- All vehicles must have a parking brake system that will give 16 per cent of vehicle holding force.

> **Did you know?**
>
> The Road Vehicle (Construction and Use) Regulations are the guidelines all vehicle manufacturers have to follow when designing a vehicle for use on the road. They also govern the vehicle condition and safety during use by the driver and passengers.

Level 2 Light Vehicle Maintenance & Repair

Most light vehicles today use split line braking systems which include a tandem master cylinder. This means that the footbrake is both the primary and secondary braking system.

The most widely used method to calculate braking efficiency is to compare the braking force to the weight of the vehicle. For example, if a vehicle weighs 1800 kg and the total braking force is 1200 kg, then the brake efficiency is 67 per cent.

$$\frac{\text{Braking force} \times 100}{\text{Weight of vehicle (kg)}} = \frac{1200 \times 100}{1800} = 67\%$$

> **Did you know?**
> The secondary brake on a split line system is controlled by the footbrake but is not testable on its own.

> **Safe working**
> Brake efficiency testing is mainly carried out on a rolling road. You must ensure the rollers are covered when this is not in use, or serious leg injuries may result from being sandwiched between the rollers.

Brake testing

When carrying out a brake roller assessment, total braking efficiency is not the only test you need to do. You also need to test for imbalance. This can occur if one side has a higher efficiency than the other. It can cause the vehicle to pull violently to one side.

You can also test for warped drums or discs and binding brakes. Where discs/drums are warped, the light application of the brakes will show up as a fluctuating reading on the tester. Binding brakes will show up where there is a reading on the tester without the brakes being applied.

Table 2.9 provides a summary of brake testing. It shows the minimum efficiencies required by law for the testing of different types of braking system.

Table 2.9 Summary of brake testing

Type of brake	Efficiency	Single line system	Dual/split line system
Primary brake	50% minimum	Footbrake	Footbrake
Secondary brake	25% minimum	Handbrake	Footbrake
Parking brake	16% minimum	Handbrake	Handbrake

Example of brake and associated components checks

Table 2.10 lists the service requirements for the different braking system components. Manufacturers will determine the exact time or mileage recommendations for the checks below and list these in the vehicle handbook. Always refer to the vehicle handbook and service schedules for accurate information on when these checks must be carried out.

Figure 2.114 Rolling road brake tester

Table 2.10 Service requirements for braking system components

Component	Service requirement
Brake fluid	Top up as required. It is generally recommended that brake fluid should be replaced every two years (depending on the manufacturer). If the level is very low or repeated topping up is required, investigate why (such as a possible leak or brake linings may be worn).
Brake pipes	Condition, security, corrosion
Brake flexi-pipes	Condition, security, cracking, bulges
Brake drums and discs	Warping, thickness, scoring, excessive heat (brakes binding), runout, ovality
Brake linings	Shoe and pad thickness, adjustment
Handbrake	Operation, security, adjustment (including linkage and cables)
Master cylinder	Condition, security, operation, leaks
Brake servo	Condition, security, operation, air leaks
Warning lights and systems	Operation

When working on braking systems you must observe the following.

- Correctly chock, lift and support the vehicle.
- Use equipment safely.
- Follow the manufacturer's procedures.
- Keep your work area clean and tidy.
- Prevent brake fluid coming into contact with paintwork.
- Do not mix different types of brake fluids.
- Observe the rules regarding the removal and disposal of waste materials.
- Remember, the brakes can get hot.

Brake fluid

You must check the brake fluid level every time a vehicle comes in for inspection or for specific brake work. Always look into any signs of loss of brake fluid. Examine pipelines for leaks, giving special attention to the flexible hoses. If there are any signs of cracking, chafing or swelling, the hose must be replaced.

Brake fluid properties

- Brake fluid is an organic oil with certain additives or it can be synthetic or man made.
- The fluid has a low freezing point, a high boiling point and maintains a constant **viscosity** across a wide temperature range.
- It is a lubricant and must resist corrosion.
- It must be compatible with rubber to prevent the rubber seals and pipes in the system from cracking. This is why organic oil is used as the main ingredient.
- Some brake fluids are **hygroscopic**.

> **Key terms**
>
> **Viscosity** – the thickness of a fluid or its ability to flow. A thinner fluid is said to have a low viscosity.
>
> **Hygroscopic** – the fluid will absorb water from the air over a period of time. The water can dilute the fluid and reduce the boiling point. This may cause brake failure. For this reason, manufacturers state that brake fluid must be changed on a regular basis (every two years).

Level 2 Light Vehicle Maintenance & Repair

Did you know?
DOT 3 and DOT 4 are mineral-based brake fluid types and DOT 5 is synthetic.

Safe working
This method must only be used to remove small amounts of air from the braking system following a brief opening of the hydraulic system to the atmosphere. For a full fluid change, a vacuum pump is attached to each brake caliper and the suction pulls air and oil fluid through the pipes when the bleed nipples are opened. Care has to be taken to make sure the master cylinder is topped up regularly.

DOT rating

Containers of brake fluid indicate a DOT rating. This refers to the properties of the brake fluid in relation to its boiling point. You should never mix brake fluids of different DOT ratings together.

The DOT ratings of brake fluid are:

- **DOT 3** – Glycol-type brake fluid with a minimum dry ERBP (equilibrium reflux boiling point) of 205 °C and a minimum wet (means the fluid is contaminated with water) ERBP of 140 °C.
- **DOT 4** – Glycol-type brake fluid with a minimum dry ERBP of 230 °C and a minimum wet ERBP of 180 °C.
- **DOT 5** – Silicone-based brake fluid with a minimum dry ERBP of 260 °C and a minimum wet ERBP of 180 °C.

Brake bleeding

If air gets into the hydraulic system, pressure from the master cylinder will not be transmitted to the hydraulic wheel actuators. The reason for this is that air may be compressed and under light compression it will not transfer force. An indication of air in the system is a very spongy or easily pressed brake pedal.

You must bleed the braking system if any component within the hydraulic line is changed or when changing the brake fluid during a service. (See Figure 7.44 on page 461.)

Bleeding the braking system (always follow manufacturer's instructions)

1. Fill the brake reservoir.	2. Attach one end of a rubber tube to the brake bleed nipple.	3. Put the other end of the tube in a jar containing fluid.	4. Ask a colleague to press the brake pedal and then you should open the bleed nipple slowly.
5. Close the bleed nipple when the brake pedal is down and ask your colleague to release the pedal.	6. Continue to follow this procedure until there is no air coming through the pipe into the jar.	7. Top up the brake fluid level and replace the cap.	8. Press the pedal until it goes hard.

Safe working
You must refer to the manufacturer's service procedures when bleeding braking systems.

Electronic anti-lock braking systems

If the brake pedal is pressed very hard in a conventional braking system, the wheels can lock before the vehicle comes to a stop. If this happens, the vehicle becomes unstable and difficult to steer.

The anti-lock braking system (ABS) controls the pressure of the brake fluid applied to each hydraulic actuator. This ensures conventional operation during normal braking conditions and prevents the wheels from locking up during hard braking or in adverse weather conditions.

2 Light vehicle chassis units & components

Features of ABS systems

- Vehicle stability can be achieved when braking in all driving conditions.
- The driver can steer the vehicle in an emergency braking situation.
- When the ABS system is in operation, a vibration is felt on the brake pedal.
- ABS systems are fail-safe (see *Emergency* page 136).
- Drivers with lower ability can control the vehicle better under difficult conditions.
- ABS systems work in conjunction with electronic traction control systems – these control the power transmitted to the wheels during vehicle acceleration. This will reduce wheel spin in icy conditions.
- The system is self-diagnosing. If a system fault is detected, the ABS indicator light situated in the instrument panel comes on. The ECU will in this case store a fault code.

> **Safe working**
> ABS systems do not reduce braking distances. In snowy conditions, ABS can actually increase the braking distance.

ABS modulator unit

The ABS modulator unit contains a number of valves operated by electrical switches. When anti-lock braking is required the switches are turned on and off by the ECU. The valves control the hydraulic pressure to each wheel and so the ECU determines when the brakes are applied and released.

Figure 2.115 ABS modulator schematic

Servo/Master cylinder

A combined master cylinder and servo assembly attached to the pedal box actuates the system in response to driver effort, as in a conventional system.

Pressure reducing valve

The pressure reducing valve (PRV) regulates pressure to the rear axle to maintain braking balance. In the event of a front circuit failure, the failure bypass PRV allows full system pressure to the rear axle.

The PRV is only used in early vehicles as more modern vehicles use electronic brake distribution (EBD). This controls the pressure to the rear brakes to prevent premature rear brake lock-up and so improves stability of the vehicle under harsh braking.

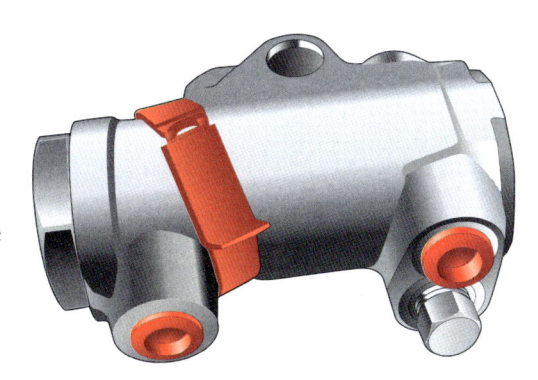

Figure 2.116 Pressure reducing valve

181

Level 2 Light Vehicle Maintenance & Repair

Electronic control unit

The electronic control unit (ECU) controls the valves by assessing the information it receives from the wheel speed sensors. In modern vehicles the ECU and modulator are now combined in one compact unit.

Wheel speed sensors and exciter rings

A sensor and toothed exciter ring are located at each wheel (see Figure 2.118). The sensor is fixed and attached to the hub, while the exciter ring rotates and is attached to the driveshaft. When the vehicle is in motion, the inductive-type sensor sends signals to the ECU to sense the speed of the wheel during rotation or lock.

Figure 2.117 Electronic control unit (ECU) combined with a modulator unit

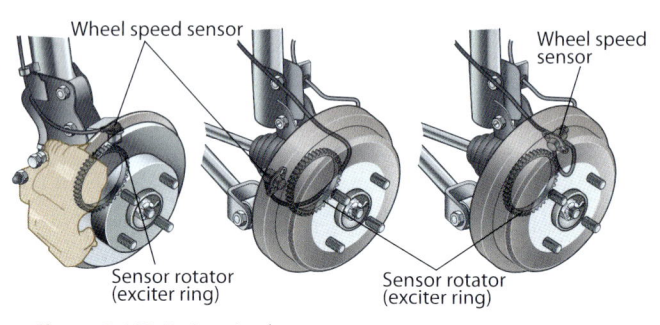

Figure 2.118 Exciter ring layout

Did you know?

The inductive-type sensor generates an electrical output signal from the breakdown of a magnetic field as each exciter tooth passes the sensor. This is in the form of an alternating current (AC) signal.

Key points

- The gap between the sensor and the exciter ring is preset and should not be changed.
- Any dirt or metallic particles attracted to the casing due to the magnetic field should be removed, to prevent any incorrect calculations occurring.
- Wheel bearings must also be in good working order, as this may affect the sensor alignment with the exciter ring.
- The voltage from the wheel speed sensor increases as the wheel speed increases.

Diagnosis

ABS is self-diagnosing. On a vehicle with ABS, faults are indicated by an ABS warning light situated in the dashboard cluster. If there is a fault the light goes on, and the fault code will be stored.

Fault codes can only be read through the use of diagnostic equipment by attaching the diagnostic connector to the vehicle. The codes are stored automatically in the Electronically Erasable Programmable Read Only Memory (EEPROM). You can only erase the fault codes by using designated diagnostic equipment. Disconnecting the vehicle's power supply will not erase fault codes from the electronic control unit (ECU) memory.

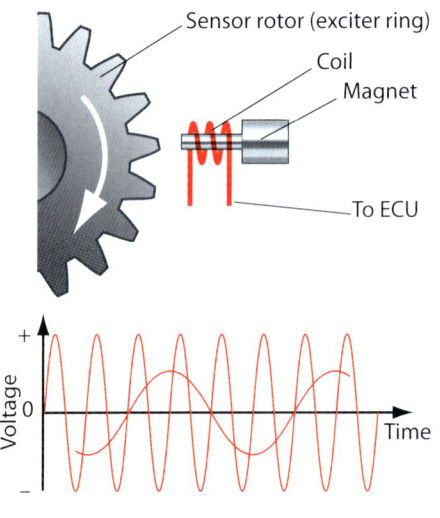

Figure 2.119 Exciter ring output

Operation

- **Normal braking mode** – During normal braking operation, all the solenoid valves remain open to allow the hydraulic fluid pressure flow from the master cylinder to the wheel actuators.
- **ABS braking mode** – When the system enters the ABS braking mode, there are four phases of operation, as shown opposite.

Key term

Traction – the friction between a body (vehicle) and the (road) surface on which it moves.

2 Light vehicle chassis units & components

1. Pressure increase (build)

Fluid is allowed to pass from the master cylinder to the modulator assembly. Fluid is then distributed to each of the inlet valves and to the respective hydraulic brake actuators. Normal braking pressure is achieved.

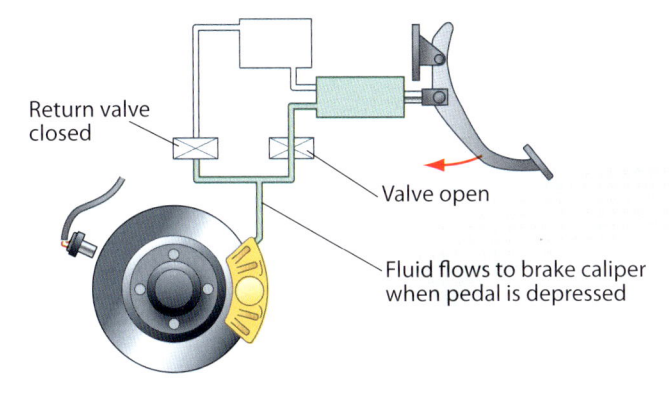

Figure 2.120 Pressure increase (build)

2. Pressure hold (maintained)

The ECU detects when the wheels are about to lock from the speed sensors. The braking effort still needs to be maintained. To achieve this, the input valve is energised, causing it to close. The outlet valve is de-energised causing this to close and fluid is now locked between both closed valves. An increase in braking pressure cannot act on the hydraulic actuator and therefore, only the initial braking pressure is maintained.

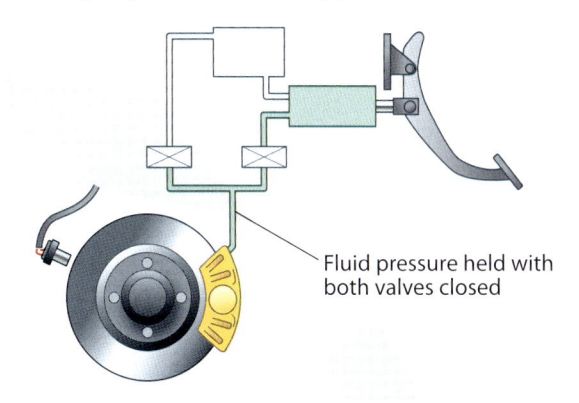

Figure 2.121 Pressure hold (maintained)

3. Pressure decrease

If the ECU calculates that a wheel is about to lock in the pressure-hold phase, then hydraulic fluid pressure is released. This is achieved by energising the outlet valve to open it and at the same time closing the inlet valve. At this point, the return pump is activated and the fluid is forced back against the pressure applied at the master cylinder.

At this stage the driver will feel a brake pedal vibration. Wheel lock is avoided, but if the vehicle is still in ABS mode the cycle will begin again. ABS operation can be at a rate of 10 times a second. This is why the driver will feel the vibration through the foot pedal.

4. Pressure increase

The vehicle starts to gain **traction** again following the pressure decrease phase. At this point, the signal from the sensor indicates that, by maintaining the pressure, the wheel speed is increasing and is not being sufficiently braked. The ECU will restore the pressure by returning to the pressure increase phase again. The cycle will continue until the vehicle has decelerated sufficiently or until the driver releases the brakes.

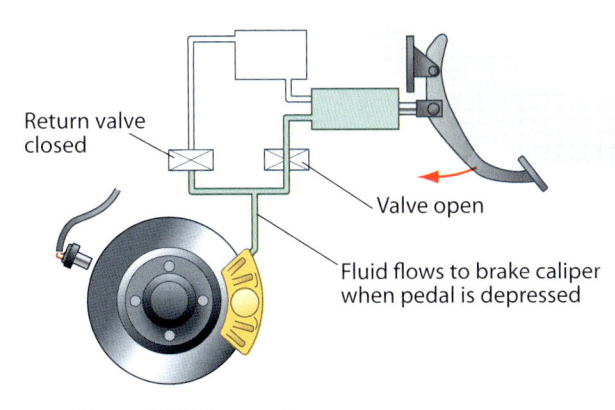

Figure 2.123 Pressure increase

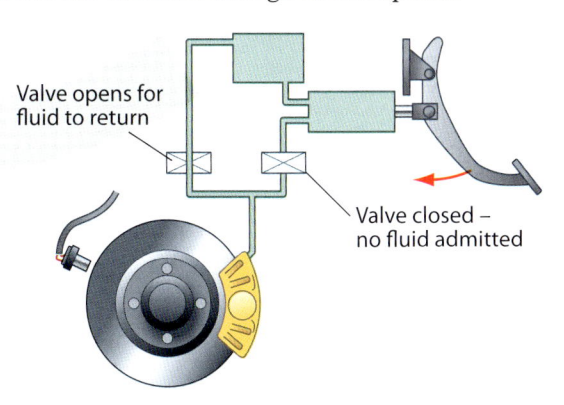

Figure 2.122 Pressure reduction

183

Level 2 Light Vehicle Maintenance & Repair

> **Did you know?**
>
> The next generation of ABS systems includes an automatic braking system, where sensors on the front of the vehicle determine how close the vehicle in front is. If the vehicle is too close, this system applies the brakes for the driver.

Active braking mode

If electronic traction control is required due to a wheel slip during acceleration, the ABS ECU determines that active braking is necessary. The ECU will operate the inlet and outlet solenoid valves to control the supply of hydraulic pressure to the individual brake actuators and slow the wheel(s).

Warning systems

A warning light (telltale) on the dashboard informs the driver about the status of the various electrical systems and components. Typical warning lights for the braking system are:

- handbrake on
- brake fluid low
- brake pads wearing low
- ABS fault.

For the exact meaning of the warning lights on a vehicle, refer to the vehicle handbook.

Figure 2.124 Dashboard warning system

Handbrake on
This consists of a small switch located just under the handbrake lever. As the lever is lifted, a circuit is closed to earth and this will illuminate a light on the dashboard.

ABS fault
The ABS warning light is situated in the dashboard cluster. It illuminates if a self-diagnosed fault occurs.

Brake fluid low
A float is situated within the master cylinder reservoir. As the level drops, two electrical contacts within the float close to earth and illuminate the light on the dashboard. This light may sometimes be combined with the handbrake warning light.

Brake pads wearing low
A wire attached to the brake pad is connected to a bulb within the dashboard cluster. As the pad wears to a predefined limit, normally 2–3 mm, the wire will earth on the brake disc through the vehicle and create a circuit to illuminate the light on the dashboard.

Brake lights

It is a legal requirement for brake lights to be in good working order, to warn other road users that a vehicle is slowing down or stopping. Light vehicles have a minimum of two brake lights. High-level brake lights are also common on today's vehicles. The brake lights must be red lights within the rear light cluster and come on when the brake pedal is depressed.

A brake light switch connected to the brake pedal operates the rear brake lights. The plunger-type switch makes or breaks an electrical connection during operation.

- **Brakes off** – The plunger switch used for operating brake lights can be pushed or released. Most brake light switches are released so no clearance exists when the pedal is not being pushed. So the brake lights are off.
- **Brakes on** – As the pedal is pressed, the plunger is released and the brake lights come on.

2 Light vehicle chassis units & components

Braking system faults

You need to be able to diagnose braking system faults and to know the causes of them and how to put them right.

Table 2.11 Braking system defects and how to correct them

Fault	Symptom	Cause	Remedy
Premature brake lining wear	• Pads/shoes need replacing frequently • Brakes sticking	• Worn drums/discs • Severe driving • Erratic driving	• Replace drum/discs • Rectify binding problem • Driver education
Unbalanced braking	• Vehicle pulls to one side when braking • Brake roller gauge needle moves on one side only	• Pad/shoe contamination • Cylinder/caliper seized	• Replace shoe/pad cause of contamination • Replace caliper/cylinder
Load related imbalance	• Rear wheels lock under normal braking	• Brake apportioning valve requires adjustment or is defective	• Adjust/replace apportioning valve
Judder/vibration, drag/grab	• Vibration during driving • Brake roller gauge needle moves up and down during light braking	• Disc warped (excessive run-out) • Oval drum	• Check run-out and ovality • Replace disc/drum
Brake fluid loss	• Fluid master cylinder needs topping up	• Leaking pipes or caliper/cylinder seals	• Replace pipe or seals
Spongy pedal	• Pedal feels spongy when pressed	• Air in the hydraulic system	• Inspect system for leaks • Bleed system or replace brake fluid
Excessive pedal/handbrake lever travel	• Pedal/handbrake lever moves further than usual when depressed/lifted	• Automatic adjusters seized • Hub end float excessive	• Adjust the brakes, clean and check operation of adjusters • Adjust hub end float
Squealing	• Brakes make squealing noise when brake pedal pressed	• Vibration between shoe and backplate/pad and caliper	• Clean and apply copper grease to shoe/backplate, pad/caliper • Fit anti-squeal shims
Brake fade and reduced efficiency	• Pedal feels hard • Loss of brake efficiency	• Overheating of brakes caused by severe, prolonged braking or binding brakes	• Allow brakes to cool; pads/shoes may need replacing
Total loss of braking	• Pedal fully depressed with little or no effect	• Fluid leak past seals or fractured pipe(s)	• Locate and rectify the fault
Brakes difficult to press	• Brake pedal hard	• Brake servo or vacuum pipe leaking	• Replace servo or vacuum pipe

Level 2 Light Vehicle Maintenance & Repair

CHECK YOUR PROGRESS

1. State the three purposes of the braking system.
2. What is the advantage of having a dual line braking system over a single line system?
3. What is meant by 'hygroscopic'?
4. What type of sensor is used to sense vehicle speed at each wheel?
5. Why is it important to keep the correct gap between the sensor and exciter ring in an ABS system?
6. State the four phases of ABS.

Light vehicle wheel and tyre systems

The requirements of a road wheel are that it must be:

- **Lightweight** – This will reduce the unsprung weight and the stress placed on the dampers. It will improve road holding. (For more on unsprung weight, see page 148.)
- **Rigid** – The wheel will resist deflection or bending during cornering.
- **Strong** – It will resist normal loads and possible accidental damage.

The wheels must be made so that the tyres do not come off the rim during cornering. They must also allow tyre removal and fitting without damage to the wheel or tyre.

To prevent 'brake fade', wheels must allow a flow of air over brake components.

Well-based rim

Light vehicle wheels usually have a well-based rim (see Figure 2.125). This makes fitting the tyre easier, as it allows one side of the tyre rim to be pushed into the well while the other side is fitted. The air pressure forces the **bead** to ride up the taper. This locks the tyre to the rim and provides an airtight seal.

Types of road wheel

There are four main types of road wheel:

1. cast alloy
2. pressed steel
3. wire (spoked)
4. detachable rims.

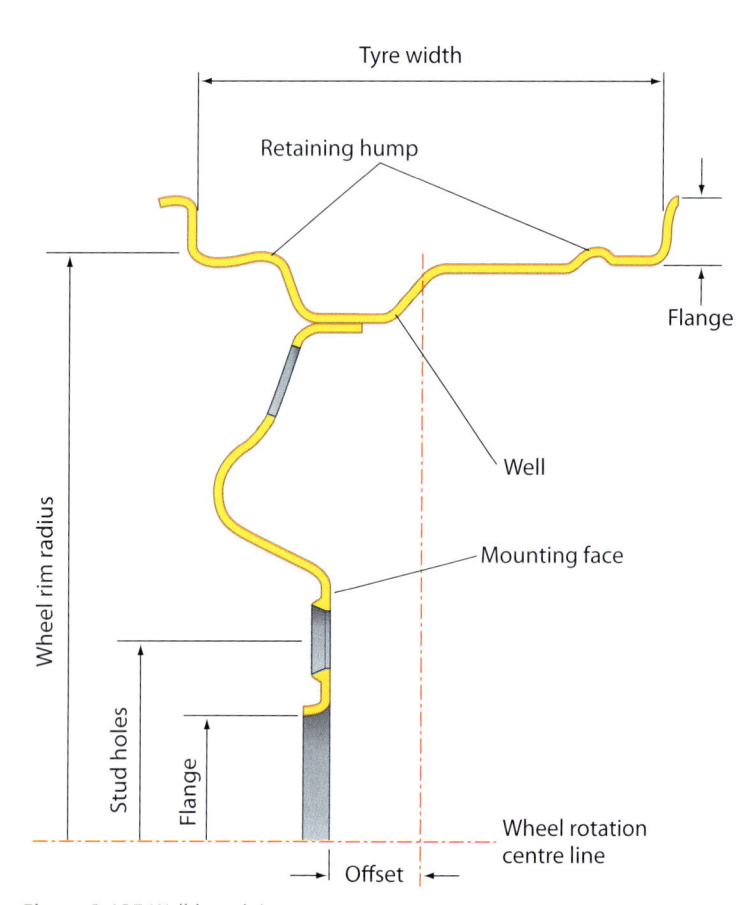

Figure 2.125 Well-based rim

2 Light vehicle chassis units & components

Cast alloy wheels

These are made from an alloy of aluminium and magnesium. The wheel is heated, poured in a cast and, when set, machined to the required **tolerances**. Cast alloy wheels have a well in the rim for easier tyre replacement. However, some alloy wheels require the tyre to be removed from the rear or damage to both tyre and wheel will occur.

The advantages of cast alloy wheels are:

- ✓ light
- ✓ good appearance
- ✓ can be machined to high tolerances
- ✓ conduct heat away from the brakes.

The disadvantages of cast alloy wheels are:

- ✗ more expensive
- ✗ lack of resistance to accidental damage and corrosion.

Pressed steel wheels

A rim, spot-welded to a pressed steel disc, gives a complete wheel (see Figure 2.126). A well is formed in the middle of the rim to allow easy removal and replacement of the tyre. To ensure an airtight fit between wheel and tyre, a seating for the tyre is tapered up to where it meets the rim. This part of the wheel, where the tyre sits when in operation, is called the flange.

The advantages of pressed steel wheels are:

- ✓ inexpensive
- ✓ easy to clean
- ✓ relatively light.

The disadvantages of pressed steel wheels are:

- ✗ do not look attractive
- ✗ can corrode quickly
- ✗ heavier than alloy rims.

All wheels have rim codes on the wheel rim to assist in tyre selection and fitting. Table 2.12 gives a few examples.

Figure 2.126 Cast alloy wheel

Safe working

Under no circumstances must an inner tube be fitted to a tyre fitted to an alloy wheel. Internal movement of the tube erodes the wheel well.

Figure 2.127 Pressed steel wheel

Key terms

Bead – wire cord that goes around the tyre where it meets the wheel rim.

Tolerance – an allowable difference from the required measurement.

Table 2.12 Examples of rim codes

Code	Rim width (inches)	Flange height (inches)	Well depth (inches)	Rim diameter (inches)
4 J 13	4.5	0.687	0.75	13
5 J 13	5.0	0.067	0.70	13
5 K 13	5.0	0.77	1.00	13
6 L 15	6.0	0.85	1.125	15

Level 2 Light Vehicle Maintenance & Repair

Did you know?

A buckled wheel is where the spokes are bent or maladjusted. The result is that the outside of the wheel does not run true to the hub or to the inside of the wheel. This can also apply to other wheels where severe damage has occurred.

Safe working

When inflating tyres on detachable rims you must use a cage. If the rim is not seated correctly, the cage will prevent the rim hitting you as a result of the force of the air pressure on the back of the rim being pushed towards you.

Key terms

Nearside – the passenger side of the vehicle for right-hand drive vehicles.

Offside – the driver's side of the vehicle for right-hand drive vehicles.

Wire (spoked) wheels

These are very rarely used today, except for bicycle wheels. They look attractive but are difficult to keep clean, are expensive, require the use of an inner tube and can buckle very easily.

Detachable rims

Detachable rims are mainly used on heavy vehicles. This is because the size and weight of the tyre makes it difficult to replace.

A variation of the detachable rim is the divided rim wheel. This is split into two halves down the well centre and the two halves are clamped together by bolts. When the tyre is deflated, the bolts are removed and the wheel can be split, to enable the tyre to be replaced.

Wheel retention

There are various methods of wheel retention. The most common method for light vehicles is to use cone-faced screws or nuts that ensure the wheel is central to the hub. The extra surface area on the cone of the nut provides a frictional locking device.

Large goods vehicles and passenger-carrying vehicles use nuts with left-hand threads on the **nearside** and right-hand threads on the **offside**. This prevents the wheel nuts becoming undone by forces generated by wheel rotation. Monitoring the tightness of the fixing on heavier vehicles is done using plastic indicators. These are set pointing to each other when the wheel fixing is tightened and any slackening can easily be observed.

Wheel locking nuts

Alloy wheels are a very desirable feature on a vehicle. To prevent theft, wheel locking nuts/bolts with a dedicated socket are used to remove and replace the wheel fixing.

Figure 2.128 Wire (spoked) wheel

Figure 2.129 Detachable rims

Figure 2.130 Wheel retention

Figure 2.131 Wheel locking nuts

Tyres

One of the most important components of the car is the tyre. Vehicle tyres have five main functions.

1. Provide a smooth ride for the vehicle's occupants

The tyre works to act as a spring to absorb some of the road shocks.

2. Support the load of the vehicle

Approximately 95 per cent of the load of a vehicle is supported by the air inside the tyre. The remaining 5 per cent is supported by the side walls of the tyre.

3. Provide traction in all-weather conditions

The tread patterns provide grip between the surface of the tyre and the ground. Channels provide a space for the water to be squeezed out and ensure that the tyre surface contacts the road. This prevents aquaplaning (see page 157), which is so dangerous that it can lead to total loss of control of the vehicle. This is another reason tyre tread depths must be above the legal limit.

4. Transmit braking and driving forces

When braking, the tyre tries to resist movement, while the momentum of the vehicle wants to keep moving the vehicle forward. A turning force is transmitted to the road surface from the rim of the wheel through the bead of the tyre, then the side wall of the tyre and on to the **tread contact patch**.

Driving forces are produced at the road surface in the same way, but the torque transmitted by the engine is to the wheel through the hub. During braking and driving, a tyre is subjected to considerable stresses.

5. Assist with the steering of the vehicle

A sideways force on the tyre exists when the steering gear of the vehicle turns the front wheels. The assistance given can vary depending on the **profile** and size of tyre.

Parts of a tyre

The main parts of a tyre are shown in Figure 2.133.

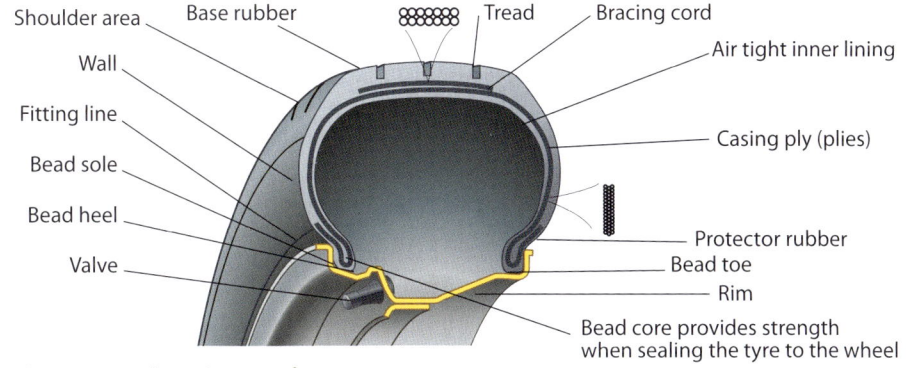

Figure 2.133 The main parts of a tyre

> **Did you know?**
> The construction of the tyre of a van differs slightly to a car in that its sidewalls are made slightly thicker to absorb the greater loads placed on them.

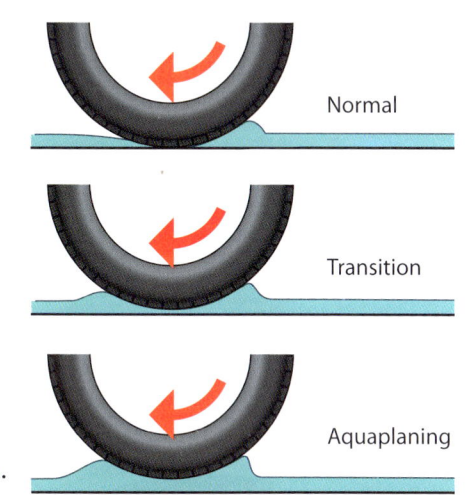

Figure 2.132 Aquaplaning

> **Key terms**
> **Tread contact patch** – the part of the tyre which is in contact with the road surface.
>
> **Profile** – the comparison between the width and height of the tyre. This is also known as aspect ratio.

Level 2 Light Vehicle Maintenance & Repair

If a tyre is sectioned it looks like a U-shape. The side walls form the sides of the shape, with the bead at the lower end and the tread on the outer edge. The internal structure of the tyre consists of wire and fabric cords. These form the strength of the tyre.

The distance from the top to the bottom of the tyre is called the section height. The distance across the tyre is called the section width.

The tread

When tyres are made, a tread pattern is moulded into the design. The grooves help the tyre move water out of the way as the tyre rolls along the road. This allows the tyre to stay in contact with the road and maintain grip. The tread pattern of a tyre is shown in Figure 2.134.

Figure 2.134 The tread

The plies

The strength of the tyre is determined by a number of fabric cords made from rayon, nylon or polyester that run through the rubber. These are called plies. They connect the beads on either side of the tyre. The angle at which they are laid affects how the tyre performs and, therefore, the way the vehicle handles.

There are three types of ply structure: radial, cross and bias belted. Most vehicles on today's roads use radial ply. On each tyre side wall there is a ply rating indicating how many layers have been added to the tyre.

Radial ply

The plies of a radial tyre are laid at an angle of 90 degrees to the tyre centre line or bead, as shown in Figure 2.135.

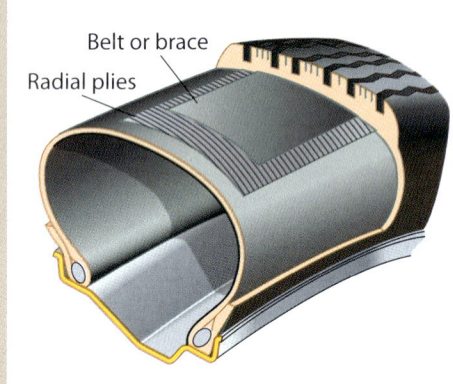

Figure 2.135 Radial ply

The advantages of the radial ply tyre structure are:

✓ It gives a more flexible tyre.
✓ This enables the tread to remain in contact with the road surface and gives good handling.

The disadvantages of the radial ply tyre structure are:

✗ The increased flexibility gives heavier low speed steering and makes the tyre noisier.
✗ Steering geometry has to be set more accurately than with any other type of tyre. This is because as the tyre flexes during cornering, driving and braking, the steering angles change.
✗ This may result in excessive tyre wear and counteract the advantage of good handling.

A strengthening layer is placed between the plies on the inside of the tread to prevent distortion of the tread as the plies fan out due to **inflation**. There are two types of radial ply tyre depending on the material used for the strengthening layer:

1. **Belted radial** – when a synthetic material is used (nylon, rayon or polyester).
2. **Braced radial** – when steel is used for this layer.

Cross ply

The plies of a cross ply tyre are laid at an angle of approximately 35 degrees to the centre line from one bead to the other (see Figure 2.136).

The advantages of cross ply tyre structure are:

- ✓ There is light steering at low speeds, as the tyre is more rigid than the radial ply tyre.
- ✓ For this reason, steering geometry does not have to be as accurate as it is for radial ply tyres.

The disadvantages of cross ply tyre structure are:

- ✗ The rigid construction means that the tyre wall does not flex well, giving poor handling when cornering.
- ✗ This will cause **tyre shuffle**.
- ✗ Tyre life is reduced due to internal friction of the crossed plies generating too much heat, softening the rubber and increasing tyre wear.

For these reasons, cross ply tyres are rarely used today.

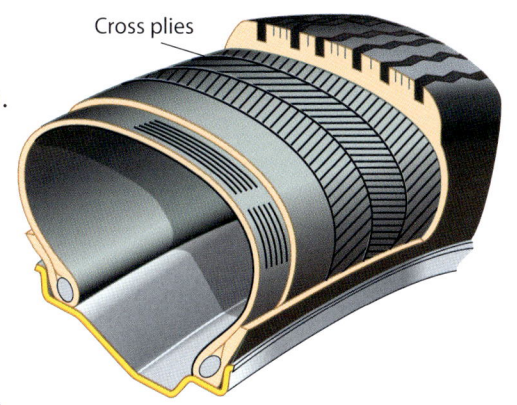

Figure 2.136 Cross ply

> **Key terms**
>
> **Inflation** – where air pressure is added to the tyre through the valve.
>
> **Tyre shuffle** – where the tyre skids sideways across the road surface.

Bias belted

The bias belted tyre consists of a layer of plies laid at approximately 35 degrees to the centre line (the same as cross ply), but it also has a belt or brace layer (as in radial ply). This gives a tyre which performs better than a cross ply tyre but not as well as a radial tyre. Bias belted tyres were mainly fitted to vehicles manufactured in the USA, and they are rarely used today.

The law regarding tyre structure

It is illegal to do the following:

- Mix radial on the front and cross ply on the rear.
- Mix bias belted on the front and cross ply on the rear.
- Mix radial on the front and bias belted on the rear.
- Mix cross ply, radial or bias belted tyres on the same axle.

Tubed and tubeless tyres

When one-piece welded rims were introduced, instead of wire wheels, there was no need to insert an inner tube into the tyre. Tubeless tyres were introduced where the seal of the tyre is achieved by lining the inside of the tyre with a soft rubber film that extends around the tyre bead. The rim of the wheel is slightly tapered. As the tyre is inflated, both the bead and rim are forced together, providing an airtight seal.

The advantages of tubeless tyres are:

- ✓ They weigh less, as there is no inner tube.
- ✓ There is less chance of the tyre deflating because the soft rubber lining seals around the object penetrating the tyre.
- ✓ Tubes cannot be fitted to low profile tyres due to the friction that is built up within the tyre and the tube.

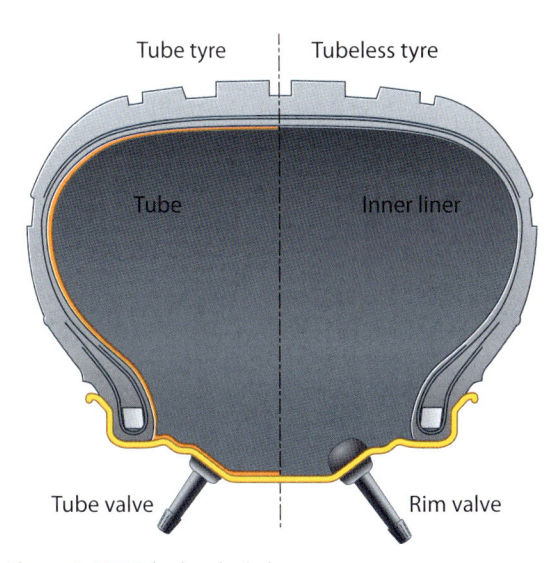

Figure 2.137 Tubed and tubeless tyres

The disadvantages of tubeless tyres are:

✗ The tyre cannot be fitted to a damaged rim.
✗ The tyre cannot fit on wire wheels.

Tyre valves

The tyre valve permits air under pressure into the tyre chamber formed between the casing and the rim when required, and it seals the air when it is in the tyre.

Emergency

On more modern vehicles, tyre valves are fitted with a sensor that measures tyre pressure (see Figure 2.137). If one tyre deflates, the sensor sends a warning via a computer to warn the driver to stop before a blow-out occurs.

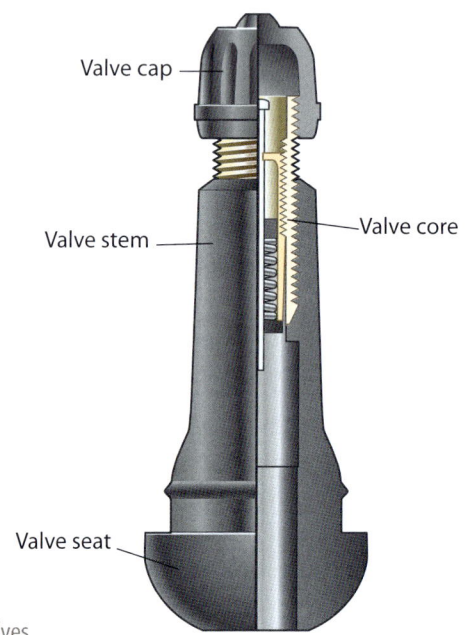

Figure 2.138 Tyre valves

Figure 2.139 Tyre pressure sensor

Tyre sidewall markings

Every tyre has a series of markings to enable the correct replacement to be fitted. Tyre size, aspect ratio, load index and speed rating are the markings that are most important. These are explained in Figure 2.140 and Table 2.13.

Figure 2.140 Tyre sidewall markings (the numbering system is explained in Table 2.13)

Table 2.13 Tyre sidewall markings

	Information given	Typical example
1	Type of tyre construction	Radial ply
2	Load index	104
3	Speed symbol	S or T
4	USA quality grading	Tread wear 160 Traction A Temperature B
5	Tread wear indicates into tread pattern	TWI or E6610356
6	Tread pattern	M+S (mud and snow)
7	Reinforcing mark	Reinforced
8	USA load and pressure specification	900 Kg (1946 LBS) at 340 KA (50 PSI)
9	Tyre size	235/70 R16
10	Tyre type	Tubeless
11	Country of manufacture	Great Britain
12	USA compliance symbol	DOT AB7C D OFF 267
13	European tyre approval	E 11 01234
14	Tyre construction	Side wall 2 plies Rayon. Tread 2 Rayon, 2 Steel
15	Manufacturer's brand name	Traction ply XYZ
16	Direction of rotation	Arrow shows forward rotational direction of tyre for correct fitment

An example of a tyre marking is **185/55 R 15 81V**:

185 – the section width measured in millimetres

55 – the profile or aspect ratio expressed as a percentage

R – radial tyre

15 – the diameter of the wheel measured in inches

81 – load index; this gives the maximum load that can be placed on the tyre (81 means 462 kg – see Table 2.14 below)

V – speed rating, this is the maximum speed the tyre can safely travel at (V means 240 km/h – see Table 2.15 on page 195)

Aspect ratio

The comparison between width and height of the tyre is known as the aspect ratio, but it is more commonly called the profile. In most tyres the section width is greater than the section height (the tyre is wider than it is high). The aspect ratio gives the section height expressed as a percentage of the section width. For example, if the height of a tyre is 60 per cent of the width, it will have an aspect ratio of 60.

Level 2 Light Vehicle Maintenance & Repair

The standard aspect ratio is 82, although many vehicles have aspect ratios of 55 or 50. Very wide tyres for performance vehicles can have an aspect ratio as low as 35.

The main advantages of low profile tyres are increased steering response and better road holding. There is also increased space inside the wheel for larger brake calipers and a greater airflow to cool the friction surfaces. The disadvantages of tyres with an aspect ratio of below 55 are reduced comfort and increased cost.

Load index

The load index gives the maximum load that can be placed on the tyre. The different load indexes are shown in Table 2.14. Where twin wheels are used for heavier vehicles, two load indexes are quoted. A load index of 114/110 means that if the tyre is a single fitment, then the load index is 114 (1180 kg). If the tyre is one of a pair on one side of an axle, then the lower load index 110 (1060 kg) is used. This prevents tyre overload if one tyre fails.

Table 2.14 Load index table

Load index	Load (kg)	Load index	Load (kg)	Load index	Load (kg)	Load index	Load (kg)	Load index	Load (kg)	Load index	Load (kg)
60	250	74	375	88	560	102	850	116	1250
61	257	75	387	89	580	103	875	117	1285
62	265	76	400	90	600	104	900	118	1320
63	272	77	412	91	615	105	925	119	1360
64	280	78	425	92	630	106	950	120	1400
65	290	79	437	93	650	107	975	121	1450
66	300	80	450	94	670	108	1000	122	1500
67	307	81	462	95	690	109	1030	123	1550
68	315	82	475	96	710	110	1060	124	1600
69	325	83	487	97	730	111	1090	125	1650
70	335	84	500	98	750	112	1120	126	1700
71	345	85	515	99	775	113	1150	127	1750
72	355	86	530	100	800	114	1180	128	1800
73	365	87	545	101	825	115	1215	129	1850

Find out

Take a look at the tyre sizes on different vehicles and check their ratings using Tables 2.14 and 2.15.

Speed rating

The speed symbol is a letter code that gives the maximum continuous speed of the tyre. This is known as the speed rating. Table 2.15 shows speed symbols and the maximum speeds in kilometers per hour.

Table 2.15 Speed symbols

Speed symbol	Speed (km/h)	Speed symbol	Speed (km/h)	Speed symbol	Speed (km/h)	Speed symbol	Speed (km/h)
A1	5	B	50	L	120	U	200
A2	10	C	60	M	130	H	210
A3	15	D	65	N	140	V	240
A4	20	E	70	P	150	VR	Over 210
A5	25	F	80	Q	160	W	270
A6	30	G	90	R	170	Y	300
A7	35	J	100	S	180	Z	Over 240
A8	40	K	110	T	190	ZR	Over 240

Legal requirements for tyres

Table 2.16 shows the legal requirements for the different features of tyres.

> **Key term**
>
> **Functional test** – where you carry out a simulation of the operation of a system.

Table 2.16 Legal requirements for tyres

Tyre feature	Legal requirement	Inspection equipment	Inspection method
Tread depth	Must be 1.6 mm across the central three-quarters of the width of the tyre and around the entire circumference. Note: tread wear indicators are moulded to the tyre to indicate when 2 mm of tread is left.	Tyre tread depth gauge	A **functional test** of four points across the width of the tyre around the full circumference
Pressure	The tyres must be sufficiently inflated to suit the use of the vehicle. Use the manufacturer's recommended pressure as a guide.	Tyre pressure gauge	Use suitable inflation methods to functionally check the pressure. Refer to the driver's handbook or workshop manual.
Type of tyre	Must be suitable for the use of the vehicle.		Visual – use the manufacturer's recommended sizes as a guide
Structure of tyre	There must be: • no breaks or cuts in the structure of the tyre that exceed 25 mm or 10% of the section width, whichever is greater • no lumps or bulges caused by separation or partial failure of the structure • no exposed plies or cords.		Visual
Sizes and aspect ratios	These must be compatible with the tyres fitted to the other wheels. Use the manufacturer's recommended sizes as a guide.		Visual – use the driver's handbook or workshop manual

Servicing tyres

The condition of the tyres and pressures must be checked weekly or before a long journey.

Tyres must be checked for uneven wear:

- around the circumference
- across the width of the tyre
- in comparison with each side
- from the front to the rear of the vehicle.

Front tyres on FWD vehicles will wear more quickly than the rear tyres. To ensure even wear, tyres must be swapped at regular intervals. This used to be done by swapping the front right tyre with the rear left, and the front left tyre with the rear right. However, more **directional tyres** are used these days. This means that they work better in one direction.

Where directional tyres are used, they must be changed from front to rear on the same side. The better tyres should always be fitted to the rear to reduce the risk of oversteer, as this is considered to be less controllable than understeer. However, all tyres must be in good condition and within the limits of the law.

Wheel imbalance

Wheels need to be correctly balanced to reduce noise and vibrations at all driving speeds. Wheel imbalance can be detected from:

- vibrations of wheel movement through the steering wheel, normally between a set speed range
- noise in vehicle panels called drumming.

Wheel imbalance can be caused by worn steering joints, wheel bearings, drive or propeller shaft joints. Out-of-balance propeller shaft and out-of-true wheels and tyres can also be a cause, although the most common cause is tyre and wheel imbalance.

There are two types of imbalance: static and dynamic.

Static imbalance

Static imbalance is caused by one area of the wheel or tyre being heavier than the rest. This imbalance causes the wheel to bounce (called wheel **tramp** or **hop**) and it also causes tyre wear. You can easily test for this by lifting the wheel off the ground, rotating the wheel and marking the bottom of the wheel where it lands. If there is a heavy spot this mark will always land at the bottom. The heavy spot is the unwanted weight that causes the imbalance (see Figure 2.141).

> **Key terms**
>
> **Directional tyres** – these have an arrow on the side wall of the tyre that indicates the direction of forward rotation. When replacing or fitting directional tyres or changing wheels, confirm that the direction of rotation is correct by checking the direction of the arrow.
>
> **Tramp** or **hop** – where the wheel and tyre assembly bounces up and down in a forward motion.
>
> **Shimmy** – where a wheel moves from side to side during rotation.

Figure 2.141 Static wheel imbalance

2 Light vehicle chassis units & components

Dynamic imbalance

Dynamic imbalance is where either side of a wheel has an unequal distribution of weight. This causes the wheel to **shimmy** as it rotates. This effect gets worse as the speed of the road wheels increases.

Figure 2.142 show the effects of incorrect balance, the location of the out-of-balance forces and the position of the balance weights to counteract them.

Figure 2.143 Balancing a wheel off the vehicle

Figure 2.142 Dynamic wheel imbalance

Wheel balancing

There are two methods of balancing wheels, either on or off the vehicle.

1. Balancing off the vehicle

The advantages of balancing off the vehicle are:

- All wheels (driven or undriven) can be easily balanced.
- The wheels can be balanced while the vehicle is not available.

2. Wheel balancing on the vehicle

The advantage of balancing the wheel on the vehicle is that out-of-balance forces of the hub, steering and suspension system are taken into consideration. The disadvantage of balancing on the vehicle is that if a wheel is removed, it must be put back in the same place on the hub.

> **Safe working**
>
> General precautions when wheel balancing:
>
> - Use the correct adhesive weights for alloy wheels.
> - Do not use steel wheel weights for alloy wheels.

Balancing a wheel off the vehicle

1. Remove the wheel.	2. Clean all mud and dirt off the wheel and the tyre.	3. Inspect the wheel and the tyre for damage.	4. Inflate the tyre to the correct pressure.
5. Ensure that the wheel is installed correctly.	6. Remove all weights.	7. Check the amount of imbalance by looking at the gauge.	8. Apply weights to the areas indicated by the wheel balancer.
9. Check the balance again to make sure both gauges read zero by turning the balancer on again.	10. Replace the wheel and torque nuts and bolts to the manufacturer's specifications.	11. Repeat the above steps for all of the wheels.	

Level 2 Light Vehicle Maintenance & Repair

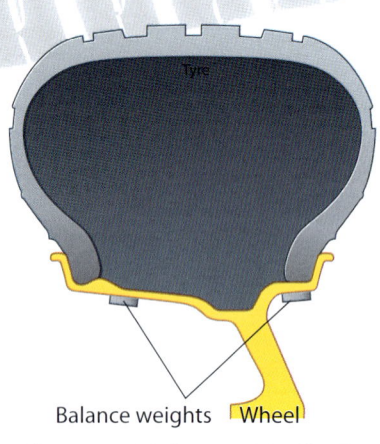

Figure 2.144 Balance weights fitted to alloy wheels

Figure 2.144 shows self-adhesive weights fitted to alloy wheels.

Tyre faults (uneven wear)

Table 2.17 shows five different types of tyre wear. It gives the causes of each type of wear and how to remedy them.

Table 2.17 Different types of tyre wear

Type of tyre wear	Cause	Remedy
Tyre wear 1	• Tyres under-inflated	Inflate tyres to correct pressure. Replace worn components.
Tyre wear 2	• Tyres over-inflated	Inflate tyres to correct pressure.
Tyre wear 3	• Track out of adjustment • Bent Panhard rod • Damaged linkages • Ride height unequal	Adjust track to correct figure. Check and replace worn or damaged components.
Tyre wear 4	• Wheel out of balance • Excessive radial run-out • Excessive braking • Worn leaking dampers • Wheel hop	Balance wheel and tyre if necessary. Check run-out and replace. Replace worn, leaking dampers.
Tyre wear 5	• Track out of adjustment • Worn suspension components • Wheel hop/bounce • Shimmy	Adjust toe to correct figure. Replace components as necessary.

CHECK YOUR PROGRESS

1 What is meant by each of the following tyre sidewall markings: 205 / 45 R 17 84 Z?
2 State the legal limit of tread depth for a light vehicle tyre.
3 What two types of wheel imbalance can occur?
4 To check that a tyre satisfies legal requirements, where should you measure the tread depth?

FINAL CHECK

1. In relation to steering geometry, what is swivel axis inclination (SAI)?
   a. the inward tilt of the swivel axis when viewed from the front or rear of the vehicle
   b. the backward tilt of the swivel axis when viewed from the side of the vehicle
   c. the forward tilt of the swivel axis when viewed from the side of the vehicle
   d. the outward tilt of the swivel axis when viewed from the front or rear of the vehicle

2. The Ackermann system steering layout produces:
   a. understeer of the front wheels during cornering
   b. oversteer of the front wheels during cornering
   c. toe-out of the front wheels during cornering
   d. centre point steering on the front wheels

3. Which of the following is an indication that the front wheels may require balancing?
   a. heavy wear around the centre section of the tyre tread
   b. heavy wear on the outer edges of both the front tyres
   c. severe feathering of both the front tyre treads
   d. steering wheel vibration that increases with vehicle speed

4. One effect of an underinflated front tyre is:
   a. both outer edges of the tyre will wear
   b. the tyre pressure will have increased
   c. the turning circle will be reduced
   d. its rolling resistance is reduced

5. The sidewall marking of a tyre is 195/65R15 81V. From this you work out that:
   a. the tyre is fitted to a 195mm diameter wheel
   b. the tyre is a cross-ply construction design
   c. the diameter of the tyre is 15 inches
   d. the section height of the tyre is 126.75mm

6. Which of the following faults will change the camber angle?
   a. unbalanced front wheels
   b. a buckled front wheel
   c. very high tyre pressures
   d. incorrect wheel alignment

7. A driver complains of steering wheel 'shimmy' that gets worse as the vehicle's speed increases. This most likely indicates:
   a. the shock absorbers need replacing
   b. the wheel alignment needs adjusting
   c. the front wheels need balancing
   d. the tyre pressures are too high

8. Brake fade is:
   a. caused by excessive moisture in the brake fluid
   b. an increase in friction at the braking surfaces
   c. caused by low temperatures at the braking surfaces
   d. caused by high temperatures at the braking surfaces

9. One advantage of disc brakes compared with drum brakes is that they:
   a. reduce the frictional force
   b. provide better heat dissipation
   c. increase the braking force
   d. increase the braking efficiency

10. The presence of air in a hydraulic braking system will cause:
    a. a very 'spongy' brake pedal during braking
    b. the brake fluid to boil under prolonged braking
    c. corrosion of the brake lines and pistons
    d. a significant reduction in braking efficiency

GETTING READY FOR ASSESSMENT

The information contained in this chapter, as well as continued practical assignments in your centre or workplace, will help you to prepare for both the end-of-unit tests and diploma multiple-choice tests.

Through reading and completing the chapter you have gained the knowledge and skills you need to remove and refit light vehicle steering, wheel, tyre and suspension components. Alongside this, you have gained an understanding of the construction of steering, suspension, braking, wheel and tyre systems. The skills you have learned will help you to prepare for working on, refitting, diagnosing and maintaining these systems on light vehicles.

You will need to be familiar with:

- The construction, operation and maintenance of steering systems
- The construction, operation and maintenance of suspension systems
- The construction, operation and maintenance of braking systems
- The construction, operation and maintenance of wheels and tyres
- How to remove and refit light vehicle wheels and tyres, steering, braking and suspension system components
- Common faults associated with light vehicle wheels and tyres, steering, braking and suspension systems
- General maintenance procedures for light vehicle wheels and tyres, steering, braking, and suspension systems

You now need to apply the knowledge you have gained in this chapter in your day-to-day working activities. For example, you need to be able to diagnose faults associated with light vehicle wheels and tyres, steering, braking and suspension systems. In this chapter you have seen examples of common faults and reasons for them. You have also learned how to remove, refit and carry out general maintenance procedures. You should now use this knowledge in your workplace, making sure that you work safely, observe the processes and regulations you have been given and communicate effectively.

This chapter has provided you with the basic knowledge that will help you with both theory and practical assessments.

Before you try a theory end-of-unit or multiple-choice test, make sure you have reviewed and revised any key terms that relate to the topics in that unit. You will need to read all the questions carefully. Take time to digest the information so that you are confident about what the question is asking you. With multiple-choice tests, it is very important that you read all of the answers carefully, as it is common for two of the answers to be very similar, which may lead to confusion.

For practical assessment, it is important that you have had enough practice and that you feel that you are capable of passing. Before you begin a task make sure you have the correct PPE, VPE, tools and equipment ready and that you have a plan to follow. You could, for example, be asked to replace a set of friction linings. For this you will need to be wearing boots, overalls, latex gloves and a dust mask. You will need to protect the vehicle using wing covers, seat covers, a steering wheel cover and floor mat covers. To carry out the task you will need a jack, wheel chocks, axle stands, a socket set, a brake piston retraction tool, brake bleeding equipment, a torque wrench and wheel torque setting information.

Re-read the chapter to confirm you have completed and understood any tasks. This will help you to be sure that you are working correctly and to avoid problems developing as you work.

When you are doing any practical assessment, always make sure that you are working safely throughout the task. You will need to observe all health and safety requirements and use the recommended personal protective equipment (PPE) and vehicle protection equipment (VPE). When using tools, make sure you are using them correctly and safely.

Good luck!

3 Light vehicle engine mechanical, lubrication & cooling system units & components

This chapter will help you develop an understanding of the construction and operation of common engine mechanical, lubrication and cooling systems. It will also cover procedures you can use when removing, replacing and testing engine mechanical systems for correct function and operation. It provides you with knowledge that will help you with both theory and practical assessments. This chapter will also help you plan a systematic approach to engine mechanical, lubrication and cooling systems inspection and maintenance.

This chapter covers:

- Safe working in light vehicle engine mechanical, lubrication and cooling system units and components
- Engine mechanical system operation
- Engine lubrication system operation
- Engine cooling, heating and ventilation systems operation
- How to check, replace and test light vehicle engine mechanical, lubrication and cooling system units and components

WORKING PRACTICE

There are many hazards associated with light vehicle mechanical, lubrication and cooling systems. Firstly, you need to give special consideration to hot components. There is also a danger of coming into contact with chemicals such as lubrication oils and coolant. You should always use appropriate personal protective equipment (PPE) when you work on these systems, to protect you from these hazards.

Personal Protective Equipment (PPE)

Safety goggles reduce the risk of small objects or chemicals coming into contact with the eyes.

Overalls provide protection from coming into contact with oils and chemicals.

Safety gloves provide protection from oils and chemicals. They also protect the hands when handling objects with sharp edges.

Barrier cream protects the skin from old engine oil, which can cause dermatitis and may be carcinogenic (a substance that can cause cancer).

Safety helmet protects the head from bump injuries when working under cars.

Safety boots protect the feet from a crush injury and often have oil- and-chemical resistant soles. Safety boots should have a steel toe-cap and steel mid-sole.

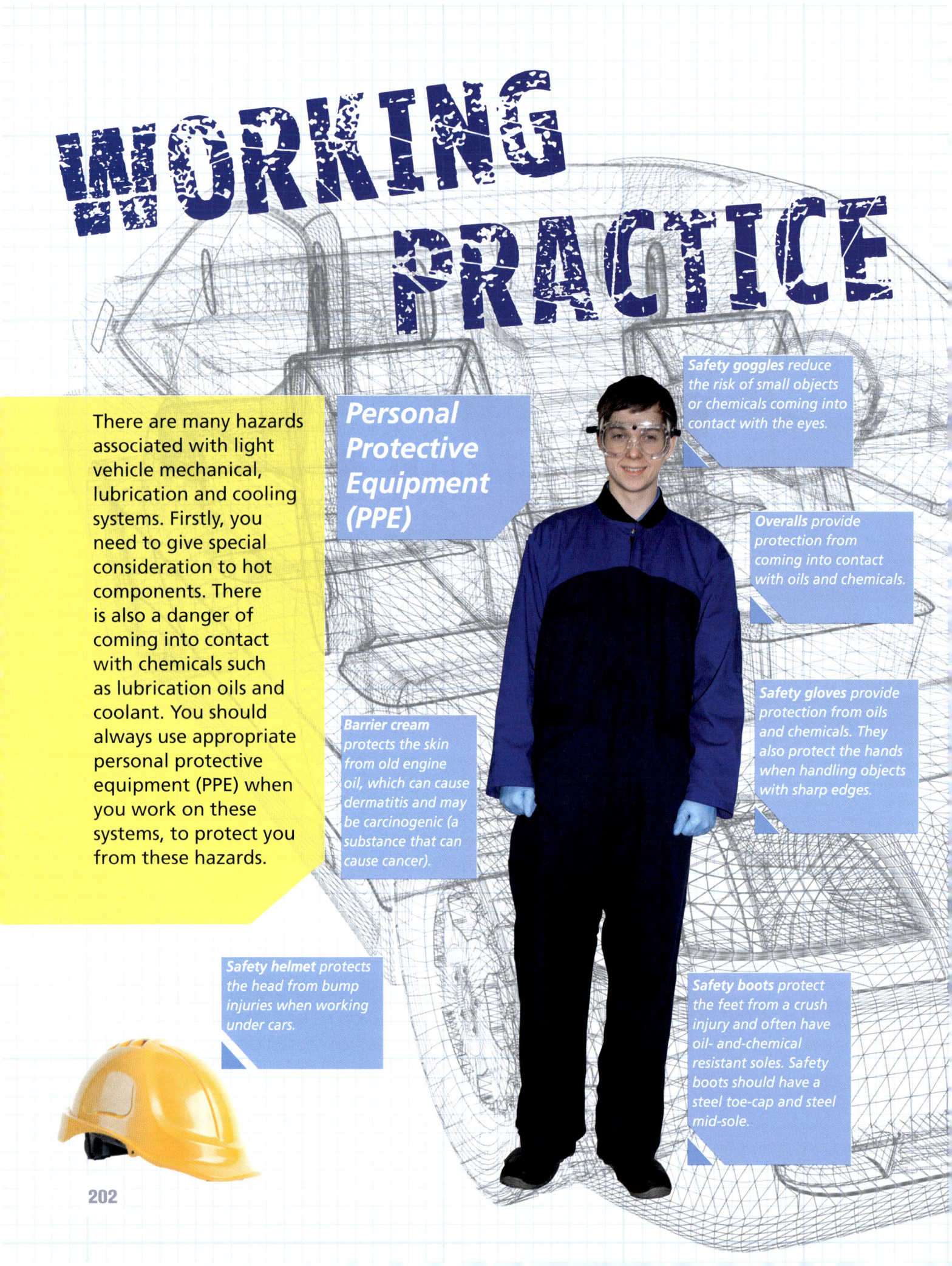

To reduce the possibility of damage to the car, always use the appropriate vehicle protection equipment (VPE):

Wing covers

Steering wheel covers

Seat covers

Floor mats

If appropriate, safely remove and store the owner's property before you work on the vehicle. Before you return the vehicle to the customer, reinstate the vehicle owner's property. Always check the interior and exterior to make sure that it hasn't become dirty or damaged during the repair operations. This will help promote good customer relations and maintain a professional company image.

Vehicle Protective Equipment (VPE)

Safe Environment

During repair and maintenance of light vehicle engine mechanical systems, you may need to drain coolant and oil. Under the Environmental Protection Act 1990 (EPA), you must dispose of these in the correct manner. They should be stored safely in a clearly marked container until they are collected by a licensed recycling company. This company should give you a waste transfer note as the receipt of collection.

All sources of ignition should be removed (for example, no smoking) when you are working on petrol fuel systems because petrol vapour is extremely flammable. The engine should be cold, you should work in a well-ventilated area, and a suitable fire extinguisher must be to hand.

To further reduce the risks involved with hazards, always use safe working practices, including:

1. Immobilise the vehicle by removing the ignition key. Where possible, allow the engine to cool before starting work.

2. Prevent the vehicle moving during maintenance by applying the handbrake or chocking the wheels.

3. Follow a logical sequence when working. This reduces the possibility of missing things out and of accidents occurring. Work safely at all times.

4. Always use the correct tools and equipment to avoid damage to components and tools or personal injury. Check tools and equipment before each use.
   - Inspect any mechanical lifting equipment for correct operation, damage and hydraulic leaks.
   - Never exceed safe working loads (SWL).
   - Check that measuring equipment is accurate and calibrated before you take any readings.

5. If engine components need replacing, always check that the quality meets the original equipment manufacturer (OEM) specifications. (If the vehicle is under warranty, inferior parts or deliberate modification might make the warranty invalid. Also, if parts of an inferior quality are fitted, this might affect vehicle performance and safety.)

6. Following the replacement of any vehicle engine, cooling or lubrication components, thoroughly road test the vehicle to ensure safe and correct operation. Make sure that all work is correctly recorded on the job card and vehicle's service history, to ensure that any maintenance work can be tracked.

Preparing the car

Tools

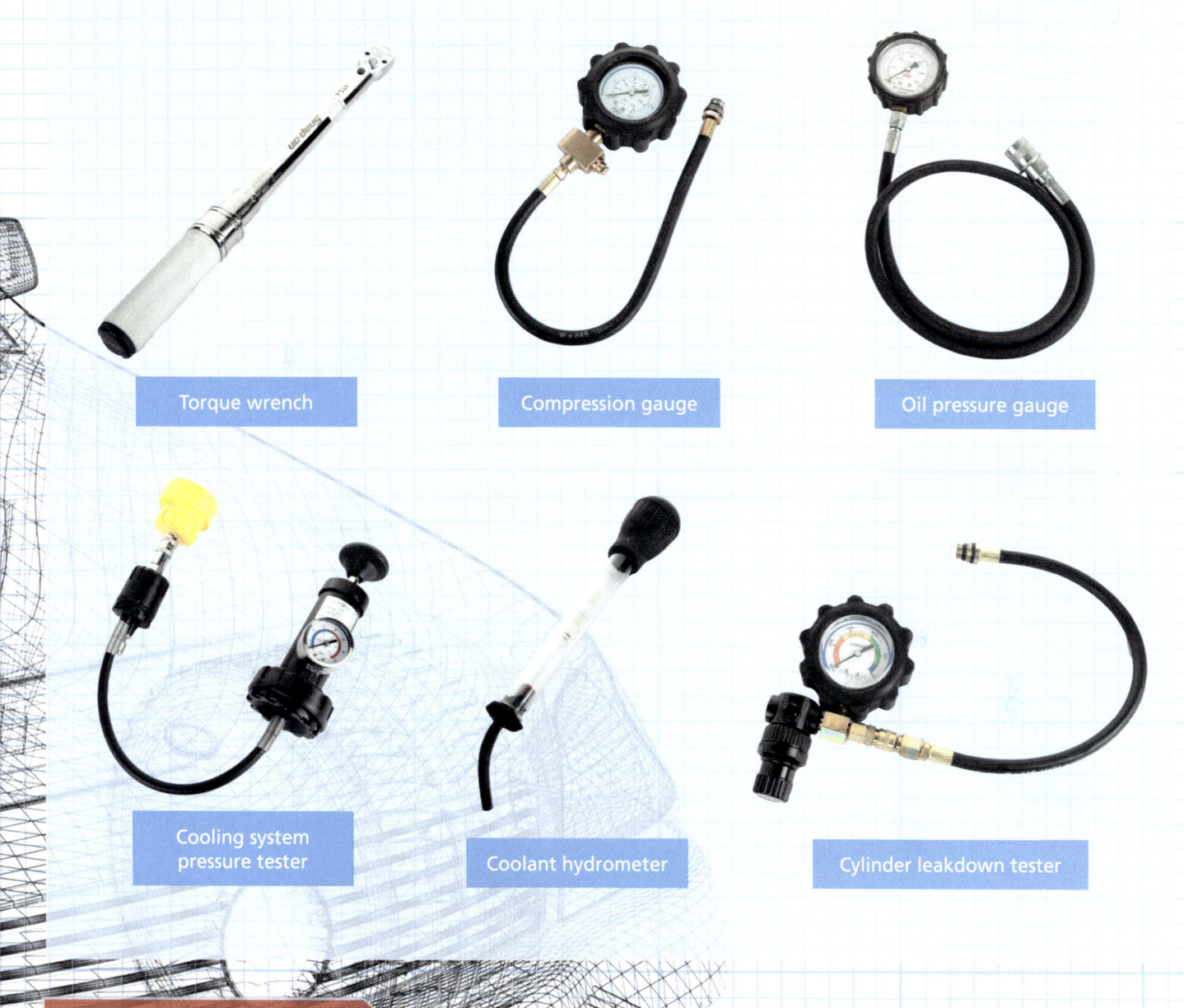

Torque wrench

Compression gauge

Oil pressure gauge

Cooling system pressure tester

Coolant hydrometer

Cylinder leakdown tester

Safe Working

- Always clean up any fluid spills immediately to avoid slips, trips and falls.
- Always make sure the engine is cool before you carry out work on it.
- Always isolate (disconnect) the electrics before you carry out work on an ignition system.
- Always remove all sources of ignition (e.g. smoking) from the area and have a suitable fire extinguisher to hand when you carry work out on petrol or diesel fuel systems.
- Always use exhaust extraction when running engines in the workshop.
- Always work in a well-ventilated area.
- Always wear safety goggles when removing and fitting exhaust system components.

Level 2 Light Vehicle Maintenance & Repair

> **Key terms**
>
> **Combustion** – the process of burning.
>
> **External combustion** – when the combustion process takes place outside the piston cylinder.
>
> **Internal combustion** – when the combustion process takes place inside the piston cylinder.

Engine mechanical system operation

The engine is the car's power plant. It provides energy for movement and to generate electricity and heat. Early vehicles, such as trains, used **external combustion** to provide the power for movement. An example is the steam engine: combustible materials were burnt and the heat created was used to boil water and make steam. This high-pressure steam was then used to drive pistons in the motor section of the engine.

A modern engine uses petrol or diesel, which it combines with air. This mixture is then burnt in a cylinder containing a piston. The heat energy given off by the burning mixture is used to drive pistons and operate the engine. As the combustion takes place directly in the cylinder containing the piston, it is known as an **internal combustion** engine.

The internal combustion engine has been around for over 150 years. There have been many developments in that time, but the basic operating principles remain the same.

Engines

Depending on the vehicle type and how it will be used, manufacturers use different engine layouts and configurations within their vehicle design.

Engine types and configurations

- **Inline** – An inline engine is one where the **pistons** are arranged in a line next to each other. This is probably the most common design, but it can take up a great deal of space because of the length of the **cylinder block** needed to hold the pistons.
- **Flat** – A flat engine is also sometimes known as a 'boxer' engine or 'horizontally opposed' engine. In this type of engine, the pistons are laid out flat on each side of the **crankshaft**. This means that if you have a four-cylinder horizontally opposed engine, there will be two pistons on each side of the crankshaft.

A flat engine provides a low centre of gravity, and the crankshaft can be kept relatively short because there are only two pistons side by side. This makes the engine compact.

Figure 3.1 A modern light vehicle engine

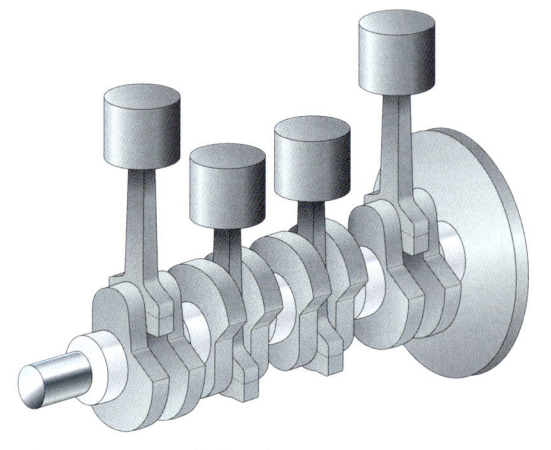

Figure 3.2 A set of inline pistons

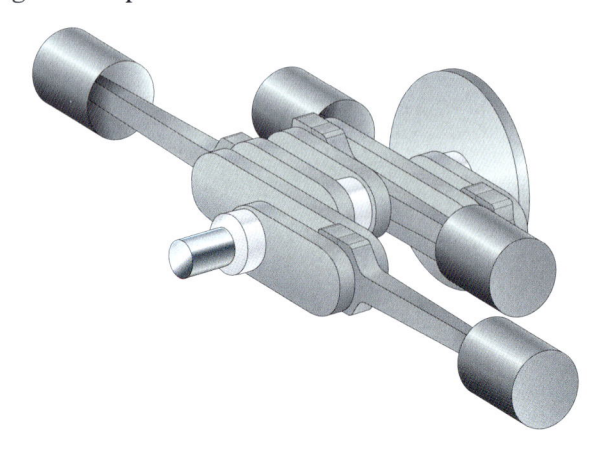

Figure 3.3 A set of horizontally opposed pistons

- **Vee** – A vee engine has its cylinders laid out in the shape of a letter V. In a similar way to the flat engine, the crankshaft can be made shorter and more compact. When the firing impulses on each **bank** occur, they help to balance each other out, providing a smooth delivery of power.

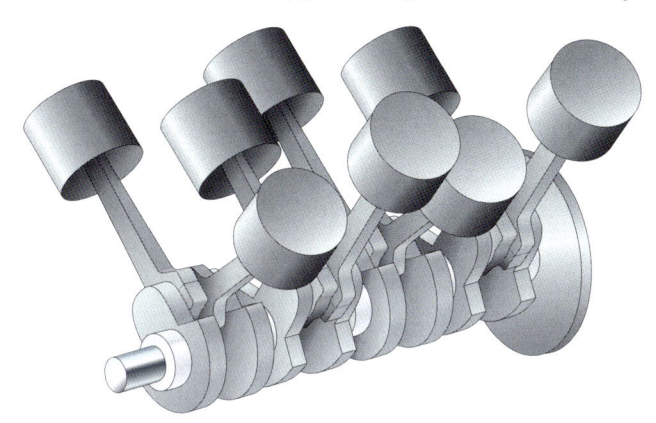

Figure 3.4 A set of pistons in a vee configuration

> **Find out**
>
> Do some research and find at least one manufacturer that uses each of the engine layouts listed above.

Single and multi-cylinder engines

In a standard four-stroke engine, there is only one combustion operation for every two **revolutions** of the crankshaft (720°). The large gap between **firing impulses** can cause vibrations, which makes engine operation harsh and noisy. To overcome this problem manufacturers use multi-cylinder engines with pistons operating at different phases of the four-stroke cycle. In this way, as one piston finishes its power stroke, another is just beginning. This produces a much smoother delivery of power and reduces the need for very large **flywheels**.

- If a two-cylinder engine is used, there is a firing impulse every 360°.
- If a four-cylinder engine is used, there is a firing impulse every 180°.
- If a six-cylinder engine is used, there is a firing impulse every 120°.
- If a eight-cylinder engine is used, there is a firing impulse every 90°

Petrol and diesel engines use different fuel types and ignition methods. However, they use many of the same components which perform the same functions.

Engine components and layouts

Engine design and layouts can vary depending on use, performance requirements, vehicle design and production costs.

Inlet and exhaust valves

Valves are used in a four-stroke engine to allow air and fuel to enter and exhaust gases to be released from the cylinder. At all other times during the combustion cycle, the valves remain closed to seal the cylinder.

> **Key terms**
>
> **Piston** – a pressure-tight plunger.
>
> **Cylinder block** – the main body of the engine containing the cylinders.
>
> **Crankshaft** – the main engine rotating shaft, driven by the pistons.
>
> **Bank** – a row of pistons in a line.
>
> **Revolution** – a complete rotation of a shaft.
>
> **Firing impulse** – another term used to describe an engine power stroke.
>
> **Flywheel** – a heavy metal disc bolted to the crankshaft which stores kinetic (movement) energy.

> **Find out**
>
> To calculate the number of degrees in the crankshaft revolution between firing impulses of a four-stroke engine, divide 720° by the number of cylinders.
>
> How many degrees between the firing impulses are there in the following engines?
>
> - Three-cylinder engine
> - Five-cylinder engine
> - Twelve-cylinder engine

Level 2 Light Vehicle Maintenance & Repair

> **Key term**
>
> **Valve train** – the components used to operate the engine's inlet and exhaust valves.

> **Did you know?**
>
> Many manufacturers are now using multi-valve arrangements in their engine designs. This is where two or more valves of each type, inlet and exhaust, are added to make best use of the space in the combustion chamber. This will also improve the amount of air/fuel that can be drawn into the cylinder and the speed with which exhaust gas can be expelled.

Single (OHC) and multi-camshaft (DOHC)

The mechanisms used to open and close the inlet and exhaust valves of the engine are collectively known as the **valve train**. These can include:

- **Camshaft drive belts, chains or gears** – used to connect camshaft to crankshaft.
- **Camshaft(s)** – used to open inlet and exhaust valves.
- **Push rods** – metal rods used to transfer movement from camshaft followers to rocker arms.
- **Camshaft followers** – used to transfer smooth operation from the cam lobe.
- **Rocker arms** – metal arms used to relay camshaft movement to the valves.
- **Hydraulic tappets** (sometimes called **lash adjusters**) – automatic adjusters used to maintain correct valve clearance.
- **Valves** – used to let air and fuel into the combustion chamber and to let exhaust gases out. (They may be called inlet valve, exhaust valve or poppet valve.)
- **Return springs** – used to close the inlet and exhaust valves.

The more direct the drive to the valve mechanism, the greater the performance that can be gained.

Older engines used a camshaft set low in the engine block (near the crankshaft). This operated push rods against rocker arms to open the valves, as shown in Figure 3.5. The design was known as overhead valve (OHV).

As engine design improved and manufacturing processes became cheaper, the camshaft was moved into the cylinder head, closer to the inlet and exhaust valves. Push rods were no longer required, and fewer

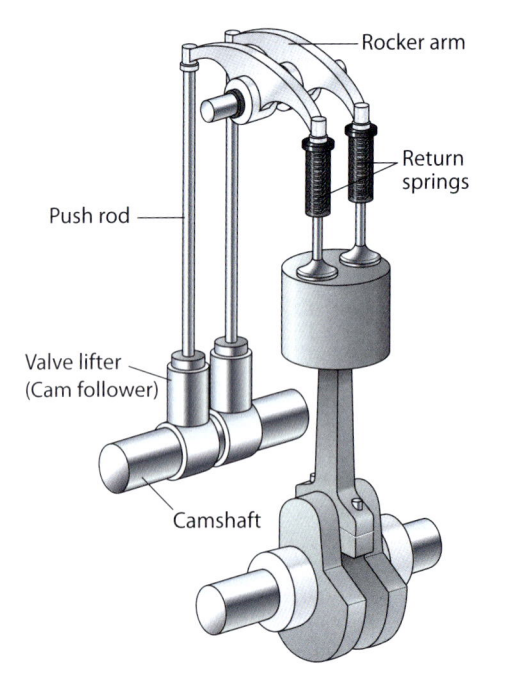

Figure 3.5 Overhead valve (OHV)

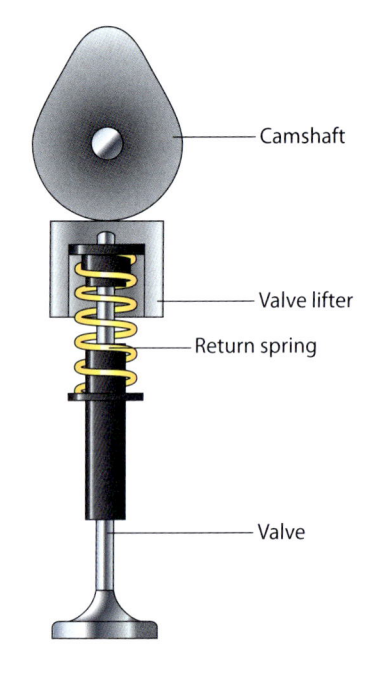

Figure 3.6 Overhead camshaft (OHC)

parts to operate led to greater performance. This set up, shown in Figure 3.6, is known as overhead camshaft (OHC).

It was soon discovered that if a separate camshaft was used to operate the two-valve mechanisms (inlet and exhaust), an even greater gain in performance could be achieved. This became known as double overhead cam (DOHC).

The construction and operation of light vehicle engine mechanical systems

The processes which make the internal combustion engine work are called cycles. These cycles of operation were patented in Germany around 1875 by a man called Nikolaus August Otto, and are therefore correctly known as the 'Otto cycle'. Other patents for internal combustion engines existed before Otto's, but his was the first to make use of four separate strokes.

An internal combustion engine needs four stages to make it work, no matter what design of internal combustion engine is used in the vehicle. These are what make up the Otto cycle. Each has an official name and a simplified name (given below in brackets):

- Induction (suck)
- Compression (squeeze)
- Power (bang)
- Exhaust (blow).

Spark ignition and compression ignition

There are two categories of Otto cycle: **spark ignition** (SI) and **compression ignition** (CI).

The operating cycle for four- and two-stroke engines

The four-stroke operating cycle of a spark ignition (SI) petrol engine

The four-stroke cycle of operations is used in **reciprocating** engines.

Induction (suck)

As the piston moves downwards in the **cylinder bore**, a low pressure (also called a 'depression') is created above it in the combustion chamber.

The camshaft turns and operates on a mechanism to open the inlet valve.

The atmospheric pressure pushes the air/fuel mixture through the open inlet valve and tries to fill the cylinder.

As the piston reaches the bottom of the induction stroke, the lobe of the camshaft moves away from the valve-operating mechanism, and a return spring is used to close it. This seals the combustion chamber.

The induction stroke of the four-stroke cycle is shown in Figure 3.7.

> **Key terms**
>
> **Spark ignition (SI)** – found in petrol engines which need a spark to ignite the fuel.
>
> **Compression ignition (CI)** – found in diesel engines which use hot compressed air to ignite the fuel.
>
> **Reciprocating** – an engine that uses pistons moving backwards and forwards, up and down.
>
> **Cylinder bore** – the area inside the engine that houses the pistons.

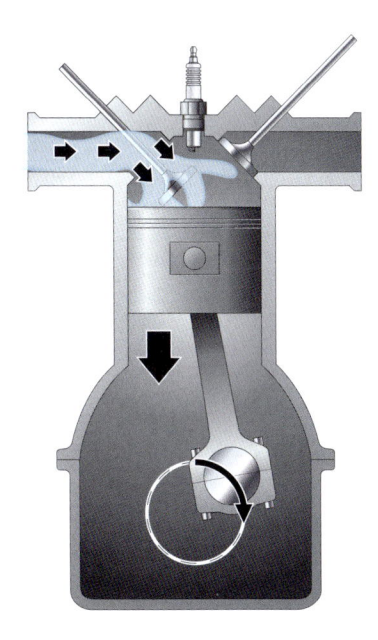

Figure 3.7 The induction stroke of the four-stroke cycle

Level 2 Light Vehicle Maintenance & Repair

Compression (squeeze)

As the crankshaft continues to turn, the piston moves upwards, squeezing the air/fuel mixture into a smaller and smaller space, which raises its pressure (see Figure 3.8). This rise in pressure will perform two functions:

1. As the pressure rises, the temperature of the air/fuel mixture increases. This rise in temperature helps to vaporise the fuel. This is important because it is the fumes (vapour) of the petrol that actually burn, not the liquid.
2. The rise in pressure means the air/fuel mixture is burnt in the confined space of the combustion chamber above the piston. The fuel releases its energy in a much more powerful manner than if it was not confined.

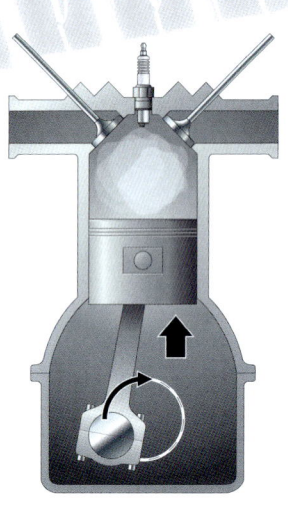

Figure 3.8 The compression stroke of the four-stroke cycle

> **Did you know?**
> When a fuel is rapidly burnt in a confined space, it releases its energy in a much more powerful manner. For example, if you set light to a small pile of gunpowder, it will burn away very quickly. But if you take the same gunpowder and seal it inside a cardboard tube, it will explode in the same way as a firework.

Just before the piston reaches the top of the compression stroke, a spark plug mounted in the combustion chamber releases a high-voltage spark. The heat from the spark is enough to start the air/fuel mixture burning. This creates a flame that spreads out evenly as it moves away from the spark plug. There is a rapid expansion of gases created by the heat.

Power (bang)

Because the gases are sealed in the confined space of the combustion chamber above the piston (see Figure 3.9), there is a rapid pressure rise. This forces the piston downwards on a bang (power) stroke.

The energy from the power stroke is transferred through the piston and connecting rod to the crankshaft, which then turns.

The rotation of the crankshaft on the power stroke is the only active operation within the Otto cycle. The other three strokes are often called 'dead' strokes because they do not help with the turning of the crankshaft.

Kinetic energy stored in the flywheel is used to keep the crankshaft turning on the dead strokes.

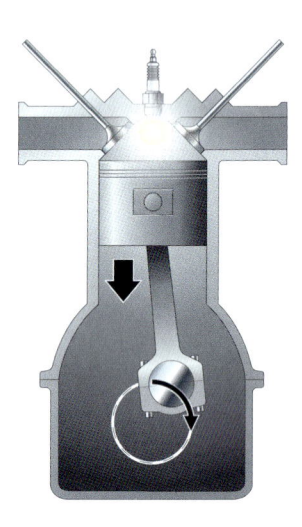

Figure 3.9 The power stroke of the four-stroke cycle

Exhaust (blow)

As the piston reaches the bottom of the power stroke, the energy in the burning air/fuel mixture is used up. This leaves waste products behind, which are known as exhaust gases.

The exhaust valve is opened by the camshaft, and as the piston travels upwards, the exhaust gases are forced out through the **manifold** and exhaust system, as shown in Figure 3.10.

> **Key term**
> **Manifold** – a set of pipes used to supply air and fuel to the engine or to direct exhaust gases away.

210

As the piston reaches the end of the exhaust stroke, the valve is closed by the return spring and the inlet valve begins to open. The whole cycle of operations starts again.

Crankshaft rotation in the four-stroke cycle of operations

To complete one full cycle of operations (induction, compression, power, exhaust), the crankshaft of a standard petrol engine must make two full revolutions. This means the piston will move up and down twice (two strokes each revolution equals four strokes in total). This will give 720° of crankshaft rotation.

The two-stroke operating cycle of a spark ignition (SI) petrol engine

Two-stroke engines are rarely used in cars because in the past they have been inefficient and highly polluting. However, advances in technology have led to improvements in the operating principles. As a result, some engine manufacturers are considering reintroducing this engine type.

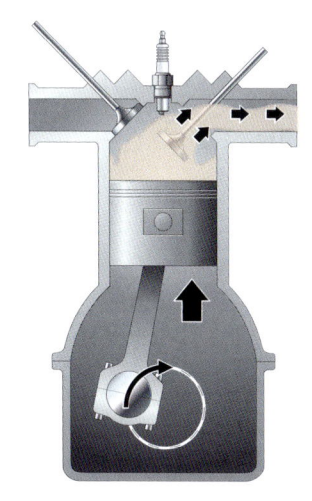

Figure 3.10 The exhaust stroke of the four-stroke cycle

A two-stroke engine completes a full cycle of operation in just two strokes (the piston moves up and down once). With a four-stroke engine, the entire process takes place above the piston, but a two-stroke engine also makes use of the area below the piston (the crankcase) to speed up the operation.

Instead of valves, most two-stroke engines use **ports** that are opened and closed (covered and uncovered) by the piston as it moves up and down (see Figure 3.11).

When you are reading the description of the two-stroke operating cycle that follows, try to imagine what is going on both above and below the piston at the same time.

Induction

Induction takes place below the piston. As the piston moves upwards, a low pressure is created in the crankcase. The piston uncovers the inlet port and an air/fuel mixture is drawn into the crankcase.

Some two-stroke engines use a passive valve or flap called a 'reed valve' to help seal the crankcase once the air/fuel mixture has been inducted.

Pre-compression

As the piston moves downwards, it closes the inlet port and squeezes the air/fuel mixture in the crankcase, raising its pressure.

Transfer

As the piston continues downwards, it opens the **transfer port** in the cylinder wall. The pressurised mixture is forced upwards above the piston. (At this point, an exhaust port machined in the cylinder wall opposite the transfer port is also open.)

> **Key terms**
>
> **Ports** – holes machined in the cylinder walls.
>
> **Transfer port** – a passageway from the combustion chamber to the crankcase that is machined in the cylinder wall.

Level 2 Light Vehicle Maintenance & Repair

Compression
The piston moves upwards (covering both the transfer and exhaust ports) and compresses the air/fuel mixture into the combustion chamber.

Power
Just before the piston reaches top dead centre (TDC), the spark plug ignites the air/fuel mixture and the high pressure created forces the piston downwards on its power stroke.

Exhaust
As the piston moves downwards on its power stroke, it uncovers the exhaust port, and the fresh incoming mixture from the transfer port helps to push the burnt gases out of the engine. This process is called **scavenging**.

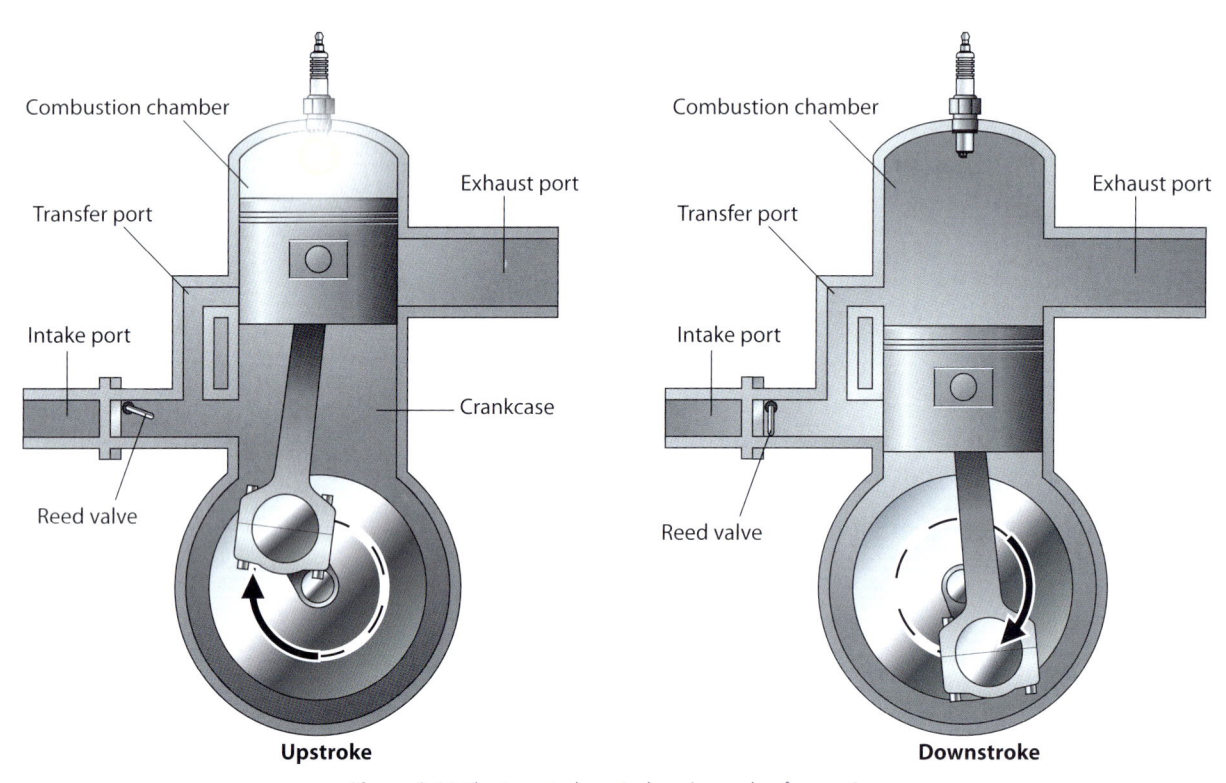

Figure 3.11 The two-stroke petrol engine cycle of operations

> **Key term**
>
> **Scavenging** – assisting the removal of exhaust gases using pressure from the incoming air/fuel mixture.

Crankshaft rotation in the two-stroke cycle of operations
To complete one full cycle of operations (induction, pre-compression, transfer, compression, power and exhaust), the crankshaft of a two-stroke petrol engine only makes one full revolution. This means the piston will move up and down once (two strokes in total). This will give 360° of crankshaft rotation.

The operating cycle of a compression ignition (CI) diesel engine
The four main elements of the Otto cycle are still used in a compression ignition (CI) engine, but the induction and the ignition processes are different from those found in a SI engine.

> **Find out**
>
> List five differences between a four-stroke engine and a two-stroke engine.

Stronger and heavier mechanical engine components are needed in a CI engine because of the higher pressures and stresses involved. But compression ignition is a more efficient way of generating energy from fuel. This means that a CI engine will run longer on diesel than a SI engine will run on the same amount of petrol.

> **Did you know?**
>
> In 1892 Rudolf Christian Karl Diesel took out a patent for his design of CI engines, and his name has become associated with these engines.

Induction

The induction stroke of a CI engine is similar to that of a SI engine. (See the explanation of the SI induction (suck) process on page 209.) However, there is a key difference:

- In the induction stroke of a CI engine, air only is forced into the cylinder. No fuel is mixed with the air at this stage.

Compression

The compression stroke of a CI engine is similar to that of a SI engine. (See the explanation of the SI compression (squeeze) process on page 210.) However, there are a number of key differences:

- Air is squashed into a smaller and smaller space as the crankshaft turns on the compression stroke.
- A much smaller combustion chamber is used in a CI engine than in a SI engine.
- Because of the higher pressure inside a CI engine, the air becomes superheated and this is used as the source of ignition instead of a spark plug.
- Just before the piston reaches the top of the stroke, a fuel injector mounted in the combustion chamber (or a pre-combustion chamber, sometimes called a 'swirl chamber') opens and sprays diesel fuel directly into the superheated air. As the small droplets of diesel fuel come into contact with this superheated air, they spontaneously combust (without the need for a spark plug) and begin to burn.
- To try to produce a smooth burn from the air/fuel mixture, the combustion chamber and/or piston are shaped to create turbulence (also known as swirl), which attempts to mix the diesel and air evenly.

Power

The power stroke of a CI engine is similar to that of a SI engine. (See the explanation of the SI power (bang) process on page 210.) However, there is a key difference:

- The droplets of diesel injected into the superheated air begin to burn, and the heat created makes the gases expand. Unlike a petrol engine, where the burn spreads out evenly away from the spark plug, diesel droplets begin to burn in a number of different positions and directions. As these burning droplets (called flame fronts) collide, a noise sometimes called 'diesel knock' is created.

Exhaust

The exhaust stroke of a CI engine is the same as that of a SI engine. (See the explanation of the SI exhaust (blow) process on pages 210–211.)

Crankshaft rotation in the standard CI engine cycle of operations

To complete one full cycle of operations (induction, compression, power, exhaust), the crankshaft of a standard CI diesel engine must make two full revolutions. This means the piston will move up and down twice (two strokes each revolution equals four strokes in total). This will give 720° of crankshaft rotation.

The operating cycle of a two-stroke compression ignition (CI) diesel engine

The two-stroke diesel engine is not currently used in the production of cars, but is in widespread use for heavy vehicles and marine engines. This type of engine is highly efficient at extracting thermal energy from fuel, which makes it an interesting concept for engine and vehicle designers.

Induction

The induction stroke of a two-stroke CI engine is similar to that of a four-stroke CI engine. (See the explanation of the CI induction process on page 213.) However, there are some key differences:

- As the piston moves downwards, it uncovers ports in the cylinder wall. No inlet valve is required.
- Some form of **supercharger** or **compressor** is needed to push the air into the cylinder through the open ports in the wall.
- As the piston reaches the bottom of the induction stroke, the piston starts to move back up the cylinder. This closes off the intake ports and seals the combustion chamber.

Compression

The compression stroke of a two-stroke CI engine is the same as that of a four-stroke CI engine. (See the explanation of the CI compression process on page 213.)

Power

The power stroke of a two-stroke CI engine is the same as that of a four-stroke CI engine. (See the explanation of the CI power process on page 213.) As with a four-stroke diesel engine, the noise known as 'diesel knock' is often produced.

Exhaust

The exhaust stroke of a two-stroke CI engine is similar to that of a four-stroke CI engine. (See the explanation of the CI exhaust process on page 213.) However, there are some key differences:

- The inlet ports in the cylinder wall will have been uncovered by the piston. So when the exhaust valve is opened by the camshaft, the high air pressure created by the supercharger forces the waste gas out through the open exhaust valve and on through the manifold and exhaust system.
- As the piston reaches the end of the exhaust stroke, the supercharger has forced in a fresh charge of air to start the cycle again.

Crankshaft rotation in the two-stroke diesel engine cycle of operations

To complete one full cycle of operations (induction, compression, power, exhaust), the crankshaft of a two-stroke diesel engine only makes one full revolution. This means the piston will move up and down once (two strokes in total). This will give 360° of crankshaft rotation.

The operating cycle of a rotary engine

Rotary engines are a form of SI petrol engine that do not use a conventional piston. The type of rotary engine used in many modern cars is based on a design made by a German mechanical engineer, Felix Wankel, in the 1950s.

A rotary engine is different from a standard reciprocating (up and down) engine because it does not have a normal piston. Instead of a piston, it has a three-sided **rotor** sitting inside a squashed oval-shaped cylinder called an 'epitrochoid'.

Because the rotor has three sides, it is effectively three pistons in one. This means that a different **phase** of the Otto cycle (induction, compression, power or exhaust) will be happening on each side of the rotor at the same time.

Unlike a standard four-stroke engine, the rotary engine doesn't have valves. Instead, it has ports in the cylinder wall that are opened and closed as the rotor turns.

The operation on one side of the rotor is as follows, and is illustrated in Figure 3.12.

Induction

As the rotor turns, the tip uncovers the inlet port. An expanding chamber is produced which creates the low pressure (or depression). Atmospheric pressure forces an air/fuel mixture through the open inlet port and tries to fill the cylinder. As the rotor reaches the end of the induction phase, the tip covers the inlet port, sealing the combustion chamber.

> **Key terms**
>
> **Supercharger/compressor** – a mechanical device that can be used to 'blow' air into the engine.
>
> **Rotor** – a triangular component used instead of a piston in the rotary engine.
>
> **Phase** – a term used instead of 'stroke' when describing the operation of a rotary engine.

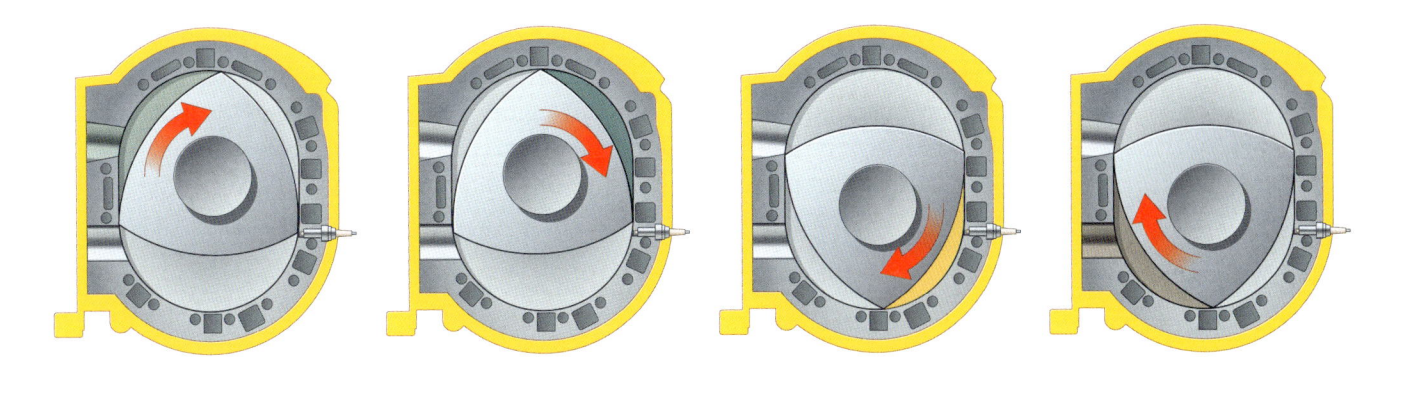

Induction — Compression — Power — Exhaust

Figure 3.12 The four phases of a Wankel rotary engine

Compression

As the rotor continues to turn, the sealed chamber starts to reduce in size, compressing the air/fuel mixture into a smaller and smaller space, which raises its pressure. This rise in pressure performs two functions:

1. As the pressure rises, the temperature also increases. This rise in temperature helps to vaporise the fuel. This is important because it is the fumes or vapour of the petrol that actually burn, not the liquid.
2. When a fuel is rapidly burnt in a confined space, it releases its energy in a much more powerful manner.

Power

At the point of highest compression, a spark plug ignites the air/fuel mixture. Because these gases are sealed in a rotor chamber, a rapid pressure rise occurs, forcing the rotor round on a power phase.

The energy from the power phase is transferred through the rotor to the crankshaft, which is turned.

The rotation of the crankshaft on the power phase is the only active operation within the Otto cycle, but as the rotor has three sides, it is quickly followed by another power phase.

This means that a smaller, lighter flywheel can be used than in a standard reciprocating engine. The crank produces a lower amount of **torque** than a standard piston engine, but can give a high power output.

Exhaust

At the end of the power phase, the rotor tip uncovers the exhaust port. The chamber continues to decrease in size, forcing the burnt exhaust gases out through the manifold and exhaust system. At this point the process repeats itself.

Crankshaft rotation in the rotary engine cycle

Because the rotary engine has a rotor with three sides, there is a power phase every 120° of revolution. Many rotary engines combine two rotor chambers and will provide a power phase every 60°.

Comparison between engine components

Spark ignition SI (petrol) and compression ignition CI (diesel) engine components are similar in operation, but built differently. Because diesel engines work with much higher pressures, the mechanical construction of their components needs to be more robust. This means that diesel engine components are much heavier. When heavy components are used inside an engine, they are unable to rev as high.

> **Key term**
>
> **Torque** – turning effort. In this case, it is the turning effort produced at the crankshaft as it is rotated. The torque will normally increase the amount of weight a vehicle can pull.

3 Light vehicle engine mechanical, lubrication & cooling system units & components

The key engineering principles involved in light vehicle engine mechanical systems

Thermal efficiency

The shape of an engine's combustion chamber affects its **thermal efficiency**. An engine that can extract more heat energy from a quantity of fuel will give better performance, be more economical and produce lower emissions. Various shapes of combustion chamber are:

- bath tub
- hemispherical (hemi)
- wedge
- penthouse roof.

The bath tub and wedge shapes (see Figures 3.13 and 3.14) are easy to design and cheap to produce. This means overall production costs can be kept down, resulting in a cheaper car. However, their thermal efficiency is low so they do not perform as well.

> **Key term**
>
> **Thermal efficiency** – how well an engine extracts heat energy from the fuel. The more efficient the engine is, the more heat energy it will extract from the fuel.

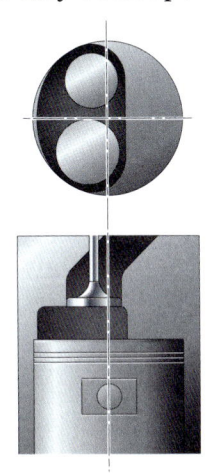

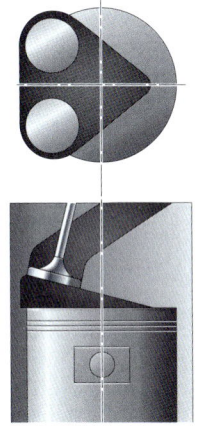

Figure 3.13 Bath-tub-shaped combustion chamber

Figure 3.14 Wedge-shaped combustion chamber

Hemispherical and penthouse roof designs (see Figures 3.15 and 3.16) improve performance but are more costly to design and build. Inlet and exhaust valves are normally positioned on either side of the combustion chamber, so with these shapes two camshafts are often required. This system is called dual overhead cam or double overhead cam (DOHC).

Figure 3.15 Hemispherical combustion chamber

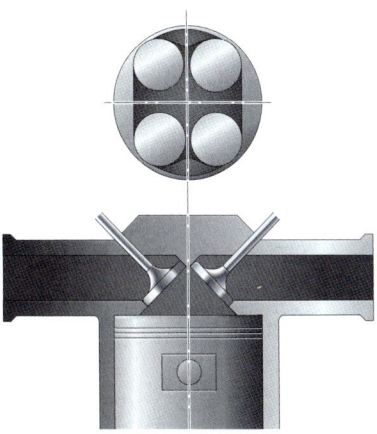

Figure 3.16 Penthouse roof-shaped combustion chamber

217

Key terms

Volumetric efficiency – how well the cylinder can be filled with air and fuel.

Valve lead – where a valve (inlet or exhaust) opens early.

Valve lag – where a valve (inlet or exhaust) closes late.

Valve overlap – the period of time when both inlet and exhaust valves are open at the same time.

Cam timing – the alignment of the camshaft in relation to the crankshaft.

Top dead centre (TDC) – this is where the piston has gone as high as it can during normal operation.

Piston crown – the top of the piston.

Bottom dead centre (BDC) – this is where the piston has gone as low as it can during normal operation.

Squish – because of the shape of the piston, air and fuel are squashed towards the centre of the combustion chamber.

The bigger the inlet and exhaust valves, the greater the performance gained from the engine. To make best use of the surface area available in the combustion chamber, many designs use multiple valve arrangements. Instead of having a single inlet and a single exhaust valve, it is common to have four valves per cylinder (two inlet and two exhaust valves). In this way a larger opening can be formed for the inlet and exhaust gases, improving **volumetric efficiency**.

Valve timing

When an engine is running, the cycle of operations is happening very fast. This doesn't allow much time for the processes of getting air and fuel into the engine, igniting it, burning it and getting rid of the exhaust gases. Because of this, certain operations take place early and finish late during the four-stroke cycle, helping to give extra time for the processes to take place.

An important example of this is valve timing. The inlet and exhaust valves are timed to coincide with piston movements. To give as much time as possible for air and fuel to enter the system and for exhaust gases to get out, **valve lead** and **valve lag** are used:

- **Valve lead** – The inlet valve is opened slightly early (while the piston is just coming to the end of its exhaust stroke) and the exhaust valve is opened slightly early (while the piston is just coming to the end of its power stroke)
- **Valve lag** – The inlet valve is closed late (just as the piston is starting its compression stroke) and the exhaust valve is closed late (just as the piston is starting its induction stroke).

Because of this valve timing process, there is a period of time when both the inlet and exhaust valves are open, called **valve overlap**.

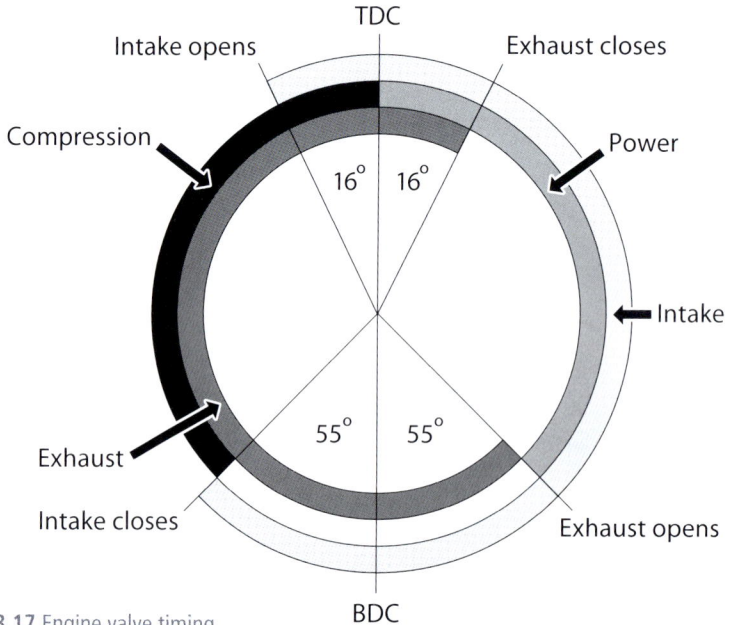

Figure 3.17 Engine valve timing

3 Light vehicle engine mechanical, lubrication & cooling system units & components

One advantage of a valve overlap period is that the incoming air and fuel help push out the exiting exhaust gas, and the exiting exhaust gas helps draw in fresh air and fuel.

The valve movement can be shown on a timing diagram, as illustrated in Figure 3.17.

To drive the valve train of the engine, the camshaft must be connected to the crankshaft. This is commonly achieved by one of three methods:

- **Gear drive** – A gear on the crankshaft is used to drive a gear on the camshaft directly.
- **Chain drive** – A chain similar to a bicycle chain is used to connect the crankshaft and camshaft.
- **Timing belt** – A toothed rubber drive belt is used to connect the crankshaft and camshaft, as shown in Figure 3.18.

Whichever drive method is used, the camshaft must turn at exactly half crankshaft speed. The gearing of the camshaft drive operates with a ratio of 2:1. This means that for every two revolutions of the crankshaft, the camshaft turns once.

It is very important that the gearing and drive between the crankshaft and camshaft are correctly aligned. This is known as **cam timing**. Most manufacturers provide timing marks that need to be set up whenever the engine is dismantled or reassembled.

If the camshaft timing is incorrect, the inlet or exhaust valves of a cylinder may be open as the piston is approaching **top dead centre** (**TDC**). These components will be travelling so fast that a collision between them will cause serious engine damage.

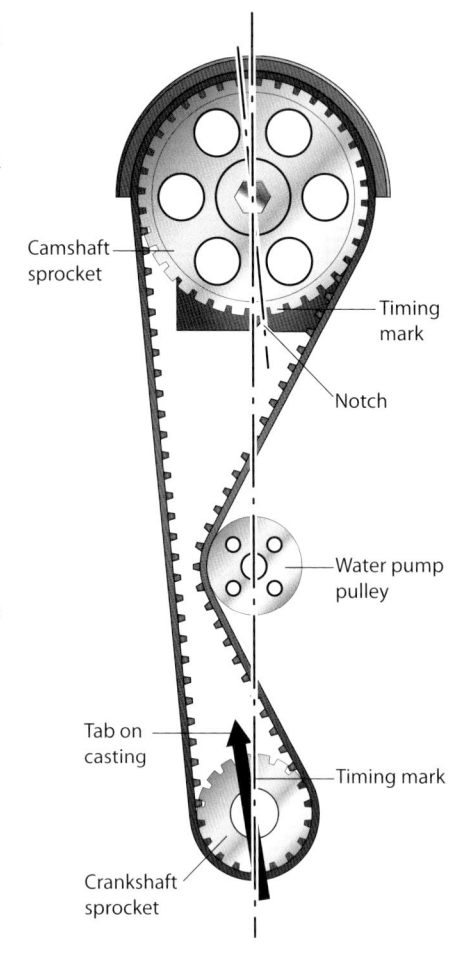

Figure 3.18 Engine camshaft timing belt with sprockets and timing marks

Working life

Mike has just finished replacing the camshaft drive belt on an engine. Before refitting the timing belt covers, he decides to double-check his work.
1. What should he be looking for?
2. How many times must the crankshaft be rotated before the timing marks are realigned?

Piston design

The design and shape of the piston can also be used to assist the smooth burning of the air/fuel mixture.

Piston crowns can be shaped to help with the mixing of the air/fuel by creating turbulence. Also, if the shape of the crown is designed to match the edges of the combustion chamber as the piston reaches top dead centre (TDC), the mixture is forced into the centre – this is known as **squish**. The shape of the piston and the combustion chamber to cause squish is shown in Figure 3.19.

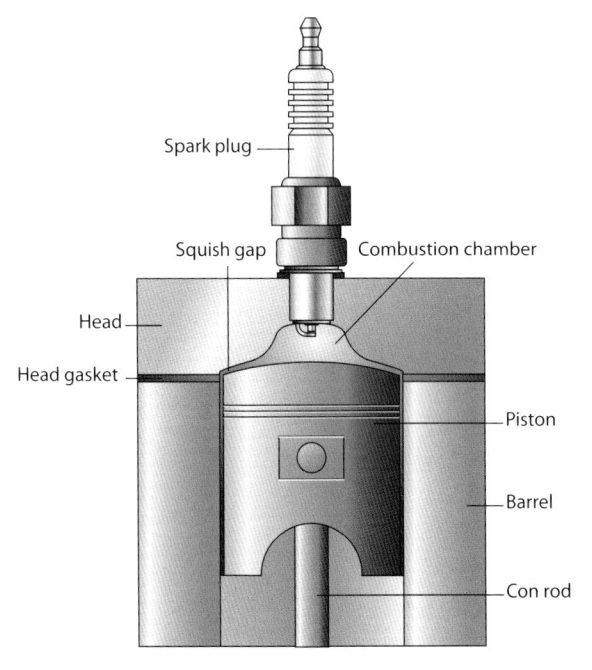

Figure 3.19 The shape of the piston and the combustion chamber to cause squish

Level 2 Light Vehicle Maintenance & Repair

Key terms

Swept volume – the amount of space which is displaced by the piston as it moves from top dead centre to bottom dead centre.

Clearance volume – the amount of space above the piston when it is top dead centre (i.e. the combustion chamber).

Bore – the circular area of the cylinder.

Stroke – the movement of the piston as it travels TDC to BDC.

Cylinder capacity

The cylinder capacity of an engine is often known as cubic capacity, or cc. It is the total of all **swept volumes** within the engine. For example, in a four-cylinder engine, it is the swept volume of cylinder 1 plus the swept volume of cylinder 2 plus the swept volume of cylinder 3 plus the swept volume of cylinder 4.

The cylinder capacity (cc) does not include the **clearance volume** of the combustion chamber. This is because it only relates to the amount of space that is displaced by the piston, and therefore the amount of air that can be taken into the cylinder. This is shown in Figure 3.20.

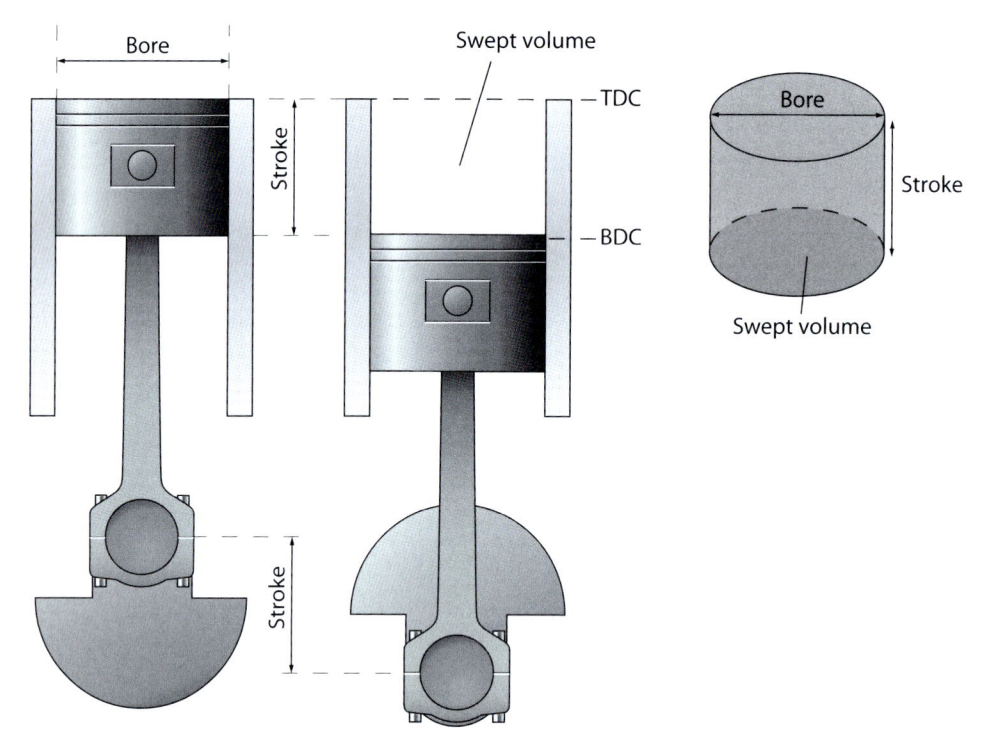

Figure 3.20 A piston and cylinder showing swept volume (cubic capacity)

To calculate the swept volume of one cylinder, follow these steps:

1. First you need to work out the area of the **bore**. To do this use the formula pi times radius squared:
   area of bore = (πr^2)
2. Once you know the area of the bore, multiply it by the length of the **stroke** to give you the swept volume of one cylinder.
3. Then multiply this volume by the number of cylinders to calculate the total cylinder capacity.

Compression ratios

The compression ratio of an engine is the difference between how much air is drawn into the engine, and the space into which it is squashed. This is illustrated in Figure 3.21.

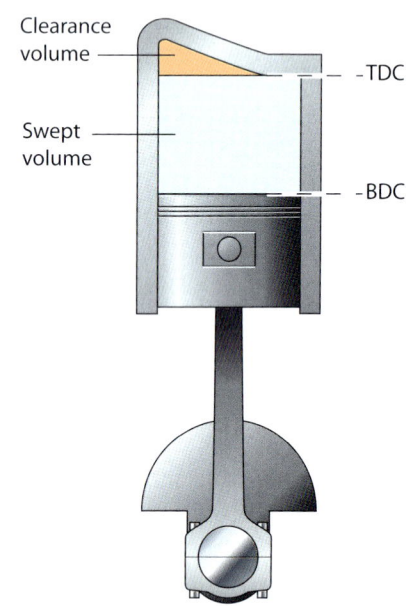

Compression ration = $\dfrac{\text{Swept volume}}{\text{Clearance volume}} + 1$

Figure 3.21 A piston and cylinder showing compression ratio

To calculate the compression ratio of an engine, you need to find how many times the clearance volume will fit into the swept volume and add 1. Therefore, swept volume divided by clearance volume plus one equals compression ratio. You can use this formula:

$$\text{Compression ratio} = \frac{\text{Swept volume}}{\text{Clearance volume}} + 1$$

For example:

> Calculate the compression ratio of an engine with a cylinder swept volume of 234 cc and a clearance volume of 26 cc:
>
> 234 ÷ 26 = 9   9 + 1 = 10
>
> This gives a compression ratio of 10 : 1.

The higher the compression ratio, the more performance or power is gained from the engine. A standard petrol engine will have a compression ratio of around 11:1 and a standard diesel engine will have a compression ratio of around 18:1.

In a petrol engine, as the compression ratio increases so does the performance. However, there is a point at which the compression ratio will make the air/fuel mixture become unstable and possibly auto-ignite (self-combust). If the fuel ignites by itself, instead of waiting for the spark plug to begin the combustion process, it is acting like diesel. This unstable ignition can lead to a misfire and engine damage.

Typical compression ratio values are:

- Petrol engines 8:1 to 12:1
- Diesel engines 16:1 to 20:1

Power and torque

Power and torque are measurements that describe the performance output from an internal combustion engine. They describe two different elements of performance. Depending on how the car will be used, different combinations of power and torque will be used.

- **Power** is the amount of energy that is efficiently released during the combustion cycle, or the rate at which work is done. This will normally give a vehicle its speed.
- **Torque** is the turning effort produced at the crankshaft as it is rotated. Torque gives a vehicle its strength and will normally increase the amount of weight a vehicle can pull.

You can think of power as being like a sprinter, who delivers a sudden burst of energy to produce speed.

> **Find out**
>
> Calculate compression ratios from the following data:
>
> 1. An engine with a total cylinder swept volume of 423 cc and a clearance volume of 47 cc.
> 2. An engine with a total cylinder swept volume of 416 cc and a clearance volume of 26 cc.
> 3. A four-cylinder engine with a total swept volume of 1600 cc and a single clearance volume of 25 cc.
> 4. A V6 engine with a total swept volume of 2982 cc and a single clearance volume of 71 cc.
>
> **Hint:** For questions 3 and 4, you will need to divide the swept volume by the number of cylinders before you do the calculation.

Figure 3.22 A sprinter represents power

Level 2 Light Vehicle Maintenance & Repair

Figure 3.23 A weightlifter represents torque

You can think of torque as being like a weightlifter, who has a great deal of strength to lift heavy weights.

The sprinter and weightlifter are both very fit athletes, but they perform two completely different disciplines, and it is the same for power and torque.

When designing a vehicle, the manufacturer must take into account what the vehicle will be used for. For example, a sports car may require a great deal of power yet because it is lightweight it does not necessarily require large amounts of torque. In contrast, a vehicle designed to carry heavy weights, such as a van or people carrier, may require more torque and less power.

Naturally aspirated and turbocharged engines

To improve the performance of internal combustion engines, air is sometimes forced into the cylinder above the piston. (The more oxygen there is contained in the cylinder, the more fuel can be burned, and therefore the more energy can be released.)

If an engine relies on atmospheric pressure to force air and sometimes fuel into the cylinder, then the engine is **naturally aspirated**.

If a pump is used to raise the pressure of air entering the cylinder, then the engine is **supercharged**.

> **Find out**
>
> Research four cars (two petrol and two diesel), each from a different manufacturer. Compare the power and torque between the different engine sizes, styles and fuel types.
>
> - How does fuel type affect torque output?
> - How does engine size affect power output?

If the air pump or compressor is mechanically driven from the engine, it is a 'parasite of power'. This means that it takes some of the power away from the engine just to drive the supercharger.

If a waste exhaust gas is used to spin a turbine, which can then drive a compressor fan, then the engine is said to be turbocharged.

Turbocharger

A turbocharger is an exhaust gas-driven compressor. It takes the incoming air and raises its pressure, forcing it through the inlet valve and into the combustion chamber. The fitting of turbochargers tends to be restricted to performance petrol engines, but is a more common design feature of diesel engines. An example of a turbocharger is shown in Figure 3.24.

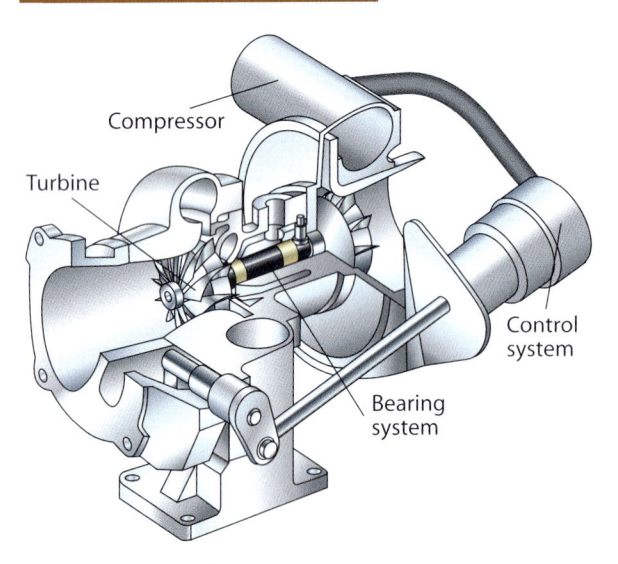

Figure 3.24 A turbocharger

Hybrid fuel engines

The need to conserve fuel and to produce lower exhaust emissions is constantly increasing. Vehicle manufacturers are looking at alternative engine design types to meet this need. Many are considering, or have already produced, cars powered fully by electric motors. These are known as battery electric vehicles (BEVs).

> **Key terms**
>
> **Naturally aspirated** – not supercharged or turbocharged.
>
> **Supercharged** – an engine which has air forced into the cylinder at above atmospheric pressure.

222

Many of the electrically powered cars have issues with:

- performance
- range of travel on a single charge
- amount of time required to recharge the batteries
- a lack of available recharging points.

Some manufacturers have built high-performance BEVs, but the cost of production has made them unsuitable for many road users. A compromise between petrol, diesel and battery electric vehicles is the hybrid.

Hybrid cars have a small-capacity internal combustion engine which runs on petrol or diesel. It also has some large-capacity batteries which run electric drive motors.

Under normal operating conditions, the car will use either the standard internal combustion engine, the electric motors or a combination of the two to provide drive.

- At low speeds with fully charged batteries, the hybrid will normally use only the electric motors and battery power.
- If the battery charge falls low, the engine can be started automatically to generate the electricity required to recharge the battery.
- As acceleration or high-speed travel is needed, the petrol or diesel engine takes over. This is assisted by the electric motors, which reduce strain on the engine and provide greater performance.

Regenerative braking

An advantage of using hybrid technology is that a system called regenerative braking can easily be included in the design. When the driver wants to slow down, he or she applies the brakes, but only some of the braking effort is supplied to the braking system. Instead, the electric motors that are usually used to drive the wheels are converted into electrical generators to charge the batteries.

The energy used to turn these motors and generate electricity provides some of the retardation force required to slow the car down. Because no fuel is used during this recharging process, the generation of electricity is essentially free.

> **Safe working**
>
> Hybrid vehicles use high-voltage electricity. When working with high voltages there is always a danger of electrocution. You should be careful when working in areas with highly insulated wiring (usually colour-coded orange).

> **Find out**
>
> Find out the names of three vehicle manufacturers that currently produce hybrid engine cars.
>
> Find out the names of two vehicle manufacturers that currently produce battery electric cars (BEVs).

CHECK YOUR PROGRESS

1. List ten engine mechanical components.
2. Describe three differences between reciprocating and rotary engines.
3. What are the differences in ignition process between petrol and diesel engines?
4. What is the difference between power and torque?
5. Explain these terms: TDC, BDC, stroke, bore.

Engine lubrication system operation

The need for lubrication

An internal combustion engine uses metal components that move against each other at very high speed. This creates friction. For example, in an engine rotating at 6000 rpm, the piston will move from the top of its stroke to the bottom of its stroke and back to the top again 100 times in a second.

Friction creates heat, and heat makes metal expand and creates wear. If left unchecked, the friction would damage engine components and cause rapid wear. The expansion of the hot metal components could be so great that they would melt and fuse together – this is known as seizure. To help overcome this problem, oil is used as a lubricant.

Engine oil has three main purposes:

1. It helps to keep the surfaces of the components apart.
2. It acts as a cooling medium, helping to take heat away from metal components.
3. It retains dirt and metal particles, helping it to clean the engine at the same time.

Oil is pumped around the engine to make sure that it reaches all components that require lubrication.

Lubrication: boundary and hydrodynamic

- **Boundary lubrication** – A thin film of oil coats the surfaces of the mechanical components. This fluid film may be only one or two molecules thick but it is very hard to penetrate. Because of this, when the two surfaces rub against each other, they slide on this thin boundary film, rarely making contact. Figure 3.25 shows how boundary lubrication works.
- **Hydrodynamic lubrication** – Oil is pumped to rotating components. As these components turn, an oil wedge is created, forcing the surfaces apart. This is similar to the way in which a tyre will aquaplane on a wedge of water. Figure 3.26 shows how hydrodynamic lubrication works.

> **Did you know?**
>
> The science of movements of liquids is called hydraulics. It states that liquids are virtually incompressible (they can't be squashed). By using oil as a lubricant, the metal components glide on a surface of liquid instead of being in contact with one another, similar to aquaplaning.

> **Key terms**
>
> **Boundary lubrication** – oil coating of a surface, sometimes under pressure.
>
> **Hydrodynamic lubrication** – the movement of engine components, forcing oil into a hydraulic wedge, which holds components apart and reduces friction.

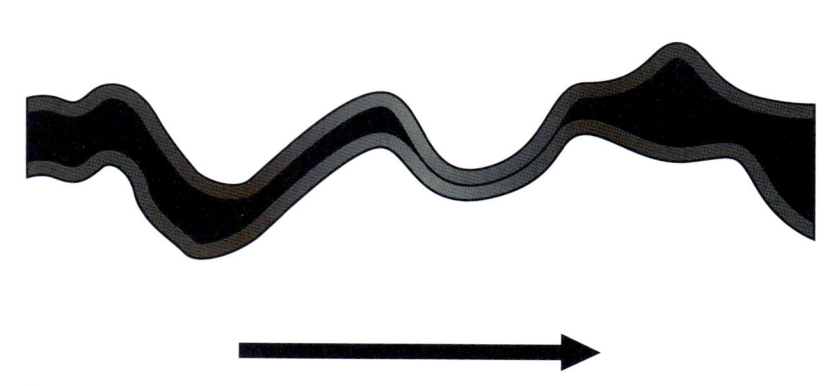

Figure 3.25 Boundary lubrication

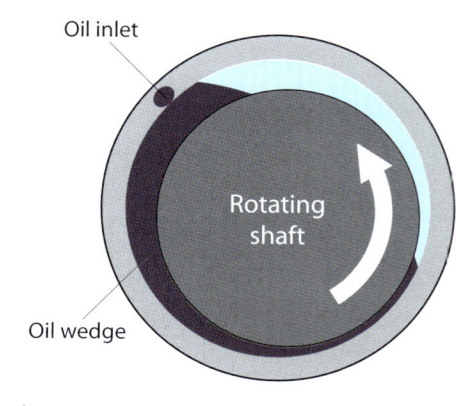

Figure 3.26 Hydrodynamic lubrication

3 Light vehicle engine mechanical, lubrication & cooling system units & components

Light vehicle engine lubrication system components

Engine sump

The sump (see Figure 3.27) is normally bolted to the lower part of the engine, and is used as a reservoir to hold the lubrication oil. It is from here that oil is picked up, pumped around the engine and returned when it has performed its function.

Some sumps contain 'baffles'. These are metal plates that help prevent the oil from sloshing around when the car is moving.

The sump is normally made from thin steel or aluminium. It is exposed to airflow under the engine to help cool the oil.

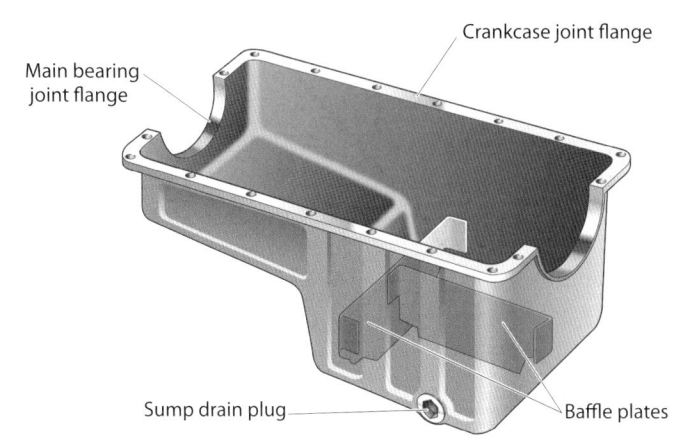

Figure 3.27 Engine sump

Oil pump and strainer

As the sump sits at the bottom of the engine, oil must be pumped upwards and around the engine under pressure. An oil strainer (a basic filter made from wire mesh) is fitted in the lowest part of the sump, and is attached to a tube which joins it to the oil pump. The oil pump, driven by the engine, draws up oil through the strainer and pickup pipe, and forces it around the engine. Figures 3.28 to 3.31 show four different types of oil pump construction.

All four pumps make use of the principle of lowering pressure inside and then forcing it out through a discharge port to provide system pressure.

Crescent oil pump

This uses a crankshaft-driven gear in the centre. This drive gear meshes with external gear teeth on a driven gear, making it turn in the same direction. In the middle of the two gears is a crescent-shaped piece of metal that separates the inlet and the outlet. As the gears rotate, a low pressure develops on the inlet side of the pump as the gear teeth move away from each other. At the other end of the crescent, the gear teeth come together, forcing oil through the outlet.

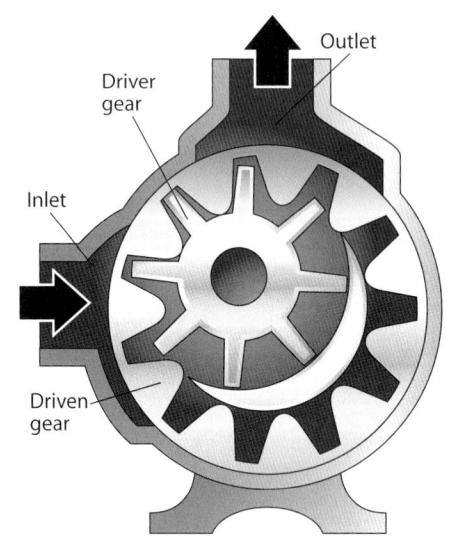

Figure 3.28 Crescent engine oil pump

Eccentric gear oil pump

This uses an engine-driven gear, which is offset (not in the middle) to the main oil pump housing. This offset gear meshes with an external gear, which contains one extra tooth section, making it turn in the same direction. As the gears rotate, a low pressure develops on the inlet side of the pump as the gear teeth move away from each other. At the other side of the pump, the gear teeth come together, forcing oil through the outlet.

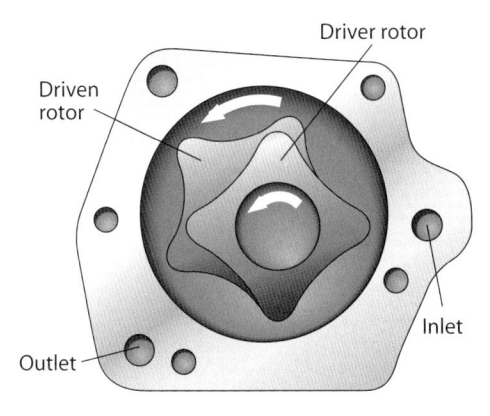

Figure 3.29 Eccentric gear engine oil pump

225

Gear oil pump

This uses two engine-driven gears. They mesh with each other and rotate in opposite directions. As the gears rotate, a low pressure develops on the inlet side of the pump as the gear teeth move away from each other. Oil is moved around the outside of the gears and, on the other side of the pump, the gear teeth come together, forcing oil through the outlet.

Vane oil pump

This uses an offset rotor with a series of paddle blades. The paddle blades are spring-loaded against the pump walls. As the rotor turns, a low pressure develops on the inlet side of the pump as the chamber expands. Oil is moved around with the paddle vanes and, on the other side of the pump, the chamber becomes smaller, forcing oil through the outlet.

When the oil has completed its function, it returns to the sump through passageways under the force of gravity.

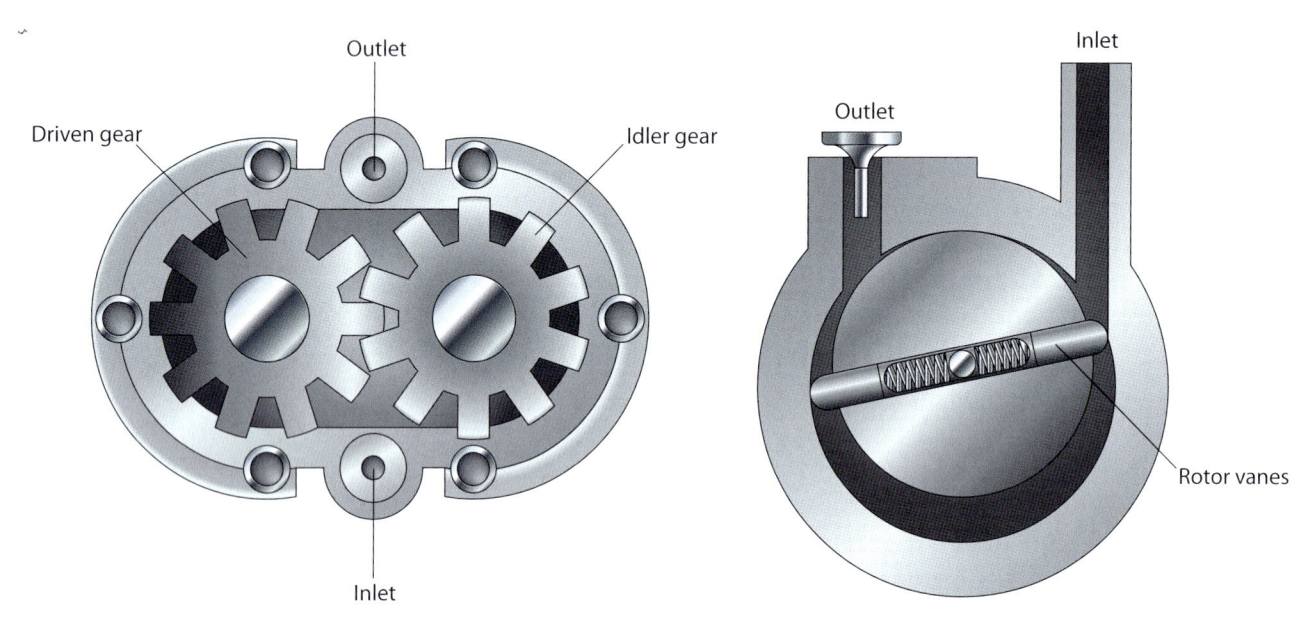

Figure 3.30 Gear engine oil pump

Figure 3.31 Vane engine oil pump

Oil galleries

Oil galleries are passageways that are cast in the cylinder block and head during manufacture. They direct oil to vital engine components and return it to the sump when finished.

Positive crankcase ventilation (PCV)

When the engine is running, it is possible for large amounts of pressure to be created in the crankcase by the pumping actions of the pistons and cylinder **blow-by gas**. If crankcase pressure is left unchecked, oil leaks and engine damage might occur.

In older engines this pressure was allowed to escape to the outside air, but the fumes from this caused environmental pollution, which is no longer legal. Instead, a system called positive crankcase ventilation (PCV) has been created.

In PCV, a series of pipes and valves are connected to the crankcase. These are able to vent pressure and fumes into the engine's inlet manifold, where they become involved in the normal combustion process. This reduces emissions released to the atmosphere.

Pressure relief valve

An oil pump is capable of providing sufficient quantities and oil pressure, even when the engine is running slowly. As the engine speeds up, oil flow and pressure rise. To prevent this becoming too much, an oil pressure relief valve is fitted. This consists of a spring-loaded plunger, which is normally fitted near the output of the oil pump. As the engine runs and the oil pump operates, spring pressure holds the plunger closed. As engine speed increases and oil pressure rises, the plunger is lifted off its seal and any excess oil or pressure is returned to the sump. This allows a constant oil pressure to be maintained under all engine operating conditions, as shown in Figure 3.32.

> **Did you know?**
> Positive crankcase ventilation (PCV) is also sometimes known as the engine's breather system.

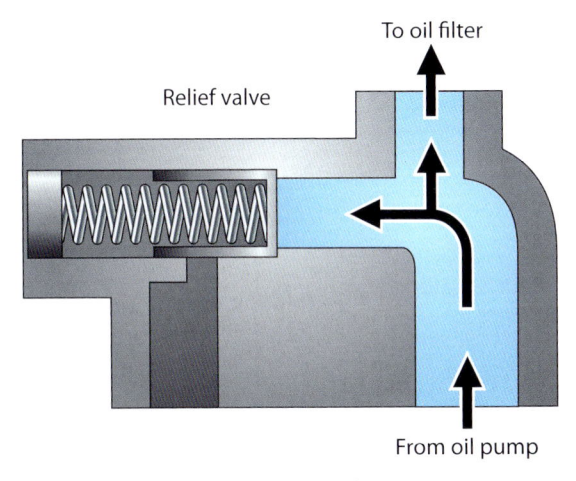

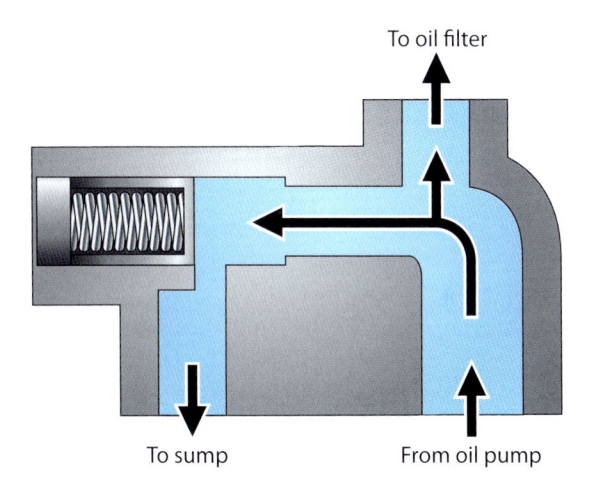

Figure 3.32 Engine oil pressure relief valve operation

Oil filter

As one of the functions of engine oil is to remove combustion **by-products** and dirt, a filter is included in the system. The job of the filter is to maintain a clean flow of oil to the engine's mechanical parts. The oil filter is a replaceable pleated paper element, designed to strain small particles of dirt and debris from the lubrication oil. It is fitted in line with the oil pump. It is usually contained in a housing designed to allow easy access, so it can be replaced during routine maintenance.

The construction and operation of light vehicle engine lubrication components and systems

Full-flow lubrication system

In a **full-flow lubrication system**, oil is pumped from its storage area through a filter element, to help remove dirt particles. It is then fed on to the rest of the engine. In this way, all the oil travelling to the engine has been filtered.

> **Key terms**
>
> **Blow-by gas** – as the extreme pressures of combustion push the piston downwards on its power stroke, a small amount of combustion gas is forced past the piston rings into the crankcase. This is known as blow-by gas.
>
> **By-products** – waste left over from the combustion process.
>
> **Full-flow lubrication system** – in this system all of the engine oil is filtered.

Over a period of time the filter element can become blocked. If this happens the engine might become starved of lubricating oil, which could lead to overheating and seizure. As a result, full-flow oil filters are equipped with a small bypass valve. If the filter becomes blocked, pressure inside the filter will rise, lifting the valve off its seat and allowing oil to bypass the filter element. In this way, the engine will still receive lubricating oil even if the filter is blocked (because dirty oil is better than no oil).

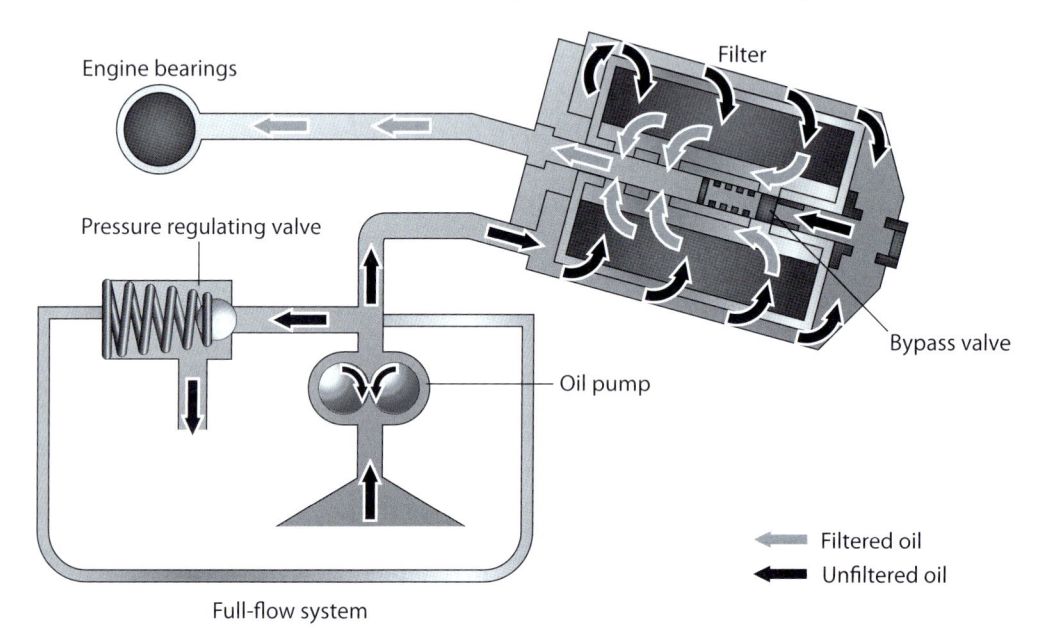

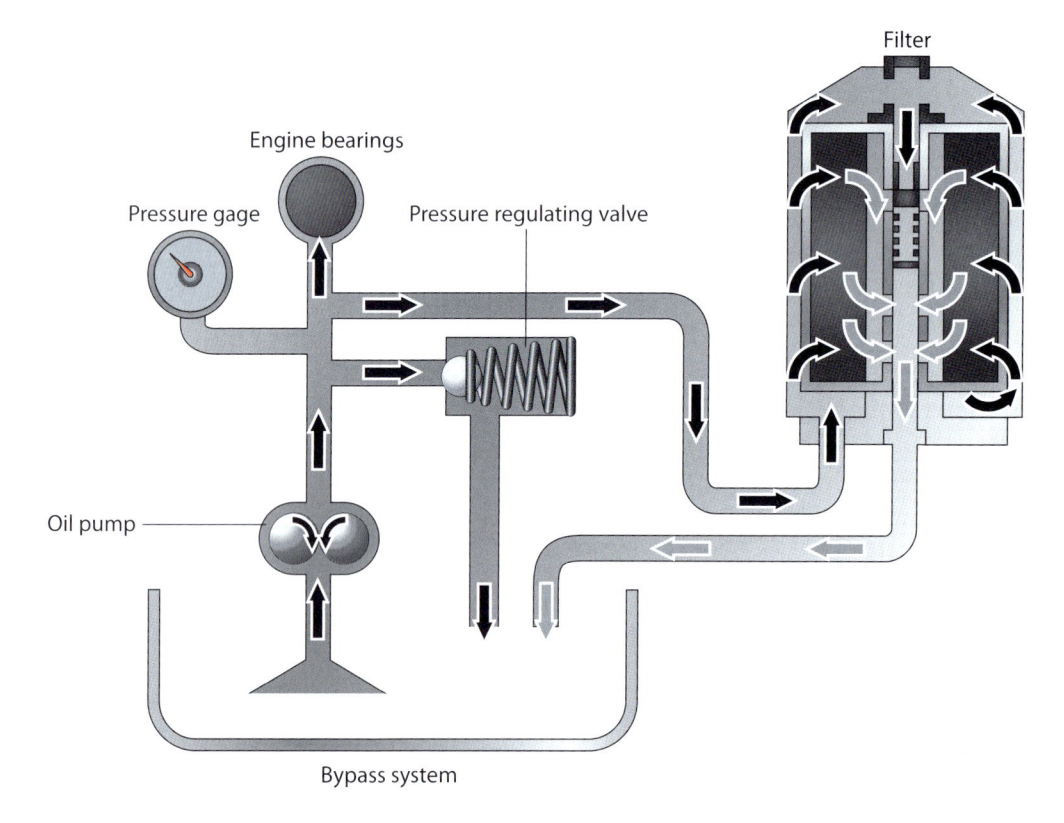

Figure 3.33 Engine oil full-flow and bypass lubrication system

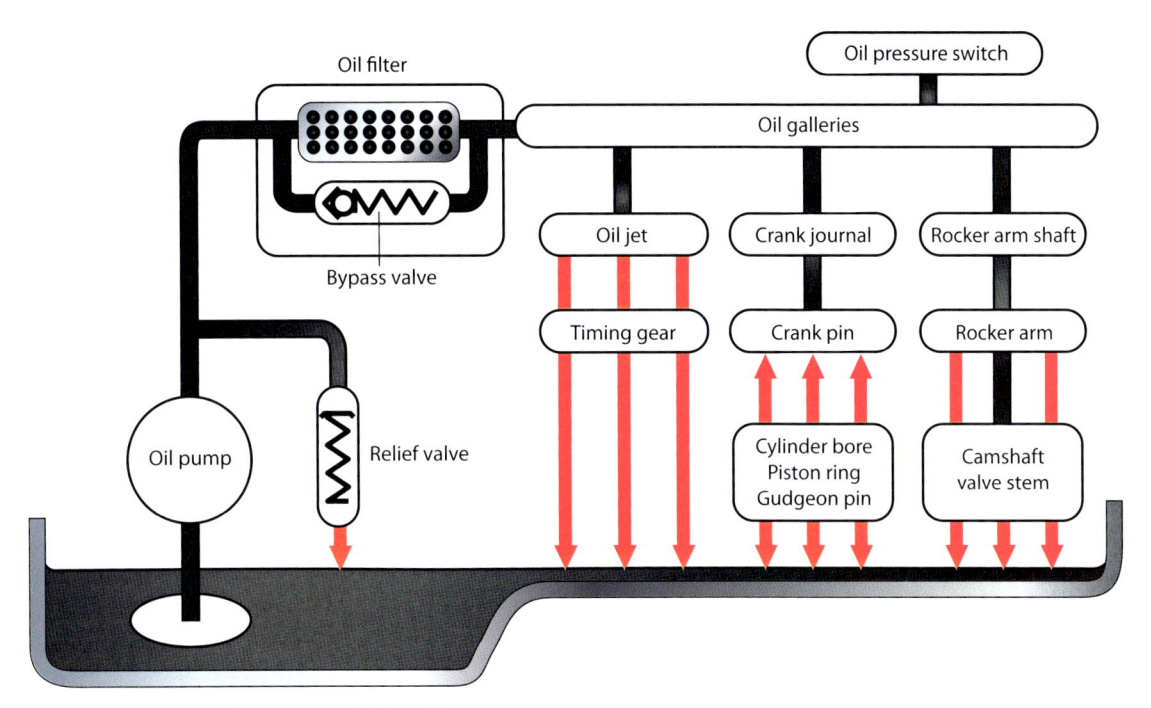

Figure 3.34 Flow of oil around the full-flow lubrication system

Bypass lubrication system

In a **bypass lubrication system**, a much finer oil filter is used, which is able to remove more dirt than in a full-flow system. It is connected to a passage parallel to the one which takes the lubricating oil to the rest of the engine. When oil is pumped from the sump, only some of the oil goes to the engine – the rest is forced through this filter element and back to the sump. Because of this, the oil within the sump is always relatively clean and, unlike the full-flow system, the bypass system does not have to filter all of the oil.

> **Key term**
>
> **Bypass lubrication system** – in this system only some of the oil is filtered.

Wet sump lubrication system

In a wet sump system, the storage area for the oil is a container at the bottom of the engine, known as a sump. When used, a pump draws oil from the sump and pushes it around the engine to where lubrication is required. When this process is complete the oil drains back to the sump for recycling.

An advantage of a wet sump lubrication system is:

✓ Because it normally sits in the air stream below the engine, it assists with the cooling of the engine oil.

The disadvantages are:

✗ Manufacturers have to allow for the wet sump area within the design of their vehicle's front suspension, steering and height. This has led to smaller and smaller sumps being used.

✗ As the size of the sump gets smaller, it can contain less oil, which means it has to work much harder to lubricate the engine's mechanical components.

Dry sump lubrication system

In a dry sump lubrication system, the storage area for the oil is a separate tank. Where the sump would normally be, an oil collection pan is used. The oil tank can be mounted away from the engine but it works in a similar way to a wet sump. Oil is pumped from the tank around the engine and, when complete, the engine oil falls back to the oil collection pan. It is then gathered up by a scavenge pump and returned to the oil storage tank for recycling.

The advantages of this type of lubrication system are:

✓ Manufacturers no longer have to design the front suspension, steering and height of the vehicle to accommodate a wet sump.

✓ A larger quantity of oil can be stored in a dry sump lubrication tank. This is an advantage because when more oil is used, the system has doesn't have to work so hard.

A disadvantage is:

✗ It is expensive to create a dry sump lubrication system.

Oil coolers

One of the functions of engine lubricating oil is to help cool the engine. The engine oil comes into contact with very hot parts of the engine (pistons, for example) that coolant does not. This heat is carried away with the oil, usually back to the sump, where it is cooled.

On some higher-performance vehicles, the heat generated is so great that another method of cooling the oil is needed. Oil coolers are used to do this. These can be either small radiators mounted in the air stream or **heat exchanger units**.

Key engineering principles that are related to light vehicle engine lubrication systems

Classification of lubricants

There are two main classifications for engine lubrication: SAE and API.

SAE (the Society of Automotive Engineers)

SAE is a classification of the lubricant's **viscosity**. The lubricant is given an SAE viscosity grade. The SAE number is calculated by measuring the time it takes for oil to flow through an accurately sized tube at a certain temperature.

The oil is normally tested at −18 °C to represent winter conditions, and its number will be followed by the letter W to show its cold viscosity.

It will also be tested at 99 °C to represent a hot running engine (no W).

The higher the number, the more slowly the oil will flow and the 'thicker' it will be.

The recommended SAE viscosity grades for different operating temperatures are given in Table 3.1.

> **Did you know?**
>
> Three main types of engine oil are available:
>
> 1. Mineral – processed from crude oil
> 2. Synthetic – created from chemicals in a laboratory or factory
> 3. Semi-synthetic – a blend of a mineral base with added synthetic chemicals.

> **Key terms**
>
> **Heat exchanger unit** – coolant is passed through pipes next to the oil and helps to transfer some of the heat away.
>
> **Viscosity** – resistance to flow. This can be described as the 'thickness' of the lubricant.

Table 3.1 SAE viscosity grades

Climate	Minimum temperature (Low)	Maximum temperature (High)	Recommended SAE grade	Viscosity
Sub-zero	−30 °C	−18 °C	0 W	Thick
Freezing	−25 °C	0 °C	5 W	
Winter	−20 °C	0 °C	10 W	
Normal	4 °C	36 °C	30	
Warm	10 °C	43 °C	40	
Hot	16 °C	43 °C	50	
Very hot	26 °C	43 °C	60	Thin

Many modern engine oils are a blend of viscosities, with additives known as viscosity improvers. This means that the oil stays relatively stable throughout a large range of temperatures. It is important to select the correct viscosity of oil for an engine:

- If it is too viscous, it will be slow to reach engine components and won't adequately lubricate.
- If it is not viscous enough, the oil will be squeezed out from between moving components. This will leave the components in contact with each other, creating friction, heat, wear and possibly seizure.

Vehicle manufacturers supply recommendations for the type of oil to be used under particular running conditions. If the engine oil is a blend of viscosities, it is known as a multigrade oil.

- The lower number indicated on the multigrade (sometimes followed by the letter 'W') shows how the engine oil will perform at low temperatures (down to −18 °C).
- The higher figure will show how the engine oil performs when hot (up to 99 °C).

If there is a large difference between these two numbers, the engine oil has a high viscosity index. This means that the viscosity changes very little in normal operating temperatures.

Find out

Using information sources available to you, research two current production vehicles (one petrol and one diesel). For each vehicle, find out:

- the recommended viscosity rating (SAE)
- the engine's oil capacity.

API (the American Petroleum Institute)

The API classification shows the quality of the engine oil being used. This classification usually consists of two letters grouped together.

- If the two letters begin with an 'S', the engine oil is designed for spark ignition (petrol) engines.
- If the two letters begin with a 'C', the engine oil is designed for compression ignition (diesel) engines.

Some engine oils are designed to be used in both spark ignition and compression ignition engines – this will be shown on the information label.

The second letter of the grouping is a designation of the quality of the engine oil. Normally, the further through the alphabet the second letter is, the higher the quality of the engine oil (see Table 3.2).

Table 3.2 API engine oil categories

Petrol engines		
Category	Status	Service
SM	Current	For all automotive engines currently in use. Introduced 30 November 2004, SM oils are designed to provide improved oxidation resistance, improved deposit protection, better wear protection, and better low-temperature performance over the life of the oil. Some SM oils may also meet the latest ILSAC specification and/or qualify as Energy Conserving.
SL	Current	For 2004 and older automotive engines.
SJ	Current	For 2001 and older automotive engines.
SH	Obsolete	For 1996 and older automotive engines. Valid when preceded by current C categories.
SG	Obsolete	For 1993 and older engines.
SF	Obsolete	For 1988 and older engines.
SE	Obsolete	CAUTION – Not suitable for use in petrol-powered automotive engines built after 1979.
SD	Obsolete	CAUTION – Not suitable for use in petrol-powered automotive engines built after 1971. Use in more modern engines may cause unsatisfactory performance or equipment harm.
SC	Obsolete	CAUTION – Not suitable for use in petrol-powered automotive engines built after 1967. Use in more modern engines may cause unsatisfactory performance or equipment harm.
SB	Obsolete	CAUTION – Not suitable for use in petrol-powered automotive engines built after 1963. Use in more modern engines may cause unsatisfactory performance or equipment harm.
SA	Obsolete	CAUTION – Not suitable for use in petrol-powered automotive engines built after 1930. Use in modern engines may cause unsatisfactory engine performance or equipment harm.

3 Light vehicle engine mechanical, lubrication & cooling system units & components

Diesel engines

Category	Status	Service
CI-4	Current	Introduced in 2002. For high-speed, four-stroke engines designed to meet 2004 exhaust emission standards implemented in 2002. CI-4 oils are formulated to sustain engine durability where exhaust gas recirculation (EGR) is used and are intended for use with diesel fuels ranging in sulphur content up to 0.5% weight. Can be used in place of CD, CE, CF-4, CG-4 and CH-4 oils. Some CI-4 oils may also qualify for the CI-4 PLUS designation.
CH-4	Current	Introduced in 1998. For high-speed, four-stroke engines designed to meet 1998 exhaust emission standards. CH-4 oils are specifically compounded for use with diesel fuels ranging in sulphur content up to 0.5% weight. Can be used in place of CD, CE, CF-4 and CG-4 oils.
CG-4	Current	Introduced in 1995. For severe duty, high-speed, four-stroke engines using fuel with less than 0.5% weight sulphur. CG-4 oils are required for engines meeting 1994 emission standards. Can be used in place of CD, CE and CF-4 oils.
CF-4	Current	Introduced in 1990. For high-speed, four-stroke naturally aspirated and turbocharged engines. Can be used in place of CD and CE oils.
CF-2	Current	Introduced in 1994. For severe duty, two-stroke cycle engines. Can be used in place of CD-II oils.
CF	Current	Introduced in 1994. For off-road, indirect-injected and other diesel engines including those using fuel with over 0.5% sulphur. Can be used in place of CD oils.
CE	Obsolete	Introduced in 1985. For high-speed, four-stroke naturally aspirated and turbocharged engines. Can be used in place of CC and CD oils.
CD-II	Obsolete	Introduced in 1985. For two-stroke cycle engines.
CD	Obsolete	Introduced in 1955. For certain naturally aspirated and turbocharged engines.
CC	Obsolete	CAUTION – Not suitable for use in diesel-powered engines built after 1990.
CB	Obsolete	CAUTION – Not suitable for use in diesel-powered engines built after 1961.
CA	Obsolete	CAUTION – Not suitable for use in diesel-powered engines built after 1959.

Properties of lubricants

The 'oiliness' of an engine lubricant is its ability to stick to surfaces. This is important in order to achieve adequate lubrication.

Methods of reducing friction

To improve the lubricating, cooling and cleaning properties of engine oil, additives are introduced:

- **Detergents** – These are used to help keep the oil clean, and disperse sludge and varnish build-up.
- **Anti-oxidants** – These are included in many additives to help reduce the possibility of the oil burning and creating a hard, tar-like substance.
- **Anti-foaming agents** – These help prevent the formation of air bubbles as oil is churned around in the engine.

- **Anti-corrosion agents** – These help counteract the damaging effects of exhaust blow-by gases (which have escaped past the piston rings and entered the crankcase). Acids from these gases can eat into metal-bearing surfaces, causing pitting.
- **Extreme pressure and anti-wear additives** – These include phosphorous and zinc, which help reduce damage where metal is in contact with metal.

> **Working life**
>
> Allan is checking the oil level on an engine. He has the front of the car jacked up and safely placed on axle stands. When he checks the dipstick, the level is below the minimum mark. Without checking the manufacturer data requirements, he gets a can of oil from the storeroom and tops it up to the maximum line.
>
> 1. What mistakes has Allan made?
> 2. Why should you always use manufacturer's data?
> 3. What are the possible consequences of Allan's actions?

CHECK YOUR PROGRESS

1. List five lubrication system components.
2. Name two types of lubrication.
3. Name two types of oil filtration.
4. What does SAE stand for?
5. What does API stand for?
6. Explain the term 'viscosity'.

Engine cooling, heating and ventilation systems operation

An internal combustion engine uses thermal energy (heat) to provide **propulsion** to push the car along the road. This stored energy is released during the combustion process when the air/fuel mixture is ignited. It creates heat and expansion to push pistons up and down or backwards and forwards.

During normal operating conditions, extremely high temperatures are created within the combustion chamber – these can exceed 2500°C. If the engine was not properly cooled, then this heat would lead to rapid expansion – this would create seizure or certain components would melt.

A large amount of the heat created during combustion is removed from the engine along with the exhaust gases. Some heat is absorbed by engine oil, but the remainder has to be taken away by the engine's cooling system.

Types of engine cooling systems

Engine cooling systems can be air based or liquid based.

Air cooling systems

In air-cooled systems, during the manufacturing process, fins are created around the engine cylinders to increase the surface area. When air moves across them, the heat is transferred into the air and then away from the engine (see Figure 3.35).

The advantages of air-cooled systems are:

- ✓ They can operate at far higher temperatures than a standard liquid-cooled system.
- ✓ They require less maintenance.
- ✓ There is no coolant to leak.
- ✓ They are often lighter in construction.

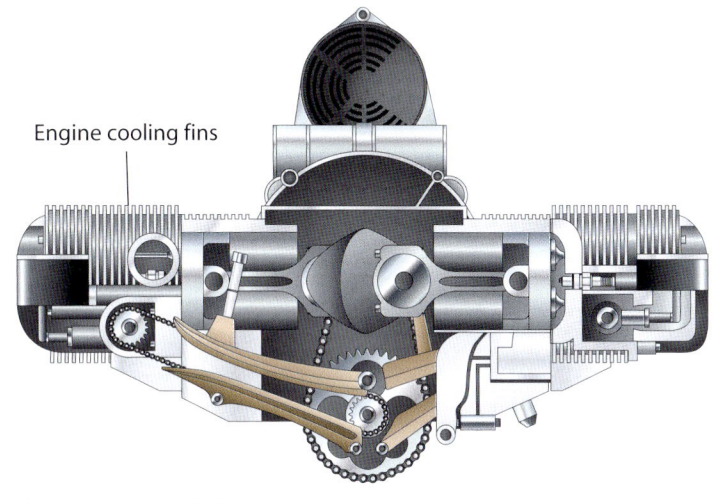

Figure 3.35 Air-cooled engine

The disadvantages of air-cooled engines are:

- ✗ The cooling of the engine is uneven.
- ✗ The manufacture of the components may be more complicated.
- ✗ The engines are often noisier than a liquid-cooled system (because the liquid provides some sound proofing).

Liquid cooling systems

To produce a liquid cooling system, galleries (waterways) are cast into the cylinder block and cylinder head during manufacture. These galleries are often called the water jacket. The engine is connected to an external radiator which can transfer the heat energy from the engine to the surrounding air. To make sure that the coolant within the system is circulated adequately, an engine-driven **water pump** is used to move the coolant around the system.

> **Key terms**
>
> **Propulsion** – a driving force.
>
> **Water pump** – an engine-driven pump used to circulate coolant around the engine.

Coolants

Early liquid-cooled systems used water as a base (as water can absorb more heat energy by volume than any other liquid). Unfortunately, water has certain properties that make it unsuitable for use in modern liquid-cooled engines:

- Water promotes corrosion inside the engine, leading to the breakdown of metal components.
- Water has a peculiar property. When most substances are heated they expand, and when they are cooled they contract. Water is the same, except that when it is cooled down to 0°C and begins to turn to ice, it begins to expand again. (This is because of the formation of ice crystals.) If left unchecked, this expansion is strong enough to cause engine damage to metal components. When sealed inside an engine cooling system, the expansion of ice crystals can crack engine casings and cylinder blocks.

Level 2 Light Vehicle Maintenance & Repair

> **Did you know?**
>
> Monoethylene glycol was a chemical added to water, used as a base for antifreeze. Many manufacturers have now replaced this with polypropylene glycol as it is biodegradable and more environmentally friendly.

Modern cooling systems use a water-based coolant with additives to reduce corrosion and the production of metal salts. **Antifreeze** is used to lower the freezing point of the liquid within the cooling system.

Many modern engine coolants come pre-mixed from the manufacturer, but if you need to mix antifreeze with water, Table 3.3 shows the amounts you should use and how much protection it should give.

Table 3.3 Percentages of antifreeze needed in engine coolant

Percentage of antifreeze added to coolant	Lowest temperature at which this percentage provides protection
25%	−10°C
33%	−15°C
40%	−20°C
50%	−30°C

The advantages of liquid-cooled systems are:

✓ They provide more even cooling of engine components.
✓ They run much more quietly than air-cooled engines.
✓ They allow an efficient heating and ventilation system for the driver and passengers.

Some disadvantages of liquid-cooled systems are:

✗ Water boils at 100°C, and when it boils it changes to steam. Once the water has changed to steam, it no longer acts as an efficient coolant.
✗ The cooling system will be complicated, with extra components such as radiators, water pumps and thermostats. Because of this, there is more to maintain, and more to go wrong.

Radiator and radiator cap

Heat is a form of energy, and energy cannot be created or destroyed, but only converted into something else or transferred. The heat absorbed by the liquid coolant from the engine mechanical components is transferred into the surrounding air using a **radiator**.

A radiator is a series of metal tubes, surrounded by cooling fins (thin pieces of corrugated metal designed to increase the surface area). The radiator is mounted away from the engine, in a position that allows airflow to pass over it and **dissipate** heat to the surrounding air. The pipework in the centre of the radiator zigzags backwards and forwards. This means that the coolant passing through it is in contact with the air for the longest time possible, which makes it efficient at getting rid of the heat.

- If the pipework is mounted top to bottom (up and down) the radiator **core** is known as **upright** (see Figure 3.37).
- If the pipework is mounted side to side, the radiator core is known as **crossflow** (see Figure 3.38).

> **Did you know?**
>
> During the manufacture of cylinder blocks and heads, holes are left in the castings to help remove sand from the water jackets. (The sand is used in the moulding process.) Small metal caps called 'core plugs' are used to block off the holes left in the cooling system. Although not part of the design process, these core plugs have been known to fall out during the freezing of a cooling system. This can help relieve pressure and save engines from damage caused by ice.

Figure 3.36 A core plug (sometimes called a welch plug)

3 Light vehicle engine mechanical, lubrication & cooling system units & components

Figure 3.37 Cooling system radiator (upright)

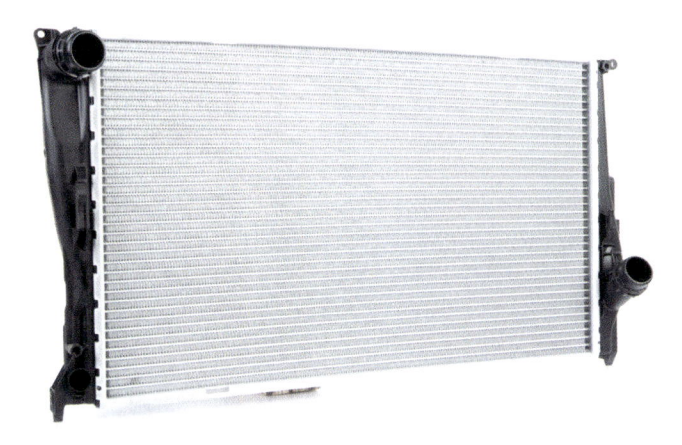

Figure 3.38 Cooling system radiator (crossflow)

Pressurising the cooling system

Antifreeze raises the boiling point of the water slightly. Under normal circumstances, this would not be enough to stop the coolant boiling at some stage within the engine system. A method used to raise the boiling point still further is to pressurise the cooling system.

Pressure has a direct effect on the boiling point of water:

- If the pressure is lowered, the boiling point is lowered.
- If the pressure is increased, the boiling point is raised.

To raise the boiling point of the cooling system, it can be pressurised by sealing it with a **radiator cap**, also called a **pressure cap**. As the coolant warms up, it tries to expand, but it has nowhere to go (because the radiator cap is sealing the system), so pressure increases. This pressure increase raises the boiling point of the liquid coolant in the system.

To make sure that the pressure does not continue increasing past safe limits, the radiator cap contains a spring-loaded valve (see Figure 3.39). When a preset pressure is reached, the valve releases, allowing some coolant to escape.

As the system cools down, pressure falls. As some of the coolant has been allowed to escape past the radiator cap, this pressure fall would create a **vacuum**, making the cooling system hoses collapse. To overcome this, the radiator cap is fitted with another valve that works in the opposite direction. As cooling system pressure falls, this valve opens, allowing the expelled coolant to be drawn back into the system and keeping it topped up (see Figure 3.39).

> **Key terms**
>
> **Antifreeze** – an additive introduced to a water-based liquid coolant to lower the freezing point of the coolant.
>
> **Radiator** – a unit designed to transfer heat energy to the surrounding air.
>
> **Dissipate** – spread out in many directions (usually associated with the movement of heat to air).
>
> **Core** – the central pipework and cooling fins of a radiator.
>
> **Upright** – where the pipework of the radiator core runs vertically (top to bottom).
>
> **Crossflow** – where the pipework of the radiator core runs horizontally (side to side).
>
> **Pressure cap/radiator cap** – a spring-loaded cap designed to raise the pressure in the cooling system, and therefore the boiling point of the coolant.

237

Level 2 Light Vehicle Maintenance & Repair

> **⚠ Safe working**
>
> If the cooling system pressure is suddenly released by opening the radiator cap, the coolant can boil. It will then gush out of the system. This could cause burns and scalding, so never open the radiator cap when the engine is hot. Once it is cool, open the cap slowly.

Pressure valve

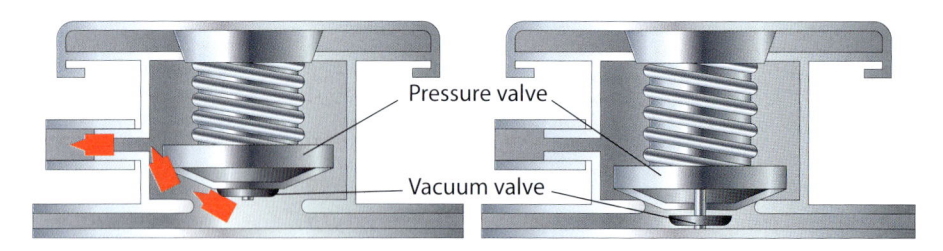

Vacuum valve

Figure 3.39 Cooling system pressure cap, with relief valve open (left) and with pressure valve open (right)

Thermostat

An engine runs most efficiently when its overall temperature is around 100°C (the boiling point of water). This means that when an engine starts from cold, high levels of fuel consumption, emissions and wear are created.

To help with the rapid warm-up of the engine, a **thermostat** is fitted in the system.

A thermostat is a temperature-sensitive valve. It is positioned in the system so that when it is closed, it will stop the flow of coolant into the radiator but will allow coolant to circulate around the engine. This restricts the circulation of water until the engine has reached a certain temperature.

A thermostat is normally filled with a gas or a liquid (usually wax).

- When the engine is cold, the thermostat valve is closed, but as the engine warms up the wax expands. This opens the valve and allows the coolant to circulate freely through the system and radiator.
- If the temperature in the cooling system falls, the wax in the thermostat contracts. This closes the valve and once again restricts the flow of coolant through the radiator.

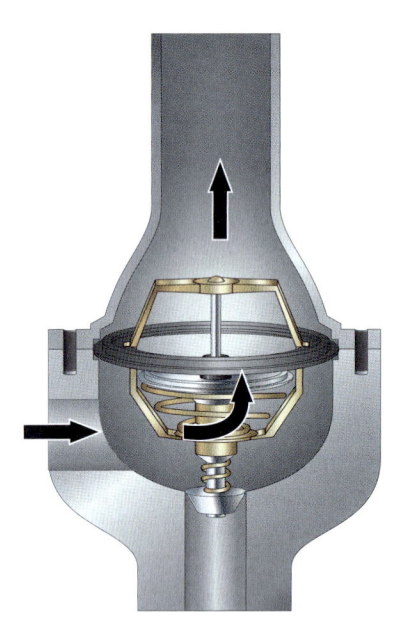

Figure 3.40 Cooling system wax-type thermostat

Expansion tank

The expansion tank is a plastic container which acts as a reservoir for excess coolant. It is connected to the cooling system by a pipe at the radiator cap valve.

- As the cooling system warms up, any coolant that is allowed to escape past the radiator valve is transferred to the expansion tank.
- As the cooling system temperature falls, any losses from the system are drawn back in from the expansion bottle, keeping the system topped up.

Because of this the cooling system is sealed, and if the coolant needs topping up, it is done via the expansion bottle. The expansion bottle normally has a maximum and minimum reading marked on the side, to indicate when there is the correct quantity of coolant. This level should be checked when the engine is cold.

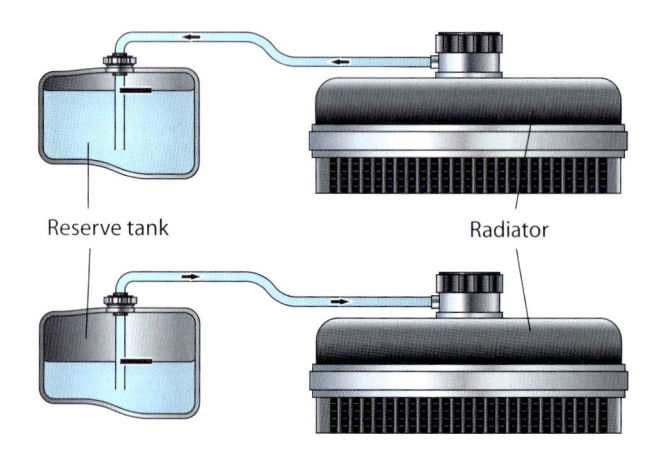

Reserve tank Radiator

Figure 3.41 Cooling system expansion tank

3 Light vehicle engine mechanical, lubrication & cooling system units & components

Pipes and hoses

To transfer coolant from the engine to the radiator, a series of pipes and hoses are used. The radiator can be mounted solidly to the vehicle body, but the engine is normally allowed to move slightly on rubber mountings. If rigid metal pipes were used between the engine and radiator, vibrations could lead to fracture and leakage. Because of this, rubber hoses are used between fixed and movable components so that any vibration is absorbed.

Gaskets and sealing rings

As with other engine components, gaskets and seals are used in cooling systems so that leaks do not occur between surfaces in contact.

Water pump and drive belt

Early engine systems used a process known as **thermosiphon** to move coolant around the engine and radiator.

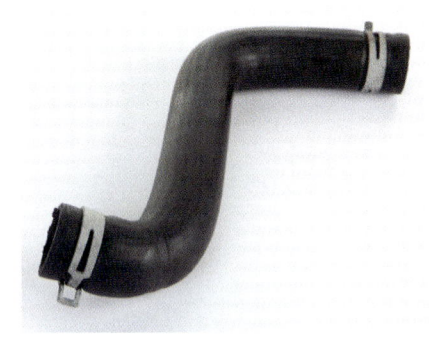

Figure 3.42 Cooling system radiator hose

Key terms

Vacuum – a negative pressure.

Thermostat – a temperature control valve which regulates system temperature.

Thermosiphon – an early type of cooling system. It uses convection currents created in the water to circulate coolant around the system.

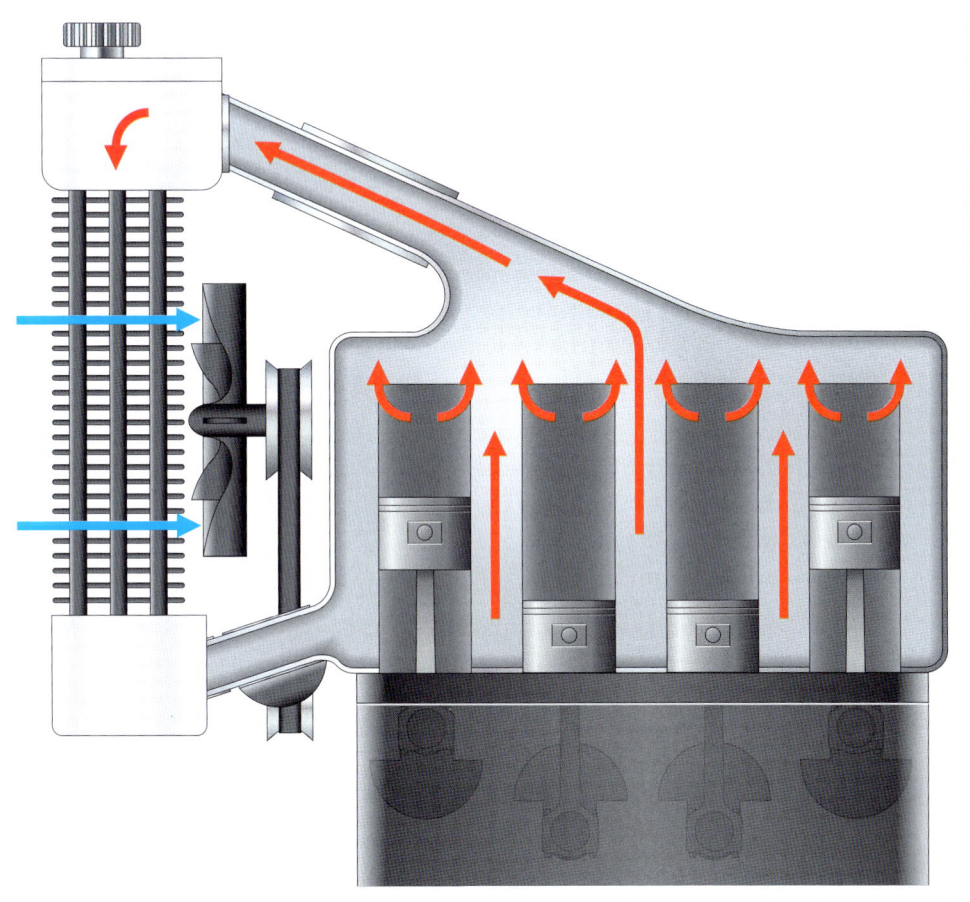

Figure 3.43 Cooling thermosiphon system – warm water rises out of the engine, cools in the radiator and re-enters at the bottom

Level 2 Light Vehicle Maintenance & Repair

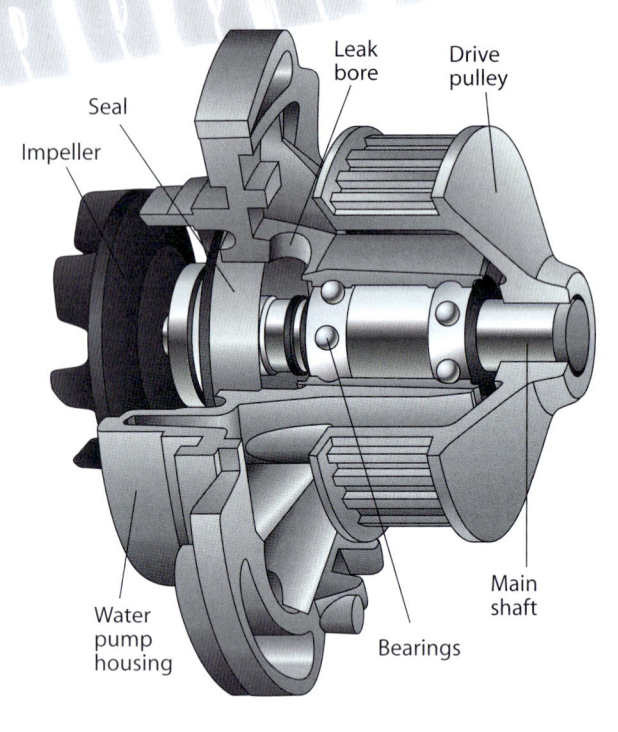

Figure 3.44 Cooling system water pump

As Figure 3.43 shows, this process relied on **convection currents**:

- As water heats up, convection currents make the warm water rise. The warm water exits the engine through a hose slanted upwards, and enters the top of the radiator.
- As the radiator cools the water, it falls to the bottom and re-enters the engine to continue the process.

To help the coolant in a modern engine system circulate, and maintain a relatively even temperature, a mechanically operated pump is used.

The water pump is normally mounted in the engine block. When it is turned, a small set of rotor blades called an impeller is used to drive coolant around the system. The impeller is connected to a shaft, and on the opposite end is a pulley which is connected to a rubber drive belt driven by the engine's crankshaft. This rubber drive belt, sometimes called the fan belt, is normally operated by the front drive pulley of the crankshaft, although some manufacturers are now using the camshaft timing belt to turn the water pump.

Mechanical cooling fans

During normal operation, there may be times when there is not enough airflow over the radiator to allow adequate cooling (for example, when a car is sitting in traffic). Manufacturers include a fan mechanism in their designs so that, in these situations, airflow can be increased by either drawing air through or blowing air over the radiator core.

Early systems had a fan mounted on the water pump pulley. As the pump was turned by the fan belt, it also operated the cooling fan. By running the cooling fan continuously, drag was created and overcooling occurred. This led to poor performance, poor fuel economy and high emissions.

The **viscous coupling** system was developed (see Figure 3.45). This used a liquid or a wax to provide varying amounts of slippage between the drive pulley and fan, depending on engine temperature.

- If the engine was cold, a lot of slippage occurred, and the fan blades hardly turned.
- As temperature increased, the viscous coupling created more drag, making the fan spin faster.

Figure 3.45 Viscous coupling

3 Light vehicle engine mechanical, lubrication & cooling system units & components

Electric cooling fans

A further development is a cooling fan that is driven by an electric motor (see Figure 3.46). This type of fan can be bolted directly onto the radiator unit, so its position is not dictated by engine shape and design.

A thermostatically controlled switch is mounted in the cooling system to control the operation of the fan depending on the engine temperature.

- When the engine temperature reaches a preset value, the switch closes and starts the electric motor which drives the cooling fan.
- As the engine cools down, the switch opens and turns off the electric motor.

Because the electrically driven cooling fan only operates when required, the load on the engine and the electrical system is reduced. This improves the overall engine efficiency, performance and fuel economy, and helps lower emissions.

Vehicle heater

A vehicle heater is created by including a second radiator (called a heater matrix), which is connected to the cooling system inside the vehicle. Hot coolant from the engine is directed by a series of valves or controls so it passes through the heater matrix. An electric fan circulates air through the heater matrix – the air warms up and is used to heat the passenger compartment. This warm air can be directed through vents inside the vehicle to angle it towards the feet, face or windows.

Air conditioning

Heating, ventilation and cooling for the vehicle's occupants can be further improved by the addition of air conditioning. Air conditioning uses an engine-driven pump called a compressor to raise the pressure of a **refrigerant** gas in a sealed system (see Figure 3.47).

The gas passes through a radiator, called a condenser, which is normally mounted just in front of the cooling system radiator. The high-pressure gas is then cooled (condensed) into a liquid. From here it is transferred into a storage unit (called a receiver drier) until it is needed.

When the driver operates controls to lower the cabin temperature of the car, the refrigerant is released through an expansion valve. As the pressure falls, the liquid refrigerant changes state in another small radiator inside the car called an evaporator. The temperature in the evaporator falls, and as the cabin air is circulated through it, heat is removed. This helps cool the air inside the car. The refrigerant is then returned to the compressor, where the whole process starts once more.

> **Safe working**
>
> If the engine is hot, the electric fan can cut in at any time without warning, even if the engine is not running.
>
> If you are working on a hot engine, it is advisable to isolate the fan electrics to reduce the risk of getting caught in the fan blades. (Don't forget to reconnect it when you have finished.)

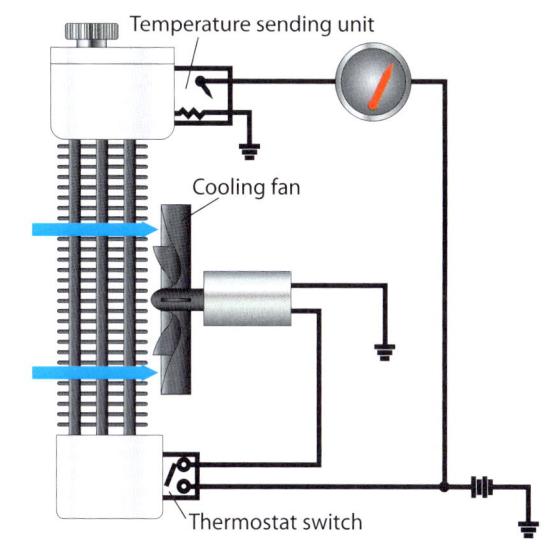

Figure 3.46 Electric cooling fan circuit

> **Key terms**
>
> **Viscous coupling** – a drive coupling that uses a fluid to transmit drive.
>
> **Refrigerant** – cooling gas.
>
> **Convection currents** – the movement of heat energy through a gas or a liquid.

Level 2 Light Vehicle Maintenance & Repair

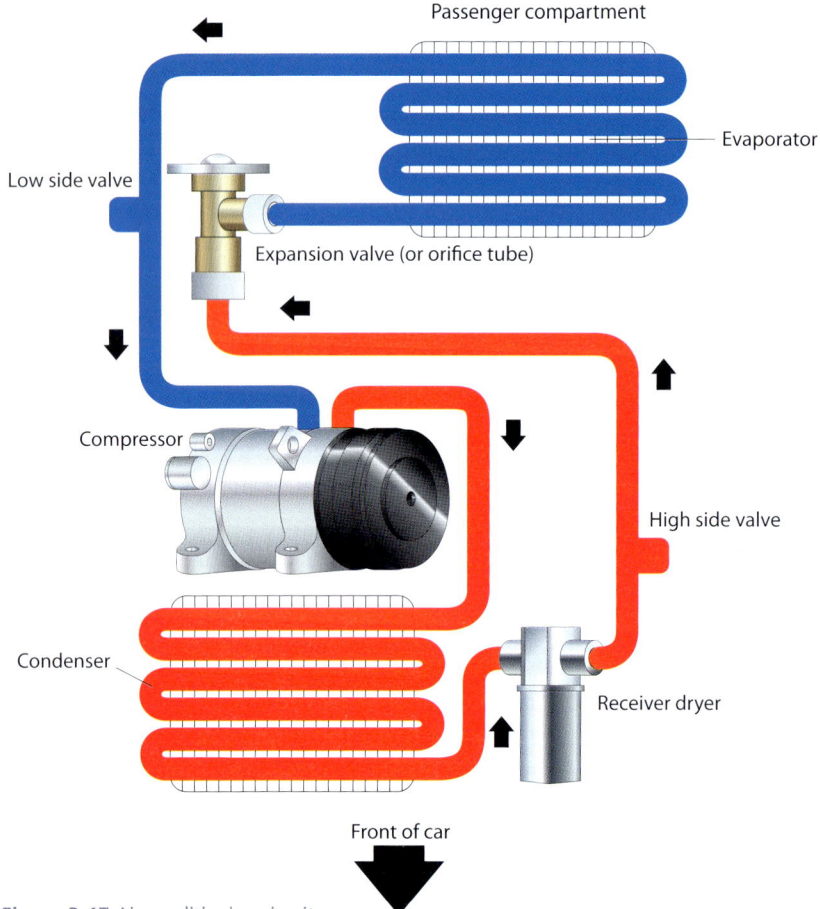

Figure 3.47 Air conditioning circuit

Key engineering principles that are related to light vehicle engine cooling, heating and ventilation systems

Heat transfer

Heat is a form of thermal energy, and it will always move from a hotter to a colder substance. There are three methods of heat transfer: conduction, convection and radiation.

- **Conduction** is the movement of heat energy through solids. For example, if you hold a metal rod at one end, and the other end is heated, the heat will travel through the solid material via conduction until it reaches your hand.
- **Convection** is the movement of heat energy through liquids or gases. For example, a convection heater in a room uses an electrical element to warm up the air. As the air is heated, it rises up to the top of room. The air cools and begins to fall to the bottom of the room, where it can once again enter the convection heater element. This circulation of warm and cool air is called convection current.
- **Radiation** is the movement of heat energy through a vacuum (or space) as electromagnetic energy. This radiation energy is often within the **infrared spectrum**, and is normally felt as heat.

Key terms

Infrared spectrum – a form of invisible electromagnetic radiation.

Sensible heat – the heat energy contained within a substance where it stays as one particular state of matter (i.e. solid, liquid or gas) and does not change to another state.

Linear and cubical expansion

As substances heat up they expand. This expansion can be linear (in a line) or cubical (in all directions at once).

Specific heat capacity

Specific heat capacity is the amount of heat energy that can be stored within a particular substance without it changing state. (The three main states of matter are: solid, liquid, gas. For example, H_2O as a solid is ice, as a liquid it is water and as a gas it is steam). Any temperature rise where a substance stays within a certain state (i.e. solid liquid or gas) is called **sensible heat**.

Boiling point of liquids

The boiling point of liquids is affected by pressure. The Celsius temperature scale is based on the boiling and freezing points of water, which freezes at 0 °C and boils at 100 °C. If the pressure around water is raised or lowered, this boiling point changes.

- If pressure is increased, the boiling point of water rises.
- If pressure is reduced, the boiling point of water falls.

> **Did you know?**
> If you climb to the top of Mount Everest and boil a kettle to make tea, because of the low air pressure, the water will not reach 100°C, but instead will boil at 69°C. No matter how long you try to boil the kettle, the water will get no hotter, because at boiling point it changes state into steam.

CHECK YOUR PROGRESS

1 List five engine cooling system components.
2 Describe how coolant is circulated to cool an engine.
3 List three differences between liquid-cooled and air-cooled engines.
4 What are the three methods of heat transfer?

How to check, replace and test light vehicle engine mechanical, lubrication and cooling system units and components

Removing and replacing engine mechanical, lubrication and cooling system units and components

Some common testing methods used to check the operation of engine mechanical, lubrication and cooling systems are: compression testing, oil pressure testing, cooling system pressure testing and antifreeze testing.

> **Safe working**
> If engine mechanical, lubrication or cooling components need replacing, where possible allow the engine to cool first. This will reduce the health and safety risks, and can also help prevent damage to engine mechanical components.

Level 2 Light Vehicle Maintenance & Repair

> **! Safe working**
>
> As you dismantle an engine, always store components carefully to keep them clean. If you don't do this, it's possible that damage will occur when the engine is reassembled and started.

Compression testing

One method of testing the mechanical operation and efficiency of an engine is to conduct a compression test. A compression gauge will measure the amount of pressure created by the piston in the cylinder.

You can do this test on both petrol and diesel engines, but you will need to use different compression gauges, because of the higher pressures involved with compression ignition engines.

Carrying out a compression test

Checklist			
PPE	**VPE**	**Tools and equipment**	**Source information**
• Steel toe-capped boots • Overalls • Latex gloves	• Wing covers • Steering wheel covers • Seat covers • Floor mat covers	• Compression gauge • Spark plug socket and ratchet • Sockets or spanners for injector removal • Torque wrench • Oil can	• Compression technical data • Spark plug torque setting information • Job card

1. Isolate the engine so that it cannot start. With a petrol engine this will normally involve disconnecting the ignition system. With a diesel engine this will normally involve interrupting the fuel supply.

2. On a petrol engine, gain access to and remove all of the spark plugs.

   On a diesel engine, gain access to and remove all of the injectors. (Make sure you clean up any spilled fuel.)

3. You can now attach the compression gauge to the engine. First fit the adapter, in place of the spark plugs or injectors.

   Now attach the compression gauge to the adapter.

4. Once the gauge is attached to a cylinder, crank the engine (with the throttle held wide open on petrol engines). Continue to crank the engine for a short period of time, until the maximum reading is shown on the gauge.

5. Record the reading and compare the figure with manufacturer's specifications. (See Figure 3.48 for an example.)

   Repeat the procedure for the remaining cylinders.

6. Once you have tested all of the cylinders, compare the results obtained against each other.

   If one or more compressions is lower than expected, further investigation might be required.

244

3 Light vehicle engine mechanical, lubrication & cooling system units & components

Technical data on the vehicle				
Make	Ford	Date	03-12-2010	
Model	Ka 1,3i	Owner		
Year	1996-1998	Registration No.		
Engine	JJB	VIN		
Variant		1. Reg. Date		

Technical item	Data
Engine	
Engine ID code	JJB
Number of cylinders	4
Number of valves	8, OHV
Capacity/ (bore/stroke)	1299 cm^3 (73,96/ 75,48)
Compression ratio (RON)	9.5: 1 (95 unleaded)
Max. output kW (din hp)/rpm	37 (50)/ 4500
Max. torque NM/rpm	97/ 2000
Engine code location	Engine block left rear side
Vehicle Identification Number location	Floor of vehicle at right front seat
Vehicle Identification plate location	Lockingplate
Production year code in VIN	Second-last character i
Valve clearance, inlet (cold/hot)	0.20 cold i
Valve clearance, exhaust (cold/hot)	0.30 cold (After 20.11.1996 – 0.50 cold) i
Compression pressure, bar	**13.0 - 16.0**
Oil pressure/rpm, bar	0.6/idle speed (1.5/ 2000) (Running temperature)
Radiator cap, bar/thermostat °C	1.2 /88° C
Clutch freeplay, mm	(Pedal travel 123 ± 2 mm)
Repair time: Clutch renewal	2 hours and 45 minutes
Timing chain:	i
Drive belt	i

> **Safe working**
> Make sure that the high-tension ignition electrics are isolated to reduce the risk of electric shock during the compression test.

Figure 3.48 Manufacturer's compression data

Wet test

It is common for compression to leak either from the top of the engine (cylinder head and valves) or at the bottom, past the pistons and rings. To help diagnose the location of compression leakage, you can use a wet test.

A wet test involves putting a small amount of oil down the spark plug or injector hole (see Figure 3.49) and repeating the compression test. The oil will normally sit around the top edge of the piston crown, against the top ring, forming a temporary seal. If the compression rises slightly after oil has been put in the cylinder bore, this can indicate that the piston rings may have worn. If the compression stays low, it can indicate that the compression loss is from the upper part of the engine (head).

Figure 3.49 Putting oil down a spark plug hole

245

Level 2 Light Vehicle Maintenance & Repair

> **Safe working**
> Always clean up any oil spills immediately to avoid the risk of any slips, trips and falls.

Oil pressure test

One method of testing the operation and efficiency of an engine lubrication system is to carry out an oil pressure test. You can do this test on both petrol and diesel engines, and the procedure and tools needed are the same.

Carrying out an oil pressure test

Checklist			
PPE	VPE	Tools and equipment	Source information
• Steel toe-capped boots • Overalls • Latex gloves	• Wing covers • Steering wheel covers • Seat covers • Floor mat covers	• Oil pressure gauge • Sockets or spanners for oil pressure switch removal • Torque wrench	• Oil pressure technical data • Oil pressure switch torque setting information • Job card

1. Locate the oil pressure switch used to control the warning light on the dashboard. (This will be in the side of the engine block or cylinder head, normally around the area of the oil filter.)

2. Remove the oil pressure switch.

3. Connect the oil pressure gauge using an appropriately sized adapter.

4. Now you can start the engine and run it at speeds recommended by the manufacturer. Take the oil pressure readings.

5. Compare the readings obtained with those found in the manufacturer's technical data. (See Figure 3.50 for an example.)

3 Light vehicle engine mechanical, lubrication & cooling system units & components

Technical data on the vehicle

Make	Opel/Vauxhall	Date	03-12-2010
Model	Corsa D 1,2	Owner	
Year	2007–2009	Registration No.	
Engine	Z12XEP	VIN	
Variant		1. Reg. Date	

Technical item	Data
Engine	
Engine ID code	Z12XEP
Number of cylinders	4 i
Number of valves	16, DOHC
Capacity, ccm	1299
Bore, mm	73.4
Stroke, mm	72.6
Compression ratio	10.5 : 1
Max. output hp/rpm	80/ 5600
Max. output kW/rpm	59/ 5600
Max. torque NM/rpm	110/ 4000
Engine code location	Engine block by flywheel
Vehicle Identification Number location	Right basement at front seat
Vehicle Identification plate location	Right B – post
Production year code in VIN	10th digit = model year i
Valve clearance, inlet (cold)	(Hydraulic)
Valve clearance, exhaust (cold)	(Hydraulic)
Cylinder head height new, mm	126.0
Cylinder head height minimum, mm	No grinding
Compression pressure, bar	16
Compression pressure, max. Difference, bar	1
Compression pressure, minimum, bar	14
Oil pressure idling. Minimum, bar	1.5
Thermostat opens, °C	92°
Clutch freeplay, mm	(Hydraulic)
Timing chain:	i
Drive belt	i

Figure 3.50 Manufacturer's oil pressure data

- **High oil pressure** – If the oil pressure is too high, check the oil pressure relief valve, as it may be stuck closed. High oil pressure may also indicate a blockage in the system.

  Example: maximum pressure at 6000 rpm 5 bar. (This will need to be checked against the manufacturer's specifications.)

- **Low oil pressure** – If the oil pressure is too low, check the oil pressure relief valve, as it could be stuck open. Low oil pressure could also indicate that excessive engine wear has taken place.

  Example: minimum pressure at 1000 rpm 1 bar. (This will need to be checked against the manufacturer's specifications.)

Pressure is created by resistance to flow. If engine components such as crankshaft main bearings or big end bearings have worn so that excessive **clearances** exist, the resistance to oil flow will fall, creating low pressure.

> **Safe working**
>
> Always allow the engine to cool before removing the cooling system pressure cap. If you remove the cap while the engine is hot, a sudden drop in pressure may cause the liquid coolant to boil, resulting in scalding or severe burns.

> **Safe working**
>
> To conduct a cooling system pressure test correctly, the thermostat should be open. If you warm up the engine with the pressure cap in place, when you undo the cap, pressure in the system will suddenly fall. This can cause the liquid to boil, which may result in scalding and severe burns.

> **Key term**
>
> **Clearance** – gap between engine components.

Cooling system pressure testing

A cooling system pressure test is used to ensure that the cooling system can be pressurised and that there are no leaks.

Carrying out a cooling system pressure test

Checklist			
PPE	**VPE**	**Tools and equipment**	**Source information**
• Steel toe-capped boots • Overalls • Latex gloves	• Wing covers • Steering wheel covers • Seat covers • Floor mat covers	• Cooling system pressure tester	• Cooling system technical data • Cooling system operating pressure from pressure cap • Job card

1. Allow the engine to cool.

   Once the engine is cool, remove the cooling system pressure cap. Then run the engine until it has reached a normal operating temperature.

2. Now attach the cooling system pressure tester in place of the system pressure cap.

3. Without operating the pump of the pressure tester, start the engine. If you see a rapid rise in pressure on the gauge, this can indicate head gasket failure.

4. Following this, operate the pump on the pressure tester until the gauge reaches the maximum pressure indicated by the vehicle manufacturer. (The maximum cooling system pressure can often be found as a marking on the radiator cap.)

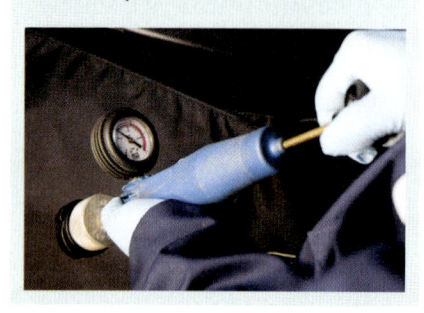

5. Now conduct a visual inspection for leaks. This can include checking the inside of the car in case the heater matrix is leaking. (You can sometimes check for this by feeling the carpet area around the driver and passenger footwell. If the heater matrix is leaking, the carpet will often be wet.)

6. If there are no visible signs of leakage, the car can be left with the cooling system pressure tester attached for a longer period of time. If the gauge pressure falls significantly during this period, there may be a system leak, which might require further investigation.

3 Light vehicle engine mechanical, lubrication & cooling system units & components

Testing the strength of coolant or antifreeze

To make sure that the correct strength of coolant or antifreeze mixture is being used in the system, you can use a hydrometer to check the **specific gravity**.

> **Key term**
>
> **Specific gravity** – the density of a liquid when compared to water.

Using a hydrometer to test strength of coolant/antifreeze

Checklist			
PPE	**VPE**	**Tools and equipment**	**Source information**
• Steel toe-capped boots • Overalls • Latex gloves	• Wing covers • Steering wheel covers • Seat covers • Floor mat covers	• Cooling system hydrometer	• Cooling system technical data • Antifreeze mixture instructions • Job card

1. Allow the engine to cool.

2. Once the engine is cool, carefully remove the radiator pressure cap.

3. Select a hydrometer for the type of coolant in the system. Two main types of coolant are used: monoethylene glycol and polypropylene glycol. Each coolant type has a different hydrometer and will provide antifreeze protection.

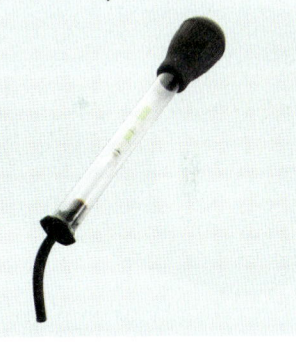

4. Insert the hydrometer into the coolant and take a sample.

   The indicator on the hydrometer will show the antifreeze strength.

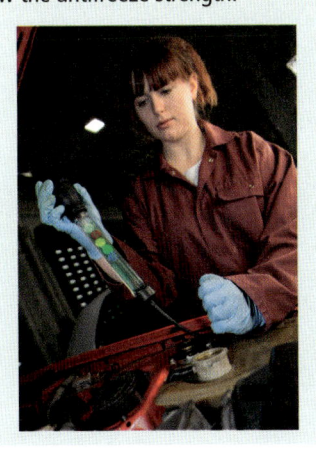

5. While you are carrying out the hydrometer test, this is also a good time to assess the condition of the coolant (check it to see if it is dirty or contaminated).

   If the coolant is below the standard required, you should drain it, flush the system and refill it with the correct quantity and type.

6. If an antifreeze and water mix is to be used, follow the manufacturer's recommendations, including the ratio of water to antifreeze. Mix the antifreeze and water in a separate container and then fill the system.

Level 2 Light Vehicle Maintenance & Repair

> **Safe working**
>
> You must always allow the engine to cool before removing the radiator pressure cap. If you remove the radiator cap while the engine is hot, a sudden drop in pressure may cause the liquid coolant to boil, causing scolding or severe burns.

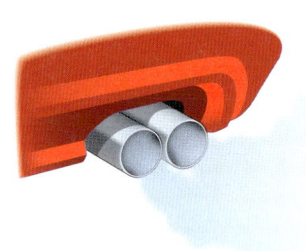

Figure 3.51 Blue smoke (from burning oil) being emitted from a car exhaust

Figure 3.52 A damaged engine cylinder

> **Key terms**
>
> **Valve stem seals** – rubber seals designed to prevent oil leaking passed the inlet and exhaust valves.
>
> **Tappet clearance** – the gap between the end of the valve stem and the operating mechanism.
>
> **Block tester** – a clear cylinder containing a special liquid that changes colour when it comes into contact with hydrocarbons present in exhaust gases. It is used to check for head gasket failure.

Common faults found in light vehicle engine mechanical, lubrication and cooling systems and their causes

Worn piston rings

Two problems can occur if piston rings wear. If the compression rings are worn, cylinder compression is reduced or lost, leading to poor performance on that particular cylinder. If the oil-control ring (the piston ring that helps keep oil below the piston) becomes worn, the engine may start to burn its own lubrication oil. When this happens, the oil level will fall and excessive blue smoke in the exhaust may occur (see Figure 3.51).

- If piston rings are the cause of oil smoke, it will usually be most noticeable during acceleration.
- If lubrication oil is leaking past the **valve stem seals**, oil smoke will usually be most noticeable on start up, or after the engine has been left idling for a short period of time.

Damage to the cylinder walls

Overheating or lack of lubrication can lead to the cylinder walls being damaged. If this happens, you will often find scoring on the piston thrust side of the cylinder. The thrust side of the cylinder is the one that the piston presses against as it moves downwards on its power stroke. The throw of the crankshaft will force the piston to one side and this is where the greatest amount of wear and damage might occur.

Excessive valve clearance

Because of expansion due to heat, inlet and exhaust valves require clearance between them and any operating mechanism (see Figure 3.53). This is often known as **tappet clearance**.

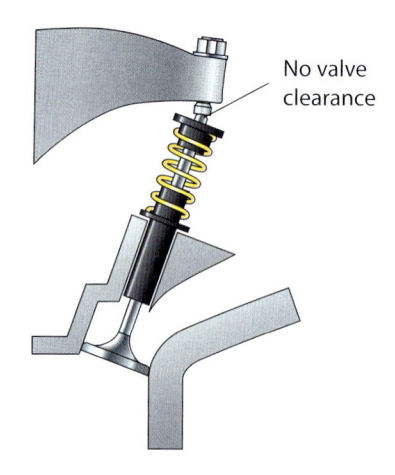

Figure 3.53 Engine valve clearance

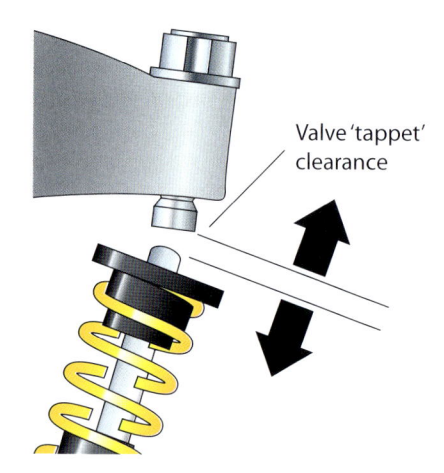

250

If the clearance is too small, two problems can occur:

1. When the valves expand due to heat, they might be held open when they come into contact with the valve-operating mechanism. This will lead to loss of compression and misfiring. Because the valves conduct their heat away through the cylinder head, if they are held open they may not cool and could burn out.
2. If the valve clearance is too small, the operating mechanism will come into contact with the valves sooner, advancing the valve timing.

If the valve clearance is too large, you can often hear a rattling noise from the top end of the engine:

1. The valves may not fully open. This reduces the amount of air/fuel mixture that can be drawn into the cylinder or reduces the amount of exhaust gas that can be expelled. This will affect the volumetric efficiency, and therefore performance.
2. If the valve clearance is too large, the operating mechanism will come into contact later, retarding the valve timing.

Some manufacturers use hydraulic tappets, which are self-adjusting. When the engine is running, oil from the lubrication system fills the hydraulic tappets, making them expand. This takes up any excess valve clearance and, because oil is virtually incompressible, the valve-opening mechanism works as normal.

Cylinder head gasket failure

A head gasket is used to ensure a good seal between the cylinder head and block. The gasket is tightly clamped between two very flat surfaces, preventing loss of compression and leakage of coolant and lubricating oil. Over a period of time, the gasket can break down when exposed to large amounts of heat, as shown in Figure 3.54. Head gasket failure can be indicated by various different symptoms, including:

- overheating
- loss of compression
- oil contamination of the coolant
- coolant contamination of the oil.

Loss of compression into the cooling system

If the loss of compression is into the cooling system, rapid pressure rises occur, often accompanied by overheating. Bubbles can sometimes be seen in the cooling system. If this occurs, it can be checked with a **block tester**:

- With the engine at normal operating temperature, insert the block tester into the neck of the radiator, in place of the pressure cap.
- Then run the engine. Fumes from the cooling system will be drawn through the liquid of the block tester.

Figure 3.54 A damaged cylinder head gasket

Safe working

Where possible, the engine should be allowed to cool before the pressure cap is removed.

- If liquid changes colour, there is a good possibility that exhaust gases are present in the cooling system, suggesting head gasket failure.

The mixing of engine oil and coolant

If oil leaks into the coolant it can often be seen in the expansion bottle – see Figure 3.55.

If coolant leaks into the oil, it can often be seen as a white emulsified, creamy substance – see Figure 3.56.

Figure 3.55 Oil contamination in engine coolant

Figure 3.56 Coolant contamination of engine oil

Oil pressure relief valve failure

The oil pressure relief valve is designed to regulate the lubrication system oil pressure. During normal operation, it should open and close at a preset value, making sure that the pressure does not rise too high or fall too low.

- If the oil pressure relief valve fails in the closed position, system pressure will rise too high as engine speed increases. This can lead to oil leaks and engine damage.
- If the oil pressure relief valve fails in the open position, most of the oil will simply return to the sump area. This means the pressure will fall too low, resulting in engine damage.

Oil pump failure

Mechanical failure of the oil pump can lead to rapid starvation of the engine's lubricating oil. If the warning system is operating correctly, the oil light should be illuminated on the dashboard so that the driver knows there is a problem. If the engine is allowed to continue to run for any period of time once oil pressure is lost, irreparable damage can occur.

Thermostat failure

The thermostat is a temperature-sensitive valve that regulates the cooling system temperature. It is designed to restrict coolant flow when the engine is cold, which assists rapid warm-up. Once the engine has reached its normal operating temperature, the thermostat should open to allow unrestricted flow of coolant to the radiator.

- If the thermostat fails in the open position, the engine will take a long time to reach its normal operating temperature, or it may not reach the temperature at all. If the engine is running cold, it will be inefficient, leading to high fuel consumption, high exhaust emissions and excessive mechanical component wear.
- If the thermostat fails in the closed position, flow to the radiator is restricted. This will lead to overheating of the engine and high cooling system pressures. High pressures will lead to leaks and damaged cooling system hoses. Overheating can also lead to problems such as head gasket failure.

How to overhaul light vehicle engine units

When assessing engines and components for serviceability (operation and usability), it is always a good idea to begin with:

- a visual inspection – look at components for wear, adjustment and condition
- an aural inspection – listen for abnormal or excessive mechanical noise.

Where possible, inspect components for excessive play and security.

Select tools that can make accurate measurements to correctly assess the following aspects of all engine components:

- condition
- size
- wear
- clearance
- levels.

You will need a good source of manufacturer's data so that you can refer to the original specifications and make judgements based on the figures given.

> **Safe working**
>
> Following the replacement of any engine components, the car should be road tested to ensure correct function and operation.

Find out

Make a list of tools and give examples of how and where you would use them when assessing engine condition and serviceability.

CHECK YOUR PROGRESS

1 List three precautions you need to take when working on engine mechanical, lubrication and cooling systems.
2 How can you test the strength of antifreeze in a car's cooling system?
3 List five typical faults that might occur in an engine's mechanical, lubrication and cooling systems.

3 Light vehicle engine mechanical, lubrication & cooling system units & components

FINAL CHECK

1. In a standard (non-turbocharged) petrol engine, the air and fuel are moved into the cylinder by:
   a pressure from the rising piston
   b a pump in the carburettor
   c atmospheric pressure
   d direct injection

2. Which of the following symptoms would suggest that the valve clearances are too great?
   a a heavy knock heard when the engine accelerates
   b a light tapping noise heard from the top of the engine
   c a high-pitched squeal heard when the engine is revved
   d a crackling noise heard when driving hard uphill

3. Oil leaking past the piston rings can be diagnosed by:
   a seeing blue smoke from the exhaust when the car is moving
   b the need to top up the engine oil at frequent intervals
   c noticing a pool of oil beneath the engine after the car has been standing for a while
   d seeing blue smoke from the exhaust when the engine is started

4. If the valve clearances are too tight, this may cause:
   a burning of the valve seats
   b wear on the valve stem
   c excessive tappet noise
   d retarded valve timing

5. Most of the wear of a cylinder bore occurs:
   a at the bottom of the bore
   b in the centre of the bore
   c on the thrust side of the bore
   d at the side of the bore

6. What is the property of a lubricant that makes it cling to a metal surface called?
   a oiliness
   b friction
   c viscosity
   d stiction

7. What does creamy or emulsified oil indicate?
   a the existence of water
   b servicing is overdue
   c oil grades have been mixed
   d loss of additives

8. In the lubrication system, what is a mesh filter fitted to?
   a main oil gallery
   b oil pickup pipe
   c filter head
   d bearing shells

9. A low viscosity rating indicates that the oil:
   a pours easily
   b pours with difficulty
   c is good-quality oil
   d is poor-quality oil

10. The radiator cap of a hot engine should be removed slowly because:
    a antifreeze solution is an acid harmful to the skin
    b there is a danger of steam gushing out
    c the air would cool the engine too rapidly
    d atmospheric pressure might damage the hoses

255

GETTING READY FOR ASSESSMENT

The information contained in this chapter, as well as continued practical assignments in your centre or workplace, will help you to prepare for both the end-of-unit tests and diploma multiple-choice tests. This chapter will also help you to prepare for working on and maintaining light vehicle engine mechanical, lubrication and cooling systems safely.

You will need to be familiar with:

- The operation and maintenance of engine mechanical systems
- The operation and maintenance of lubrication and cooling systems
- The properties of engine oils
- The properties of coolant
- How to remove and refit light vehicle engine mechanical, cooling and lubrication components
- Common faults associated with light vehicle cooling, lubrication and engine mechanical systems
- General maintenance procedures for light vehicle fuel, cooling, lubrication and engine mechanical systems

This chapter has given you an introduction and overview to the maintenance and repair of light vehicle engine mechanical, lubrication and cooling systems, providing the basic knowledge that will help you with both theory and practical assessments.

Before you try a theory end-of-unit test or multiple-choice test, make sure you have reviewed and revised any key terms that relate to the topics in that unit. Take time to read all the questions carefully and digest the information so that you are confident about what the question is asking you. With multiple-choice tests, it is very important that you read all of the answers carefully, as it is common for two of the answers to be very similar, which may lead to confusion.

For practical assessments, it is important that you have had enough practice and that you feel that you are capable of passing. It is best to have a plan of action and work method that will help you. Make sure that you have enough technical information, in the way of vehicle data, and appropriate tools and equipment. It is also wise to check your work at regular intervals. This will help you to be sure that you are working correctly and to avoid problems developing as you work.

When you are doing any practical assessment, always make sure that you are working safely throughout the test. Engine mechanical systems are dangerous and your precautions should include:

- Isolate ignition systems when conducting a compression test to reduce the possibility of electric shock.
- Wear latex gloves or barrier cream when conducting an oil pressure test to reduce the possibility of dermatitis.
- When you are working on cooling systems, allow them to cool down where possible before you start work.
- When running engines, make sure you use exhaust extraction or work in a well-ventilated area.

Make sure that you observe all health and safety requirements and that you use the recommended personal protective equipment (PPE) and vehicle protection equipment (VPE). When using tools, take care to use them correctly and safely.

Good luck!

4 Light vehicle fuel, ignition, air & exhaust system units & components

This chapter will help you develop an understanding of the construction and operation of common fuel, ignition, air and exhaust systems. It will also cover procedures you can use when removing, replacing and testing systems for correct function and operation. It provides you with knowledge that will help you with both theory and practical assessments. Finally, this chapter will help you plan a systematic approach to fuel, ignition, air and exhaust system inspection and maintenance.

This chapter covers:

- Safe working on light vehicle fuel, ignition, air and exhaust system units and components
- How light vehicle fuel systems operate
- Ignition system construction and operation
- Air and exhaust system construction and operation
- How to check, replace and test light vehicle engine fuel system units and components

WORKING PRACTICE

There are many hazards associated with light vehicle fuel, ignition, air and exhaust systems. You need to give special consideration to hot components, high voltages and the possibility that you will come into contact with chemicals such as petrol and diesel. You should always use appropriate personal protective equipment (PPE) when you work on these systems to protect you from these hazards.

Personal Protective Equipment (PPE)

Safety helmet protects the head from bump injuries when working under cars.

Overalls provide protection from coming into contact with oils and chemicals.

Safety gloves provide protection from oils and chemicals. They also protect the hands when handling objects with sharp edges.

Barrier cream protects the skin from old engine oil, which can cause dermatitis and may be carcinogenic (a substance that can cause cancer).

Safety goggles reduce the risk of small objects or chemicals coming into contact with the eyes.

Safety boots protect the feet from a crush injury and often have oil- and chemical-resistant soles. Safety boots should have a steel toe-cap and steel mid-sole.

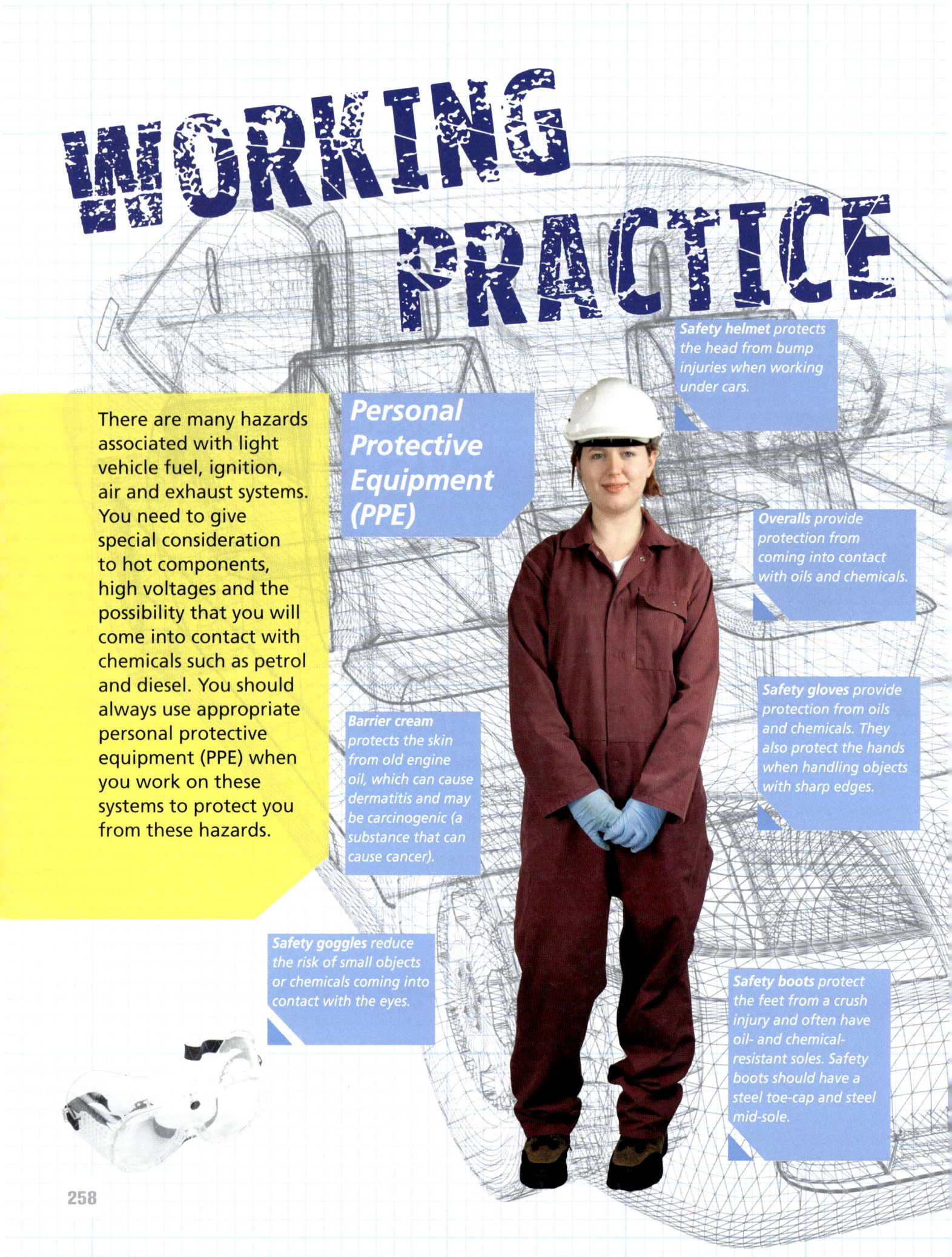

To reduce the possibility of damage to the car, always use the appropriate vehicle protection equipment (VPE):

Wing covers

Steering wheel covers

Seat covers

Floor mats

If appropriate, safely remove and store the owner's property before you work on the vehicle. Before you return the vehicle to the customer, reinstate the vehicle owner's property. Always check the interior and exterior to make sure that it hasn't become dirty or damaged during the repair operations. This will help promote good customer relations and maintain a professional company image.

Vehicle Protective Equipment (VPE)

Safe Environment

During the repair or maintenance of light vehicle ignition, fuel, air and exhaust systems, you may need to drain petrol or diesel. Under the Environmental Protection Act 1990 (EPA), you must dispose of these fuels in the correct manner. They should be stored safely in a clearly marked container until they are collected by a licensed recycling company. This company should give you a waste transfer note as the receipt of collection.

It is important to remember that the storage of fuel waste and contaminated rags poses a high fire risk.

All sources of ignition should be removed (for example, no smoking) when work is being carried out on petrol fuel systems because petrol vapour is extremely flammable. The engine should be cold, you should work in a well-ventilated area, and a suitable fire extinguisher must be to hand.

To further reduce the risks involved with hazards, always use safe working practices, including:

1. Immobilise the vehicle by removing the ignition key. Where possible, allow the engine to cool before starting work.

2. Prevent the vehicle moving during maintenance by applying the handbrake or chocking the wheels.

3. Follow a logical sequence when working. This reduces the possibility of missing things out and of accidents occurring. Work safely at all times.

4. Always use the correct tools and equipment to avoid damage to components and tools or personal injury. Check tools and equipment before each use.
   - Inspect any mechanical lifting equipment for correct operation, damage and hydraulic leaks.
   - Never exceed safe working loads (SWL).
   - Check that measuring equipment is accurate and calibrated before you take any readings.

5. If ignition, fuel, air or exhaust components need replacing, always check that the quality meets the original equipment manufacturer (OEM) specifications. (If the vehicle is under warranty, inferior parts or deliberate modification might make the warranty invalid. Also, if parts of an inferior quality are fitted, this might affect vehicle performance and safety.)

6. Following the replacement of any vehicle engine, cooling or lubrication components, thoroughly road test the vehicle to ensure safe and correct operation. Make sure that all work is correctly recorded on the job card and vehicle's service history, to ensure that any maintenance work can be tracked.

Preparing the car

Tools

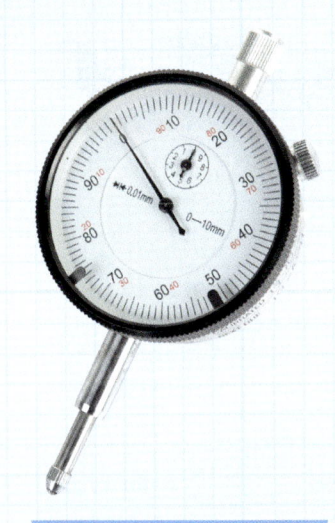

Dial test indicator (DTI)

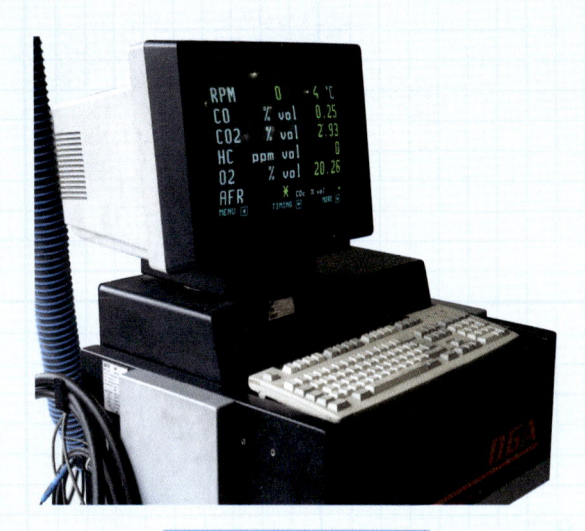

Exhaust gas analyser

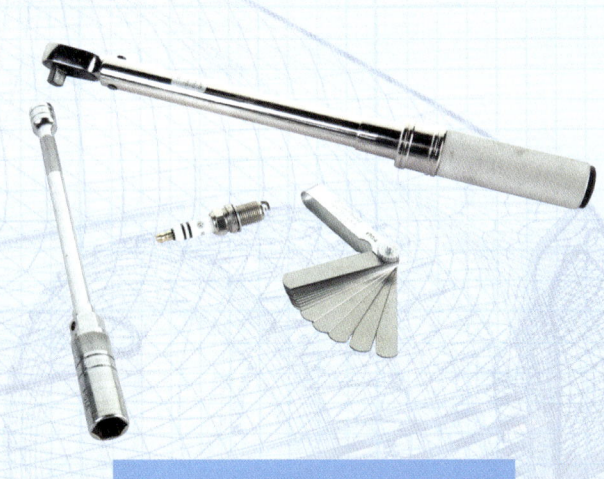

Tools used to adjust spark plugs

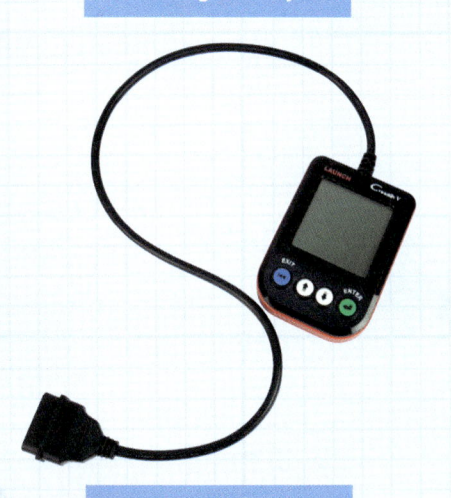

Diagnostic scan tool

Safe Working

- Always clean up any fluid spills immediately to avoid slips, trips and falls.
- Always make sure the engine is cool before you carry out work on it.
- Always isolate (disconnect) the electrics before you carry out working on an ignition system.
- Always remove all sources of ignition (e.g. smoking) from the area and have a suitable fire extinguisher to hand when you carry work out on petrol or diesel fuel systems.
- Always work in a well-ventilated area.
- Always use exhaust extraction when running engines in the workshop.
- Always wear safety goggles when removing and fitting exhaust system components.

How light vehicle fuel systems operate

Cars use internal combustion engines, which burn hydrocarbon fuel inside a combustion chamber. (See Chapter 3, pages 206–234, for detailed information on how the internal combustion engine works.)

Internal combustion engines need three key elements to work. They are:

- oxygen
- fuel
- heat.

Properties of fuels

Petrol and diesel fuels are made from crude oil. These fuels are distilled in an oil refinery, where the oil is heated and different chemicals boil off at different temperatures, as shown in Figure 4.1.

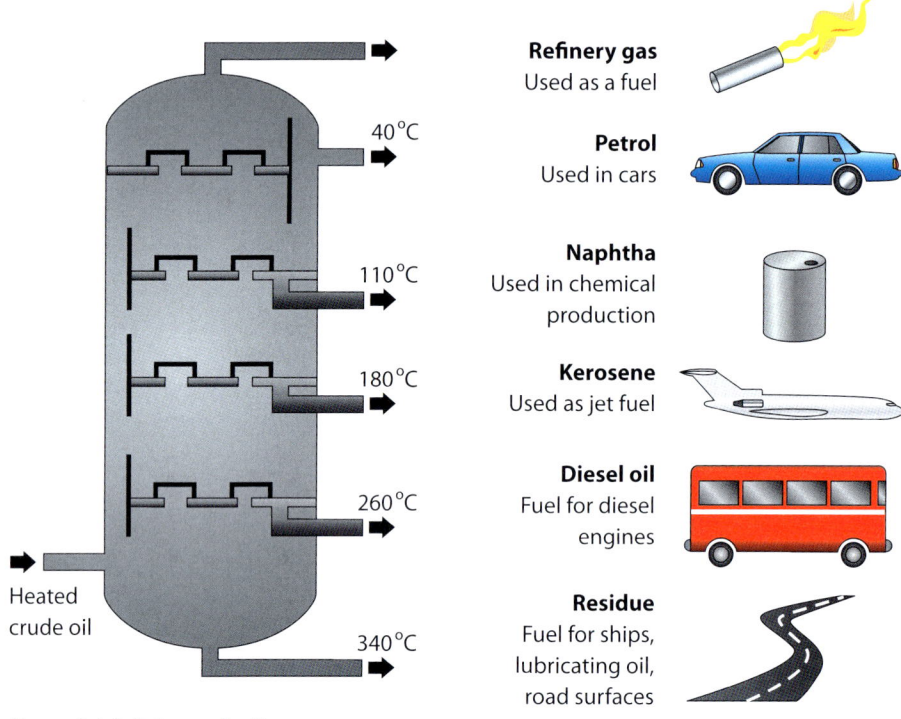

Figure 4.1 Refining crude oil

The main basis of these fuels is hydrogen and carbon, and that is why these fuels are called hydrocarbon (HC) fuels.

Whether a fuel is petrol or diesel depends on how much hydrogen and carbon there are in the fuel. This is based on the number of carbon molecules in the hydrocarbon chain. The amounts can differ slightly depending on how the fuel is refined, but as a general rule:

- When the chain has between 5 and 9 carbons, the hydrocarbon is petrol (gasoline).
- When the chain has about 12 carbons, the hydrocarbon is diesel.

Figure 4.2 shows a petrol hydrocarbon chain containing 7 carbon molecules, and a diesel hydrocarbon chain with 14 molecules.

> **Find out**
>
> Using sources of information available to you, list five petrol companies.
>
> Do all of the companies that you have found refine their own petrol and diesel?

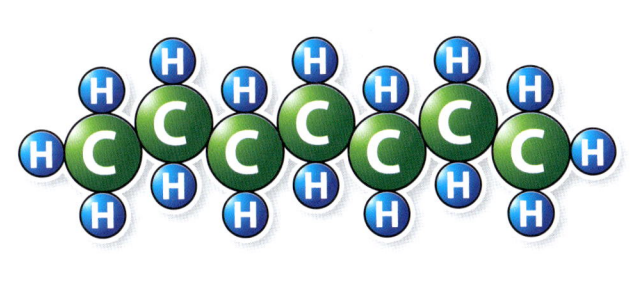

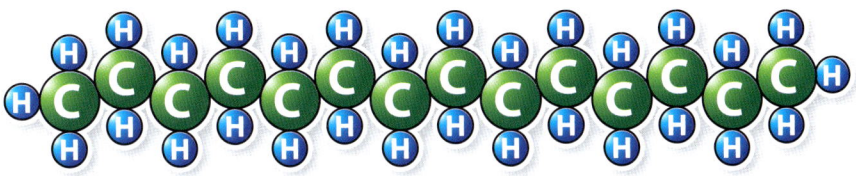

4 Light vehicle fuel, ignition, air & exhaust system units & components

Figure 4.2 Petrol (top) and diesel (bottom) hydrocarbon chains

Each fuel has a **calorific value**, which is the energy stored within it. (This is measured in the same way as the calories in food.) When the fuel is burnt, this energy is given up in the form of **thermal energy**, which can be used to provide motion power to the vehicle. The internal combustion engine is not very good at converting heat stored into other forms of energy. As a result, only about 20 per cent of the energy produced by burning petrol is used. This means that 80 per cent of the stored thermal energy is wasted.

Diesel is slightly more efficient, with a theoretical energy use of over 50 per cent. This is why a car can travel so much further on a litre of diesel than it could on the same quantity of petrol.

Working life

Carl is stripping the fuel system of a petrol car that has just driven 5 miles to reach the garage workshop. He must take off the fuel delivery pipes under the bonnet.

1. What safety precautions should he take?
2. How could he depressurise the fuel pipes?
3. Would a diesel engine be safer? If so, why?

Light vehicle fuel system construction and operation

Depending on engine design and fuel type (petrol or diesel), two different methods of introducing oxygen, fuel and heat are used – one occurs outside the **combustion chamber** and one inside.

Method one: petrol engine

With a petrol engine, air and fuel are normally mixed outside the combustion chamber and enter the cylinder through valves. Once compressed inside the combustion chamber, the source of heat is provided by a high-voltage spark from the plug. This burning mixture then rapidly expands, forcing the piston downwards on its power stroke.

Safe working

Be aware that when working with fuel:

- Petrol and diesel float on the surface of water and may travel long distances, eventually causing danger away from the place where it escaped.
- Petrol and diesel vapour can also travel long distances. It tends to sink to the lowest possible level and may collect in tanks, drains, pits or other enclosed areas. This means that there is always a risk of fire or explosion if a source of ignition is present.

Key terms

Calorific value – the amount of energy stored within a fuel.

Thermal energy – energy stored as heat.

Combustion chamber – the area inside the cylinder where air and fuel are burnt.

Safe working

Working on petrol or diesel light vehicle fuel systems presents certain problems.

- Where possible, allow the engine to cool before starting work.
- Clean up any accidental spills immediately to avoid slips, trips and falls.
- Remove all sources of ignition.

263

Method two: diesel engine

With a diesel engine, the air and fuel are normally mixed inside the combustion chamber. This means that only air is drawn along the inlet tract, entering the cylinder through valves. The compression of this air will create large quantities of heat, so that when diesel fuel is injected and **atomised** into the combustion chamber, it will spontaneously combust. This creates a rapid rise in pressure, forcing the piston downwards on its power stroke.

Petrol fuel delivery (the carburettor)

Until the 1990s the carburettor was one of the main methods used to mix petrol and air in cars. As exhaust emission regulations were tightened, the carburettor was replaced by fuel injection (see pages 267–280).

A carburettor uses the movement of air and different air pressures to mix air and fuel in the correct quantities.

The main components of a carburettor are:

- venturi
- float chamber
- float and needle valve
- jets
- butterfly valve.

Venturi

The **venturi** is designed to lower air pressure in the inlet tube. It works in a similar way to an aircraft wing.

As shown in Figure 4.3, an aircraft wing has a curved upper surface, and a flat lower surface. When propelled forwards, the air is split at the front edge of the wing, with some passing over the top and some passing underneath.

- The air passing above the wing and the air passing below the wing both have to arrive at the rear edge at the same time.
- Because the top edge of the wing is curved, the air passing over it has to travel further.
- As a result, the air moving above the wing travels faster.
- As air speed increases, pressure drops.

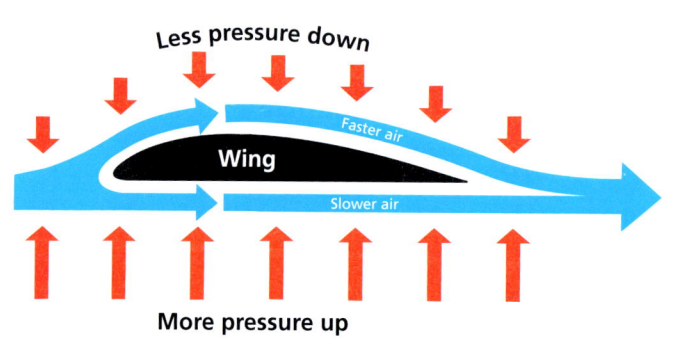

Figure 4.3 Air travelling over an aircraft wing

> **Did you know?**
>
> Around 1896 Frederick William Lanchester, from Birmingham, produced the first petrol-engined car in England. To mix the petrol and air, he experimented with a wick system that worked in a similar way to an oil lamp.
>
> A wick was submerged in petrol, and the upper part exposed to the incoming air of the inlet tract. The amount of air entering the engine was controlled by a valve and fuel could then be adjusted manually. Depending on engine speed, this could be done by winding the wick up and down. To ensure correct operation, a careful balance between air and fuel had to be maintained.
>
> An improvement on this design was the carburettor, which was invented by two Hungarian engineers, Donát Bánki and János Csonka. It was later patented by the German Wilhelm Maybach.

With an aircraft wing, this means that low pressure is created above the wing, which allows the high pressure below to create lift.

In Figure 4.4, you can see that the narrowing of the air intake into the venturi is similar to the upper curve of an aircraft wing on both sides of the tube. As air passes through this venturi restriction, it speeds up and air pressure drops.

Figure 4.4 The venturi section of a carburettor

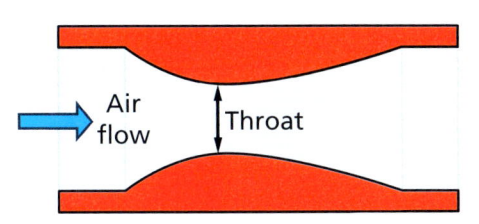

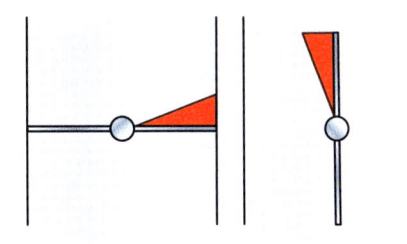

Figure 4.5 Throttle butterflies

> **Key terms**
>
> **Atomise** – to break up liquid fuel into small droplets or a spray.
>
> **Venturi** – a narrowing of the air intake (usually in a carburettor) designed to speed up air flow and lower air pressure.
>
> **Butterfly** – a mechanical flap in the intake system which controls the flow of air.
>
> **Main jet** – an accurately sized tube that regulates the amount of petrol being drawn through the carburettor.

The amount of air passing through the venturi at any one time is controlled by a mechanical flap called the throttle **butterfly** (see Figure 4.5).

The engine speed of a petrol engine is controlled by the amount of air entering the combustion chamber. The more air that enters the engine, the faster the engine will run. The butterfly is connected to the accelerator pedal (either with a cable or electronically), so that the driver is able to control the amount of air and, as a result, the speed of the engine.

The float chamber

To one side of the venturi is the float chamber. This is a small reservoir of petrol that holds an accurate amount of fuel ready to be mixed with the air in the carburettor. It is kept at a constant level by a float and needle valve.

An accurately sized tube, called the **main jet**, is angled upwards out of the float chamber, and sticks out into the narrowest section of the venturi. The top of the float chamber will be vented to the atmosphere, so that atmospheric pressure is acting on the fuel in the float chamber.

When the engine is running, a low pressure (depression) is created by the downward movement of the piston drawing air in through the venturi of the carburettor. As the air passes through the restriction, it speeds up and air pressure drops, creating a pressure lower than that of atmosphere.

The pressure difference between the venturi and the float chamber pushes petrol out of the main jet and the fast moving air breaks it up into small droplets. This is called atomisation.

Figure 4.6 on the next page shows a basic downdraft carburettor.

> **Did you know?**
>
> The float and needle valve of a carburettor works in a similar way to a ballcock in the cistern of a toilet.

> **Find out**
>
> Cut two thin strips from an A4 piece of paper.
>
> Hold them approximately 30 mm apart, and blow down between them.
>
> - What happens to the two pieces of paper?
> - Why does this happen?

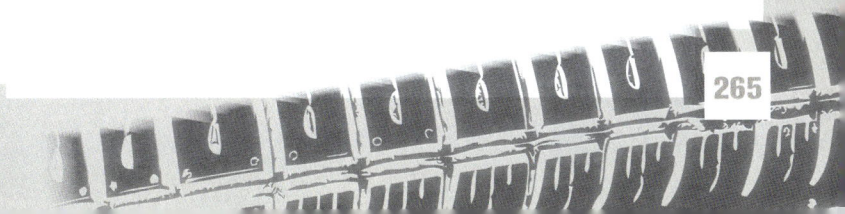

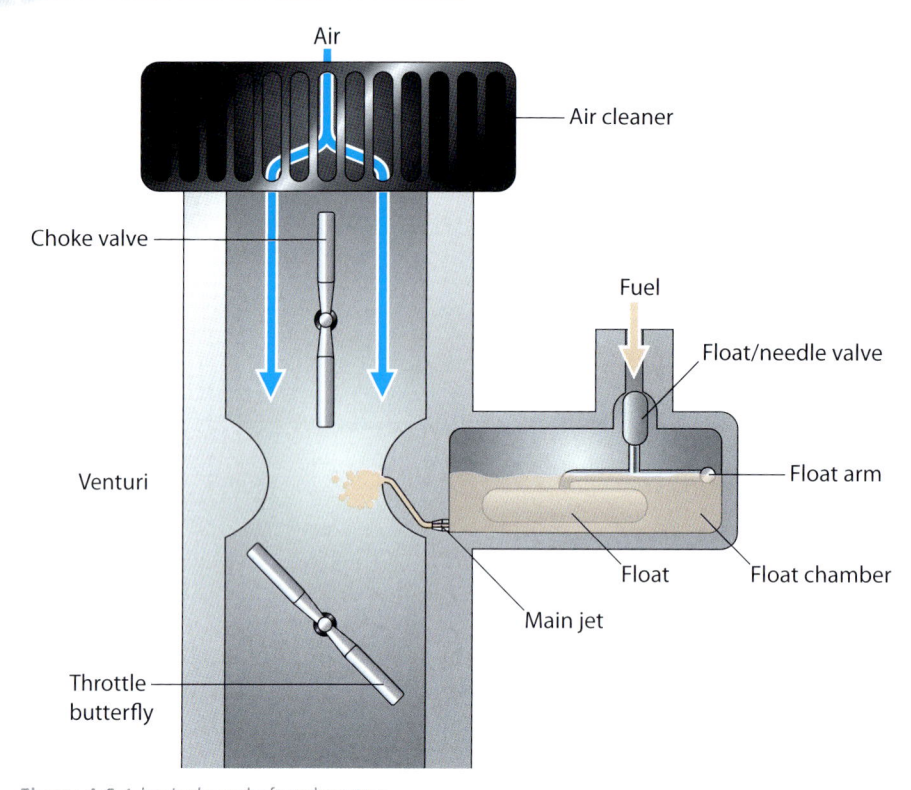

Level 2 Light Vehicle Maintenance & Repair

Figure 4.6 A basic downdraft carburettor

Light vehicle engine petrol and diesel fuel system components

The tank

The tank is a metal or plastic container, designed to safely store fuel away from the engine until it is needed (see Figure 4.7).

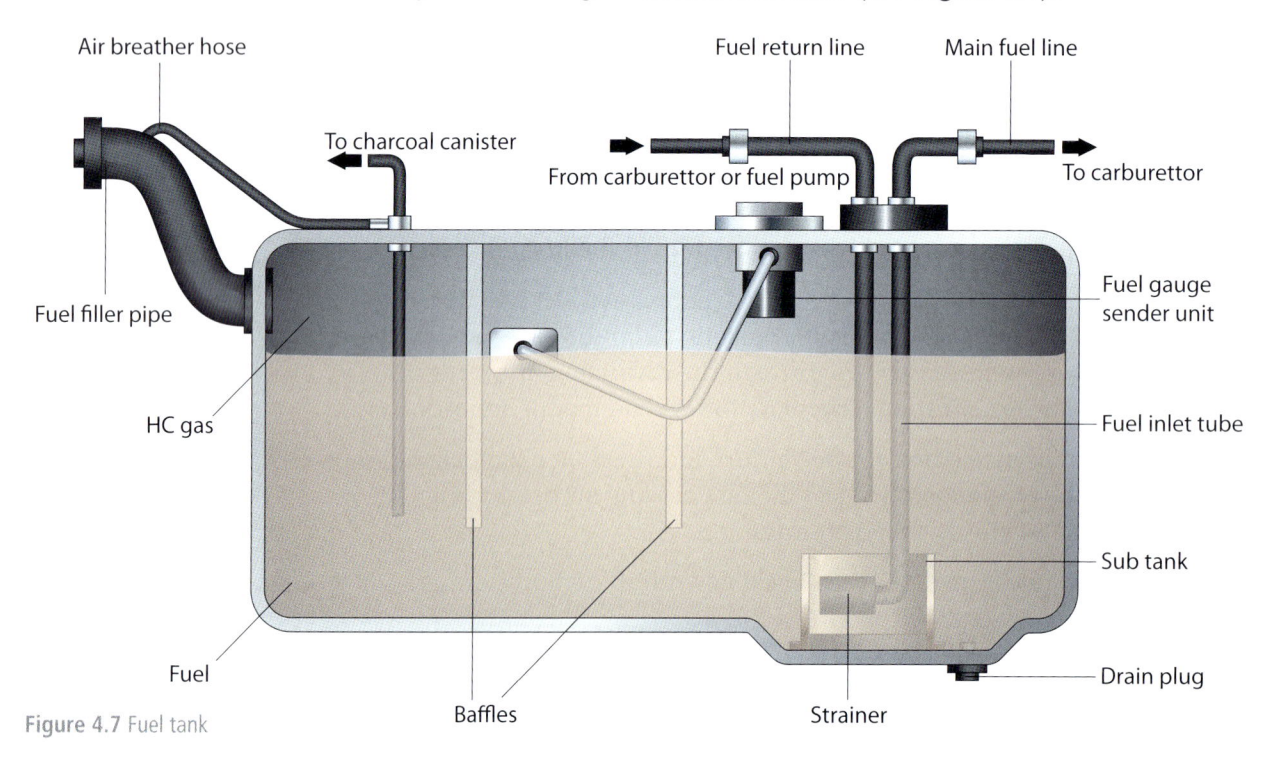

> **Key term**
>
> **Baffles** – a series of metal or plastic separators, designed to reduce the amount of fuel sloshing around inside the tank while driving.

Figure 4.7 Fuel tank

Baffles

A series of **baffles** is often used to reduce the amount of fuel movement inside the tank during normal driving operations.

Pipes and hoses

A network of pipes and hoses is used to transfer the fuel from the tank to the **engine bay**, where it can be mixed with air before being burnt inside the cylinder. These pipes and hoses are called fuel lines.

Pump

A fuel pump is needed to help transfer the fuel from the tank through the fuel lines.

Petrol pump

- In a carburettor petrol system, a mechanically driven pump is often used.
- In a petrol fuel injection system, an electric pump is required because of the high quantities and pressures needed.

Diesel pump

In a diesel fuel injection system, a high-pressure mechanical engine-driven pump is used to inject fuel directly into the combustion chamber. Because of this, extremely high pressures are required:

- on a standard diesel engine, around 60 times higher pressure than petrol
- on a **common rail diesel** engine, around 600 times higher pressure than petrol.

Fuel filter

To help make sure that the fuel delivered to the engine is clean, a fuel filter is often included in the design.

Petrol fuel delivery (injection)

Once petrol has reached the engine, methods used to mix air and fuel include:

- carburettors (see pages 264–266)
- single point fuel injection
- multipoint fuel injection.

Petrol fuel injection systems

Petrol fuel injection systems have been around for a long time. Early systems were mechanical and relied on fuel distributed to the correct injectors under pressure. This was similar to the way that fuel is injected in a standard diesel engine (although the fuel was injected into the **intake manifold**, rather than directly into the cylinder as found on diesel engines). These early mechanical fuel injection systems were an improvement on carburettors, but were still inaccurate in many ways.

Key terms

Baffles – a series of metal or plastic separators, designed to reduce the amount of fuel sloshing around inside the tank while driving.

Engine bay – the area under the bonnet that houses the engine.

Common rail diesel – an extremely high-pressure electronic fuel injection system used on modern diesel engines.

Intake manifold – a series of pipes designed to duct air into the engine.

Safe working

The high fuel pressures used with diesel injection systems can force fuel through your skin, which can cause death. You should depressurise these systems before you work on them. You can normally do this by stopping fuel supply to the pump and starting the engine. When the engine cuts out, fuel pressure should have dropped to a safe level, but still be careful when removing a fuel pipe.

- On a standard diesel pump, disconnect the fuel shut-off solenoid.
- On a common rail diesel, you can often remove the fuel pump fuse.

Single point injection

Early developments of electronic fuel injection included single point fuel injection (also called throttle body fuel injection). A **throttle body** is used, which is of similar size, shape and position to a carburettor. This contains a butterfly valve and is mounted in the air inlet tract. Instead of a difference in pressure drawing fuel from a float chamber, a single large **fuel injector** is mounted above the throttle butterfly, as shown in Figure 4.8.

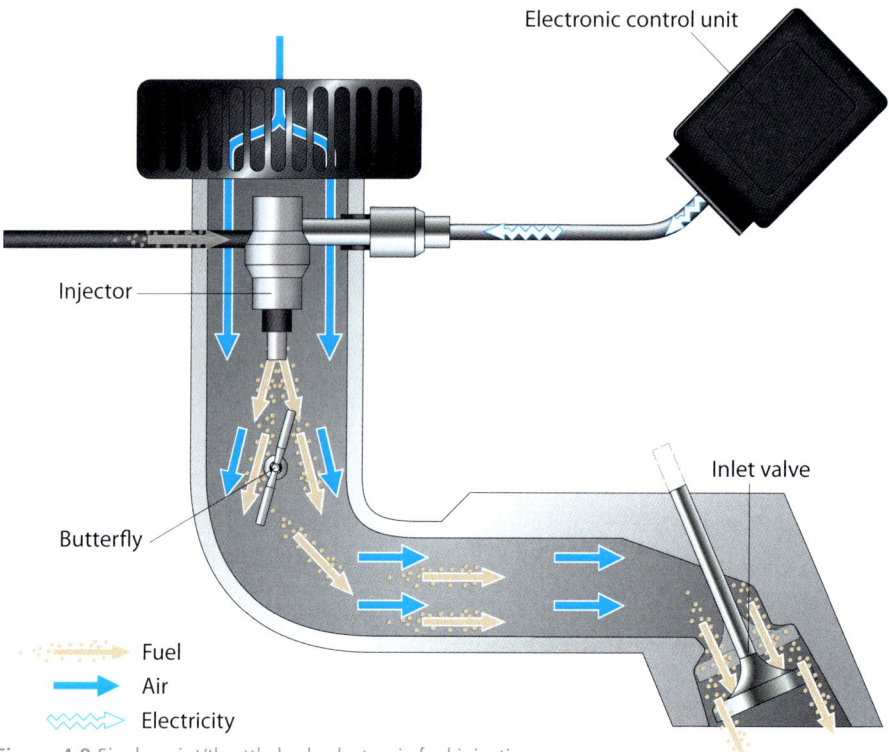

Figure 4.8 Single point/throttle body electronic fuel injection

Intake air is regulated by the throttle butterfly, and a **solenoid**-type fuel injector is used to atomise varying amounts of fuel into the air stream. This mixture of air and fuel travels along the intake manifold and into the cylinder. As with carburettors, the air and fuel are mixed in an early part of the air inlet tract. As a result the mixing can be poor and the amount of fuel/air entering the cylinder is inaccurate because of pumping losses and intake robbery. This means that a single point fuel injection system is inefficient and can lead to high emissions output.

Intake robbery

Because only one fuel injector is used to feed all cylinders, it is possible for a cylinder in good condition to steal air and fuel from the others. This means that as the engine wears, a cylinder with greater efficiency can end up with larger amounts of fuel than cylinders with poor compression/depression. This uneven fuel distribution between cylinders can lead to poor running, poor performance and high emissions output.

As emissions regulations became tighter and tighter, single point fuel injection became unsuitable.

Multipoint electronic fuel injection (EFI)

The next stage in the development of petrol fuel injection was to introduce multipoint fuel injection. This is where each cylinder has its own fuel injector mounted in the inlet manifold just before the intake valve. The intake tract now only draws air. Fuel is injected at the last moment before it enters into the cylinder. By mounting the injector behind the inlet valve (as shown in Figure 4.9), as fuel is injected it only has to travel a short distance before entering the combustion chamber. This helps to reduce the possibility of intake robbery, and because of this, the metering of fuel is more accurately controlled.

> **Key terms**
>
> **Throttle body** – an area of the inlet tract containing a throttle butterfly.
>
> **Fuel injector** – an electronically controlled valve that sprays fuel into the air stream.
>
> **Solenoid** – a linear electric motor sometimes used to control a fuel injector.
>
> **Actuator** – a mechanism that puts something into action.

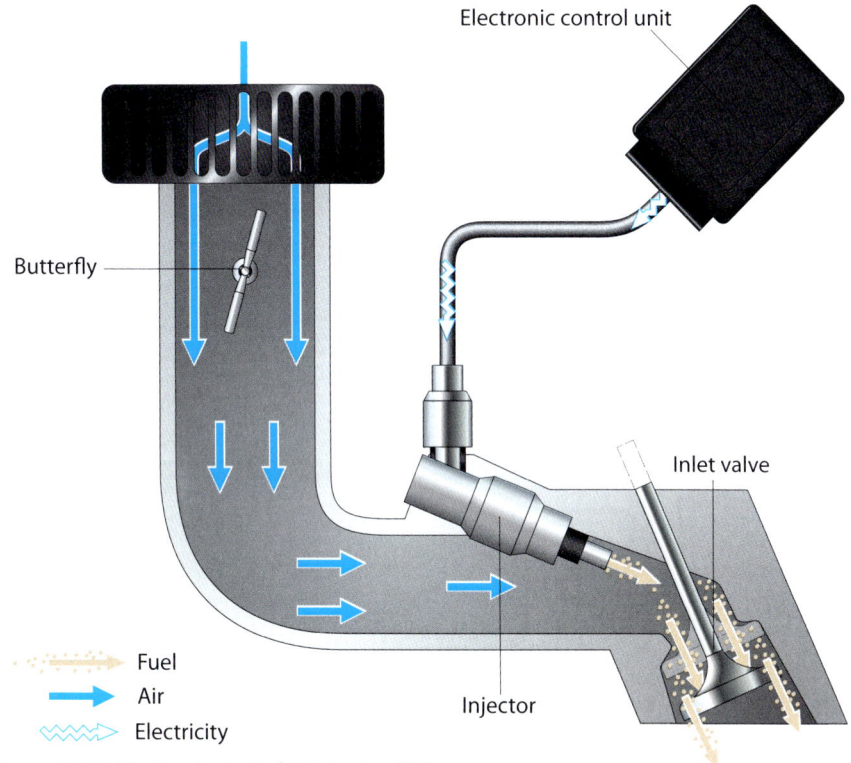

Figure 4.9 Multipoint electronic fuel injection (EFI)

A multipoint electronic fuel injection has three main functions:

- air intake and metering (measurement)
- fuel supply and delivery
- electronic processing and control.

System control

The electronic fuel injection system is controlled by computer. This computer can be called an ECU (electronic control unit), an ECM (electronic control module) or a PCM (power train control module).

Input signals produced by sensors are processed within the ECU, which then sends signals to the **actuators** (i.e. fuel injectors, auxiliary air valves and throttle controllers), providing engine control.

Figure 4.10 Engine management electronic control unit (ECU)

The basic operation of an ECU is similar to a calculator. Various engine sensors send information to the ECU as voltage signals. The ECU is programmed to carry out a mathematical calculation – to add, subtract, multiply or divide these voltage signals. When a result is obtained, the answer will correspond with a setting to turn one of the actuators of the fuel injection system on or off. Many ECUs operate as **earth switches**, completing the electrical circuit of an actuator and allowing it to operate.

Because the ECU completes the electric circuit for many of the engine management actuators, it is important that it has a good earth. Any resistance created where it attaches to the vehicle body can lead to poor performance or running issues.

Petrol fuel supply

In a petrol engine, the main components making up the petrol fuel supply circuit are:

- fuel tank
- fuel pump
- fuel filter
- fuel rail.
- fuel injectors
- fuel pressure regulator
- overflow return to tank

> **Key term**
>
> **Earth switch** – a switch on the negative side of an electric circuit.

> **Did you know?**
>
> Manufacturers are now producing returnless fuel injection systems. The pressure regulator and filter have been moved inside the fuel tank and a pressure sensor is mounted on the fuel rail instead. This stops engine heat being transferred into the fuel that is returned to the tank and reduces evaporation.

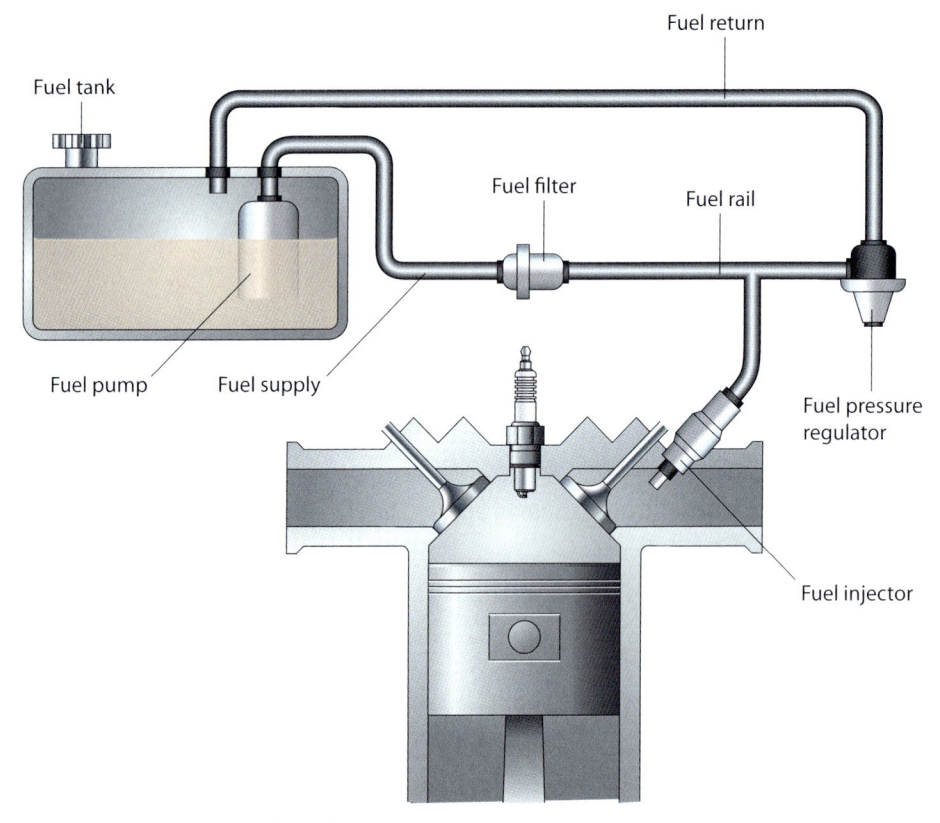

Figure 4.11 Multipoint EFI fuel delivery system

Fuel tank

The fuel tank is the main reservoir and supply of fuel for the vehicle fuel injection system. Petrol fumes are not allowed to escape into the surrounding air. Because of this, fuel tanks are sealed to reduce the possibility of hydrocarbon emission due to evaporation. If the tank has nowhere to store these evaporated fumes, high pressures could build up in the tank, causing leaks or damage.

The evaporation of fuel continues as with any other fuel system, but the vapours must be captured, stored and disposed of in an environmentally friendly way.

Fuel emission vapours are stored in a charcoal canister. Under certain engine running conditions, they are vented into the intake manifold through purge pipes and burnt along with the rest of the air/fuel mixture. This is called an evaporative emission system or EVAP.

Fuel pump

A large quantity of fuel, supplied under pressure, is needed for the electronic fuel injection system to function efficiently. This is the job of the fuel pump.

Many fuel pumps are of the roller cell type. A small electric motor is used to drive a roller cell (as shown in Figure 4.12). The roller cell acts as a pumping mechanism to draw fuel from the tank and feed it on to the fuel injection rail and injectors.

> **Working life**
>
> While servicing a car, Jan is replacing the air filter. When the casing is reassembled, he accidentally traps the vacuum and system purge pipes from the EVAP charcoal canister.
>
> 1 What effect might this have on:
> - the fuel system
> - emissions?

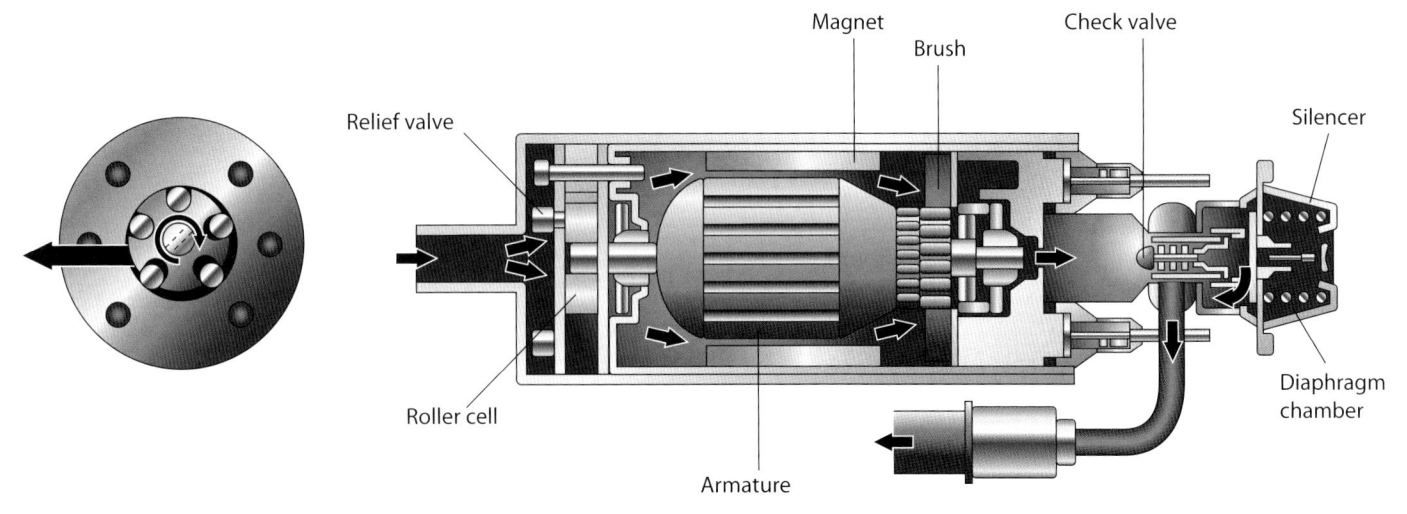

Figure 4.12 Roller cell electric fuel pump

The fuel pump can be mounted in the main fuel line under the car, or submerged in the fuel tank itself. (Many vehicle manufacturers are choosing to submerge the fuel pump inside the tank, as this assists with cooling and noise suppression.)

Even if the fuel pump is mounted in line under the vehicle, fuel flows through the middle of the electric motor, and this will help with cooling.

As with all electric motors, fuel pumps create sparks under normal operation. However, the risk of ignition is very low because fuel vapour (which is the part that burns) is minimal. This means that any liquid fuel passing through the pump will put out any sparks and should not ignite.

Features of the fuel pump

- **Fuel supply** – A standard electronic fuel injection fuel pump is capable of supplying 4 to 5 litres of fuel per minute at a pressure of approximately 8 bar (120 psi).
- **Pressure relief valve** – Inside the pump there is a pressure relief valve. This is designed to lift off its seat if system pressure goes higher than 8 bar (in case of a blockage, for example of the fuel filter).
- **Non-return valve** – At the outlet end of the fuel pump is a non-return valve, to stop fuel flowing backwards through the pump. When the fuel pump is switched off, it closes and keeps pressure within the system.

Fuel pump relay

Many electric fuel pumps are controlled by a remote electromagnetic switch called a relay. This helps to reduce load on the vehicle's electrical system. It allows a small current supplied by the ECU to switch the much larger current required by the fuel pump.

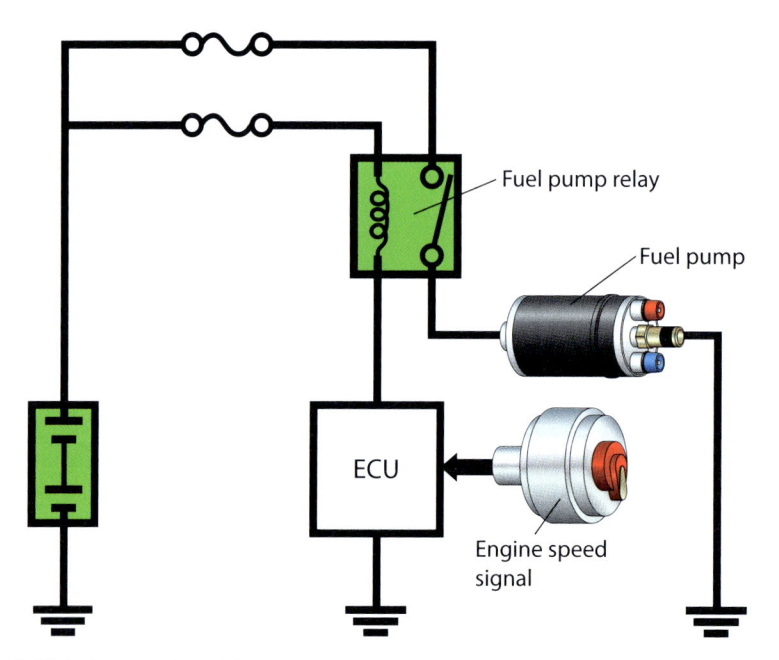

Figure 4.13 Fuel pump relay wiring

The fuel pump relay can also act as a safety device, cutting power to the electric fuel pump if an excessive leak occurs in the system. An engine speed signal is sent to the ECU. When this is received, the fuel pump relay is switched on and the pump begins to run.

If an excessive leak occurs in the fuel system, the engine will stall and the ECU will no longer receive a speed signal. At this point, the fuel pump relay is switched off, stopping the fuel pump running. This is particularly useful if the car is involved in an accident – it means that the high risk of fire can be reduced because the pump is not forcing fuel out of any broken pipes.

Fuel filters

Because of the very small **tolerances** within an electronic fuel injection system, tiny amounts of dirt would easily block fuel injectors and other components. This means that an inline fuel filter should be fitted.

Inline fuel filters are service items, but are often overlooked when diagnosing fuel starvation problems. If services are missed, then these fuel filters may not be replaced for some considerable time and can become blocked.

Fuel pressure regulators

Very few electronic fuel injection systems have methods for measuring fuel flow. If a constant pressure can be maintained in the fuel rail, the electronic control unit can calculate how much fuel is being injected. The constant pressure is controlled by the fuel pressure regulator.

The fuel pressure regulator works as in the following example:

- Imagine a large number of tennis balls sitting on top of a trapdoor. They are all exactly the same size and weigh exactly the same amount.
    - If the trapdoor is opened for one second, 10 tennis balls fall through.
    - If the trapdoor is open for 10 seconds, 100 tennis balls fall through.
- Because the tennis balls are all the same size and weight, you don't need to count the number of balls that fall through the trapdoor. You can calculate the exact number of tennis balls from the amount of time the trapdoor is open.

The process is very similar with the pressure in the fuel rail. The ECU does not have to measure the flow or quantity of petrol. It simply has to know how long to hold the fuel injector open to calculate the amount of fuel being delivered.

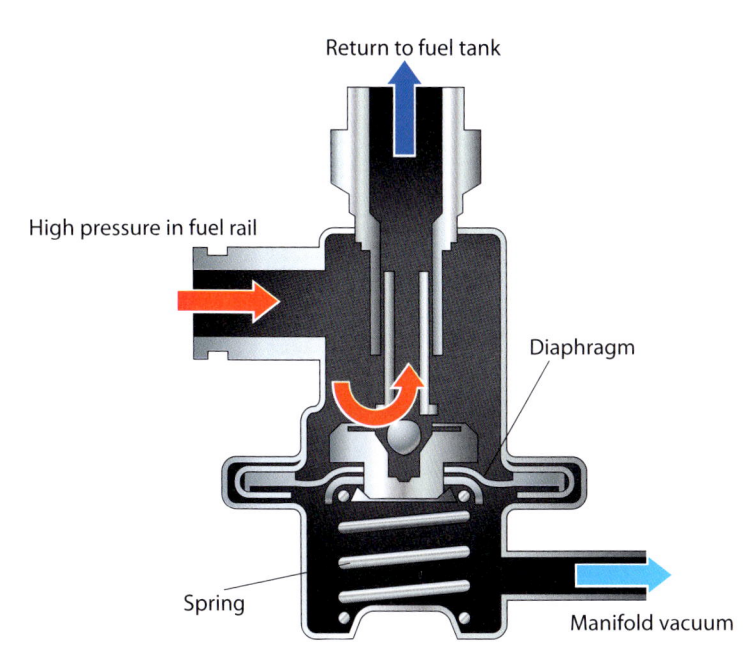

Figure 4.14 shows a vacuum-operated fuel pressure regulator.

> **Did you know?**
>
> Some manufacturers use a shock-sensitive switch called an inertia switch to shut off the fuel pump if the car is involved in a collision. If this switch is disturbed, it must be reset before the engine will run.

> **Key term**
>
> **Tolerance** – an allowable difference from the required measurement.

Figure 4.14 Vacuum-operated fuel pressure regulator

Level 2 Light Vehicle Maintenance & Repair

Fuel pumps are designed to deliver high volumes of fuel, at a far higher pressure than needed by the fuel injection system. In this way a surplus of fuel should always be available under all engine running conditions.

Fuel from the pump is delivered to one end of the fuel rail from which the injectors feed. At the other end of the fuel rail is a return system to the petrol tank. How much fuel returns to the tank is controlled by the fuel pressure regulator.

A spring-loaded valve is fitted inside the fuel pressure regulator that will lift off its seat when a preset pressure is reached (normally around 3 bar). So, for example, if 10 bar of fuel pressure is delivered to the rail, the regulator will maintain 3 bar in the rail, and allow 7 bar to return to the tank.

Under different running conditions low pressures are created in the inlet manifold and will affect the amount of fuel injected. (A low manifold pressure can draw too much fuel through an injector.)

The pressure regulator is often connected to the manifold by a vacuum pipe, which will automatically adjust the spring force in the regulator to compensate for pressure variations in the manifold.

Fuel injectors

Fuel injectors are of the solenoid type. A small coil of wire is wound around a movable **armature** (see Figure 4.15). When current is applied to this coil, a magnetic field is created, which draws the armature through the middle of the **winding**. When current is removed from the coil, the magnetic field collapses, and a return spring is used to move the armature back into its original position.

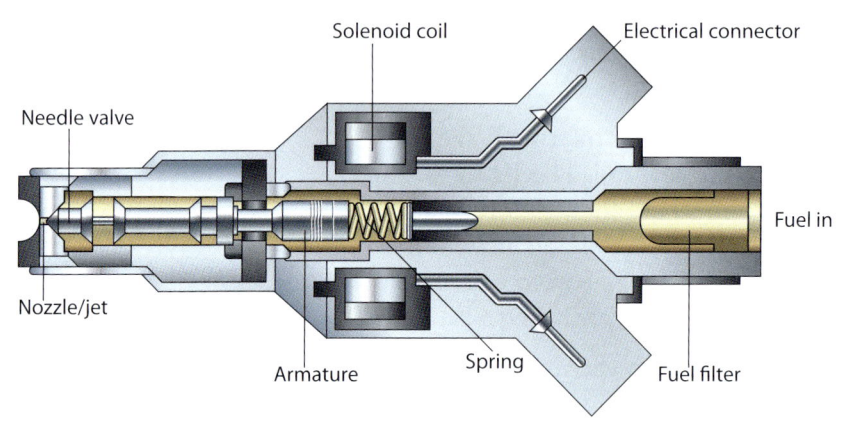

Figure 4.15 Solenoid-type electronic fuel injector

The armature of a fuel injector forms a needle valve at one end, and fuel pressure is supplied behind the needle valve. When the winding of the solenoid is supplied with current and the needle of the armature lifts from its seat, fuel can be sprayed under high pressure into the intake manifold.

Air induction

In a petrol engine, speed and load depend on the position of the throttle butterfly. As the throttle butterfly is opened, more air is allowed into the engine, and so the engine speeds up.

The throttle butterfly can be mechanically operated by an accelerator cable. On many vehicles they are now what are known as 'drive by wire'. This means that a sensor picks up the position of the accelerator pedal, and motors drive the throttle butterfly into the correct position (no direct connection).

To enable the ECU of the electronic fuel injection system to calculate engine demands (load and speed), throttle position and air quantity must be measured.

Air quantity measurement

In order that the engine management system can mix the correct quantities of petrol and air, it needs to know the amount of air entering the engine. The air is usually measured using one of three methods:

- airflow
- manifold pressure
- air mass.

Airflow meters

Early electronic fuel injection systems used airflow meters to measure the amount of air entering the inlet tract.

An airflow meter consists of a small spring-loaded door or flap (often called the vane) that is pushed open by the incoming air. The more air entering the system, the further open the flap will move. On top of the spring-loaded door is a **potentiometer** (a type of variable **resistor**) that can accurately measure position. This gives an indication of the amount of air entering.

> **Key terms**
>
> **Armature** – the central shaft of an electric motor.
>
> **Winding** – a small coil of electrical wire.
>
> **Potentiometer** – variable resistor.
>
> **Resistor** – an electronic component that slows down the flow of electricity.

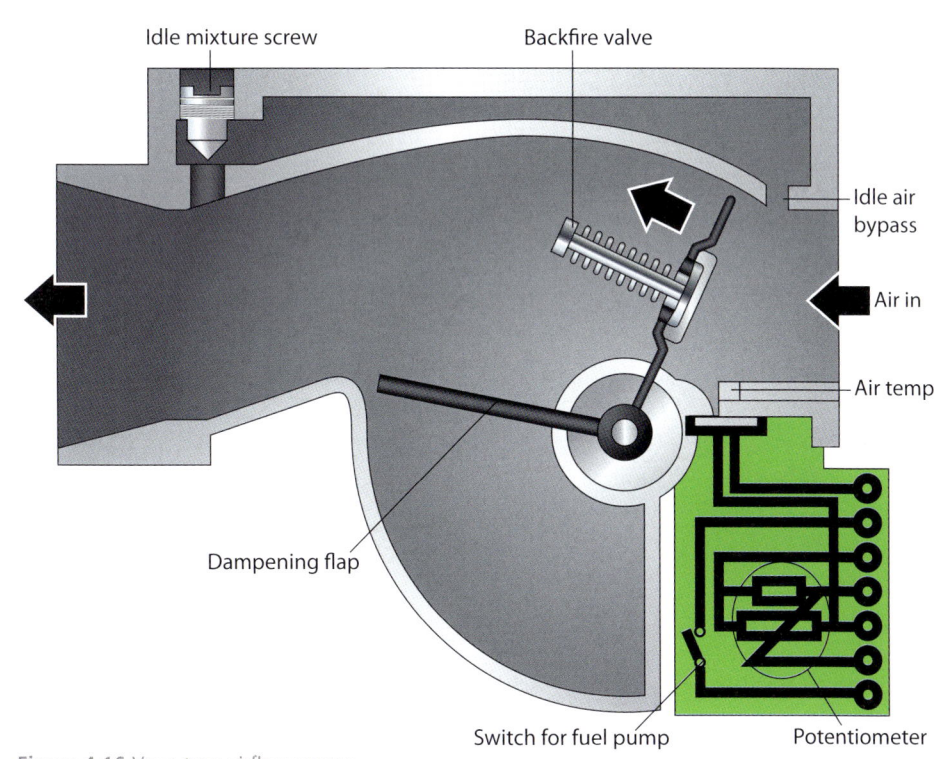

Figure 4.16 Vane-type airflow sensor

If the ECU of the vehicle is connected to the sliding contact of the potentiometer (similar to that found in a toy slot car), the varying voltage signal can be used to show the exact position of the spring-loaded door.

When the engine is running, pressure in the inlet manifold created by the engine's pistons can make the airflow sensor door pulsate. A second door is attached at a right angle to the first. During operation, this door is moved in a damping chamber (like a shock absorber) and helps prevent incorrect signals being sent to the ECU.

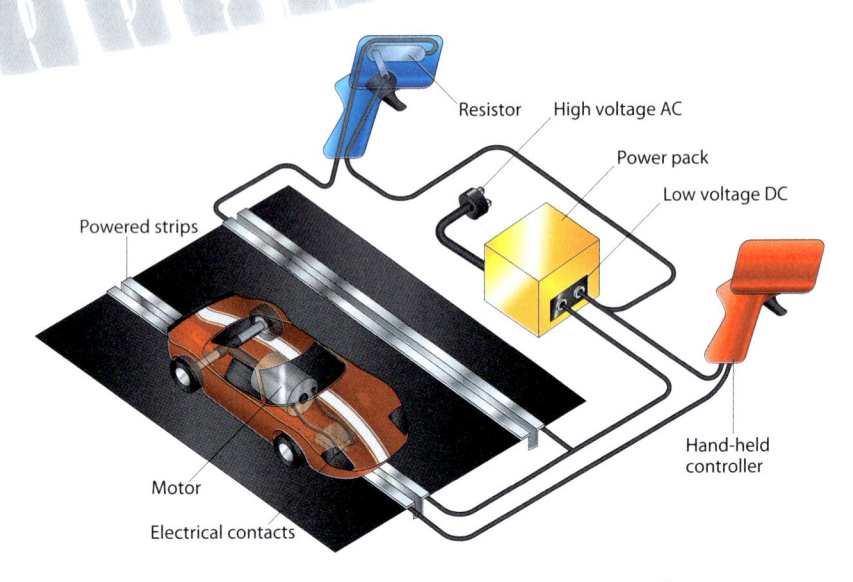

Figure 4.17 A toy slot car with a variable resistor potentiometer controller

Did you know?

The operation of a potentiometer is similar to the controller of a toy slot car. A slot car controller slides an electrical contact along a resistor (see Figure 4.17). This changes the amount of electrical voltage being transmitted to the track. The more voltage, the higher the electrical pressure and the faster the slot car motor will run.

Manifold absolute pressure sensors (MAP)

A manifold absolute pressure sensor (MAP) can also be used to calculate air intake for the engine. In addition, it gives a good indication of engine load. For any given size of inlet manifold, if the amount of vacuum or low pressure can be measured, the ECU can calculate roughly how much air is being drawn.

The MAP is connected to the intake manifold, either directly or by a vacuum pipe. The low pressure (depression) created in the manifold will be sensed by a diaphragm. This is converted into a signal and sent to the ECU for calculation.

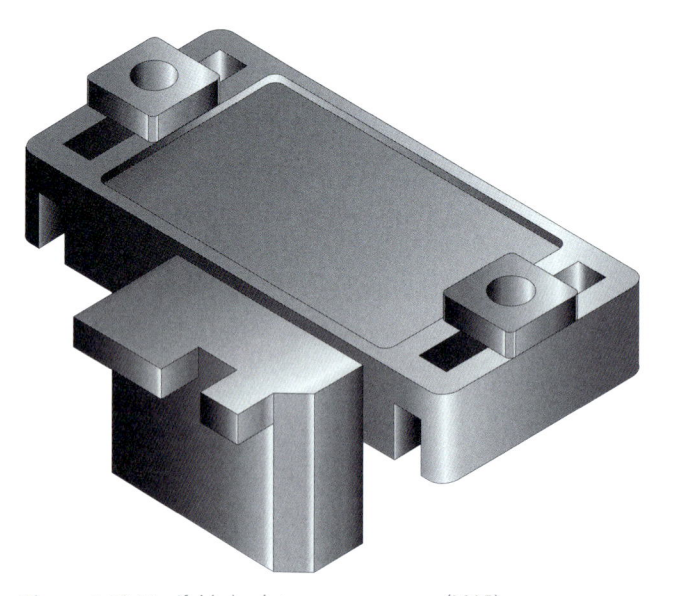

Figure 4.18 Manifold absolute pressure sensor (MAP)

Hot wire mass airflow sensors (MAF)

A hot wire mass airflow sensor (MAF) is a very accurate way of measuring the amount of air entering an engine. Unlike a vane-type airflow sensor, there is very little restriction to the air intake (the air doesn't have to push a little flap open) and so performance is improved.

Air flow

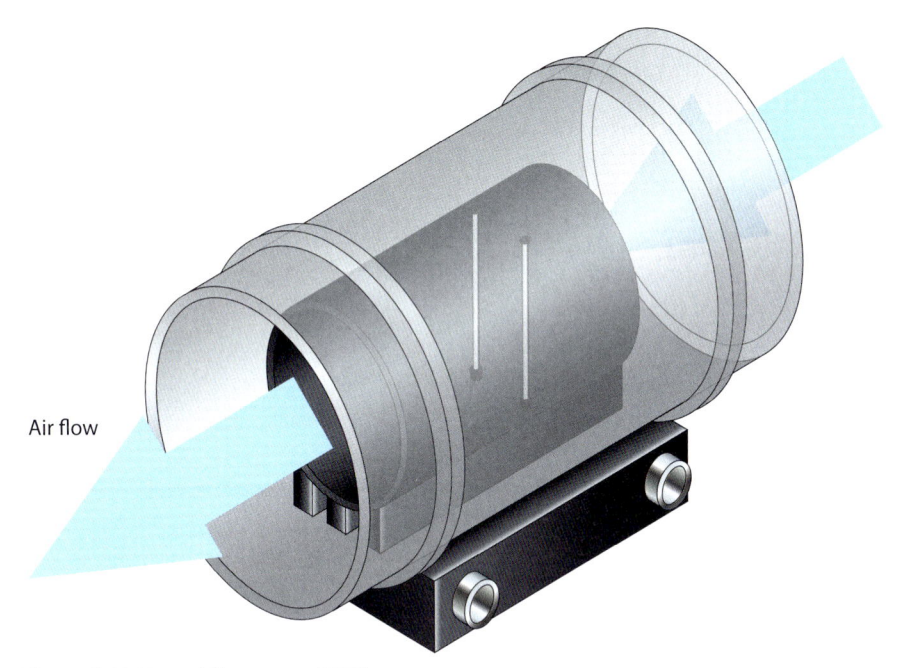

Figure 4.19 Mass airflow sensor (MAF)

The MAF is normally mounted before the throttle butterfly. It contains wires or heating elements (sometimes called hot film). When supplied with an electric current, the wires get hot. As the intake air passes over the wires or elements, they are cooled. To maintain a constant heat, the ECU must supply more electric current as the amount of air increases. The amount of current needed to keep the wires at a constant temperature is in proportion to the mass (weight) of air.

Throttle position sensors

Throttle position sensors are potentiometers and work in the same way as described on pages 275–276. A sliding contact moves along a variable resistor, sending the ECU information about throttle position. They can be mounted at the pedal, throttle body or both.

If the system is driven by wire, there is no accelerator cable attached to the throttle butterfly. Instead, the butterfly is controlled by small motors, signalled from the ECU.

The accelerator pedal is attached to a potentiometer, which signals the position of the driver's foot. The ECU then moves the butterfly using motors, and its exact position can be checked using the throttle position sensor.

Level 2 Light Vehicle Maintenance & Repair

> **Did you know?**
>
> Stepper motors are sometimes used as engine control devices, helping maintain smooth idle for example. A stepper motor is able to operate in small stages or 'steps' which allows it to be set to various positions, making it ideal for controlling certain mechanical operations.

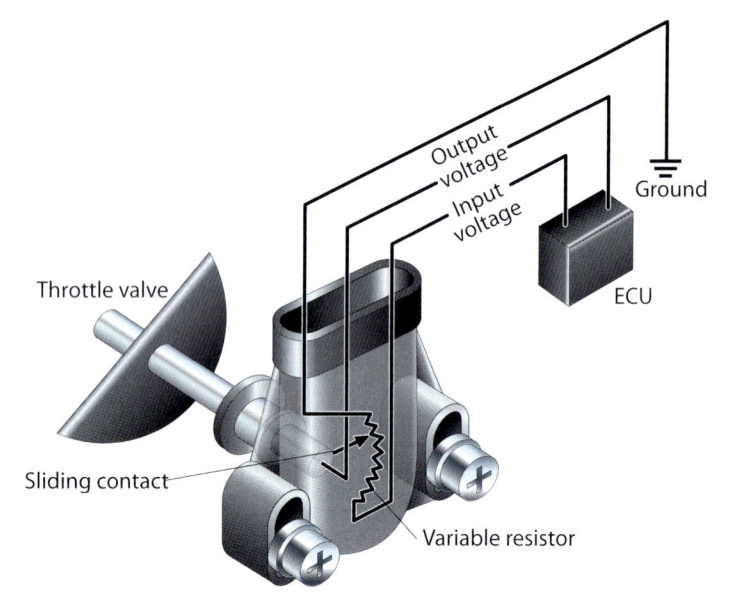

Figure 4.20 Throttle position sensor (TPS)

> **Working life**
>
> Sunil has been working on a multipoint fuel injection system. When he reassembles the throttle position sensor, he notices that it is adjustable and can be mounted in a number of different positions. He didn't mark its original position, so decides to guess.
>
> 1. Does it matter where the throttle position sensor is mounted?
> 2. What information does the ECU get from the throttle position sensor?
> 3. What symptoms might be seen if the throttle position sensor is mounted in the wrong place?

Auxiliary air valves

As electrical loads are switched on and off, the demands on the alternator and charging system try to slow the engine down. If the engine is on **tickover**, **idle speed** can fall to a level where it can't be maintained and the engine will stall.

If small quantities of air are allowed to bypass the throttle butterfly, the tickover speed can be increased when needed. In many cases this air bypass is controlled by a small valve or solenoid, as shown in Figure 4.21. The valve can be opened and closed by the ECU when an rpm sensor shows that idle speed has fallen below a set value. By regulating the amount of air that passes the throttle butterfly, correct idle can be maintained.

> **Find out**
>
> Run a car engine until it reaches its normal operating temperature. (Remember to observe health and safety, including exhaust extraction.)
>
> - Make a note of the idle speed.
> - Switch on various electrical components and see if a difference occurs in idle speed and tickover.
> - Check the speeds against the manufacturer's specifications.

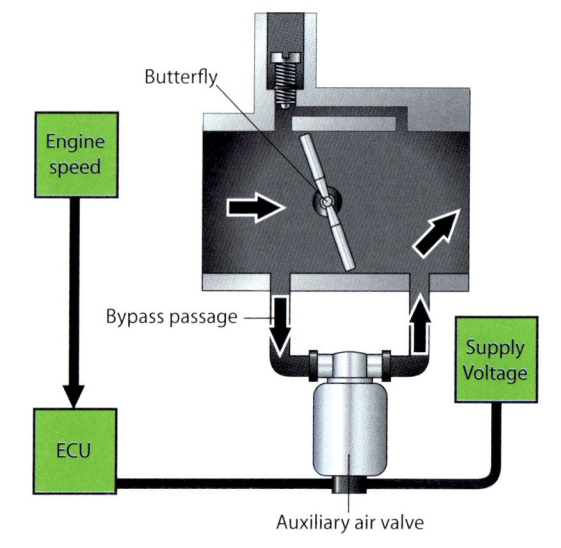

Figure 4.21 Auxiliary air valve

Plenum chambers

The length of the air inlet tract (from intake near the air filter to the inlet valve) will vary from manufacturer to manufacturer. The ideal length of the air inlet tract depends on engine load and speed demands. This means that performance will be affected if the inlet tract is too long or too short.

To help overcome this, some manufacturers use a plenum chamber (see Figure 4.22) in the design of their intake manifolds. This is a chamber that can be used as a reservoir of air. The cylinders can draw from this, which means that the intake air does not have to travel the entire length of the inlet tract.

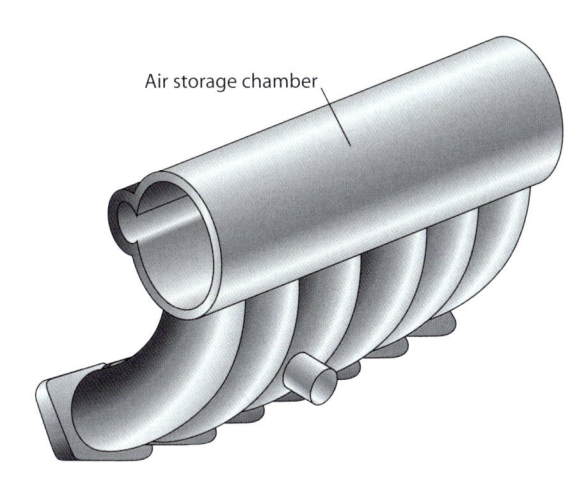

Figure 4.22 Plenum chamber

Engine coolant temperature sensors (ECT)

Because fuelling demands vary according to engine temperature (for example, a cold engine needs more fuel, similar to applying a choke on a carburettor), the engine's ECU needs to measure engine temperature so that it can control fuelling as required.

The engine coolant temperature sensor (ECT) is a heat-sensitive variable resistor (**thermistor**).

Two types of ECT are in common use: negative temperature coefficient (NTC) or positive temperature coefficient (PTC).

- With an NTC thermistor, as engine temperature rises, resistance falls.
- With a PTC thermistor, as engine temperature rises, resistance also rises.

This change in resistance can be signalled to the ECU to give an indication of engine temperature.

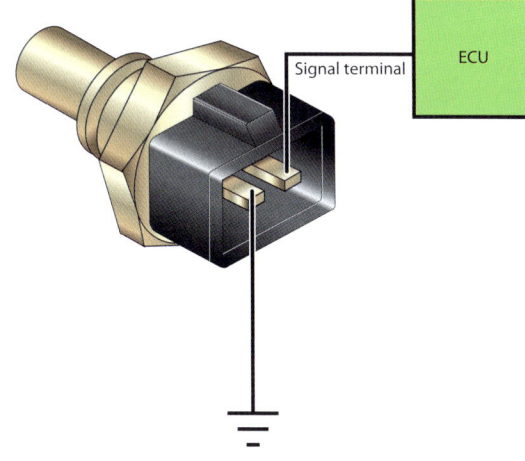

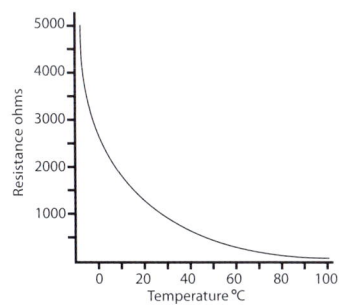

Figure 4.23 Engine coolant temperature sensor

Air temperature sensors

The **density** of oxygen changes when the temperature changes, so the ECU must be able to measure air temperature in order for it to compensate for this. (On a cold day the air charge will be denser and contain more oxygen.) If this is not allowed for, **stoichiometric** values may be incorrect.

Air temperature sensors can be NTC or PTC thermistors. They work in exactly the same way as an engine coolant temperature sensor (ECT).

> **Key terms**
>
> **Tickover/idle speed** – when the engine is running at its lowest set speed.
>
> **Thermistor** – a temperature-sensitive resistor.
>
> **Density** – how closely packed the molecules of a substance (gas, liquid or solid) are.
>
> **Stoichiometric** – a balanced chemical reaction used to describe the ideal air/fuel ratio.

Level 2 Light Vehicle Maintenance & Repair

Engine speed sensors

A number of methods can be used to sense engine speed. The main ones are crankshaft position sensing and ignition primary trigger signals.

Ignition primary trigger signals

These are very rarely used now, as distributorless ignition systems (see page 297) are common. This means that the switched signal to the primary circuit of the ignition coil is an output from the ECU and not a sensor input of engine rotation speed.

Crankshaft position sensor

Engine speed and crankshaft position can be detected by using a sensor normally mounted close to the flywheel. A toothed gear is rotated past the crankshaft sensor, creating a small alternating current. The frequency of this current can be used to calculate crankshaft speed.

If a gap in the gear teeth is left at a position representing top dead centre (TDC), the ECU will know when cylinder number one and its corresponding piston are at the top of their stroke. (On a four-cylinder engine this will be one and four.) By knowing when the pistons have reached TDC, the ECU can more accurately meter the amount of fuel to improve performance, and lower fuel consumption and emissions.

Figure 4.24 Crankshaft position sensor

> **Key terms**
>
> **Sequential** – in order, one after another (in sequence).
>
> **Reluctor** – a toothed ring used together with an inductive sensor.

Camshaft position sensor

Some multipoint fuel injection systems operate individual injectors according to engine firing order. This is called **sequential** injection. So that this can happen, the ECU also needs to know which cylinder is on its induction stroke. A camshaft position sensor (see Figure 4.25) is often used to detect this. It works in a similar way to a crankshaft position sensor, but instead of a number of gear teeth rotating, a single **reluctor** tooth is normally used. Because the camshaft turns at half the speed of the crankshaft, if its signal is combined with the crankshaft position sensor, the ECU can work out induction strokes and firing orders.

Figure 4.25 Camshaft position sensor

CHECK YOUR PROGRESS

1. As an engine warms up, what happens to the resistance of an NTC engine coolant temperature sensor?
2. Why are plenum chambers used in some air intake systems?
3. What does the acronym MAP mean?

4 Light vehicle fuel, ignition, air & exhaust system units & components

Fuel delivery in diesel engines

A diesel compression ignition system uses heated air to ignite injected fuel. As with petrol engines, fuel is stored in a tank and is transferred to the engine through fuel lines and pipes (see Figure 4.26).

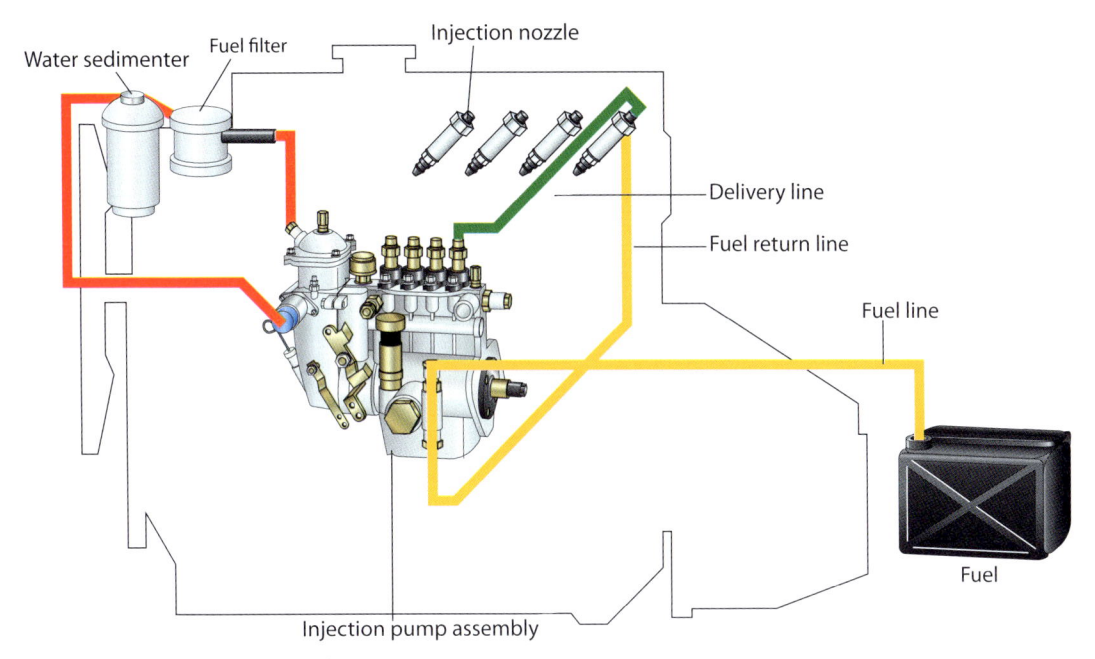

Figure 4.26 Fuel system layout of a compression ignition engine

A fuel injection pump is used to pressurise diesel in the system until it reaches a point where it can overcome the force of a spring inside a mechanical injector. When this happens, a needle in the injector lifts off of its seat and sprays (atomises) fuel into the combustion chamber.

The two main types of injection pump commonly used are inline and rotary.

Inline fuel injection pump

An inline pump (see Figure 4.27) uses an engine-driven camshaft to operate a set of plungers in cylinders. These act as the pumping elements and create pressure. The pumping elements are arranged in a line, one after another, which gives this pump its name.

As the plungers move on their pumping stroke, fuel pressure rises until the injector opening pressure is reached. At this point fuel is injected into the combustion chamber. Then, so that injection can be immediately stopped, a port opens in the chamber which allows excess fuel to escape and pressure to drop.

When the port in the chamber opens, it is called spill. To vary the amount of fuel that is injected, this spill port can be moved through a mechanism connected to the accelerator. In this way the engine speed can be increased and decreased. These fuel injection pumps are very accurate but are heavy and expensive. As a result, they are rarely used on smaller diesel cars.

> **Did you know?**
>
> Early diesel fuel pumps were connected to the throttle by a cable. This cable was replaced in later systems by sensors and actuators to regulate the pump, and became known as electronic diesel control EDC. Modern common rail direct injection systems use processes that are very similar to electronic petrol fuel injection when controlling the amount of fuel delivered to the engine.

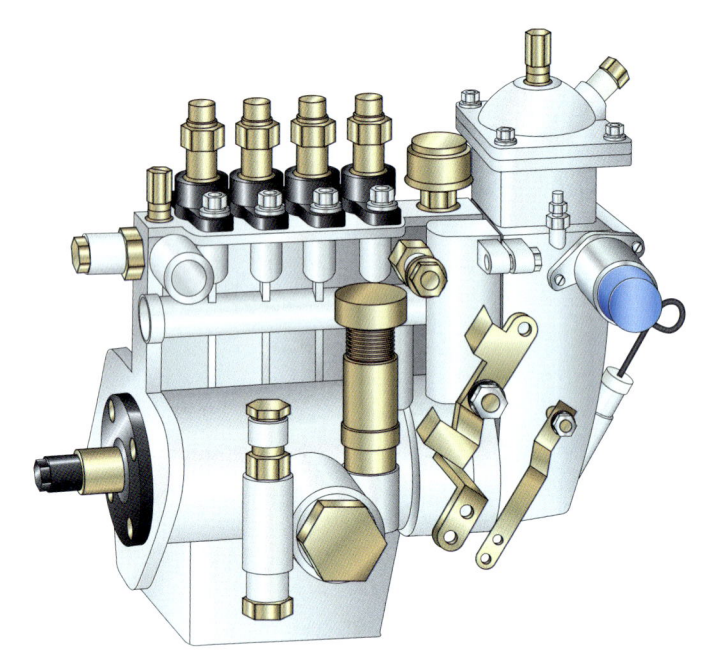

Figure 4.27 Inline diesel injection pump

Rotary fuel injection pump

A rotary injection pump (see Figure 4.28) is smaller, more compact and cheaper to produce than an inline pump, making it more popular for use on smaller diesel cars.

- The engine-driven pump rotates a **cam ring**.
- Pumping plungers are mounted inside this cam ring and operate as a pair.
- The pressure created overcomes the spring force inside the injectors.

As with the inline pump, the pumping chamber can be opened with a spill port connected to the accelerator, controlling engine speed.

Just like a spark ignition system, the process of combustion has to start before the piston has reached TDC on its compression stroke. To make sure that diesel fuel is injected at exactly the right time, the pumping elements can be advanced by setting the angle of the fuel pump in relation to the crankshaft rotation.

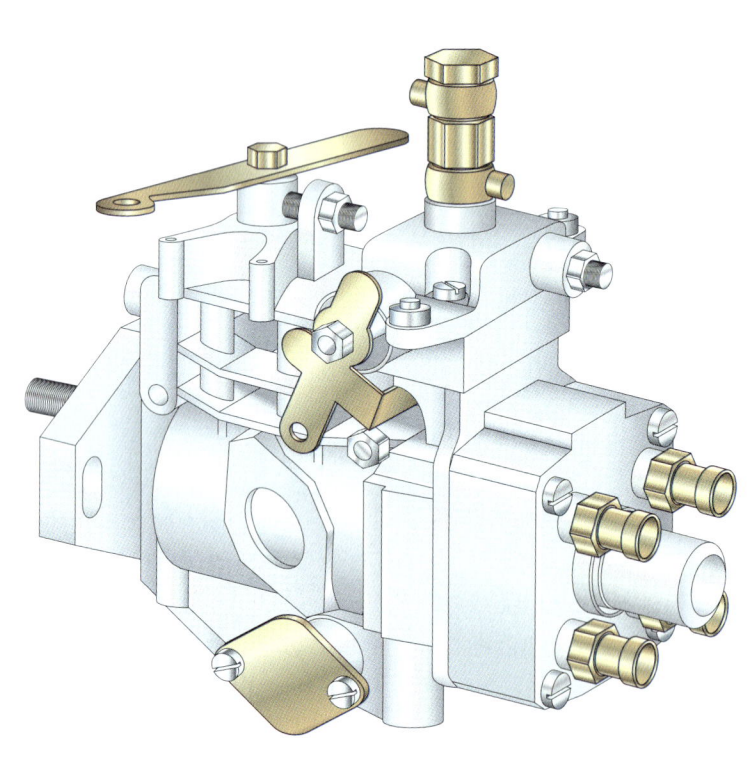

Figure 4.28 Rotary diesel injection pump

To time the injection pump statically (without the engine running), you can mount a dial test indicator (DTI) against the pump plunger, as shown in Figure 4.29. Following the manufacturer's routines, you rotate the injection pump until you obtain the correct readings.

With some systems it is also possible to carry out injection pump timing dynamically (while the engine is running). To do this you fix a special clamp containing a **piezoelectric crystal** around number one fuel injection pipe. As diesel is pumped through the pipe, pulsations are picked up by this clamp and converted into a signal which can be used to power a strobe timing lamp. With the engine running, you can use the timing lamp to align timing marks by rotating the injection pump in relation to the crankshaft.

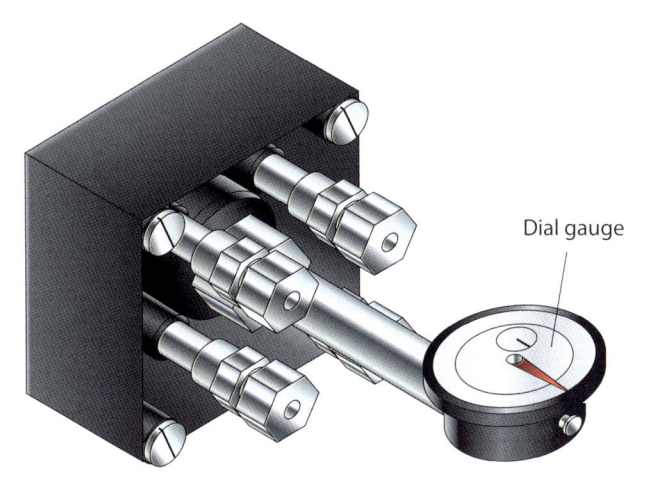

Figure 4.29 Injection pump timing

Diesel speed control

Unlike a petrol engine, a compression ignition engine controls speed by the amount of fuel it injects. As more fuel is injected, revs increase, and this would continue until the engine self-destructs. To stop this happening, a way of restricting fuel injection is needed. A mechanical or electronic rev limiter, called a **governor**, is used with both types of pump to prevent damage.

Because no spark plugs are used, a mechanical diesel system will continue running all the time it is supplied with fuel. To stop the engine, the fuel must be cut off. This is normally done with a small valve controlled by an electric solenoid, as shown in Figure 4.30.

- All the time the ignition switch is on, the solenoid holds the valve open, which allows fuel to flow and the engine to run.
- When the ignition key is switched off, power is cut to the solenoid and a spring closes the valve. This stops the fuel and makes the engine cut out.

Fuel injectors

Fuel injectors used with both inline and rotary pumps are usually mechanically controlled. A central needle is held firmly against a specially machined seat in the nozzle of the injector by strong spring force. These springs are accurately calibrated. As the injection pump raises the pressure of fuel inside the injectors, force acting on this needle valve increases until it reaches a point where it is able to lift off of its seat.

> **Key terms**
>
> **Cam ring** – a component with cam lobes inside a circular metal ring.
>
> **Piezoelectric crystal** – a piece of crystal that produces electricity when placed under pressure.
>
> **Governor** – a device used to limit the speed of a diesel engine.

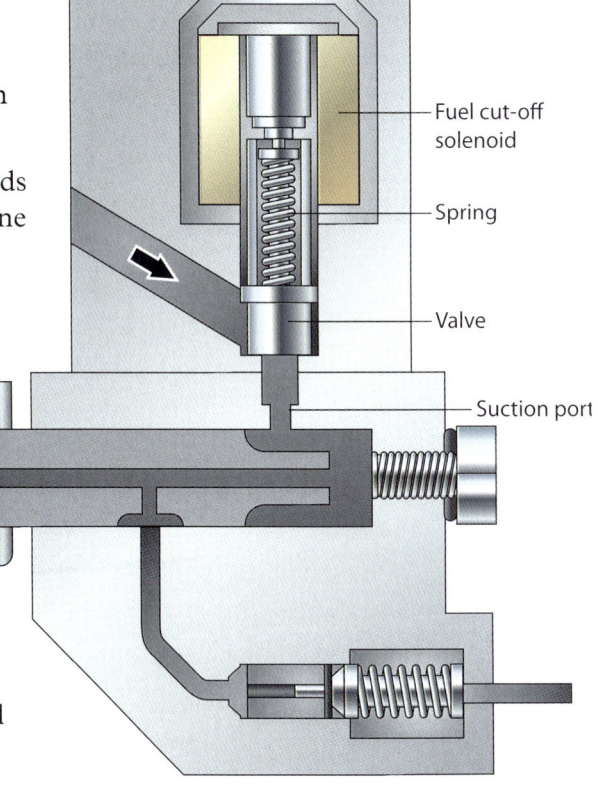

Figure 4.30 Fuel cut-off solenoid

Depending on the design of the injector nozzle, fuel is atomised in a particular spray pattern. Examples of injector nozzle types and spray patterns are shown in Figure 4.31.

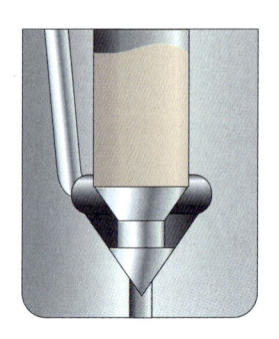

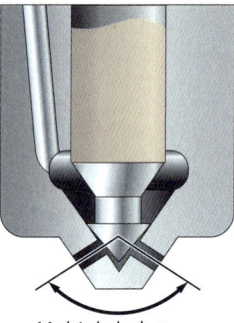

Single hole type | Multiple hole type | Pintle type | Throttle type

Figure 4.31 Injector nozzle types

At the end of the injection period, it is important that the fuel injector stops immediately. When the spill port of the injection pump opens, injection pressure rapidly falls and the return spring forces the injector needle back against its seat.

On a multicylinder engine, fuel injectors are calibrated as a matched set, so they all work at the same pressures and deliver the same quantity of fuel.

Diesel injector feed pipes

The feed pipes between the injection pump and the fuel injector are made of rigid metal to prevent pressure losses. Each injector feed pipe is carefully shaped and bent so that it fits round other engine components. If these metal feed pipes were straightened out, they would all be exactly the same length. By having these pipes all the same length, the timing of the fuel injection can be accurately controlled.

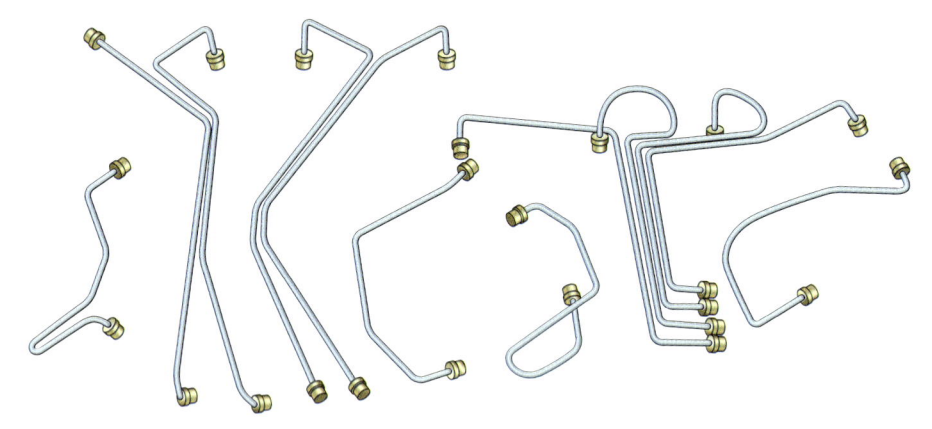

Figure 4.32 Metal diesel injector feed pipes

Diesel combustion chambers

The combustion chamber of a compression ignition engine is much smaller than in a spark ignition engine. This provides the very high pressures required to heat the air, so that when fuel is injected it will spontaneously combust.

Direct injection

The combustion chamber can be manufactured in the cylinder head or in the piston crown. The shape of the combustion chamber will be specifically designed to produce a swirling action of the incoming air (see Figure 4.33). This swirling action mixes the air and fuel, as it is injected, in an even manner. If the air and fuel are not evenly mixed as the fuel is injected, droplets of diesel will start to ignite in a number of different positions at the same time. As the fuel burns and the flame spreads, a collision of flame fronts produces a distinctive noise called 'diesel knock'.

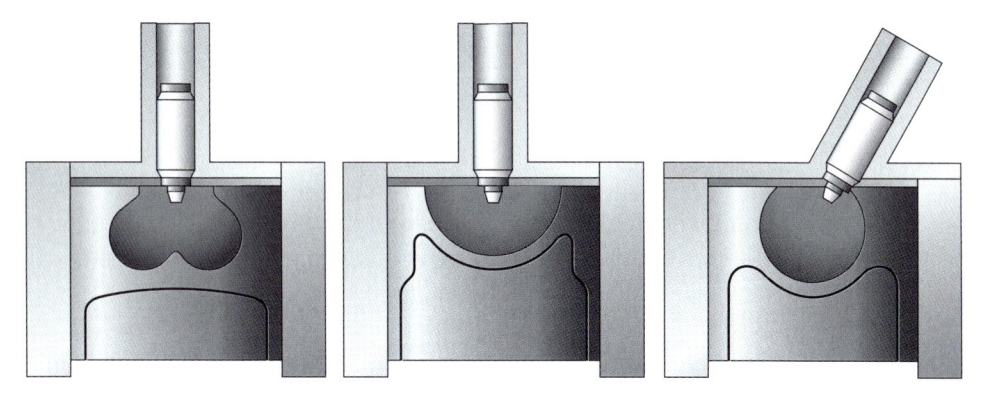

Figure 4.33 Different types of piston crown combustion chambers

Indirect injection

Another method used to try to produce an even burn is indirect injection. With indirect injection, instead of diesel fuel being atomised directly into the combustion chamber, the injector is mounted in a small pre-combustion or swirl chamber, as shown in Figure 4.34. As the fuel starts to burn, it is contained within the pre-combustion chamber and the flame spreads out in a far more even manner. This provides the pressure inside the cylinder to operate the piston on its power stroke.

An indirect injection engine usually uses an injector of the 'pintle' type (see Figure 4.31 on page 284) and is much quieter in operation than a direct injection engine.

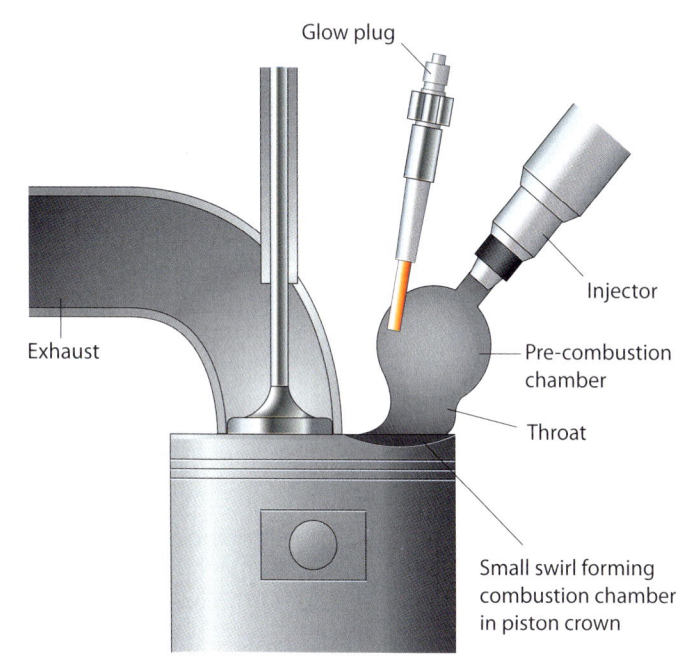

Figure 4.34 Swirl chamber

Cold-starting

Because a compression ignition engine uses the heat of air to initiate combustion, when an engine is started from cold, temperatures may be too low and starting can be difficult. (This can be a particular problem on indirect injection diesel engines.) Glow plugs are therefore used to pre-heat the air in the combustion chamber to assist with cold starting.

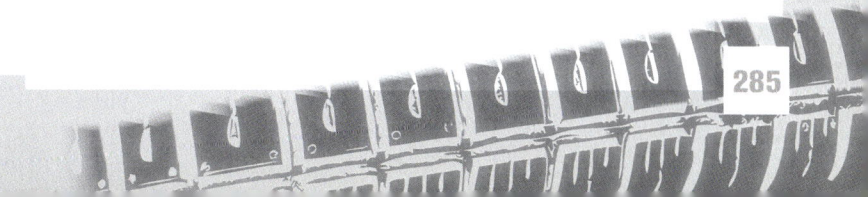

> **Find out**
>
> Measure the length of the fuel injection pipes of a standard diesel engine in your workshop.
>
> - Are they all the same length?
> - What type of injection pump is fitted?

When the ignition key is first switched on, a timer circuit and relay is used to supply electric current to a set of heating elements called glow plugs (see Figure 4.35). These plugs are used only for a short period at start up and should not be confused with spark plugs.

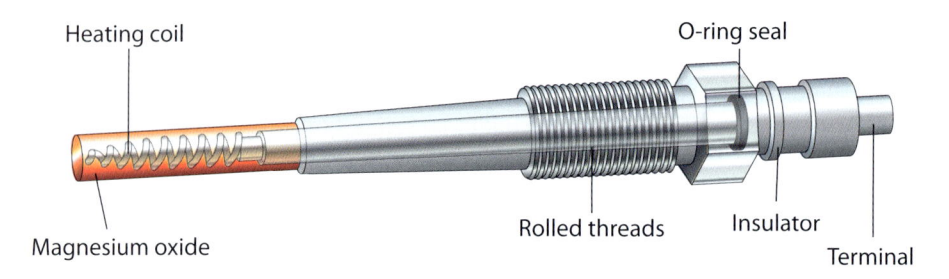

Figure 4.35 Diesel engine glow plugs

Some diesel engines can also use a fuel enrichment device. This can be automatically operated when the engine is cold, to assist with starting.

Turbocharging

A turbocharger uses waste exhaust gases to rotate a **turbine** (like a small windmill) which is attached by a shaft to a set of **compressor vanes**. As exhaust gases exit the engine, the turbine spins at speed, driving the compressor and forcing air into the cylinder at higher than atmospheric pressure.

Two problems occur with the use of turbochargers:

1. Exhaust gases need to be at a relatively high pressure to turn the turbine fast enough to produce any useful boost. This means that until the engine is revving at a fairly high rpm, no performance increase from the turbocharger is felt. This is known as 'turbo lag'.
2. As engine revs continue, **turbo boost** and, therefore, performance continue to increase until a point where overboost can cause engine damage. Because of this, a safety valve is needed to release this extra boost if it becomes too much. Should this happen, a pressure or vacuum-operated valve is used to bypass exhaust gases away from the turbine if this happens. This is called a 'wastegate'.

A turbocharger is shown in Figure 3.24 on page 222.

> **Key terms**
>
> **Turbine** – a bladed rotor.
>
> **Compressor vanes** – a set of turbine blades used to compress air.
>
> **Turbo boost** – an increase in engine performance given by a turbocharger.
>
> **Air charge** – the air entering the engine during induction.

Intercoolers

Intercoolers are often used together with turbochargers. When the turbocharger raises air intake pressure, it also increases the air temperature. This is because of the transfer of heat from the exhaust gases used to drive the turbo and the compression of the air. This rise in temperature reduces the density of the oxygen in the **air charge**.

After exiting the turbocharger, the air can be passed through a radiator called an intercooler, before it enters the combustion chamber. In this way some of the heat can be dissipated to atmosphere, thereby improving the air density.

Comparison of petrol and diesel system components

Table 4.1 gives a summary of the differences and similarities between a petrol and a diesel fuel system.

Table 4.1 Comparison between petrol and diesel system components

Component	Petrol system	Diesel system
Fuel tank	• Similar in design and construction to that found on a diesel	• Similar in design and construction to that found on petrol
Fuel pump	• Either low-pressure mechanically driven or electrically operated for fuel injection • Delivery pressure approximately 3 bar	• Usually high-pressure mechanically driven from the engine • Delivery pressure approximately 180 bar to 1800 bar
Fuel filter	• Designed to remove dirt • Does not contain a water separator or sedimenter	• Designed to remove dirt and water • Normally contains a sedimenter, which can be drained to remove excess water
Speed control	• An air flap is connected to the accelerator, and is designed to regulate the amount of air entering the engine and therefore engine speed	• No throttle butterfly: the accelerator is connected directly to the fuel pump or engine management system and regulates the amount of fuel being injected. The amount of fuel being injected controls engine speed.
Fuel injectors	• Usually of an electronic solenoid type, injecting fuel before it enters the combustion chamber • Opening period controlled by an electronic control unit (ECU) • Operates at a pressure of approximately 3 bar	• Can be mechanically operated or an electronic solenoid type • Fuel is usually injected directly into the combustion chamber of the engine • Mechanical injectors operate at a pressure of approximately 180 bar • Electronic injectors operate at a pressure of approximately 1800 bar
Ignition	• Achieved by the spark plug	• Achieved by raising the air temperature with high compression pressures

Key terms related to petrol and diesel fuel

Volatility

In a petrol or diesel engine, it is not the liquid fuel that burns but the vapour. The volatility of a fuel is a measurement of how well it turns into vapour at different temperatures.

- Petrol is an extremely volatile fuel, turning to vapour easily at room temperature. The fumes given off by petrol are extremely flammable and are easily ignited, making it suitable for use with spark plugs.
- Diesel fuel is less volatile. Its temperature needs to be raised considerably before it starts to turn into a vapour. High compression pressures are needed inside a diesel engine so that the atomised

fuel can be vaporised as it is injected into the combustion chamber. If pressures and temperatures are too low, diesel fuel is relatively stable and is not easily ignited. It is also hygroscopic, which means it absorbs moisture from the surrounding air.

The flashpoint of a fuel is the lowest temperature at which it will vaporise to form an ignitable mixture. When fuel reaches its flashpoint, if supplied with a source of ignition, it will combust (burn).

Octane value

In a petrol engine, the flashpoint of the fuel must be kept relatively stable to prevent spontaneous combustion. In other words, you don't want the fuel igniting until the spark plug fires.

To help stabilise petrol, chemicals are added to 'suppress detonation'. Originally, a chemical substance called tetraethyl lead was added to petrol (this petrol was called 'leaded petrol'). As science and medicine improved, it became clear that lead is a toxic substance which is hazardous to health. As a result, leaded fuel has been phased out, although it is still available in some countries.

When tetraethyl lead was removed from petrol, other chemicals were added to suppress detonation. The stability of a fuel is graded and given an **octane number** – also called a research octane number (RON). The higher its octane number, the more stable the fuel, meaning that it can be used in higher compression, higher performance engines.

Cetane value

Whereas the octane number shows the stability of petrol fuel, a **cetane value** is used to show how easily diesel fuel will ignite. Because a diesel engine uses compression ignition, it is important that as the injector atomises fuel into the combustion chamber and it mixes with the superheated air, the ignition and combustion process is smooth and rapid.

After the fuel has been injected, three stages of combustion occur:

1. **The delay period** – This is the short period of time between when the fuel is first injected and ignition in the heated air of the combustion chamber starts to take place. The length of the delay period is related to the cetane number of the fuel. Normally, the higher the cetane number, the shorter the delay period.
2. **Flame spread** – This is where the atomised fuel has vaporised and starts to combust, hopefully with the flames spreading out in a relatively even manner using the oxygen in the combustion chamber air. Because no single source of ignition exists, such as the spark produced with a plug, atomised fuel can begin to ignite in a number of different places at once. As the flames spread, and the flame fronts collide, it produces a distinctive knocking noise.
3. **Complete burn** – This is where the fuel has used up the energy contained in the hydrogen to create a very high temperature and pressure, which forces the piston down on its power stroke.

> **Key terms**
>
> **Octance value** – number which shows the stability of petrol fuel. The higher the number, the more stable the fuel.
>
> **Cetane value** – a number which shows how well diesel fuel burns. The higher the number, the better it burns.
>
> **Stoichiometric value** – the ideal air to fuel ratio, which is approximately 14.7 : 1 by mass.
>
> **Lambda window** – the engine operating range which achieves an air to fuel ratio of 14.7 : 1 by mass. Lambda is a Greek letter used to represent the stoichiometric value.

Combustion processes

Three things are needed for combustion: a source of fuel, a source of heat, and oxygen. The source of fuel comes from either petrol or diesel, but it is only the hydrogen in the hydrocarbon fuel that is burnt. This means that the carbon that is found in petrol and diesel is a waste product.

The constituent components of air are shown in Figure 4.36. They are: nitrogen at 78 per cent, oxygen at 21 per cent and 1 per cent other gases (known as trace gases). As it is only the oxygen that is used in the combustion process, the nitrogen is a waste product. Nitrogen is an inert gas, which means that it doesn't burn or support combustion.

When oxygen and hydrogen are brought together and are supplied to the source of heat via the spark at the plug, they burn and release their energy as heat. In order to minimise the amount of waste products given off by the combustion process, just the right ratio of oxygen to hydrogen is required, so that very little is left over after combustion. This is known as a **stoichiometric value** or a balanced chemical reaction. Under normal operating conditions the stoichiometric value should be 14.7 : 1 by mass (14.7 parts air to 1 part fuel, by weight).

Exhaust gas constituents

Exhaust gases consist of chemical components made up after the combustion process has taken place.

The main chemical elements taken into the engine are oxygen and nitrogen in air, and hydrogen and carbon in fuel. Theoretically, ideal combustion (sometimes called the **Lambda window**) would only produce carbon dioxide, nitrogen and water (H_2O), as shown in Figure 4.37. Unfortunately, the combustion process is never perfect and so exhaust pollutants are also produced.

Following the combustion process the elements are rearranged to form the following substances:

- **Carbon dioxide** – With effective combustion, the chemical elements carbon and oxygen combine to form carbon dioxide (CO_2). These emissions should normally be higher than 14 per cent by volume. Carbon dioxide is a greenhouse gas and is considered an environmental pollutant.
- **Hydrocarbons** – If fuel passes through the combustion process with no chemical change, hydrocarbons (HC) are given off. These should be kept as low as possible, usually under 200 parts per million at idle. Hydrocarbons are considered harmful to health and may cause lung damage or cancer.
- **Carbon monoxide** – During the combustion process, if the burning fuel goes out (maybe due to lack of oxygen or rapid cooling), carbon monoxide (CO) is produced. Carbon monoxide is a product of incomplete combustion. Carbon monoxide is harmful to health. It is colourless, odourless and tasteless, but if inhaled it is poisonous.

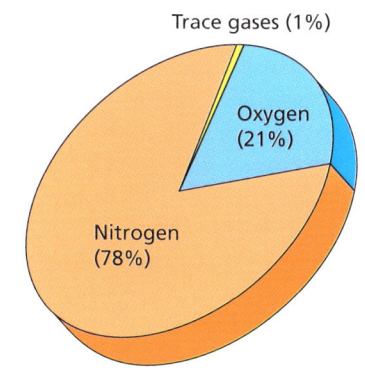

Figure 4.36 The main constituents of air

Did you know?

Strict controls are placed on the amount and type of exhaust emissions that can be released to atmosphere from cars. In the UK these will be tested annually when the car has its MOT. Vehicles produced for use in Europe must meet Euro emission standards before they are approved for sale.

Safe working

Some exhaust pollutants are hazardous to health. When working on a running engine you should avoid breathing in exhaust fumes by using special exhaust fume extraction equipment. If this is not available you must make sure that the area is well-ventilated.

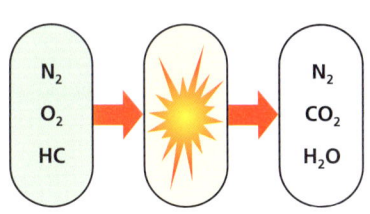

Perfect combustion - stoichiometric

Figure 4.37 Ideal combustion

> **Did you know?**
>
> If a piece of wood is burnt completely it will turn to ash. But if the piece of wood goes out halfway through its burning process you will be left with a lump of charcoal. This is an example of incomplete combustion.

It replaces oxygen in the blood and starves the organs of required oxygen.

- **Oxides of nitrogen** – Most of the combustion process takes place at temperatures between 2000 and 2500 °C. In this extreme heat, the oxygen and nitrogen in the incoming air are combined to produce a pollutant called oxides of nitrogen (NO_x). Unfortunately, the better the combustion process is, the more oxides of nitrogen are produced. Because of this, specific methods of control are required, such as exhaust gas recirculation (EGR) – see pages 302–303. Oxides of nitrogen can cause health issues for the lungs and may also damage plant life and reduce visibility.

CHECK YOUR PROGRESS

1. List five petrol or diesel fuel system components.
2. Explain two differences between single and multipoint injection systems.
3. How do the fuel pumps differ between petrol and diesel systems?
4. Which exhaust gas emissions are environmental pollutants?

Ignition system construction and operation

To ignite the air and fuel in a petrol engine, a high-voltage spark is required to start the burn.

With an engine speed of 6000 rpm the piston is travelling from top dead centre (TDC) to bottom dead centre (BDC) and back again 100 times a second (6000 rpm divided by 60 seconds = 100). If a spark is required every fourth stroke, that is 25 sparks per second for each cylinder.

An ignition system must take 12 volts from the battery and turn it into many thousands of volts, many times a second. Any slight deviation in this process can lead to a misfire or the engine not running at all – near enough is not always good enough!

Ignition system components

A standard ignition system is made up of a number of components. These include:

- battery
- ignition switch
- ignition coil
- switching mechanism (such as contact breakers).
- distributor
- high-tension leads
- spark plugs

> **Key term**
>
> **Electrodes** – the electrical conductive tips of a spark plug.

Spark plugs

Spark plugs are used in the internal combustion petrol engine. They ignite the fuel and start the rapid burn at precisely the correct time. The gap at the tip of the spark plug creates an open circuit. This means that no electric current can flow until a high enough voltage is produced to overcome the resistance of the air gap.

To produce this voltage, two **electrodes** are used. As shown in Figure 4.38, these are:

- a centre electrode, normally fed from the high-tension ignition lead
- an earth electrode, normally manufactured as part of the spark plug shell.

The tips of these electrodes are machined to precise shapes.

High-tension sparks find it easier to jump from a sharp point on an electrode. The thinner and sharper an electrode, the better it will function. However, due to high temperatures and spark erosion, narrow, pointy electrodes do not last very long. For this reason, the electrodes found on a standard spark plug are machined flat with sharp edges, allowing sparks to be created at a number of points. Over a period of time the sharp edges will begin to erode and round off. This reduces the performance of the spark plugs, and they must be replaced during normal maintenance procedures.

A number of precious and rare metals such as platinum and iridium are used in the production of some spark plugs. This means that they are less prone to spark erosion and can be manufactured with smaller, sharper electrodes.

Spark plug dimensions

Many spark plugs are manufactured so that they can be used across a range of vehicles. When designing engines, vehicle manufacturers take into account size and shape (dimensions) but also the operating heat range.

Thread diameter

Spark plugs are manufactured with a number of different thread diameters. The most common are 10 mm, 12 mm, 14 mm and 18 mm. This is the size indicated on a spark plug socket. (Note that it is not measured across the flats of the nut surface as found on standard sockets used for nuts and bolts.)

Spark plug reach

The reach of a spark plug is the length of the threaded section that is screwed into the cylinder head. Spark plugs can have either long reach or short reach. This will depend on the engine design and spark plug position.

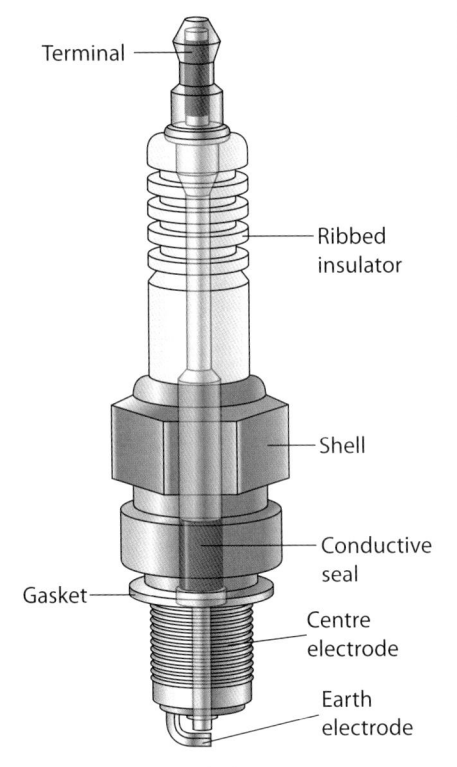

Figure 4.38 Spark plug

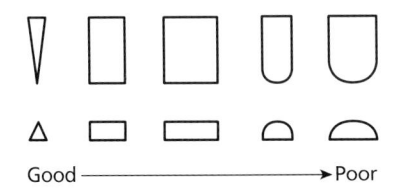

Figure 4.39 Spark plug electrode shapes

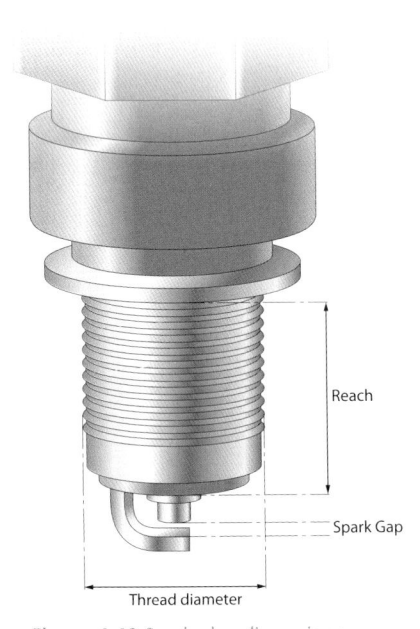

Figure 4.40 Spark plug dimensions

Spark plug gap

The air gap found between the electrodes at the tip of the spark plug is accurately calculated by the manufacturer during the engine design process. Each engine requires a spark plug with a different gap to produce the correct quality and timing of sparks used to ignite the petrol inside the combustion chamber.

To set the spark plug gaps, you should follow the manufacturer's recommendations and use feeler gauges to accurately measure the gap setting. Wire gauges are available to set the spark plug gaps, and these have multiple earth electrodes.

If spark plug gaps are incorrectly set, this can lead to poor running and even misfiring.

- Spark plug gaps set larger than recommended by the manufacturer require higher firing voltages. This high voltage will take longer to produce and the timing will be retarded (too late).
- Spark plugs that have a smaller gap than recommended by the manufacturer require lower firing voltages. This low voltage will take less time to produce and the timing will be advanced (too early).
- If a multicylinder engine has spark plugs fitted with different gap sizes, then each individual cylinder will be receiving a firing voltage of varying strength and timing. This will lead to uneven running.

> **Did you know?**
>
> New spark plugs do not come 'ready gapped' from the spark plug manufacturer. This means they need setting before they are fitted to a particular engine.

> **Working life**
>
> Sam is replacing the spark plugs on a car during a scheduled service. She asks her supervisor what the correct plug gap should be. Her supervisor says, 'As long as you get the correct part number, it doesn't matter, as they come ready gapped.'
>
> 1. Is this correct?
> 2. How should Sam check the gaps?
> 3. What could happen if the plug gaps are incorrect?

Spark plug heat range

During normal operation, spark plugs are designed to operate within a certain heat range (an ideal range is between 500°C and 850°C).

During the design process of new engines, manufacturers test different spark plug heat ranges to make sure that the correct spark plug is selected for the vehicle. This will be included in the vehicle's technical data.

- If the temperature of the spark plug is too low, carbon and combustion chamber deposits may not be burnt off.
- If the temperature of the spark plug is too high, electrode and/or piston damage can occur (components may melt).

The heat range of a spark plug is determined by its shape and design. Heat **dissipation** occurs in the following processes:

- Approximately 20 per cent of the heat is taken away by the incoming air on the intake stroke.
- Approximately 58 per cent of the heat is absorbed into the cylinder head through **conduction**.
- Approximately 2 per cent goes into the HT leads.
- The remaining 20 per cent is taken up by the spark plug insulator and side walls.

> **Key terms**
>
> **Dissipation** – the spreading of heat into the surrounding air.
>
> **Conduction** – the transfer of heat through solid materials.
>
> **Mutual inductance** – an electric current created in two coils of copper wire by magnetic fields.

The amount of heat dissipated determines the heat range of the spark plug. This depends on two things:

- the shape of a ceramic insulator in the tip of the spark plug around the electrode
- the amount of air allowed to circulate.

The length of the ceramic insulator will help to make the spark plug run either hot or cold (see Figure 4.41).

- A long insulator around the electrode means that heat has to travel much further before it can be transferred to the spark plug walls and into the cylinder head. As a result, the spark plug will run hot.
- A short insulator around the electrode means that heat does not have to travel so far before it can be transferred into the spark plug walls and into the cylinder head. As a result, the spark plug will run cold.

The size, shape and heat range are normally found as part of the spark plug's identification number. Recommendations for these can be found within the manufacturer's data.

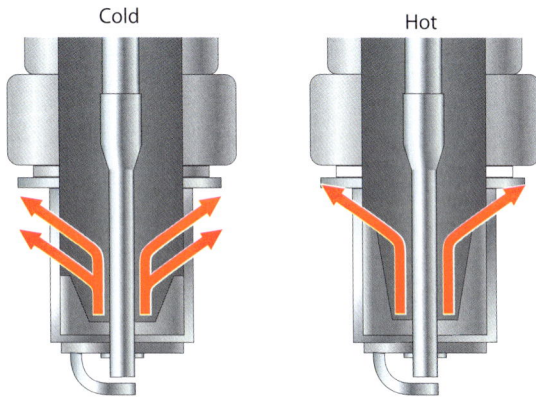

Figure 4.41 Cold spark plug and hot spark plug

Ignition coil

Many modern ignition systems are based on the principle of **mutual inductance** that was originated by Charles Franklin Kettering. For this reason they are often referred to as Kettering ignition systems.

A standard ignition coil contains two sets of copper windings: a primary and a secondary winding (see Figure 4.42).

- The primary winding is connected to the low-tension or 12-volt circuit of the car. It contains thick copper wire wound loosely with several hundred turns.
- The secondary winding contains a very thin copper winding (connected at one end to the primary circuit) with many thousands of turns. The secondary winding fits inside the primary winding.

A soft iron core is inserted in the centre of these two windings. This helps to increase the magnetic field produced when current is passed through the primary winding.

Level 2 Light Vehicle Maintenance & Repair

The primary and secondary windings and the soft iron core are mounted inside a cylinder. The cylinder is filled with oil to help cooling, and is then sealed. The top of the ignition coil has connectors to join it to the positive and negative of the **low-tension** circuit. One end of the secondary winding has a terminal for a **high-tension** discharge lead (normally known as the coil or **king lead**).

When the low-tension or 12-volt circuit is switched on, current flows through the primary winding. The current generates a magnetic field that also encloses the secondary winding and the soft iron core.

When current is flowing through the primary circuit, a magnetic field is building/charging – this is called the **dwell** period.

When the current flow to the primary winding is switched off, a rapid collapse of the magnetic field occurs. As this collapse passes through the secondary winding, it induces a high-tension voltage of many thousands of volts.

The voltage in the secondary winding builds and builds until the pressure is great enough to jump the air gap at the plug and produce a spark.

Conventional ignition systems

What is an ignition system?

In a conventional ignition system (see Figure 4.43), a mechanical switch, called contact breaker (points), is used to make and break the low-tension circuit to the ignition coil. These contact breakers are normally driven from their own camshaft, which turns at half crankshaft speed. The camshaft has a lobe for each cylinder needing a spark. When the engine is cranked, the contact breakers open and close.

When closed, current flows through the primary winding of the ignition coil. As soon as the contact breakers open, current flow stops. The magnetic field in the primary circuit breaks down, a high voltage is induced in the secondary winding and the spark is produced.

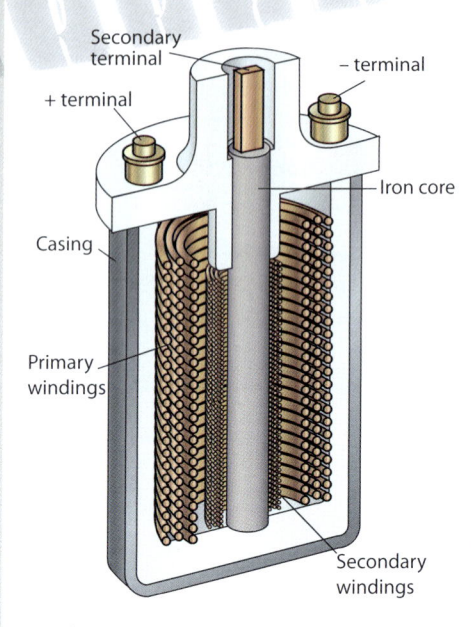

Figure 4.42 Ignition coil

> **Key terms**
>
> **Low tension** – the low voltage side of an ignition circuit (sometimes referred to as primary). It is normally measured at around battery voltage (12 V).
>
> **High tension** – the high voltage side of an ignition circuit, normally measured in many thousands of volts.
>
> **King lead** – the main high-tension lead from the ignition coil to the distributor cap.
>
> **Dwell** – the charge time of an ignition coil. (Also the period of time contact breakers are closed in a conventional ignition system.)
>
> **Condenser/capacitor** – a temporary storage of electricity.
>
> **Transistor** – electronic switch with no moving parts.

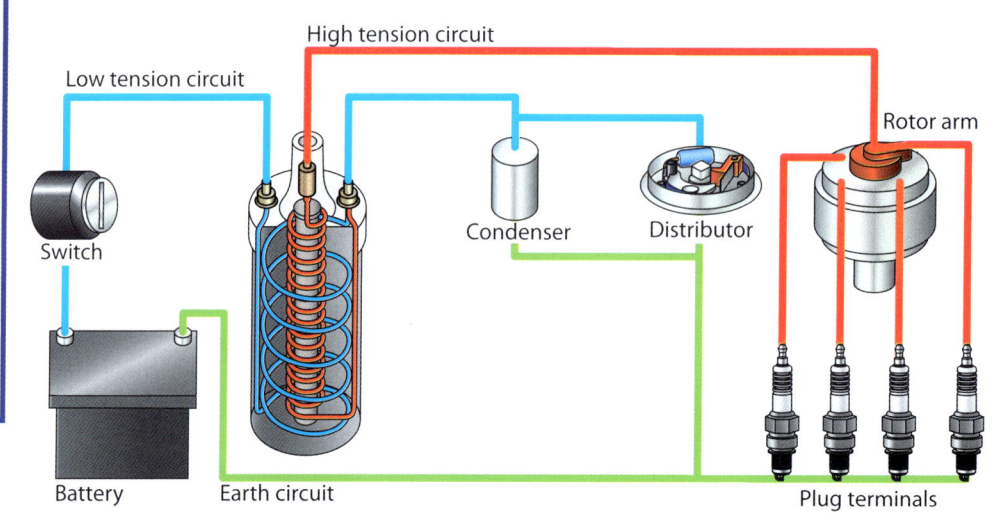

Figure 4.43 Conventional ignition circuit

294

4 Light vehicle fuel, ignition, air & exhaust system units & components

Condensers

As contact breaker points open and close, sparks are created which will damage the points and lead to misfiring and eventually breakdown. Conventional ignition systems are fitted with a **condenser** (also callled a **capacitor**) so that sparks at the points are prevented.

As the contact breaker points start to open, the current flowing in the primary winding of the ignition coil can no longer find an easy route to earth. Instead the electricity starts to fill the condenser. As soon as the points are wide enough and a spark can no longer jump the gap, the condenser discharges back through the primary winding of the ignition coil. This reverse current reacts against any current left in the primary winding, and stops it abruptly. This makes the magnetic field collapse even more quickly, which improves its efficiency.

Distributor cap and rotor arm

Once a high-tension spark has been produced in the secondary winding of the ignition coil, it leaves through the king lead and travels to the distributor cap. The high-tension spark normally enters the cap through a centre contact, which touches a rotating arm within the distributor. As this rotor arm turns, it points in sequence to the correct spark plug lead. This ensures that each spark plug receives the high voltage in the correct order.

Electronic ignition systems

Distributor ignition systems

An electronic ignition system uses many of the same processes as a conventional system. The main difference is that the mechanical contact breakers have been replaced by an electronic switch with no moving parts (in most cases a **transistor**).

The advantage of using an electronic system to switch the primary circuit of the ignition coil is that no mechanical wear is produced. It operates far more quickly and efficiently than contact breaker points, meaning the ignition timing is more accurate and reliable.

Ignition amplifier

The small signals produced at the trigger sensors are not enough to switch the primary circuit of the ignition coil, and this is where the ignition amplifier comes in.

The ignition amplifier effectively takes the place of the mechanical contact breakers that are found in a conventional ignition system. The ignition amplifier contains transistors, which are switches with no moving parts.

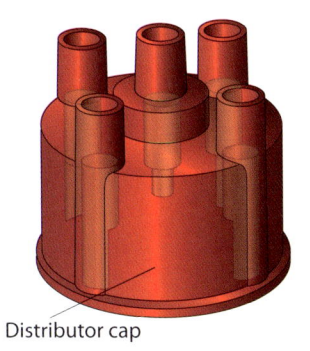

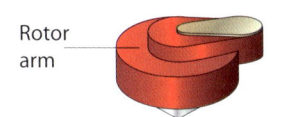

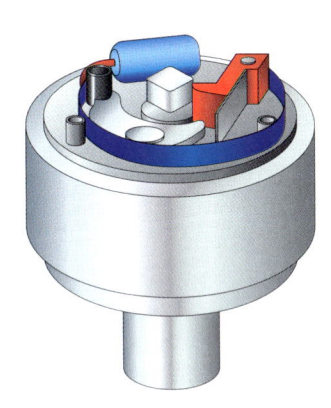

Figure 4.44 Distributor cap and rotor arm

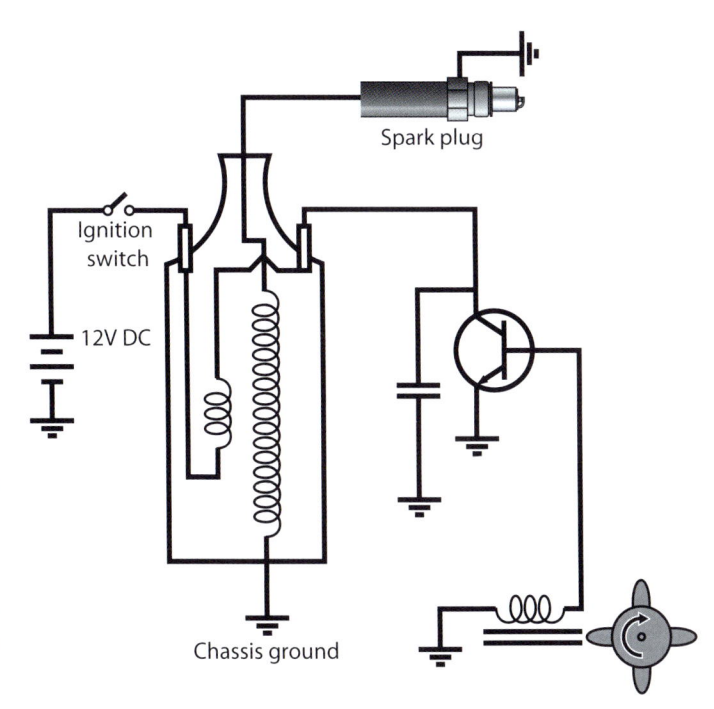

Figure 4.45 Electronic ignition system

295

Figure 4.46 Electronic ignition amplifier unit

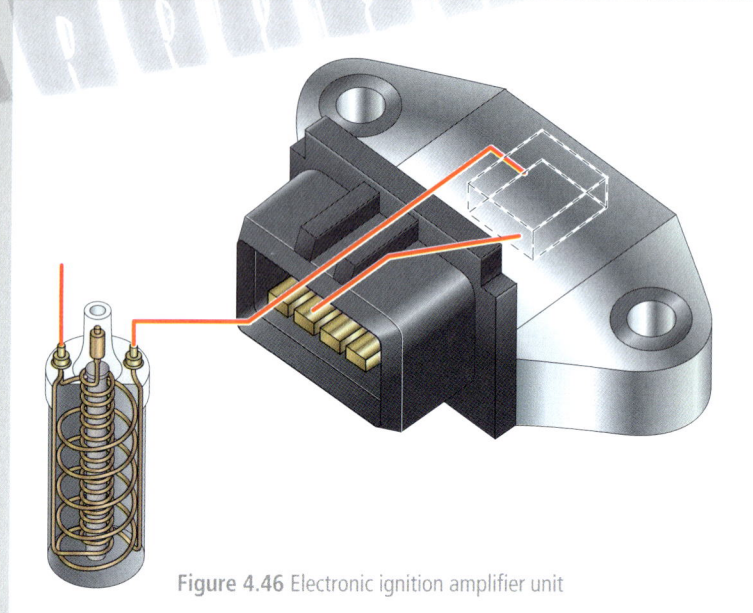

When connected to an ignition circuit, the **base** of the transistor is normally joined to the trigger sensor and the **collector** and **emitter** are joined to the ignition coil primary circuit. In this way, when a small trigger signal is produced, the primary circuit can be switched on and off.

Trigger signals

In order that the transistor knows when to switch the ignition coil on and off, a system is needed to trigger its operation. This is normally achieved using either:

- inductive sensors
- Hall effect sensors
- optical sensors.

Inductive sensors

An inductive sensor uses magnetism to produce a small electric current in a similar way to a generator. This small electric current can be used to signal the transistor unit, allowing it to switch on and off at the correct time. Because it makes its own electric current, it is often known as 'active'.

A rotating shaft (normally turning at half crankshaft speed) has a number of protrusions (or fingers) corresponding to the number of cylinders in the engine – this is called a **reluctor ring**.

Mounted closely to the reluctor ring is a permanent magnet with a small coil of copper wire wound around it – this is called the **pickup** (see Figure 4.47).

As the reluctor ring rotates and one of the fingers moves towards the pickup, the magnetic field produced by the permanent magnet is disrupted. As this magnetic field moves across the coil winding of the pickup, a small voltage will be induced.

- As the reluctor finger comes level with the pickup, disruption to the magnetic field falls and voltage within the coil of wire also falls to zero.
- As the reluctor finger moves away from the pickup, the magnetic field is once again disrupted, but this time in the opposite direction so that a negative voltage is produced.

This creates a small alternating current within the pickup, as shown in Figure 4.48.

> **Safe working**
>
> Many ignition amplifiers are sealed in the factory during production and you should not attempt to open these up. During the manufacturing process dangerous chemicals are used (such as hydrofluoric acid), which have the potential to cause serious harm.

> **Key terms**
>
> **Base, collector and emitter** – the electrical connections on a transistor.
>
> **Reluctor ring** – a toothed ring used in conjunction with an inductive sensor.
>
> **Pickup** – the sensor unit producing a signal created by the reluctor ring.

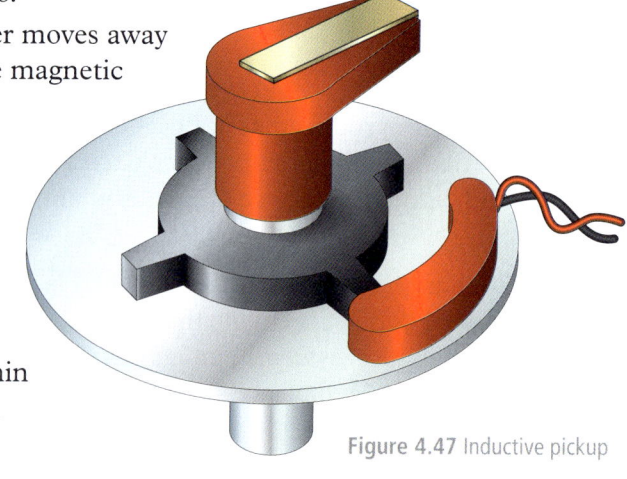

Figure 4.47 Inductive pickup

This small alternating current is not powerful enough to switch the ignition coil primary circuit on and off. Instead, the small signal is used to switch the base of the transistor inside the ignition amplifier unit (see Figure 4.46 opposite). This then switches the primary circuit of the ignition coil, which creates a dwell or charge period and then a secondary high-tension discharge to create the spark.

Hall effect sensors

The Hall effect sensor (see Figure 4.49) is an example of a **passive sensor**. It relies on an electric current that is normally supplied by the vehicle's ECU.

The Hall effect sensor is a unit that produces a magnetic field when supplied with electric current. This magnetic field is sensed by a small **integrated circuit** and a signal is produced. A rotor drum with slots and panels (doors and windows) rotates through the middle of the Hall effect sensor, interrupting the magnetic field. The integrated circuit is connected to a signal wire, which produces an on and off pulse. The on and off pulsing signal is used to switch the base of a transistor in the ignition amplifier, in a similar manner to the inductive sensor. The ignition amplifier is then able to switch the primary circuit of the ignition coil, creating a dwell period or charge time and then a high-tension spark.

The number of slots in the rotor drum corresponds to the number of engine cylinders and, therefore, the number of sparks required.

Optical pickup system (OPUS)

An optical pickup system (see Figure 4.50) uses a light source (usually a **light-emitting diode**) to shine on to a photo-sensitive receiver. A metal disc with slots cut in it (called a chopper plate) is rotated by the distributor shaft and interrupts light to the photo-sensitive receiver. As the receiver picks up or loses its light source, a small signal is sent to an ignition amplifier to switch the primary circuit on the ignition coil on and off in relation to ignition timing.

Distributorless ignition systems

Ignition systems have improved to minimise exhaust emissions, while maintaining good performance and increasing fuel economy.

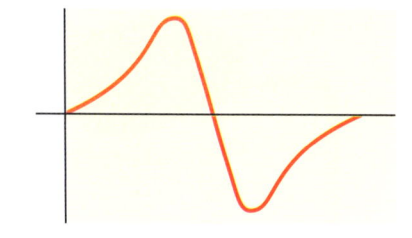

Figure 4.48 Voltage produced by an inductive sensor

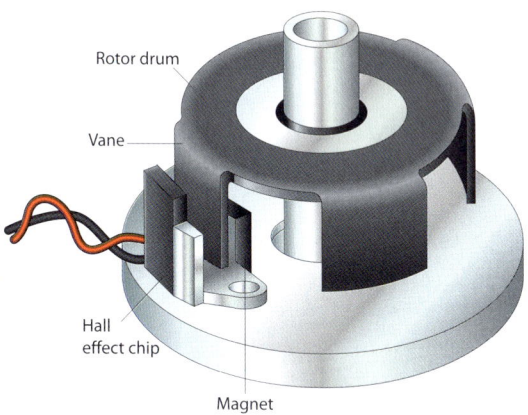

Figure 4.49 Hall effect sensor

Figure 4.50 Optical pickup system (OPUS)

Key terms

Passive sensor – a sensor that does not generate its own electrical current.

Integrated circuit – a miniature electric circuit normally contained on a silicon chip.

Light-emitting diode (LED) – an electronic component that produces light when current is passed through it in one direction.

A distributorless ignition system (DIS), as the name suggests, is one that no longer requires a distributor in order to send the spark to the correct spark plug at the right time.

To replace the distributor many systems have a single ignition coil with a high voltage output for each cylinder. Some systems mount an individual ignition coil on top of each spark plug. (This is called coil on plug or COP.)

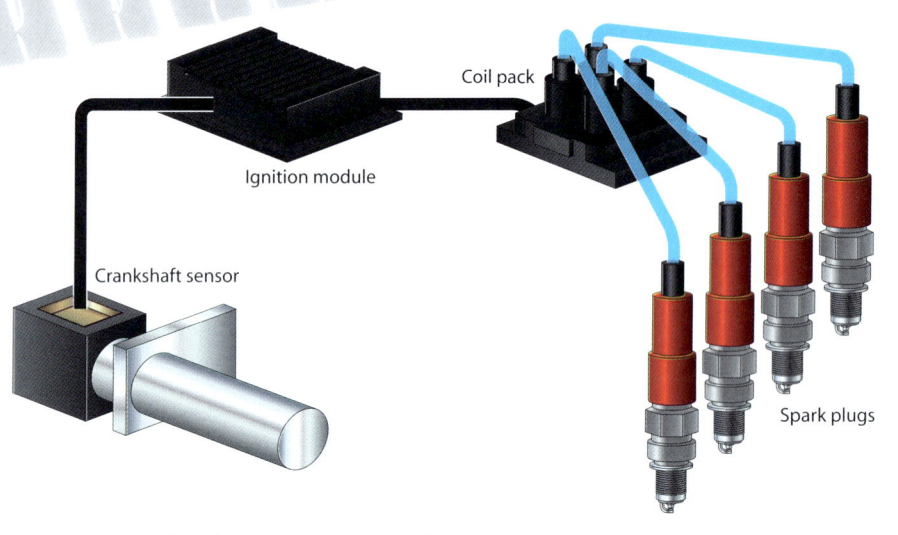

Figure 4.51 Distributorless ignition system (DIS)

The advantages of the DIS are:
- fewer moving parts
- no mechanical timing
- less maintenance
- no mechanical load on the engine
- increased coil saturation (charge).

Computer-controlled electronic ignition

A further advance in the electronic ignition system is to have the entire operation controlled by the car's ECU.

The switching mechanism for the primary circuit is normally still transistorised and engine position is calculated from crankshaft or camshaft sensors. The advantage of putting ignition control into the hands of the engine management ECU is that other sensor information can now also be included, such as engine temperature, speed, load and knock sensing.

Wasted spark ignition system

As most engines have an even number of cylinders, the cylinders normally operate in pairs (two go up and down together). As a pair of companion cylinders moves upwards, one will be on its compression stroke while the other will be on its exhaust stroke. The cylinder on the compression stroke will require a spark, while the one on the exhaust stroke will not. It is normally the job of the distributor to choose which cylinder to supply with a high tension spark.

Wasted spark works on the following principle:
- If a spark is provided to both companion cylinders at the same time, the one which is on its compression stroke will ignite the fuel and start the power stroke, while the one on the exhaust stroke will be wasted.
- The next time around, after 360 degrees of crankshaft revolution, the other cylinder will be on its compression stroke and ignite the air/fuel mixture, while the companion cylinder will be on its exhaust stroke and the spark will be wasted.

By using this process, the timing of a pair of companion cylinders can be reduced by half, so there is no need for a distributor and the associated components. (Normally in this type of system, a paired ignition coil is used to fire a set of companion cylinders – see Figure 4.52).

Operation of the wasted spark ignition system

As both spark plugs fire at the same time, the system does not need to know which spark plug needs the spark. It just needs to know when.

Figure 4.52 DIS ignition coil

- A spark plug is connected to each end of the secondary winding on a pair of companion cylinders. The cylinder head acts to complete the circuit between the two spark plugs.
- When the primary circuit of the ignition coil is supplied with power, it charges. This creates an invisible magnetic field which cuts across the secondary circuit of these paired spark plugs.
- When the primary circuit is switched off, the magnetic field collapses. This induces a high voltage in the secondary circuit of the companion cylinders. High voltage flows instantly through the secondary winding, spark plugs and cylinder head, and produces the spark.
- As current only flows in one direction, one spark plug receives a positive spark (jumping from the centre to the earth electrode) and the other receives a negative spark (jumping from the earth electrode to the centre electrode) to complete the circuit. Because of this, spark erosion and wear at the tips of the spark plug may be different. Also, if one spark plug fails (becomes open circuit), the other spark plug can also fail.

Key engineering principles relating to light vehicle engine ignition systems

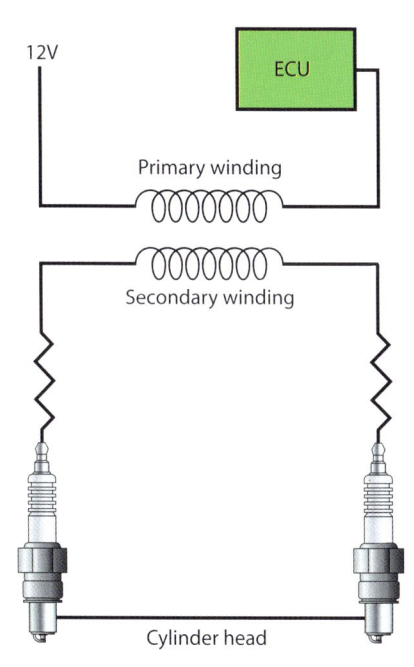

Figure 4.53 Wasted spark ignition system

Flame travel

Flame travel describes how the combustion process spreads out, away from the spark plug as it ignites the air/fuel mixture. An even flame spread gives good performance and fewer emissions.

With diesel engines, flame spread can be inaccurate. This is because fuel is normally atomised inside the combustion chamber, is ignited by the superheated compressed air and may start to burn from a number of different positions. As the flame fronts collide, a distinctive knocking noise can be heard which is common to most diesel engines.

Ignition timing

In order to obtain the maximum amount of power from the combustion of the air/fuel mixture, the piston should not be at top dead centre (TDC) when full pressure build-up occurs. This is because with the piston at TDC, the expansion of gases would push on the piston crown

Level 2 Light Vehicle Maintenance & Repair

> **Did you know?**
>
> The crankshaft of an engine is like the pedals of a bicycle. If you push down on the pedal when it is right at the top (12 o'clock), most of your effort will be wasted. If you push on the pedal just after it has passed the top (1 o'clock) your effort is turned into torque (see Figure 4.54).

when the connecting rod was vertical. This would mean that a large proportion of the torque (turning effort) would be lost due to lack of leverage at the crankshaft. Because of this, manufacturers design the engine so that the greatest amount of pressure build-up on the piston crown occurs at approximately 10° to 12° after TDC (see Figure 4.55).

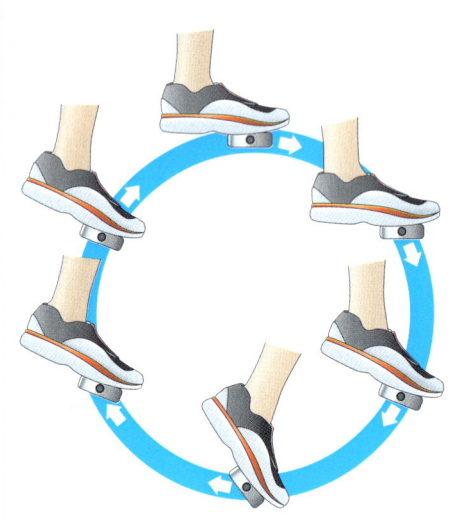

Figure 4.54 A set a bicycle pedals being rotated to represent the production of torque by an engine crankshaft

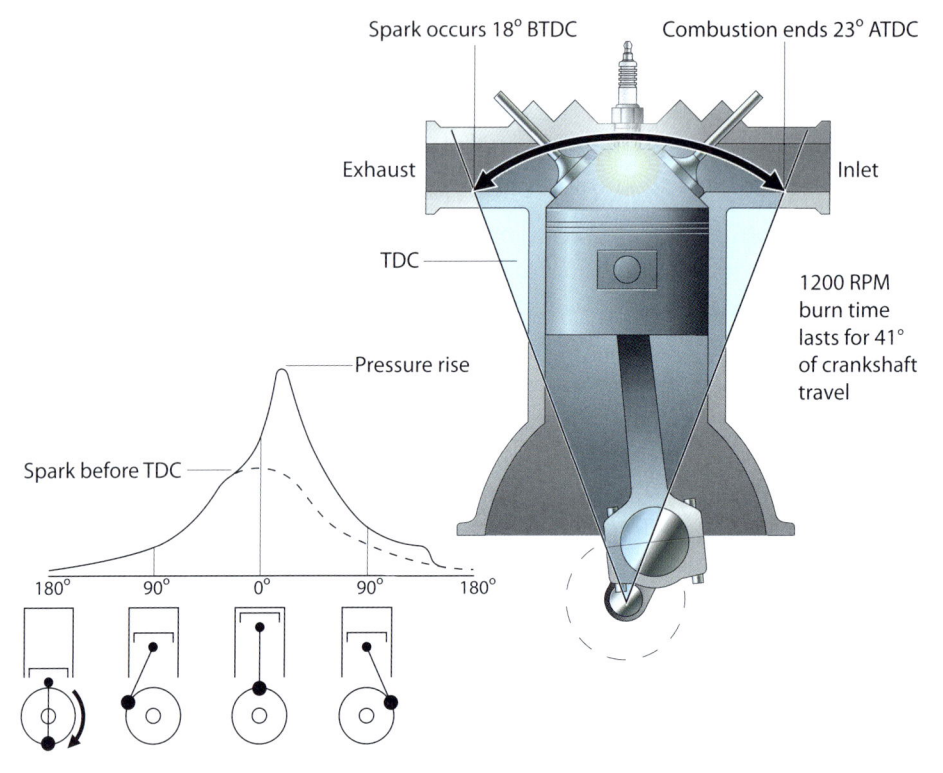

Figure 4.55 Ignition spark advance

Although the burning of the fuel and flame spread are rapid, they are not quick enough to keep up with the fast-moving pistons. This means that as the piston rises on its compression stroke, the spark will normally occur early (before TDC). The fuel is ignited and the flame spreads as the piston continues to travel upwards, past TDC, and begins to move downwards on the power stroke. When it reaches approximately 10° to 12° after top TDC, the full pressure build-up pushes on the piston crown, forcing the connecting rod to turn the crankshaft and provide the greatest amount of torque.

> **Key terms**
>
> **Timing advance** – the spark occurs earlier in the cycle as engine speed increases.
>
> **Pinking** – a metallic knocking noise created inside the engine, normally caused by ignition occurring too early.

The time that it takes for air and fuel to burn does not vary a considerable amount. Because of this, the faster the engine turns, the earlier in the cycle the spark must occur (this is known as **timing advance**).

If the spark occurs too early in relation to crankshaft revolution, it is possible that the connecting rod will be completely vertical. If the fuel is ignited and pressure builds up on the piston crown, turning effort (torque) is lost, and a high-pitched knocking noise called **pinking** can occur.

CHECK YOUR PROGRESS

1 List five ignition system components.
2 Explain two differences between a distributor and a distributorless ignition system.
3 What are the advantages of an electronic ignition system when compared with a conventional ignition system?
4 Why is ignition advance required?

Air and exhaust system construction and operation

To enable a four-stroke engine to run effectively:

- air must be allowed to enter the engine with the least amount of restriction
- exhaust gases must be allowed to escape with the least amount of restriction.

To reduce the possibility of noise pollution, silencers are fitted to exhaust systems. The two main types of silencer are absorption-type and expansion-type.

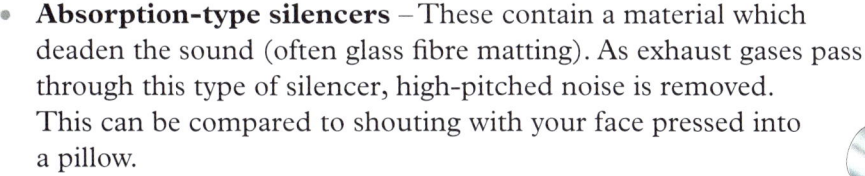

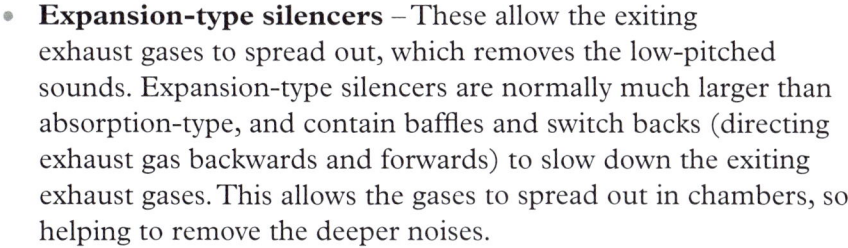

- **Absorption-type silencers** – These contain a material which deaden the sound (often glass fibre matting). As exhaust gases pass through this type of silencer, high-pitched noise is removed. This can be compared to shouting with your face pressed into a pillow.
- **Expansion-type silencers** – These allow the exiting exhaust gases to spread out, which removes the low-pitched sounds. Expansion-type silencers are normally much larger than absorption-type, and contain baffles and switch backs (directing exhaust gas backwards and forwards) to slow down the exiting exhaust gases. This allows the gases to spread out in chambers, so helping to remove the deeper noises.

Figure 4.56 Absorption-type exhaust silencer

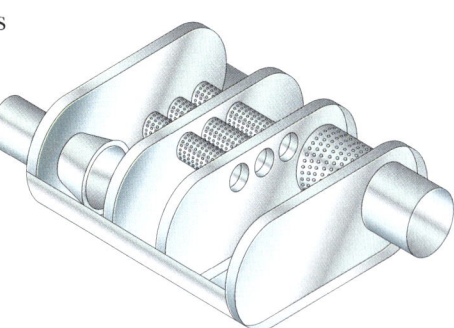

Figure 4.57 Expansion-type exhaust silencer

Supercharging

During normal operation, a standard engine relies on atmospheric pressure to push air into the cylinder (along with fuel in the case of petrol engines) to fill or 'charge' the combustion chamber. Because of the rapid induction process and the pressures involved, a standard naturally aspirated engine is not able to fill the cylinder to 100 per cent full (a normal figure is around 80 per cent). The ability to fill the cylinder well is known as **volumetric efficiency**. The better the cylinder can be filled, the more performance is gained from the engine.

> **Key term**
>
> **Volumetric efficiency** – how well the engine cylinder fills with air and fuel.

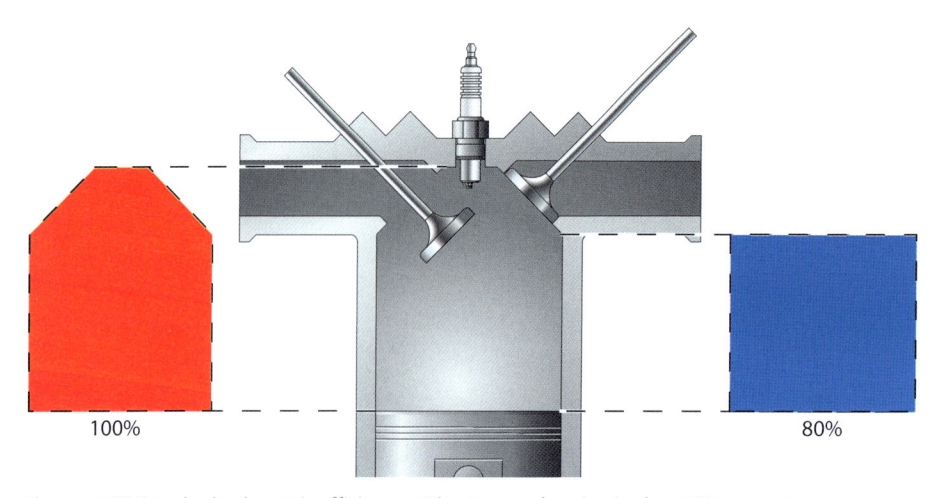

Figure 4.58 Standard volumetric efficiency without supercharging is about 80 per cent

If air can be forced into the cylinder at higher than atmospheric pressure, then the cylinder can be overfilled (with volumetric efficiency being greater than 100 per cent). This is known as 'supercharging'.

Most superchargers consist of an engine-driven pump or compressor which takes the incoming air and raises its pressure above atmospheric, thereby forcing it into the engine.

Although the supercharger increases performance in the engine, it is also an engine **power parasite**.

Turbochargers

Another way of pressure charging an internal combustion engine is to use a turbocharger (see page 286). A turbocharger is an exhaust gas-driven compressor. It takes the incoming air and raises its pressure, forcing it through the inlet valve and into the combustion chamber.

Unlike the engine-driven supercharger, it is not a power parasite as it uses the waste exhaust gas to power the turbine. The turbine must be spinning at considerable speed to provide the engine with any noticeable power increase. This means that engine and exhaust speed must also be high. A condition known as 'turbo lag' occurs when no boost is felt at the start of acceleration, but it cuts in as speed increases.

The fitting of turbochargers tends to be restricted to performance petrol engines. They are a more common design feature of diesel engines.

Exhaust gas recirculation (EGR)

In the extreme heat of combustion (temperatures above 1800 °C), the oxygen and the nitrogen from the incoming air combine to produce a pollutant called oxides of nitrogen (NO_x) – see page 290. This occurs in both petrol and diesel engines.

Normally, the better the combustion process, the higher the level of NO_x produced. Exhaust gas recirculation (EGR) can be used as an emission control system to reduce the production of NO_x.

4 Light vehicle fuel, ignition, air & exhaust system units & components

EGR is a method for taking some of the burnt exhaust gases and directing them through pipes back into the inlet manifold. There are two main reasons for this, and they are slightly different for petrol and diesel.

Petrol engines

By introducing exhaust gas back into the inlet manifold of a petrol engine, the amount of incoming air and fuel that reaches the cylinder is reduced. This means that the performance output of the engine is also reduced, and so engine temperatures will be lower. This helps prevent the production of oxides of nitrogen.

It is important the performance is not affected during most of the operating conditions under which a car is driven. A device calle an **EGR lift valve** is fitted so that exhaust gas is only recirculated when outright performance or acceleration are not required. For example, when cruising on a motorway at a steady speed with your foot resting only lightly on the accelerator pedal, exhaust gas can be recirculated and the production of oxides of nitrogen reduced.

Figure 4.59 Exhaust gas recirculation (EGR) system

Diesel engines

In a diesel engine, it is very difficult to achieve ideal air/fuel ratios (14.7:1, for example) because diesel engines draw the same amount of air every time, and it is the amount of fuel that is injected that dictates how fast the engine will run. This means that when a standard diesel engine is operating under normal conditions, the air/fuel ratio will be extremely weak at low revs and idle, or rich during acceleration.

These extremes of air/fuel ratio create emission and pollution problems. By introducing a quantity of burnt exhaust gases back into the intake manifold of a diesel engine, the amount of fresh air entering the cylinder can be more accurately controlled, resulting in better air/fuel ratios. (Exhaust gas will still compress and raise the temperature in the cylinder for ignition of the fuel.)

Secondary air injection

A secondary air injection system is sometimes fitted to vehicles to help with emission-related issues. A small air pump controlled by the car's computer (ECU) can blow air into the exhaust manifold close to where it exits the cylinder head. By introducing extra oxygen to the hot exiting exhaust gases, unburnt hydrocarbons can continue the combustion process and be burnt away.

As with blowing on the embers of a fire or the charcoal of a barbecue, introduced air helps heat the exhaust gases. If the vehicle is fitted with a catalytic converter, this will assist with the converter's warm-up period, making it operate efficiently sooner.

> **Key terms**
>
> **Power parasite** – anything that takes power away from the engine. (Because the supercharger is driven by the engine, it takes some power away to produce a higher performance.)
>
> **EGR lift valve** – valve that regulates the amount of exhaust gas directed back into the engine.

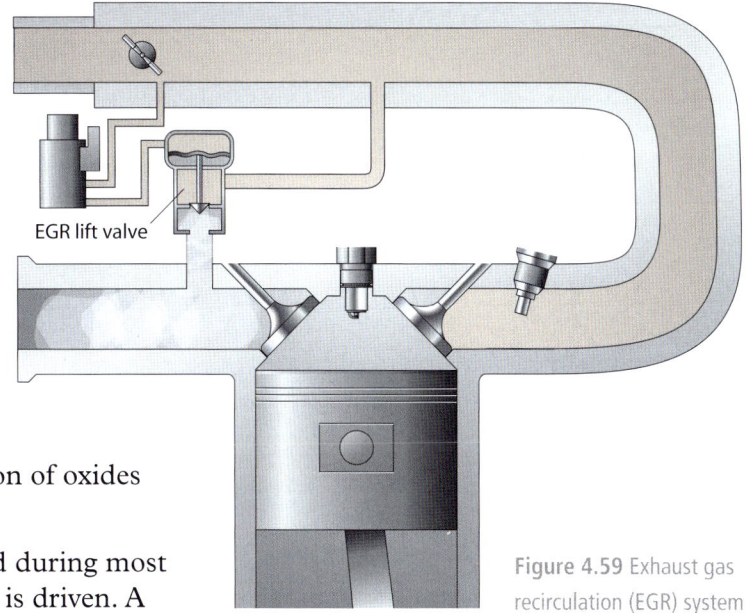

Level 2 Light Vehicle Maintenance & Repair

> ⚠ **Safe working**
>
> Be careful when working around a catalytic converter. They can get extremely hot and cause severe burns.

Catalytic converters

Catalytic converters are fitted to some vehicles to help reduce the amount of harmful pollutants present in exhaust gases. A catalytic converter is normally mounted in the exhaust system very close to the inlet manifold where the exhaust gases exit the cylinder head. (The closer the catalytic converter is to the exiting exhaust gases, the quicker it heats up and begins to operate in an efficient manner.)

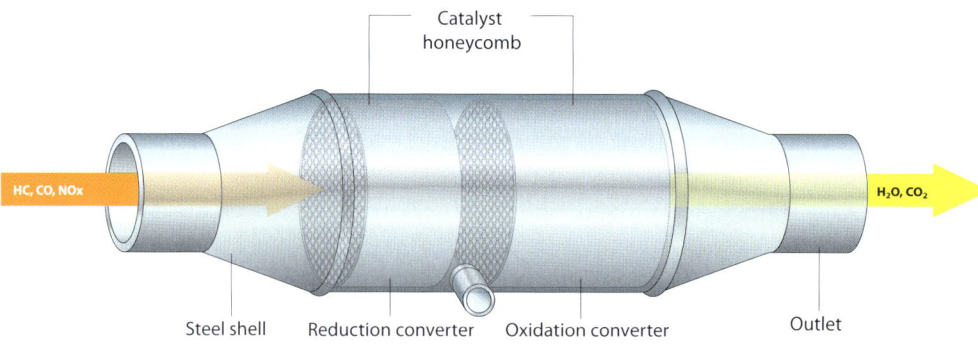

Figure 4.60 Catalytic converter

From the outside, a catalytic converter may look similar to an exhaust system silencer. On the inside, it is normally made up of a matrix-style mesh which contains precious metals such as platinum and rhodium. The mesh-style matrix increases the surface area exposed to the exiting exhaust gases. When the gases pass over the precious metals, a chemical/catalytic reaction takes place (a catalyst is anything that makes a situation or substance change).

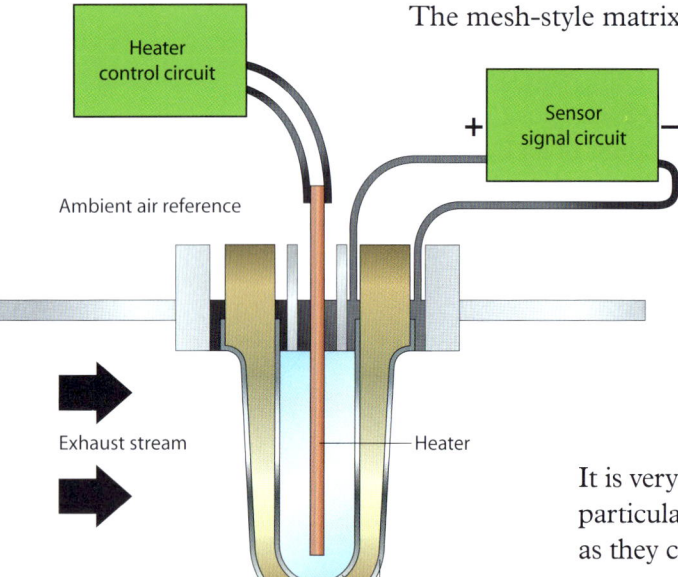

The exhaust pollutants are not removed from the exhaust gases, but are converted to form less harmful pollutants. To work efficiently, catalytic converters require the engine to operate within the ideal stoichiometric values (see page 289). This means that accurate engine management is essential.

It is very important that the correct type of fuel is used for a particular engine. Fuels containing lead should be avoided, as they can cause damage to the catalytic converter.

To make sure that the correct air/fuel ratio is achieved from the engine management system, an oxygen sensor (also called a Lambda sensor) is fitted in the exhaust system before the catalytic converter (see Figure 4.61). The Lambda sensor measures the oxygen content of the exiting exhaust gases, and instructs the ECU if the engine is running too rich or too weak. (Too much oxygen in the exhaust gas and the engine is running weak; too little oxygen in the exhaust gas and the engine is running rich.)

Figure 4.61 Exhaust system oxygen sensor (Lambda sensor)

- During warm-up periods of the catalytic converter and Lambda sensor, signals from the oxygen sensor to the ECU are ignored. When this happens, the engine management system is running **open loop**.

- As soon as the catalytic converter and Lambda sensor are up to the operating temperature, signals produced are used to correct fuel injection for the ideal air/fuel ratio. When this happens, the engine management system is running **closed loop**.

To make sure that the catalytic converter is operating correctly, some manufacturers include a second Lambda sensor mounted after the catalytic converter. The signal produced by the two Lambda sensors (pre-cat and post-cat) are compared. If they are found to be similar, no chemical reaction is taking place within the converter and the catalyst has failed.

> **Key terms**
>
> **Open loop** – when the engine management ECU operates on pre-programmed values from its memory.
>
> **Closed loop** – when the engine management ECU operates with signals supplied by sensors.

CHECK YOUR PROGRESS

1. What does EGR stand for?
2. What is the main difference between a turbocharger and a supercharger?
3. Where are the fuel and air usually mixed in a:
   - petrol engine
   - diesel engine?
4. What is the purpose of a catalytic converter?

How to check, replace and test light vehicle engine fuel system units and components

Since the mid-1980s, many car manufacturers have been including a self-diagnosis facility in the design of their engine management systems. The ECU monitors sensor signals and actuator outputs, looking for voltages or readings that are outside a specified range. If a problem is detected, the ECU will store a **diagnostic trouble code** (**DTC**) (sometimes called a fault code) in its memory. It can also illuminate a **malfunction indicator lamp** (**MIL**) on the car's dashboard.

If a fault code is stored, the engine management system is sometimes able to bypass the affected component and use pre-programmed values from its own memory. When this happens the car is on a **limited operating strategy** (**LOS**), sometimes called limp home. The ECU is often able to do this so effectively that the driver may not even notice a change in performance – the only indication will be the illumination of the malfunction indicator lamp.

> **Key terms**
>
> **Diagnostic trouble code (DTC)** – a fault code stored in the ECU memory.
>
> **Malfunction indicator lamp (MIL)** – a warning light on the dashboard of the car to inform the driver that the engine management system has detected a fault.
>
> **Limited operating strategy (LOS)** – a system used when a fault occurs, so that sensor inputs can be ignored and pre-programmed values used to maintain acceptable operation. It is also called 'limp home'.

Diagnostic scan tool

To retrieve fault codes, you will often need a specialist scan tool. You connect the scan tool to a diagnostic connector (data link connector or DLC), which will interrogate the ECU and display the diagnostic trouble code on a liquid crystal display screen (see Figure 4.62).

Contrary to popular belief, a diagnostic trouble code does not tell you what the fault is. It only points you towards the area of the system where the fault is. It is then up to you to carry out the correct diagnostic procedures to find the fault.

Using a scan tool to retrieve diagnostic trouble codes

Checklist			
PPE	VPE	Tools and equipment	Source information
• Steel toe-capped boots • Overalls • Latex gloves	• Wing covers • Steering wheel covers • Seat covers • Floor mat covers	• Diagnostic scan tool	• Data link connector location • Diagnostic trouble codes • Technical data • On-screen instructions

Figure 4.62 Diagnostic scan tool

1. Locate the data link connector (DLC). This should be located inside the car, between the centre line and the driver's side (although this is not always the case). You can use the manufacturer's vehicle repair data to help you find the socket.

2. Connect the scan tool and follow the on-screen instructions to retrieve any stored diagnostic trouble codes.

3. Record any stored codes.

4. Clear all fault codes.

5. Road test the car and rescan for any trouble codes.

6. Concentrate diagnostic procedures on any fault codes that have returned. (Electrical diagnostic procedures are covered in Chapter 6.)

Checking the emission system

To check the function and operation of the vehicle's emission system, use an exhaust gas analyser (see opposite). This calculates the constituents of the exhaust gases so that you can make decisions about the operation and efficiency of the induction and ignition system.

Legal limits exist for exhaust pollutants, which are normally checked annually during the vehicle's MOT. Manufacturers also include specific emission limits in their technical data. If the gas analyser is connected to the exhaust system, readings of carbon dioxide, carbon monoxide, hydrocarbons and oxygen can be used to calculate if the engine is running within the Lambda window (see page 289).

Using an exhaust gas analyser

Checklist			
PPE	**VPE**	**Tools and equipment**	**Source information**
• Steel toe-capped boots • Overalls • Latex gloves	• Wing covers • Steering wheel covers • Seat covers • Floor mat covers	• Exhaust gas analyser • Exhaust extraction equipment	• Manufacturer's technical data • On-screen instructions

> **Safe working**
>
> If you are running the engine in the workshop, you must use exhaust extraction equipment to protect you from harmful pollutants. Many exhaust extraction units include a specially designed hole for use with the exhaust gas analyser probe.

1. Switch on the exhaust gas analyser and allow it to go through its preset warm-up and calibration procedure.

2. Connect the exhaust extraction equipment, start the engine and run up to normal operating temperature.

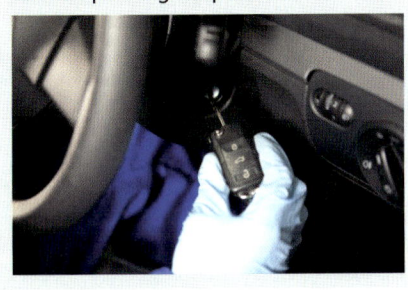

3. Insert the exhaust probe at the tail pipe.

4. Allow the digital reading to settle on the analyser (this may take up to 15 seconds) and compare with the manufacturer's recommendations.

Table 4.2 Expected exhaust gas readings for a vehicle with a catalytic converter

Exhaust gas	Reading
Carbon monoxide (CO)	Less than 0.3%
Hydrocarbons (HC)	Less than 200 ppm (parts per million)
Carbon dioxide (CO_2)	Higher than 14%
Oxygen (O_2)	Approximately 0.5%

Because repairs to fuel, ignition and exhaust systems can affect the operation of engine management and emission systems, maintenance is usually restricted to the direct replacement of components. If manufacturers allow adjustment, it is possible that the legal emission outputs might be exceeded. You should always follow any manufacturer specific instructions for removal and refitting.

You can find specific information on ignition and fuel system settings in vehicle data books or using computer-based sources. Figure 4.63 shows an example of a vehicle's technical data.

Technical data on the vehicle

Make	Ford	Date	04-12-2010
Model	Escort 1,6i	Owner	
Year	1990-1992	Registration No.	
Engine	LJF	VIN	
Variant		1. Reg. Date	

Technical item	Data
Engine Management	
Engine management system	Ford E - DIS - 4 (Ford/Weber EFI)
Spark plug	Motorcraft AGPR 22 CD1
Electrode gap, mm	1,00 ± 0,05
Firing order	1 - 3 - 4 - 2 (Cylinder 1 at timing gear)
Ignition timing (BTDC)	Control 10° ± 2°/ 900 ± 50 rpm
Diagnostic connector	In engine room
Max. timing advance (max. rpm)	(Electronic)
Timing mark location	Belt pulley
Ignition coil	DIS - coil
Ignition coil: Primary/secondary resistance	0,50 ± 0,05 ohm/13,50 ± 2,50 kohm
Fuel pressure, w/wo vacuum, bar	2,5 ± 0,2
Holding pressure, bar	2.0 ± 0.2 after 1 minute
Injector resistance, ohm	In pairs 8.0 ± 3.0 (Pressure 2.5 ± 0.2 bar)
Min. idle manifold vacuum, mbar	590
Coolant temperature sensor 20°/ 80°C	37,50 ± 2,50 kohm/ 3,75 ± 0,75 kohm
Intake air temperature sensor 20°/ 80°C	37,5 ± 2,5 kohm

Remarks

GDS Diagnostics   Order No.:
Address   **Mechanic**

Figure 4.63 Vehicle technical data for a Ford Escort

Common faults found in light vehicle fuel, air supply and exhaust systems

Because of the reliability of many engine management systems, engine running and performance problems have been considerably reduced. If problems do occur, they are often caused by lack of maintenance or electrical/electronic failure.

Two components that should be replaced during regular maintenance are:

- **The air filter** – If the air filter is not replaced at regular intervals, it will slowly become blocked and reduce the flow of air into the intake system. Many engine management computers now compensate for the reduced airflow, but performance will be affected because less fuel is injected and volumetric efficiency falls.
- **The fuel filter** – If the fuel filter is not replaced at regular intervals, it will gradually become blocked and reduce the flow of fuel to the injection system. As a fuel filter slowly becomes blocked, the quantity of petrol that reaches the fuel rail will fall, giving a symptom known as starvation. Starvation is normally shown when trying to accelerate. At slow engine revs or tickover, the engine will run smoothly. As the car is driven much harder, the engine will start to falter and 'hold back'.

Symptoms and faults associated with fuel systems

Diesel fuel system

Air in fuel system

If air is allowed into a diesel fuel system, bubbles can form that will cause a **vapour lock**, particularly in the pumping elements of the system. This can prevent the engine from starting or running. If an air leak exists in the diesel fuel system while the engine is running, this can lead to erratic idle or stalling.

> **Key term**
>
> **Vapour lock** – air bubbles forming in the fuel system and stopping the flow of fuel.

Filter blockage

In order that a compression ignition engine can work properly, it needs a clean supply of fresh fuel. As with a petrol engine, diesel is passed through a filter element (see Figure 4.64) to help remove particles of dirt.

If the fuel filter is not replaced during regular service or maintenance, it can become blocked, leading to poor performance due to fuel starvation.

Water in fuel

Like brake fluid, diesel fuel is hygroscopic, meaning that it absorbs moisture from the surrounding air. Over a period of time, diesel fuel can be contaminated with enough water to give running problems or prevent the engine from starting. If left, this could create a misfire because water is not combustible. Also, the metal components of fuel injectors and pumps can corrode.

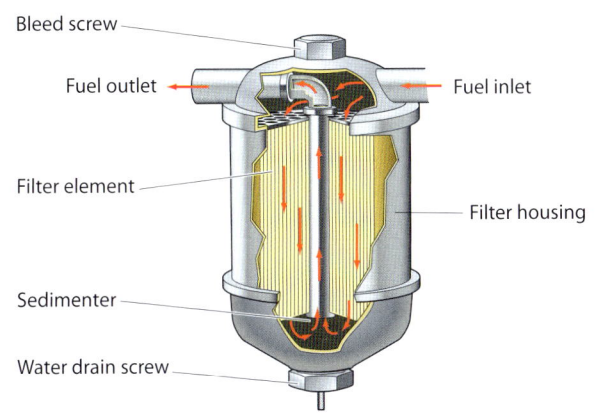

Figure 4.64 Diesel engine fuel filter element

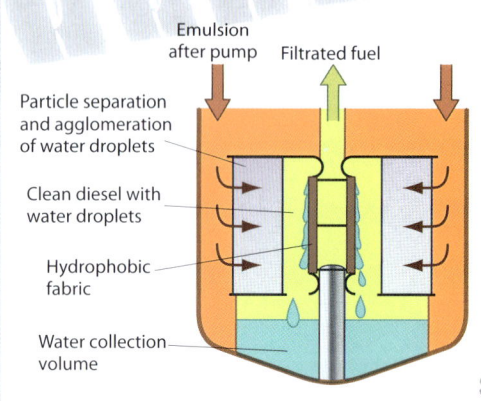

Figure 4.65 Sedimenter

Figure 4.66 Diesel filter

Safe working

As fuel is bled from the system, it is important that you clean up any spilt diesel to reduce the risk of accidents.

To help overcome this problem, many compression ignition fuel systems use a sedimenter. This is a small container mounted in the delivery fuel line. Some manufacturers produce this as part of the fuel filter housing. Diesel fuel will normally float on top of the water. As fuel flows through the sedimenter, water particles tend to settle out in the bottom of the container, which allows a clean supply of fuel to the injection system, as shown in Figure 4.65. Many manufacturers include a facility to drain the sedimenter during routine maintenance procedures.

System bleeding

If the diesel fuel system has been opened (for example, to change a fuel filter or drain a sedimenter), it is possible for air to enter the system. When this happens, trapped air can prevent the engine from starting or running.

Once the system has been reset it will normally require bleeding. Many modern systems are self-purging. This means that as the engine is cranked the air is pumped through the system and back to the tank via the fuel return line.

Some systems have to be bled using a hand 'priming' pump. Bleed screws may be found on the fuel filter housing or fuel pump, as shown in Figure 4.64. You need to open these screws and operate the hand pump until you see clean, air-free fuel.

Difficult starting

The most common problem with diesel engines is poor starting from cold. This is usually due to the failure of one or more glow plugs. The symptoms are normally that the car will start badly first thing in the morning, but can run acceptably for the rest of the day. Once the car has started, until it warms up slightly, it might run badly and produce large quantities of smoke.

A diesel engine exhaust system can emit three main types of smoke (see Figure 4.67):

- Black smoke is usually an indication of overfuelling.
- White smoke is usually an indication of water contamination, either in the fuel or, more commonly, coolant that may have entered past the head gasket.
- Blue smoke (seen as a light blue haze, particularly in sunlight) is usually an indication that lubrication oil is being burnt.

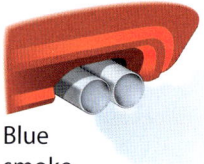

Black smoke

White smoke

Blue smoke

Figure 4.67 Types of diesel exhaust smoke

Engine knock

Engine knock on a diesel engine is usually an indication that fuel with a lower cetane value than recommended is being used. Diesel knock can also be created by incorrect fuel injection timing and leaking injectors.

Turbocharger failure

Turbochargers fitted to diesel engines can fail in a number of ways. Air leaks and seizure of the wastegate are common. Because a turbocharger spins at very high rotational speeds, it is important that a good supply of lubricating oil is available to the spindle bearings. If these bearings fail the turbo can become noisy or stop working altogether.

If the vehicle is fitted with a turbocharger, it is important to let the engine slow down to tickover for a short period of time before turning the engine off. If this is not done, oil pressure to the turbocharger bearings is lost, but the turbo may be spinning at speed, which creates wear in the bearings.

Petrol fuel system

Petrol leaks

Because fuel-injected petrol engines run under pressure, it is important that no leaks exist. Petrol fuel is extremely volatile – if it is allowed to leak around hot engine components, it could easily ignite and cause a fire.

Hunting

Erratic running, particularly around idle speed, is called 'hunting'. If this occurs, it is commonly caused by an air leak in the intake system, which creates a weak mixture (too much air and not enough fuel).

System running too rich

If the fuel injection system is running too rich (too much fuel and not enough air), this can lead to poor performance, excessive fuel consumption and high emissions. Some common causes of this are:

- blocked air filters
- leaking fuel injectors
- high fuel pressure.

The inspection and replacement of serviceable fuel, ignition and exhaust system components can be found in the manufacturer's service schedule. Check sheets are available to allow the technician to record any replacement or recommendations. An example is shown in Figure 4.68.

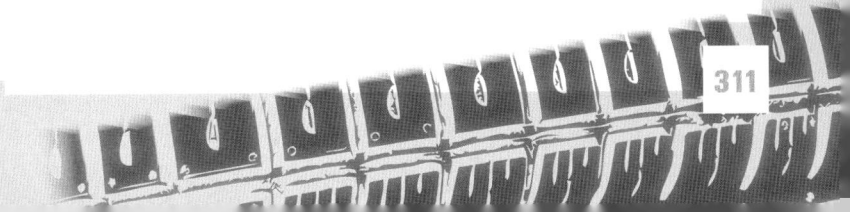

Level 2 Light Vehicle Maintenance & Repair

Make	Opel/Vauxhall
Model	Astra G 1,7 DT
Year	1998-2000
Engine	X17DTL
Maintenance	30,000 km/2 years

Date:	04-12-2-2010
Owner:	
Registration No.	
Mileage	
1. Reg. Date	

Engine compartment — OK — Remarks

1	Change air filter	
2	Change engine oil and filter	
3	Change fuel filter	
4	Change pollen filter	
5	Check cooling system and fluid level and enter freezing point: _____ °C	
6	Check fuel pipes and hoses (Positioning, condition and impermeability)	
7	Check oil level in power steering	
8	Check/Adjust driving belts	
9	Check/Adjust valve clearance (X15DT, X17D, X17DT engine)	

Body/Cabin — OK — Remarks

10	Check body/paint and undercoating (Inform customer of damage/defects)	
11	Check front windscreen and headlamp wiper/washer system (Remember to check fluid level)	
12	Check lamps and signalling lamps (Remember license plate lights)	
13	Check/Adjust clutch pedal freeplay (Only models with front wheel drive)	
14	Control/Adjust front lights	
15	Lubricate door - locks - hinges - stop, engine bonnet lock	

Chassis/Underbody — OK — Remarks

16	Change Automatic Transmission Fluid, 4 step type AF 13, every 45,000 km/3 years (ONLY CORSA B AND TIGRA, WITH X14XE ENGINE)	
17	Check all gaiters	
18	Check brake pipes and hoses	
19	Check disc brakes front and rear (Pads/discs and calipers, change when necessary)	
20	Check engine, transmission and, if applicable, rear axle for leakage (Check level and inform customer)	
21	Check exhaust system and mounting (also Control heat shield)	
22	Check oil level in automatic transmission and read out any error codes for gearbox (TE CH1)	
23	Check rear brakes (Cylinder/linings and condition of drums)	
24	Check tyres, adjust inflation pressure and enter tread depth LF: RF: LR: RR: SPARE:	
25	Check/Adjust ALB-valve (Found for instance on all st. wagons, Corsa B, Astra F)	

Testdrive — OK — Remarks

26	Clean steering wheel and handles after service	
27	Stamping of service manual	
28	Test drive with brake Control and final Control of all vehicle systems	

Additional points — OK — Remarks

29	Change timing belt (Interval: see notice/important)	

We have noted the following points, of which you should be aware, while examining your vehicle:
_____
_____
_____
_____
_____
_____

GDS Diagnostics
Address

Yours sincerely

_____
Date   Mechanic

Figure 4.68 Service inspection check sheet

CHECK YOUR PROGRESS

1 List three common faults found in light vehicle fuel, ignition, air and exhaust systems.
2 Name a piece of diagnostic test equipment that could be used to test a fuel, ignition, air or exhaust system.

FINAL CHECK

1 What is the chemically correct air to fuel ratio for complete combustion?

  a 10 : 1
  b 14.7 : 1
  c 20 : 1
  d 25 : 1

2 Diesel engine speed is controlled by:

  a the amount of air entering the engine
  b the amount of fuel entering the engine
  c a wick system
  d a butterfly valve

3 In petrol fuel injection, which of the following is a typical injector delivery pressure?

  a 1 bar
  b 3 bar
  c 8 bar
  d 12 bar

4 In a four-cylinder wasted spark ignition system, a spark occurs:

  a once every cylinder cycle
  b twice every cylinder cycle
  c three times every cylinder cycle
  d four times every cylinder cycle

5 Diesel is injected into:

  a the manifold
  b the plenum chamber
  c the combustion chamber
  d the throttle body

6 Spark plugs with a long insulator nose:

  a run hot
  b run cold
  c need small plug gaps
  d need large plug gaps

7 What is sequential fuel injection?

  a all injectors pulse simultaneously
  b the injectors pulse with the engine firing order
  c fuel is injected at the start of the air intake
  d injectors are pulsed in pairs

8 Which of the following is not a symptom caused by a blocked fuel filter?

  a high fuel consumption
  b fuel starvation
  c non-starting
  d poor performance

9 A multipoint fuel injection system has:

  a an injector for each cylinder
  b multiple outlets in each injector
  c multiple injectors for each cylinder
  d an injector in place of the carburettor

10 With single point fuel injection, the injector is fitted:

  a in the manifold, just after the air filter
  b directly into the combustion chamber
  c in the manifold, immediately before the valve port
  d next to the spark plug, pointing towards the valve

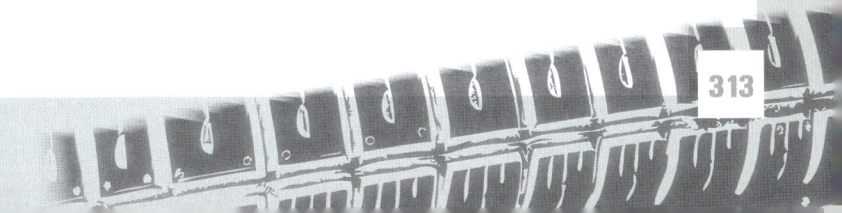

GETTING READY FOR ASSESSMENT

The information contained in this chapter, as well as continued practical assignments, will help you prepare for both the end-of-unit tests and diploma multiple-choice tests. This chapter will also help you to prepare for working on and maintaining light vehicle ignition, fuel, air and exhaust systems safely.

You will need to be familiar with:

- The operation and maintenance of air and fuel supply systems
- The operation and maintenance of ignition systems
- The constituents of air and fuel
- Air to fuel ratios
- Exhaust pollutants produced by combustion
- How to remove and refit light vehicle fuel, ignition, air and exhaust components
- Common faults associated with light vehicle ignition and fuel systems
- General maintenance procedures for light vehicle fuel, ignition and, air exhaust systems

This chapter has given you an introduction and overview of the maintenance and repair of light vehicle ignition, fuel, air and exhaust systems, providing the basic knowledge that will help you with both theory and practical assessments.

Before you try a theory end-of-unit or multiple-choice test, make sure you have reviewed and revised any key terms that relate to the topics in that unit. You will need to read all questions carefully. Take time to digest the information so that you are confident about what the question is asking you. With multiple-choice tests, it is very important that you read all of the answers carefully, as it is common for two of the answers to be very similar, which may lead to confusion.

For practical assessments, it is important that you have had enough practice and that you feel that you are capable of passing. It is best to have a plan of action and work method that will help you. Make sure that you have enough technical information, in the way of vehicle data, and appropriate tools and equipment. It is also wise to check your work at regular intervals. This will help you to be sure that you are working correctly and to avoid problems developing as you work.

When you are doing any practical assessment, always make sure that you are working safely throughout the test.

- Isolate electrical systems when working on ignitions to reduce the possibility of electric shock.
- Remove all sources of ignition when working with fuel, and have an appropriate fire extinguisher handy.
- When running engines, for example when taking exhaust gas analyser readings, make sure you use exhaust extraction or work in a well-ventilated area.
- When working with exhaust systems, remember that they can be very hot, especially catalytic converters.
- Do not operate diesel injectors when removed from the engine, as fuel can penetrate the skin and cause death.
- Make sure that you observe all health and safety requirements and that you use the recommended personal protective equipment (PPE) and vehicle protection equipment (VPE). When using tools, take care to use them correctly and safely.

Good luck!

5 Light vehicle transmission & driveline units & components

This chapter will help you to develop an understanding of the principles, construction and testing methods involved with common light vehicle transmission and driveline systems. It will also cover procedures you can use when removing, replacing and testing transmission systems for correct function and operation. This chapter provides you with knowledge and skills that will help you with both theory and practical assessments. Finally, it will help you plan a systematic approach to transmission inspection and maintenance.

This chapter covers:

- Safe working on light vehicle transmission and driveline units and components
- Transmissions and drivelines
- Light vehicle clutch systems
- Light vehicle gearbox operation
- Light vehicle driveline systems
- How to test and maintain light vehicle transmission and driveline systems

WORKING PRACTICE

There are many hazards associated with light vehicle transmission and driveline systems. You need to give special consideration to the possibility of:
- crush or bump injuries
- coming into contact with chemicals such as lubrication oils.

You should always use appropriate personal protective equipment (PPE) when you work on these systems. Make sure that your selection of PPE will protect you from these hazards.

Personal Protective Equipment (PPE)

Safety helmet protects the head from bump injuries when working under cars.

Overalls provide protection from coming into contact with oils and chemicals.

Safety gloves provide protection from oils and chemicals. They also protect the hands when handling objects with sharp edges.

Barrier cream protects the skin from old transmission oil, which can cause dermatitis and may be carcinogenic (a substance that can cause cancer).

Safety goggles reduce the risk of small objects or chemicals coming into contact with the eyes.

Safety boots protect the feet from a crush injury and often have oil- and chemical-resistant soles. Safety boots should have a steel toe-cap and steel mid-sole.

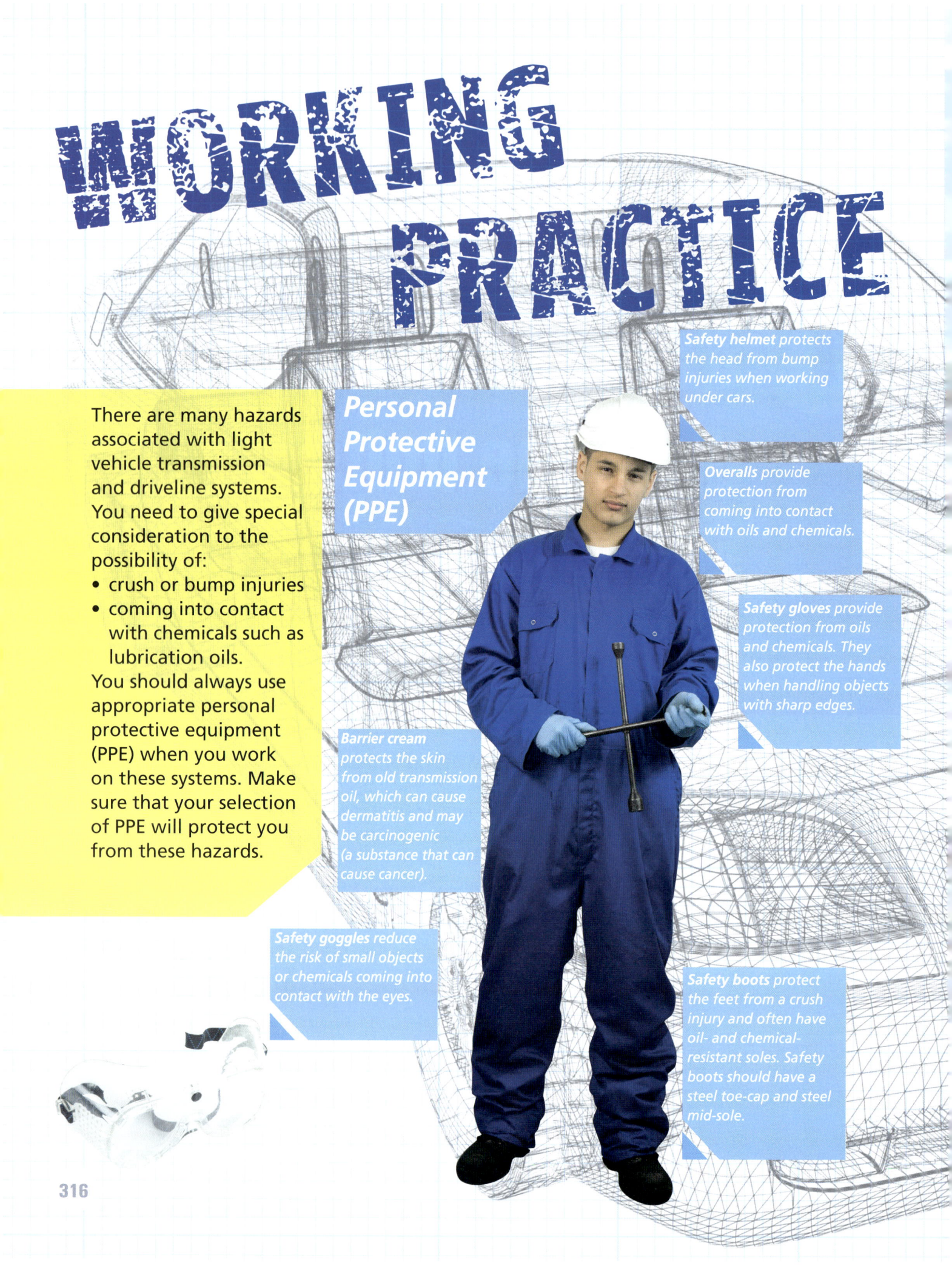

To reduce the possibility of damage to the car, always use the appropriate vehicle protection equipment (VPE):

Wing covers

Seat covers

Steering wheel covers

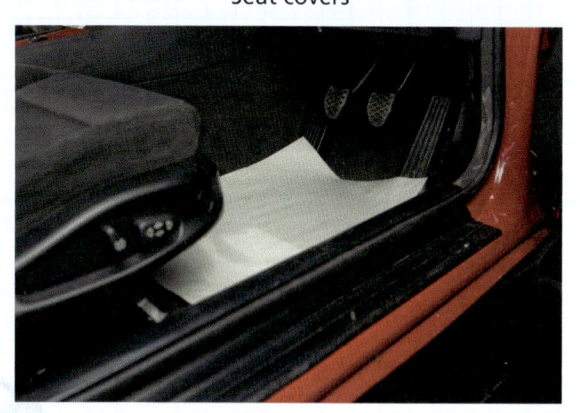

Floor mats

If appropriate, safely remove and store the owner's property before you work on the vehicle. Before you return the vehicle to the customer, reinstate the vehicle owner's property. Always check the interior and exterior to make sure that it hasn't become dirty or damaged during the repair operations. This will help promote good customer relations and maintain a professional company image.

Vehicle Protective Equipment (VPE)

Safe Environment

During the repair or maintenance of light vehicle transmission and driveline systems, you may need to drain lubrication oil. Under the Environmental Protection Act 1990 (EPA), you must dispose of oils in the correct manner. They should be stored safely in a clearly marked container until they are collected by a licensed recycling company. This company should give you a waste transfer note as the receipt of collection.

To further reduce the risks involved with hazards, always use safe working practices, including:

1. Immobilise the vehicle by removing the ignition key. Where possible, allow the engine to cool before starting work.

2. Prevent the vehicle moving during maintenance by applying the handbrake or chocking the wheels.

3. Follow a logical sequence when working. This reduces the possibility of missing things out and of accidents occurring. Work safely at all times.

4. Always use the correct tools and equipment to avoid damage to components and tools or personal injury. Check tools and equipment before each use.
   - Inspect any mechanical lifting equipment for correct operation, damage and hydraulic leaks.
   - Never exceed safe working loads (SWL).
   - Check that measuring equipment is accurate and calibrated before you take any readings.

5. If transmission or driveline components need replacing, always check that the quality meets the original equipment manufacturer (OEM) specifications. (If the vehicle is under warranty, inferior parts or deliberate modification might make the warranty invalid. Also, if parts of an inferior quality are fitted, this might affect vehicle performance and safety.)

6. Following the replacement of any vehicle driveline or transmission components, thoroughly road test the vehicle to ensure safe and correct operation. Make sure that all work is correctly recorded on the job card and vehicle's service history, to ensure that any maintenance work can be tracked.

Preparing the car

Tools

Sockets and ratchets

Spanners

Screwdrivers

Hammers

Clutch aligner

Transmission jack

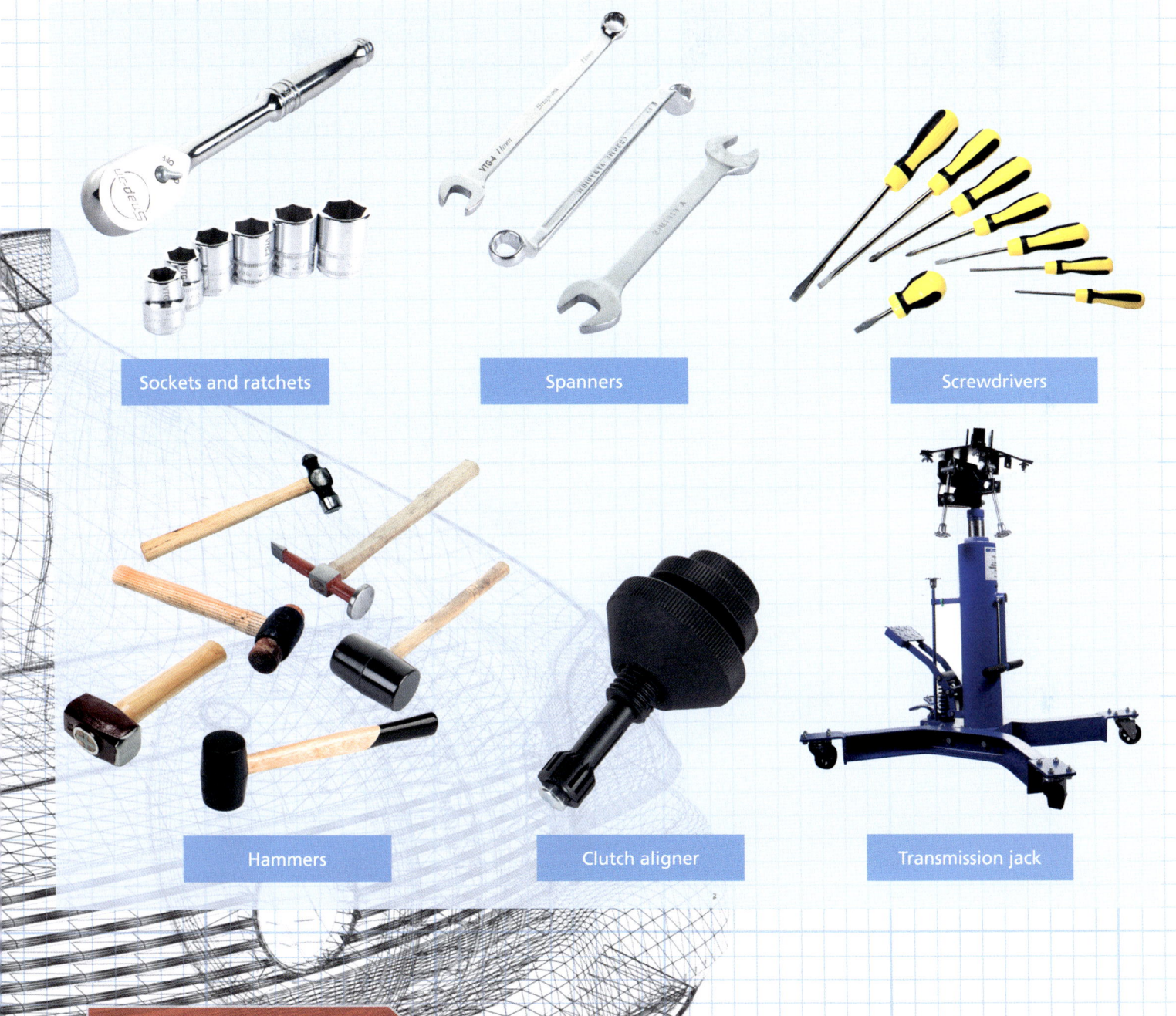

Safe Working

- Always clean up any fluid spills immediately to avoid slips, trips and falls.
- Always isolate (disconnect) the electrics before you remove connections at the gearbox, such as starter motors.
- Always use correct manual handling techniques when you remove and refit heavy transmission components.
- Always work in a well-ventilated area.
- Always wear safety goggles and a particle mask when you remove and fit clutch components.

Transmissions and drivelines

The engine is the car's **power** plant. It provides the mechanical energy needed for movement. Once this mechanical energy has been generated, it must then be transferred to the driving wheels. This is the job of the transmission system.

A number of components make up the vehicle's **driveline**, depending on the layout and the type of **transmission** that is used (i.e. manual or automatic).

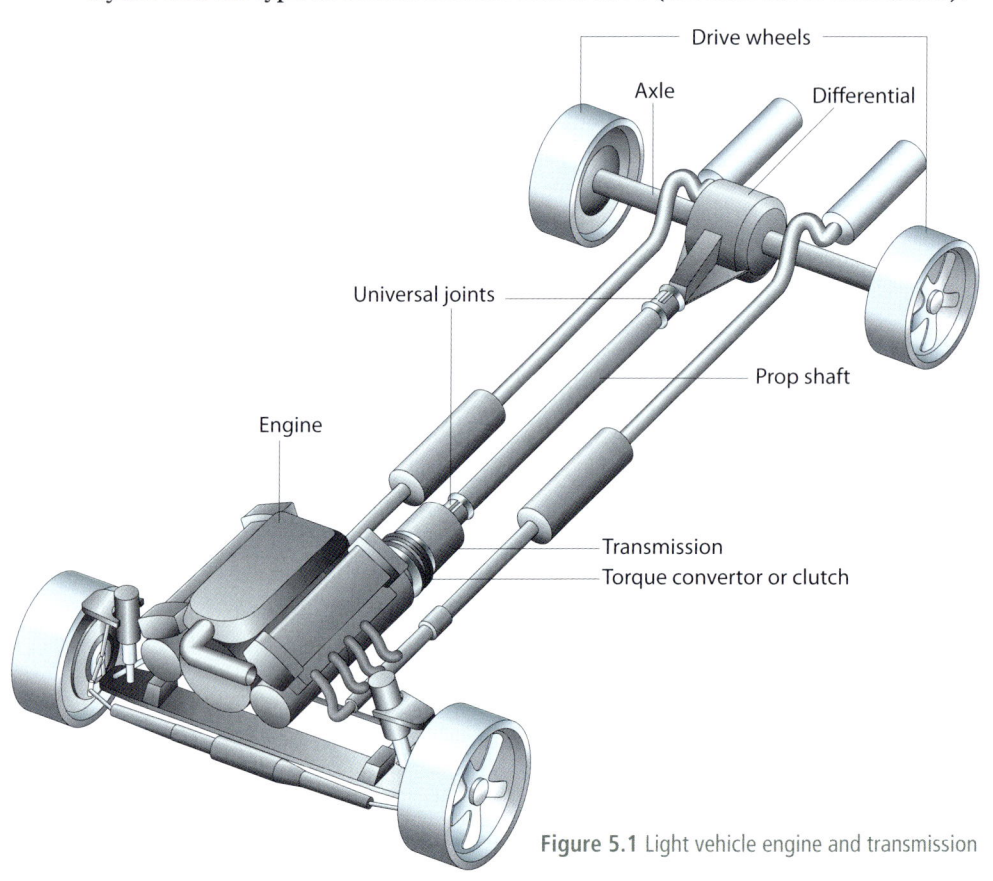

Figure 5.1 Light vehicle engine and transmission

A standard manual transmission system contains:

- clutch
- gearbox with manually selectable gear ratios
- final drive gear reduction
- differential unit
- propeller shafts and drive shafts.

A standard automatic transmission system contains:

- torque converter
- gearbox with automatically selectable epicyclic gear ratios
- final drive gear reduction
- differential unit
- propeller shafts and drive shafts.

You will learn about all these components in this chapter.

Key terms

Power – the rate of doing work. It is measured using a standard unit called the watt (W).

Driveline – the collective term for the components that transfer engine crankshaft movement all the way to the wheels.

Transmission – the collective term for the components (such as clutch, gearbox and drive shaft) which transmit power from the engine to the driven wheels.

Light vehicle clutch systems

A **clutch** is needed to connect and disconnect the engine from the transmission and driveline system.

The clutch must perform three main functions:

1. Provide a temporary position of **neutral** (so that the car can be brought to a stop without taking it out of gear).
2. Disconnect the engine from the gearbox so that a smooth gear change can be achieved (allowing gear speeds to be **synchronised**).
3. Give a smooth take-up of drive from rest (so that the car can pull away without **juddering**).

There are two main types of clutch: friction clutches and dog clutches.

The friction clutch

The friction clutch is most common type used for connecting and disconnecting drive from the engine to the gearbox, as it will give a smooth take-up of drive. The clutch is fitted at one end of the engine and is bolted to the engine's flywheel.

The surfaces of the clutch components are held together with very strong spring pressure. The **friction** from the surfaces in contact and the pressure with which they are held together provides the grip to lock the rotating components and transmit drive.

The clutch is made up of three main sections, constructed like a sandwich, as shown in Figure 5.2. All of the component parts are squashed together to provide grip.

- When the three sections are held tightly together by strong spring force, they all spin round at the same time and the clutch is **engaged**.
- When the driver depresses (pushes down) the clutch pedal, the parts separate so that there is no connection between the engine flywheel and the input shaft of the gearbox. The clutch is now **disengaged**.

The three main components of a friction clutch are shown in Figure 5.2. They are:

- **Flywheel** – a heavy metal disc used to store kinetic (movement) energy from the engine.
- **Friction plate** – the driven component of a friction clutch, which is attached to the input shaft of the gearbox. It is also known as a drive plate.
- **Pressure plate** – a metal surface used to clamp the components of a friction clutch together.

Key terms

Clutch – a mechanism for connecting and disconnecting two moving components.

Neutral – where no drive is transmitted through the gearbox.

Synchronised – when gears are turning at the same speed.

Juddering – a vibration felt when trying to pull away.

Friction – a measurement of grip or how hard it is to slide one surface against another.

Engaged – all parts of the clutch are held together and drive is transmitted.

Disengaged – the clutch components are separated and no drive is transmitted.

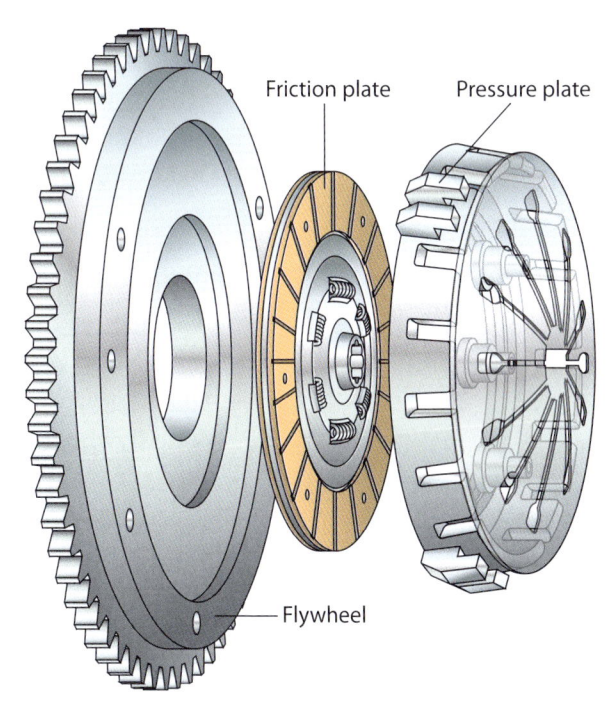

Figure 5.2 The three component parts of the clutch

A friction clutch is what is known as a gradual engagement type. This means that:

- When disengaged the components are separated, with no drive being transmitted.
- As the clutch is slowly engaged, friction between the surfaces is slowly increased, so that they eventually rotate as one complete unit.

The component parts of the clutch operating mechanism are shown in Figure 5.3.

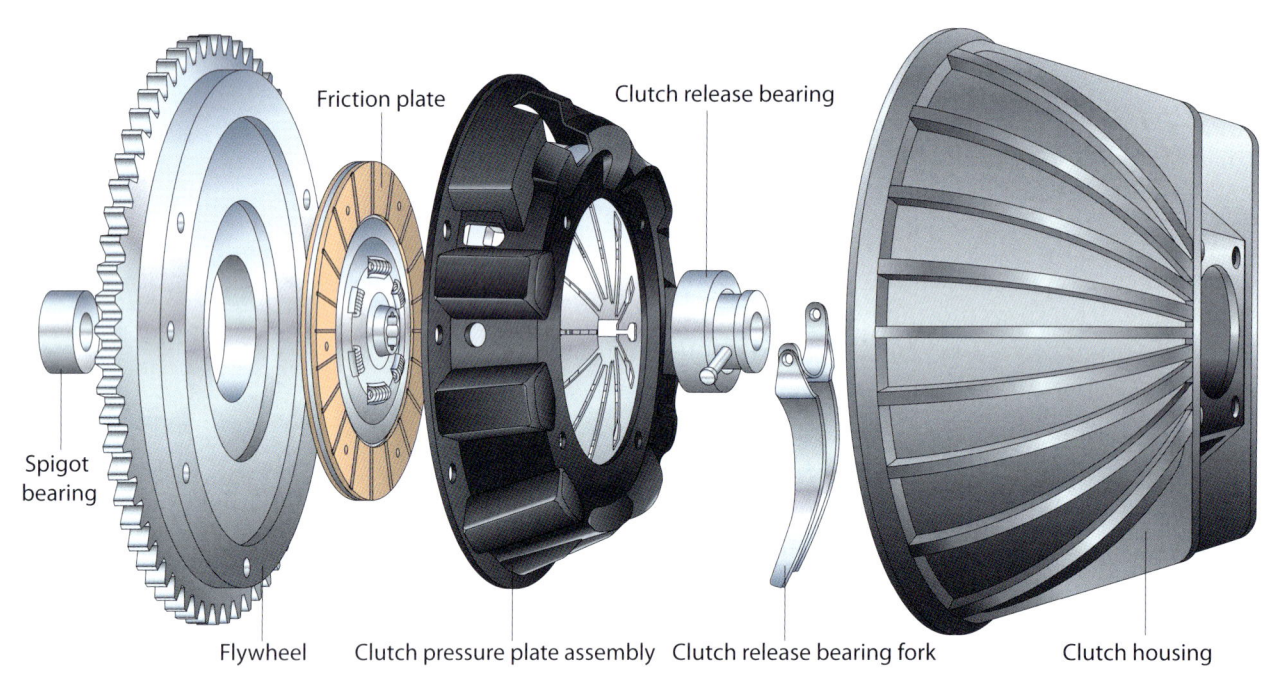

Figure 5.3 Clutch operating mechanism

Engaging the clutch

To engage the clutch, lift your foot slowly off the clutch pedal. Springs push the pressure plate against the drive plate, which in turn presses against the flywheel. This process is shown in the flow chart in Figure 5.4.

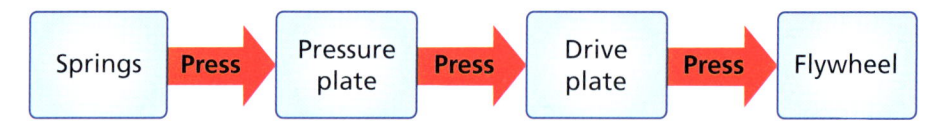

Springs → Press → Pressure plate → Press → Drive plate → Press → Flywheel

Figure 5.4 Engaging a friction clutch

The flywheel and pressure plate are attached to the engine crankshaft, and the friction plate is attached to the input of the gearbox.

As the parts of the clutch are slowly engaged, friction and grip begin to drag the drive plate around with the flywheel and pressure plate. To begin with they will be turning at different speeds, which is known as **slip**.

As full engagement is reached, friction and grip increase to the point where all three parts are turning at the same speed. Drive from the engine is now transferred to the input shaft of the gearbox.

Friction clutch components

In order that the clutch can be operated while the engine is running, a device is needed to connect the pressure plate springs to the clutch pedal operated by the driver's foot. This is usually done by a clutch release or 'throw out' bearing, which is controlled by a lever called a release fork.

When the clutch pedal is depressed (pushed down to the floor) it operates as a lever on the end of a clutch cable or hydraulic master cylinder (which pushes fluid through a series of pipes to a slave cylinder). This then operates on the clutch release bearing. This takes pressure off the springs that are providing the clamping force to drive the friction plate. Once spring pressure has been released, the friction plate is free to move independently from the flywheel and pressure plate. This allows gears to be selected or the car to be brought to a stop.

Drag and slip

- A clutch must fully disengage with no **drag**.
- A clutch must fully engage with no slip.

The amount of **torque** produced by an engine can have a considerable effect on how the clutch operates. If the torque from an engine can be increased, the car's performance can be improved. A friction clutch must be able to transmit all of this engine torque to the rest of the transmission system without slip.

The amount of torque that can be transmitted by the clutch can be increased by the following methods:

- **Increasing the diameter or 'mean radius' of the clutch** – By making the diameter bigger, the size of the surface area is increased, which increases the amount of friction. This means that the more torque an engine produces, the larger the clutch must be.
- **Increasing the forces squeezing the surfaces together** – This is normally done by using stronger springs in the clamping mechanism. The stronger the spring force is, the more effort is required to operate the system. This can mean that the clutch mechanism feels hard to operate for the driver.
- **Increasing the number of surfaces in contact** – These are often referred to as **multi-plate clutches**. By increasing the number of surfaces in contact, the overall size of the clutch (particularly the diameter) can be kept fairly small, but the surface area is increased. This creates more friction and so transmits more drive.

Differences in friction clutch design

The main differences in the construction of light vehicle clutches are in the design of the springing mechanism that clamps the surfaces together.

> **Did you know?**
> Although the engine flywheel is used as part of the clutch mechanism, its real purpose is to store kinetic (movement) energy, which will help keep the engine crankshaft turning on non-power strokes.

> **Key terms**
> **Slip** – when there is a loss of grip and two surfaces slide over each other.
> **Drag** – when friction surfaces rub against each other.
> **Torque** – turning effort normally created by some form of leverage.
> **Multi-plate clutch** – a number of clutch friction plates mounted in one unit.

Coil spring clutch

A number of early clutches used coil springs to provide the clamping effort (see Figure 5.5). Due to wear through operation, coil spring tension and length will change over a period of time. This can result in an uneven clamping force being produced, which can lead to clutch drag, slip or vibrations caused as take-up of drive is required (often referred to as clutch judder).

Diaphragm spring clutch

To overcome the problems created by coil springs, a spring mechanism known as a **diaphragm** was developed. The diaphragm spring is a single metal plate, manufactured into a series of sprung steel fingers (see Figure 5.6). It is shaped like a shallow dish, and when one end of the fingers is pressed, they pivot about a **fulcrum**. This moves the opposite end of the diaphragm fingers in the other direction.

When the ends of the fingers of the diaphragm are released, the sprung nature of the steel returns them to their original position. Because the spring diaphragm fingers are made from one single piece of metal, an even clamping force can be produced.

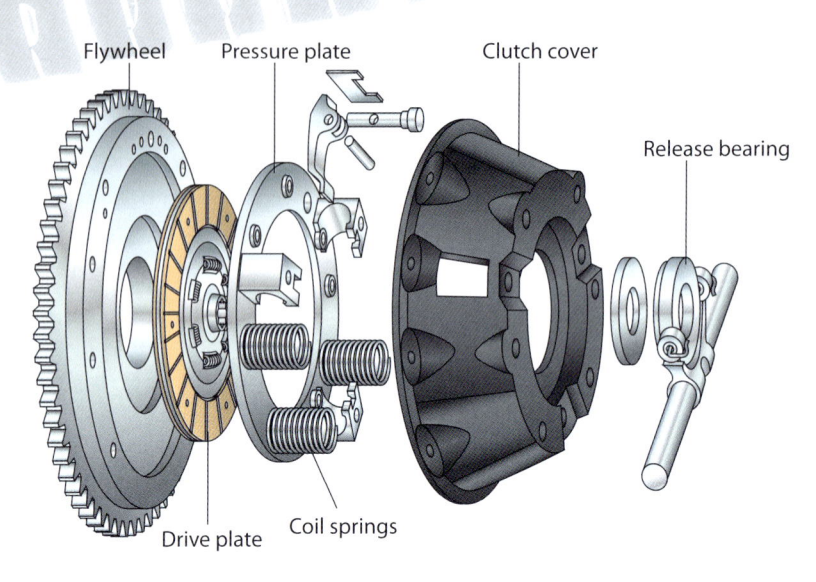

Figure 5.5 Coil spring clutch

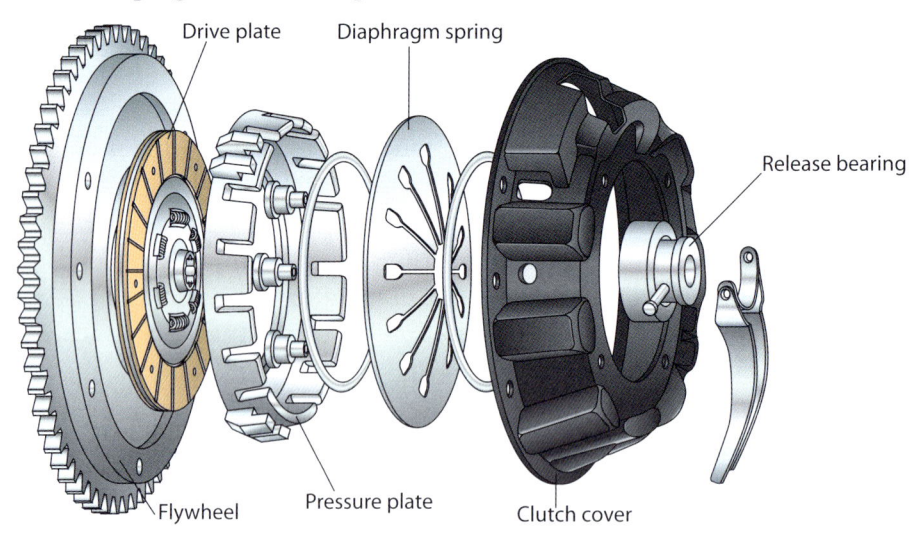

Figure 5.6 The component parts of a diaphragm spring clutch assembly

Friction plate construction

The friction plate of a clutch is made up from a number of different parts (see Figure 5.7).

- **Central hub** – This has an internally splined hole which mates with the **splines** on the input shaft of the gearbox. These splines stop the friction plate spinning on the gearbox input shaft.

> **Key terms**
>
> **Diaphragm** – a metal plate shaped like a shallow dish.
>
> **Diaphragm clutch** – clutch that uses sprung steel fingers instead of traditional coil springs.
>
> **Fulcrum** – a pivot point (like on a seesaw).
>
> **Splines** – slots or grooves machined into a round shaft or hole.

- **Outer plate** – The central hub is connected to an outer plate, which carries the friction linings.
- **Torsion dampers** – Mounted between the central hub and the outer plate are a series of springs that act as dampers against rotational forces as drive is taken up.
- **Friction linings** – Two sets of friction linings are attached to the outer drive plate (sometimes bonded but more usually riveted).
- **Wave plate** – The two friction linings are held apart by a thin sprung steel spacer called a cushion or wave plate. As drive is taken up, the wave plate allows the friction linings to be squashed together gradually, allowing a smoother engagement.

Thin grooves are manufactured as part of the friction material. These grooves assist with cooling, but also allow air between the friction material, flywheel and pressure plate. As the clutch is disengaged, the grooves help the components separate quickly and easily, reducing the possibility of drag.

If a large amount of torque needs to be transmitted, a number of friction plates can be used. This keeps the overall size of the clutch mechanism to a minimum.

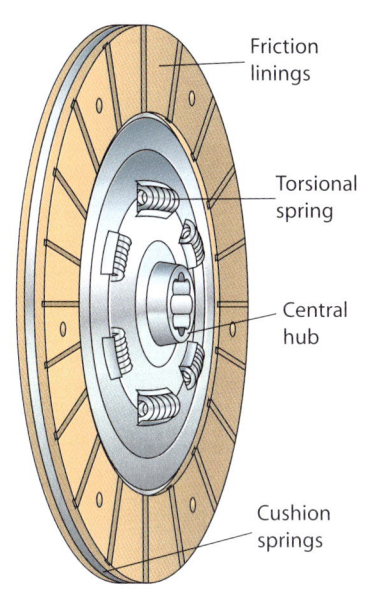

Figure 5.7 Clutch friction plate assembly

Clutch operating mechanisms

Movement from the clutch pedal can be transferred to the clutch mechanically or hydraulically.

Mechanical operation

In a mechanical system, such as the one shown in Figure 5.8, a clutch cable is attached to one end of a pivoting clutch pedal. When the pedal is pressed, it pulls on this cable and operates the clutch fork at the gearbox end. The clutch fork also pivots, usually pressing the release bearing against the fingers of the diaphragm spring, which pushes them inwards towards the flywheel.

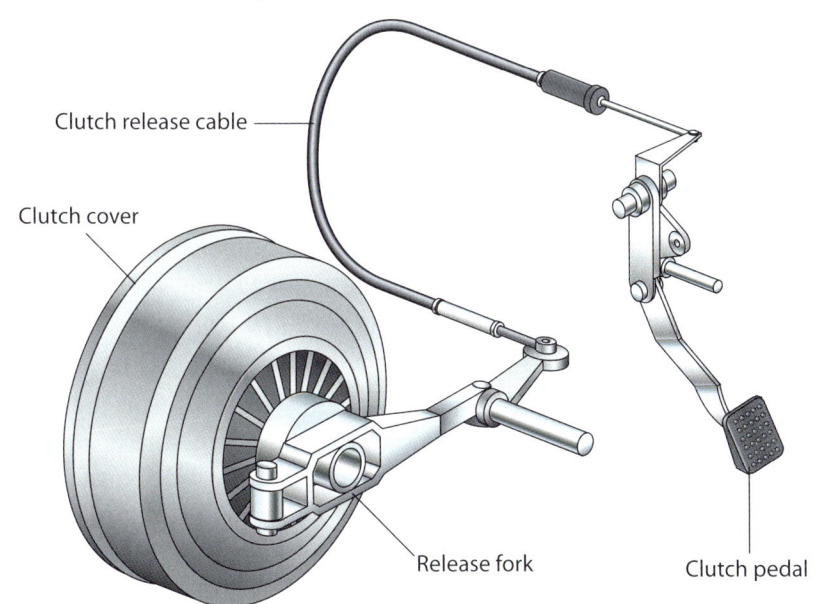

Figure 5.8 Cable-operated clutch assembly

Did you know?

The snapping of clutch cables happens more often when a vehicle has been originally designed as a left-hand drive. On these models, the clutch cable comes from the end of the pedal mechanism and is joined almost directly onto the back of the gearbox.

When manufactured to be used as a right-hand drive vehicle, the engine and gearbox remain in the same position, but the pedal mechanism has to be moved to the right-hand side. To allow clutch operation, the cable is simply made longer. As a result there have to be more bends in its route from the pedal to the clutch. This creates greater friction and wear, which can lead to premature failure.

Level 2 Light Vehicle Maintenance & Repair

> **Key term**
>
> **Mechanical advantage** – where effort is increased by leverage.

> **Did you know?**
>
> Clutch slave cylinders can be mounted inside or outside of the clutch bell housing: these are known as 'internal' or 'external'. Many manufacturers are now using internal slave cylinders as they act directly against the release bearing, reducing wear and free play between components.

> **Did you know?**
>
> The same type of hydraulic fluid is often used to operate both brake and clutch systems.

As the diaphragm fingers pivot against their fulcrum, the outer end of the diaphragm springs moves the pressure plate away from the friction plate, which disengages the clutch.

A **mechanical advantage** can be gained by the leverage produced at the clutch pedal and release arm. This leverage means you will notice a difference in the amount of movement at the pedal and release arm. (The pedal moves a long way, while the release arm and bearing only move a short distance.) Mechanical advantage makes it easier for the driver to operate the clutch.

In a mechanical system, the cable has to transmit drive through a number of different angles. This creates drag (making clutch operation harder) and wear (which can lead to premature failure of the clutch cable mechanism as the clutch cable is prone to snapping).

With a mechanical cable-operated system, as the clutch begins to wear a method of adjustment is needed to keep the clutch running and operating properly. This can be done manually using a screw thread, or automatically using a pedal ratchet mechanism.

Hydraulic operation

A hydraulic system, such as the one shown in Figure 5.9, uses the same principle that is used in the braking system (see Chapter 2, pages 172–177). The clutch pedal is attached to a master cylinder. When the pedal is depressed (pushed down), this forces a piston inside the master cylinder to push fluid through a series of pipes and hoses. The fluid is used to operate a slave cylinder piston at the clutch release end.

The advantages of a hydraulic system are:

✓ It doesn't matter how long the pipes are, or whether they move around corners and bends. Hydraulic pressure acts instantly in all parts of the system at the same time, so the slave cylinder operates as soon as the pedal is pressed.

✓ Because there is no clutch cable to route, there is no cable to snap.

✓ In a hydraulic system, as the clutch friction plate wears, hydraulic fluid takes up the space created. This makes the system self-adjusting.

✓ Just like a mechanical system, a hydraulic clutch operating system can give a mechanical advantage, making it easier for the driver to use.

Pressure at the clutch end of the system can be increased by using pistons of different sizes in the master and slave cylinders. Because pressure is a force acting over an area, if a small piston is used in the master cylinder and a large piston is used in the slave cylinder, pressure can be multiplied. This multiplied pressure means you will notice a difference in the amount of movement at the pedal and release arm. (The pedal moves a long way, while the release arm and bearing only move a short distance.)

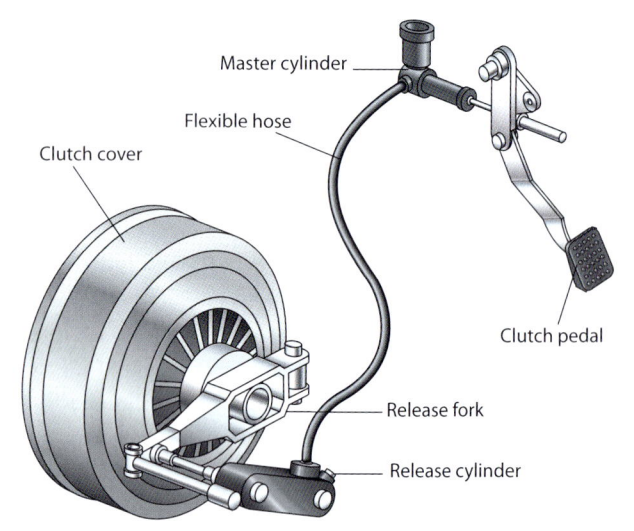

Figure 5.9 Hydraulically operated clutch assembly

5 Light vehicle transmission & driveline units & components

> **Find out**
>
> Examine the cars in your workshop and make a note of:
> - one make and model that has a mechanical clutch operating mechanism
> - one make and model that has a hydraulic clutch operating mechanism.
>
> How is the mechanical clutch operating mechanism adjusted – manually or automatically?

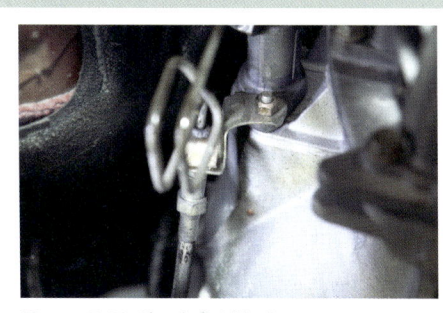

Figure 5.10 Clutch fluid leak

Problems with the hydraulic system

Just like the hydraulic system used in brakes, air bubbles must be avoided. This is because air is a gas and therefore compressible. If you have replaced any components, you should bleed the system so that no air exists. If you do not, operation of the clutch mechanism might not be possible and the clutch will drag (it won't be fully disengaged, so the friction surfaces will still be rubbing against each other).

> **Safe working**
>
> - Do not spill clutch fluid onto vehicle bodywork as it will damage paintwork.
> - Clean up any spilt fluid immediately to avoid slips trips and falls.
> - If the car has to be raised, make sure that you don't exceed the safe working load of lifting equipment.

Clutch bleeding

Checklist			
PPE	**VPE**	**Tools and equipment**	**Source information**
• Steel toe-capped boots • Overalls • Latex gloves	• Wing covers • Steering wheel covers • Seat covers • Floor mat covers	• Clutch/brake fluid • Spanners • Bleed bottle and pipe	• Technical data • Manufacturer's recommended bleeding procedure

1. Locate and top up the clutch fluid reservoir with fresh clutch/brake fluid.

2. Connect a bleed bottle and pipe to the clutch slave cylinder.

3. Open the bleed screw. With the aid of an assistant inside the car, ask them to push the pedal down.

4. Close the bleed nipple and ask your assistant to release the clutch pedal.

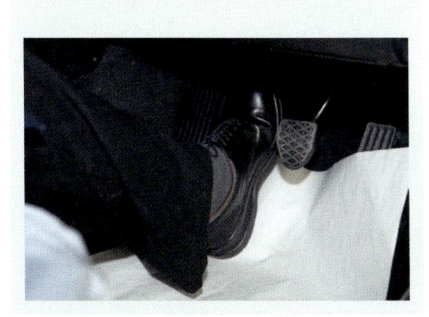

5. Repeat steps 3 and 4 until all air has been removed from the system.

6. Top up the clutch fluid reservoir and clean up any spilt fluid.

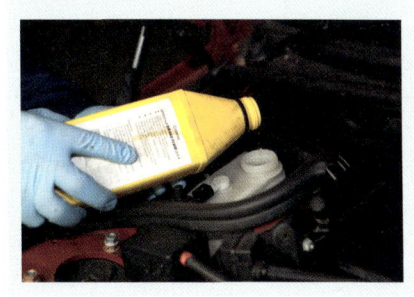

7. Road test the vehicle and check for correct function and operation.

327

Level 2 Light Vehicle Maintenance & Repair

> **Find out**
>
> Can you suggest possible clutch faults that produce the following symptoms?
>
> - A vibration that is felt as the clutch is engaged and the car starts to move
> - Difficulty getting the car into gear (especially first gear and reverse gear)
> - Partial loss of drive as the car pulls away up hill
> - A sudden jolt as the clutch is engaged and the car starts to move

> **Working life**
>
> Omar is road testing a customer's car wearing dirty overalls. He has placed a plastic cover over the driver's seat to stop it getting dirty.
>
> In accordance with the legal requirements, Omar wears a seatbelt during the road test.
>
> 1. What might happen when the customer collects their car?
> 2. What could Omar have done to prevent this problem?

When the clutch drags, gear selection can be difficult, or it might not be possible to provide a temporary position in neutral. Also, because the system uses a hydraulic fluid, if a leak occurs at the clutch end, the fluid could contaminate the friction surfaces. This will reduce the amount of friction provided and make the clutch slip.

Key engineering principles related to light vehicle clutch systems

Principles of friction

Friction is a force that occurs whenever two surfaces rub against each other, as shown in Figure 5.11. In some areas of motor vehicle design, for example the engine, friction is unwanted and must be reduced using lubrication. In other areas, such as the clutch, friction is positively encouraged.

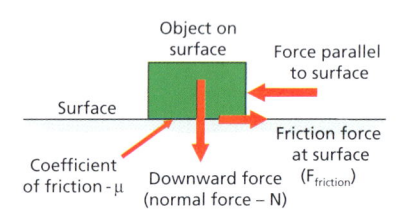

Figure 5.11 Friction forces

The amount of friction depends on:

- the materials coming into contact with each other
- how much surface area is in contact
- the pressure with which the materials are held together.

Principle of levers

Without multiplication of mechanical effort, the operation of a clutch mechanism would be very difficult. Leverage is able to multiply force by using the principle of a fulcrum (or pivot point) set at a position along the leverage line (see Figure 5.12). By increasing the length of the lever, the amount of force applied can be multiplied. This means that the amount of input force is increased at the output.

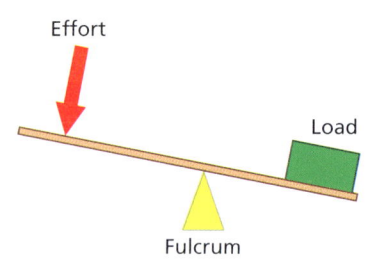

Figure 5.12 Principles of leverage

> **Did you know?**
>
> The famous ancient Greek scientist Archimedes said, 'Give me a lever long enough and a fulcrum on which to place it and I shall move the world.'

Torque transmission

The largest amount of torque (turning effort) produced by the engine is only available at certain running speeds. The clutch must be able to handle this torque without slippage. During the design process, manufacturers must choose a clutch system that is capable of transmitting the entire available engine tuning effort with no slip.

Dog clutches

A dog clutch has teeth which, when engaged, are able to physically lock two rotating shafts together, (see Figure 5.13). This allows drive with no slippage.

The disadvantage of dog clutches is that they need to be stationary or moving at the same speed to be engaged and disengaged. Any difference in speed means that the teeth will collide, causing noise and damage.

Power take-off

A power take-off can be used to connect additional equipment to a vehicle so that the engine is able to power it.

For example, a worker might need to attach a grass-cutting tool to a tractor to mow a school playing field. While driving to the school, the mowing attachment should not be operated, as this would be highly dangerous. When it reaches the playing field, the engine is stopped and a shaft from the back of the tractor's motor is connected to the mowing equipment via the power take-off clutch (which is a dog clutch). Now, when the tractor engine is started and the tractor begins to move, the mowing tool will also operate and will cut the grass.

The use of dog clutches in transmissions

A common use for dog clutch mechanisms is found inside the gearbox of a car. The gears of the transmission are able to spin freely on shafts, but when the driver selects a gear, a dog clutch is used to lock drive to the output shaft of the gearbox. Once locked, this will transmit the turning motion from the engine to the wheels.

A result of using dog clutches inside the transmission means that the gears have to be synchronised (running at the same speed). If this isn't done the teeth can collide, which causes noise and damage. (This is the grinding noise that is sometimes heard when incorrectly changing gear.) To help overcome this problem, a friction clutch is used to connect and disconnect drive from the engine to the gearbox. Once disconnected, the speed of the dog gears can be synchronised.

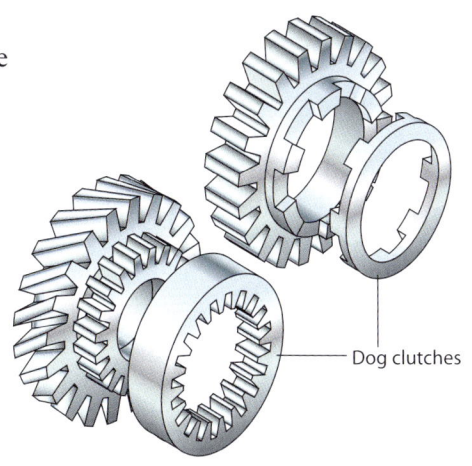

Figure 5.13 Dog clutch

> **Did you know?**
>
> Dog clutches are used as a power take-off (PTO) in commercial vehicles.

Automatic transmission torque converters

A major difference between automatic and manual transmission is that automatic transmission does not use a conventional style clutch. However, a method of providing a temporary position of neutral and giving a smooth take-up of drive is still needed. This is done by using a component called a **torque converter**. It is mounted in a similar position to a standard clutch, but instead of using friction surfaces clamped together to provide drive, fluid is forced between **turbine** blades. This creates drag, which will rotate the input shaft of the gearbox.

A torque converter consists of three main components (see Figure 5.14):

- impeller
- turbine
- stator.

These components are sealed inside a casing. The casing is pressurised with automatic transmission fluid (ATF) from a crankshaft-driven oil pump. When the torque converter is spun by the engine crankshaft, fluid is taken into the impeller blades and thrown outwards by **centrifugal force**.

> **Key terms**
>
> **Torque converter** – a form of fluid flywheel which is designed to connect drive to the gearbox in place of a standard clutch.
>
> **Turbine** – a bladed rotor.
>
> **Centrifugal force** – a force that makes rotating objects move outwards.

The fluid strikes the blades of the turbine, making it spin. The spinning turbine is connected to the input shaft of the gearbox, which now also turns.

The hydraulic fluid (ATF) now leaves the turbine and strikes the blades of the stator, which direct it back into the impeller at high speed. This force helps to multiply the torque provided by the crankshaft.

If the car is held stationary, using the brakes, the turbine is also held still while the impeller spins. As the brakes are released, the hydraulic action of the transmission fluid striking the turbine blades makes them turn and provides a smooth take-up drive.

The largest amount of torque multiplication (see page 339) happens when there is a difference in speed between the impeller and the turbine. As the speed of the impeller and the turbine begin to synchronise, torque multiplication falls to zero. This is called **coupling point**. Drag and slip between the turbine and impeller can reduce performance, so as the torque converter reaches coupling point, a hydraulically operated lock-up clutch can be used. This holds all internal components together and prevents slip.

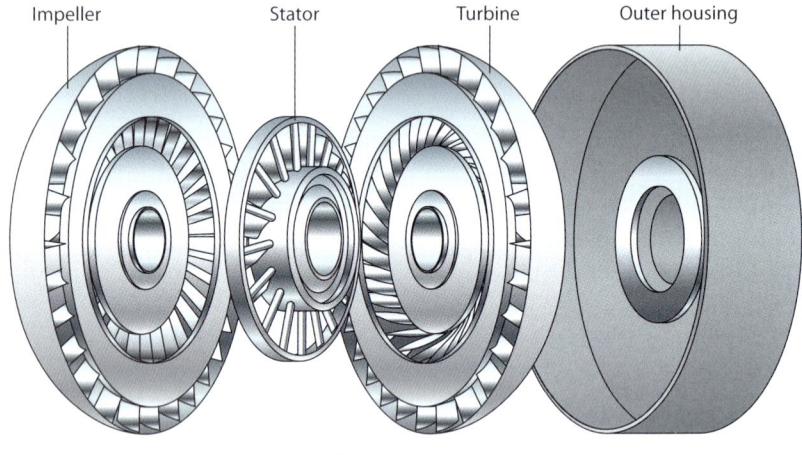

Figure 5.14 The three main sections of an automatic transmission system torque converter

Did you know?

A feature of the torque converter is that when the brake pedal is released and fluid from the impeller begins to react against the turbine, drive to the gearbox begins. This means that the vehicle will start to move unless held stationary by the brake. This movement is often referred to as 'creep'.

Safe working

If the vehicle will be left stationary for any period of time with the engine running, you must take it out of gear and select the neutral or park position. The neutral position will simply remove the connection to the gearing inside the box. The park position has the added advantage of locking a lever mechanism into the gearing to physically prevent any further movement.

As an added safety feature, an electrical cut-out is often included on the starter system so that it cannot operate if the automatic transmission is in gear. This **inhibitor switch** is designed to prevent accidental movement of the vehicle due to creep if the automatic transmission is started in a **drive range**.

Key terms

Coupling point – the point when the turbine and the impeller are spinning at roughly the same speed.

Inhibitor switch – safety cut-out switch for the starter motor when the car is in gear.

Drive range – forward or reverse gears.

Find out

Using sources of information available to you, choose a make and model of automatic car. Find out:

- the recommended grade and quantity of automatic transmission fluid (ATF) that is used in the gearbox
- the manufacturer's procedure for checking and topping up the gearbox ATF.

5 Light vehicle transmission & driveline units & components

> **CHECK YOUR PROGRESS**
>
> 1 Name four clutch system components.
> 2 State three reasons why a clutch is required.
> 3 What is the difference between a clutch and a torque converter?
> 4 Explain two ways of increasing the amount of torque that can be transmitted through a clutch.

Light vehicle gearbox operation

Depending on the design of an engine, it delivers torque (turning effort) in a very narrow rev band. This means that the greatest effort being produced by the engine is only available when the engine is running at certain speeds.

Without a **gearbox** the engine would struggle and find it very hard to pull away, carry heavy weights, tow a caravan or go uphill.

Manual transmission gearbox

The torque produced by the engine needs to be increased so that it is able to turn the wheels of the car easily. This is called **torque multiplication** and it improves the car's performance. A manual transmission system uses gears which will multiply torque using leverage. Leverage is able to increase the effort supplied, making it easier for the engine to turn the wheels. This can be explained using the example shown in Figures 5.15 and 5.16.

If you look at the gears shown in Figure 5.16, one is larger than the other. If the outside edges of the gears were removed and laid out flat, you would see that the larger gear gives a long line and the smaller gear gives a short line. These lines can be compared to the length of a lever bar. If you increase the length of a lever, you can multiply the mechanical effort.

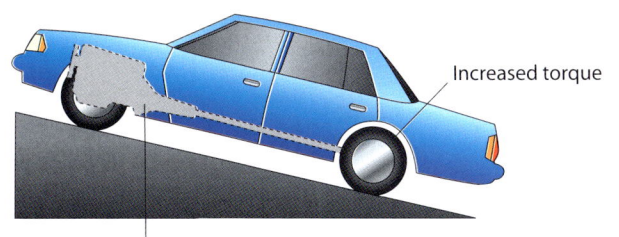

> **Key terms**
>
> **Gearbox** – the housing which contains gears, shafts, bearings and selectors.
>
> **Torque multiplication** – increasing the torque produced by the engine so that it can turn the wheels of the car more easily.

Figure 5.15 A transmission system is needed to increase the torque (turning effort) produced by the engine to help the car go up the hill

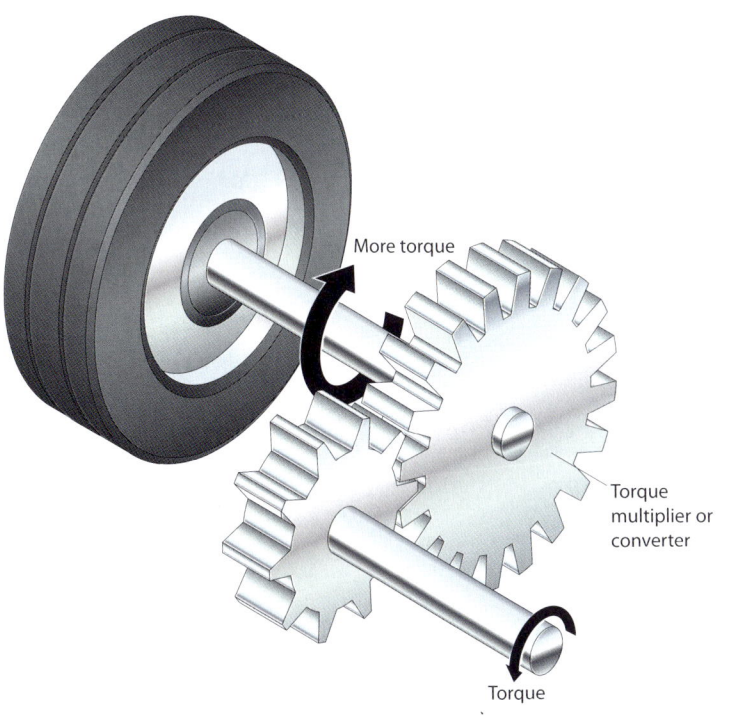

Figure 5.16 Using gears to multiply torque

331

Level 2 Light Vehicle Maintenance & Repair

Unfortunately, you don't get something for nothing. When using a small gear to turn a large gear, torque is increased but speed is reduced. This trade-off of torque multiplication for speed can be seen in the operation of a gearbox.

When the car needs to pull away from a standing start, the torque effort required to turn the wheel is large. A low gear ratio is selected by the driver (small gear turning a large gear). This will give a large turning effort but a slow speed (first gear, for example).

Once the car is moving, it now has **inertia** and because of this requires less torque. This means that higher gears can now be selected to multiply (increase) the speed of the vehicle. But as speed increases, torque goes down.

> **Key term**
>
> **Inertia** – a force that holds the body steady or keeps it moving at a constant speed and direction.

Light vehicle manual gearbox system components

Gears

Gears are circular wheels with teeth cut around the edge. They can be brought into contact to help increase torque. For example, an input gear is turned by the engine. When it is meshed with another gear, the torque will be increased by this output gear to help turn the wheels.

When meshed together, if the first gear turns in a clockwise direction, then the second gear will turn in an anticlockwise direction, as shown by Figure 5.17.

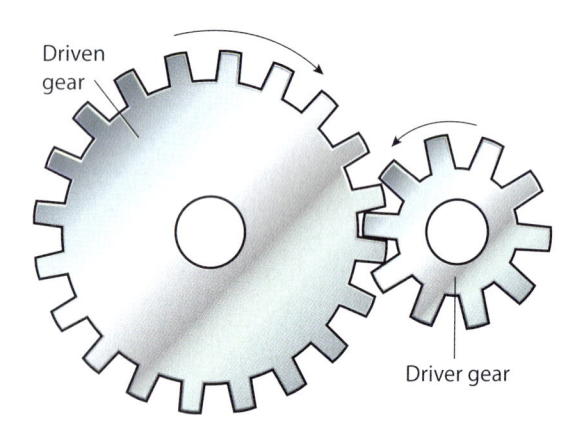

If an intermediate gear called an idler gear is placed between the two gears, then the direction of the output is reversed (input gear turns clockwise, idler gear turns anticlockwise, output gear turns clockwise). The idler gear has no overall effect on the gear ratio between the input and the output, it simply changes the direction of rotation. This is how reverse gear is achieved (see Figure 5.18). (You will learn about gear ratios on pages 339–341.)

Figure 5.17 Two gears in mesh, turning in opposite directions

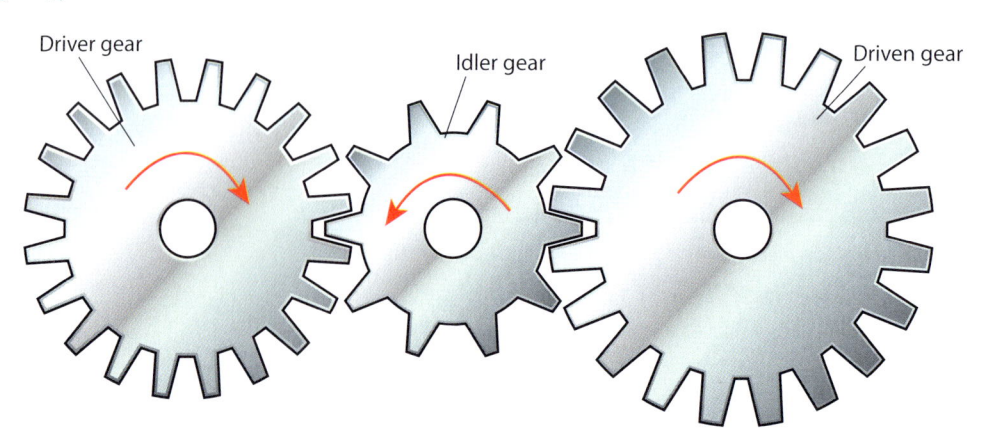

Figure 5.18 Reverse gear is obtained using an idler gear

5 Light vehicle transmission & driveline units & components

The gears inside a transmission are mounted on round shafts that rotate as the gears are turned. The input shaft is connected to the clutch drive plate and the output shaft is connected to the final drive, which turns the wheels. In a rear-wheel drive car, a third shaft is often used to help transfer movement between the gears. These shafts are often called:

- **First motion shaft** – the input shaft of the gearbox.
- **Counter shaft or lay shaft** – the second motion shaft that will turn in the opposite direction on a rear-wheel drive gearbox.
- **Main shaft** – the output shaft of the gearbox.

When the driver wants to select a gear, he or she moves a lever inside the car which pulls and pushes on a series of rods or cables. These rods or cables relay this movement to **selector forks** inside the gearbox. The selector forks then move internal gearbox components (such as gears and selector hubs) into contact with one another to provide drive.

Once a gear has been connected, a method is often needed to help keep it in place. A spring-loaded ball bearing is pushed into a groove machined on the selector shaft, which creates resistance to movement. This helps prevent the car jumping out of gear. This spring-loaded ball bearing (see Figure 5.19) is known as a **detent** (from the word 'detention', which means being detained or held back).

Two main types of gear are used in a standard manual transmission system: spur gears and helical gears.

> **Key terms**
>
> **Selector fork** – horseshoe-shaped component used to slide the selector hubs in and out of mesh with the gears.
>
> **Detent** – method used to help lock selected gears in place.
>
> **Spur gear** – gear with straight-cut teeth.
>
> **Mesh** – in contact by interlocking or fitting together.

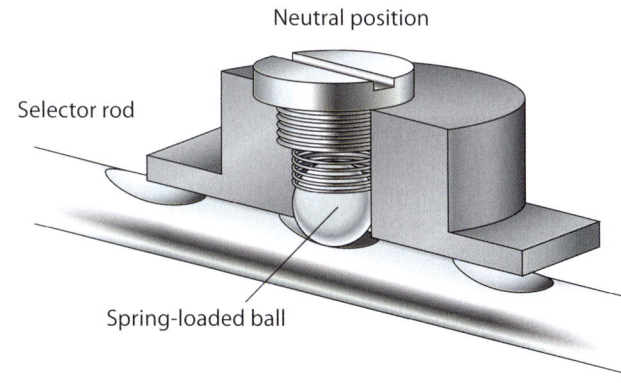

Figure 5.19 Detent mechanism

Spur gears

A **spur gear** has straight-cut teeth, as shown in Figure 5.20. These are direct-acting and create low amounts of drag, which helps improve overall performance. However, they can be noisy in operation.

A spur-cut gear can be slid in and out of **mesh** with another spur-cut gear, making them ideal for use as a reverse gear idler. (The characteristic whine when reversing a car is caused by the design of the teeth on the spur-cut gear.)

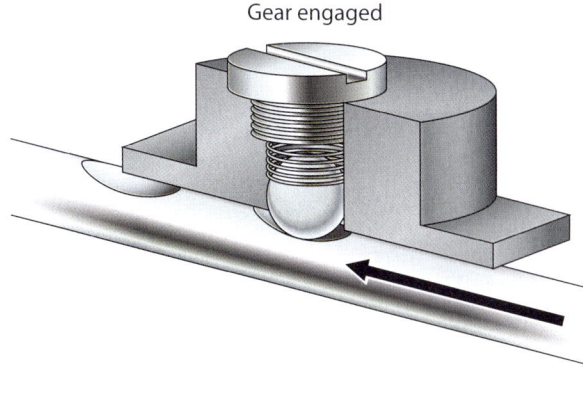

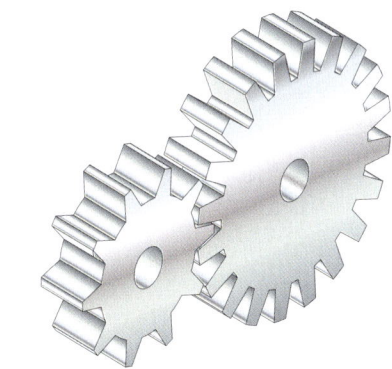

Figure 5.20 Spur gears with straight-cut teeth

Helical gears

A **helical gear** has teeth which are cut on an angle, as shown in Figure 5.21. The word 'helical' comes from the word 'helix', which is a form of spiral (think of the spiral of a coil spring). This means the teeth are not just cut on a diagonal, but if they were extended around a cylinder they would actually be shaped like a screw thread. Helical-cut gear teeth have a large surface area, making them very strong, less prone to wear and quiet in operation.

The design and shape of helical-cut gears means that they cannot be slid in and out of mesh like a spur gear. For this reason they are used in a **constant mesh** type of gearbox, where all the gear teeth on all the different gears are always engaged with each other. Since all of the gears in this type of gearbox are continuously rotating, another method of selecting the appropriate gear is needed, and a selector hub is used, which is splined to the output shaft of the gearbox.

The selector hub sits between the helical-cut gears. When a certain gear ratio is selected by the driver, a selector fork slides the selector hub towards an appropriate gear setting, as shown in Figure 5.22. Dog teeth on the side of the selector hub locate with dog teeth on the side of the appropriate gear, locking the two components together.

Figure 5.21 Helical gears with teeth cut at an angle

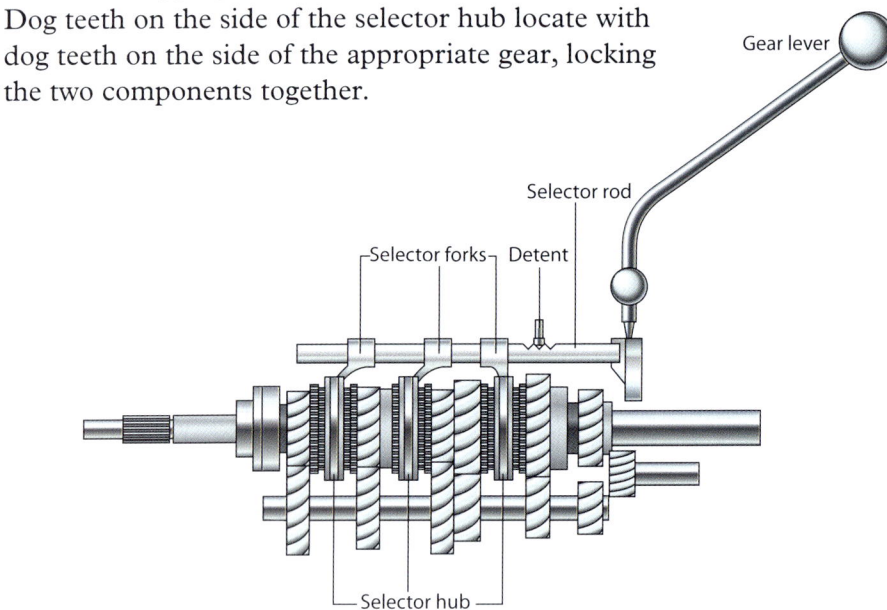

Figure 5.22 A selector hub and fork mechanism

Figure 5.23 shows the process of gear selection.

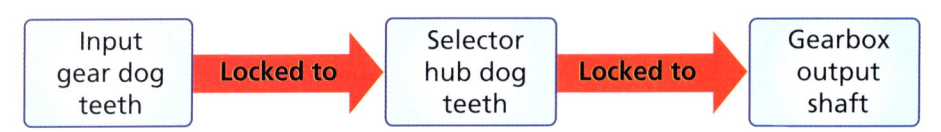

Input gear dog teeth → **Locked to** → Selector hub dog teeth → **Locked to** → Gearbox output shaft

Figure 5.23 Selector hub operation

In this way the selected gear is positively attached (via the dog teeth) to the selector hub, and the selector hub is positively attached (via the splines) to the output shaft of the gearbox, so that engine torque is transmitted.

Interlock mechanism

It is possible that when the driver operates the gear selector, two selector rods move at the same time. If this happens the selector forks inside the gearbox may try to lock two gears to the output of the transmission simultaneously. The gears inside the gearbox are different sizes and they are turning at different speeds, so if both gears were selected at the same time, the gearbox would lock solid.

To try to prevent this, a system known as an **interlock** is provided in the selection mechanism. The interlock commonly uses a series of balls, plungers or rods that lock the selector shafts when operated, so that only one shaft is able to move at any one time (see Figure 5.24).

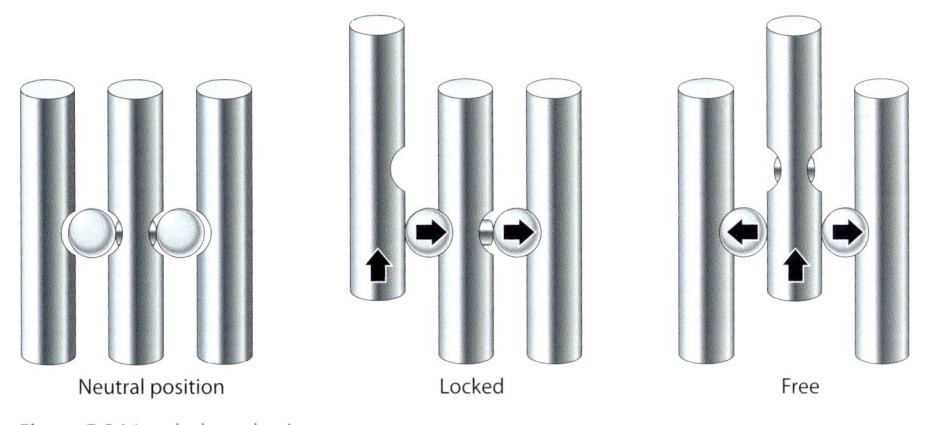

Neutral position   Locked   Free

Figure 5.24 Interlock mechanism

> **Key terms**
>
> **Helical gear** – gear with teeth cut on a helix (or spiral).
>
> **Constant mesh** – gearbox where the gears are always engaged with each other.
>
> **Interlock** – a mechanism designed to prevent the selection of two gears at the same time.
>
> **Synchromesh** – component used to equalise the speeds of two different gears.

Synchromesh

Gears of varying sizes are used inside the gearbox, so they will all be travelling at different speeds. Since dog teeth or dog clutches are used, when the driver wants to select a certain gear, the speed of these teeth must be synchronised before they can be engaged. This means that they must be sped up or slowed down so they are all turning at the same speed. If this does not happen, severe noise and damage may occur.

To achieve this synchronicity, a system sometimes called the **synchromesh** is used in combination with the selector hub (see Figure 5.25). As the selector hub is moved towards the gear to be used, two surfaces come into contact with each other and act as a friction clutch. This friction clutch provides grip which brings the speed of the selector hub either up or down, so that it is spinning at the same speed as the gear (synchronising their speed).

Once the gears are spinning at the same speed, the selector hub dog teeth and gear dog teeth are able to lock together without noise or damage, providing a definite engagement.

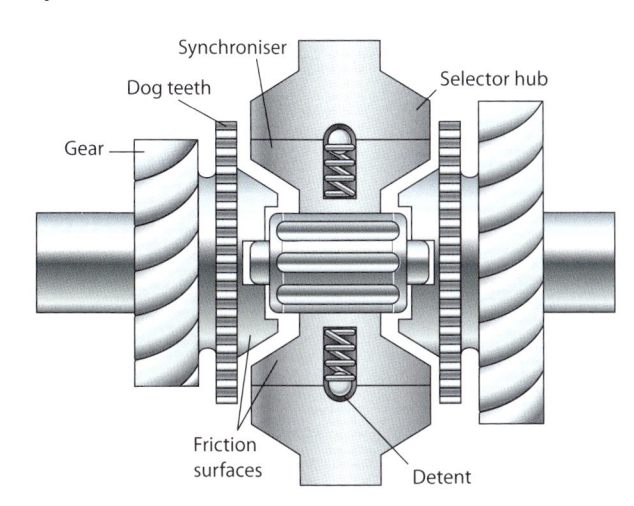

Figure 5.25 Synchromesh gear selector hub

> **Key terms**
>
> **Baulk ring** – a blocking ring that sits between the selector hub and the gear to be selected.
>
> **Splash feed** – a method of lubrication used in some gearboxes, where oil is dragged around with the movement of the gears.

Baulk ring

It is possible to outrun the speed of the synchronisation (by forcing the gear selector lever too quickly). Because of this another component was developed for the synchromesh, to prevent gear selection if the selector hub and the chosen gear speeds were not synchronised. This component is known as a **baulk ring** or blocking ring (see Figure 5.26).

A baulk ring has teeth that are the same size and shape as the dog teeth on the gear and selector hub. The baulk ring is able to move slightly when compared to the gear and selector hub. This means that if the teeth on all three components (selector hub, baulk ring and gear) don't line up, then gear selection is blocked.

As the driver pushes the gear lever, the baulk ring acts like the drive plate of a friction clutch, and is sandwiched between the selector hub and the gear. As pressure is applied, friction makes it grip and bring all three components up to the same speed. Once all three components are travelling at the same speed, the teeth on the baulk ring line up. This allows the selector hub to slide across, locking the dog teeth to those on the side of the chosen gear.

A manual gearbox is lubricated by **splash feed**. This means that as the gears turn they scoop up oil and drag it around, lubricating the gearbox components. Because friction is necessary between the selector hub, the baulk ring and the appropriate gear, a system is needed to allow small amounts of friction to exist (cutting through the lubricating oil). For this reason, many baulk rings are manufactured with small grooves in the friction surface that act like the tread found on tyres (see Figure 5.26). Just as tyre tread is designed to cut through water tension to prevent aquaplaning, the grooves that are found on the synchromesh baulk ring perform a similar function with the lubricating oil in the gearbox.

Figure 5.26 Synchromesh baulk ring

Over a period of time the grooves on the baulk ring begin to wear. This can lead to slippage of the friction surfaces, which might mean that complete synchronisation of the gear speeds is not possible. If this happens the dog clutch teeth may be travelling at different speeds and will strike each other, leading to noise and gear teeth damage.

Automatic gearbox

A manual gearbox is one where the driver selects the appropriate gear ratio, relying on their own judgment regarding the load placed on the engine.

An automatic gearbox is another option used by vehicle manufacturers. Engine speed and load are detected automatically and the transmission system itself chooses the most appropriate gear.

The main differences between automatic gearboxes and manual gearboxes are the type of gear used and the way in which the gears are selected. In an automatic transmission, instead of the normal spur or helical gears which are engaged and disengaged, a system known as an epicyclic gear train is used.

Epicyclic gear train

An epicyclic gear train (see Figure 5.27) uses gears that are constantly in mesh and consist of:

- a large outer ring gear, often called the 'annulus'
- a central gear, often called the 'sun gear'
- a series of intermediate gears (that sit between the sun gear and the annulus), called the 'planet gears'. These are supported on spindles attached to a planet carrier.

The gears of an epicyclic gear train are selected by holding one of the components stationary (sun gear, planet gears or annulus). This means the remaining two gears become input and output.

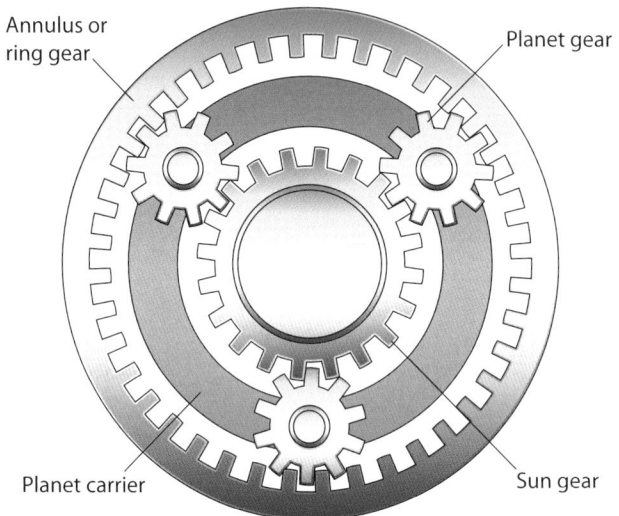

Figure 5.27 Epicyclic gear mechanism

The differing numbers of teeth on the input and output are called gear ratios. The gear ratios show how much the torque is increased (multiplied). (Gear ratios are explained fully on pages 339–341.)

Table 5.1 shows how different gear ratios are obtained using an epicyclic gear mechanism. Only two of the gearing components are used at any one time. The third is held stationary and takes no part in the torque multiplication.

- The first column shows which part is held still.
- The second column shows which part is turned by the engine.
- The third column shows which part will turn the wheels.
- The fourth column shows by how much the torque is multiplied and which direction the output turns.

Table 5.1 Example of epicyclic gear selection, torque increase and direction of travel

Stationary	Input	Output	Ratio and direction
Annulus	Sun gear	Planet carrier	3.4 : 1 (forward)
Sun gear	Planet carrier	Ring gear	0.71 : 1 (forward)
Planet carrier	Sun gear	Ring gear	2.4 : 1 (reverse)

As all three components are constantly in mesh, the locking of one of the components is done using brake bands or clutch packs.

Level 2 Light Vehicle Maintenance & Repair

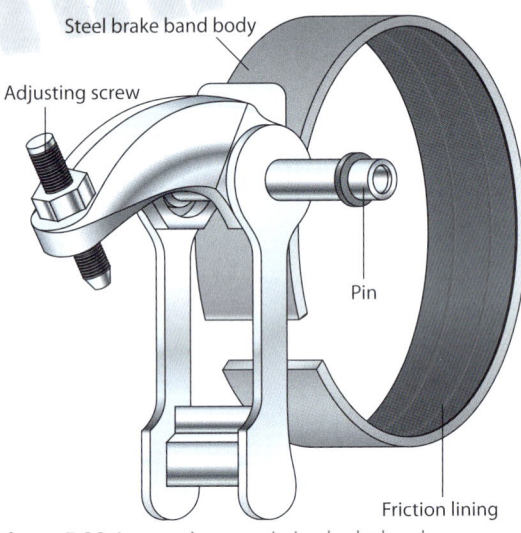

Figure 5.28 Automatic transmission brake band

- **Brake band** – This is like a belt that is wrapped around one of the components. When the brake band is pinched together, it holds that component stationary against the side of the gearbox casing (see Figure 5.28).
- **Clutch pack** – This is a series of small multi-plate clutches which are operated by hydraulic pressure from the automatic transmission fluid (ATF). They are used to connect and disconnect two rotating epicyclic components inside the automatic gearbox.

Working life

Colin is trying to move an automatic car into the garage workshop. The car won't start so he checks that the battery is fully charged. The battery is fine.

1. What is stopping the car from starting?
2. What does Colin need to do to start the car?

Continuously variable transmission (CVT)

A **continuously variable transmission (CVT)** gearbox is a form of automatic transmission. Instead of using mechanical gear sets, such as an epicycle, a drive belt is held between two pulleys, as shown by Figure 5.29, in a similar way to the chain and **sprockets** on a bicycle.

A bicycle is able to vary its gearing by changing the size of the sprocket on which the drive chain runs. The principle of CVT is similar. By changing the size of the drive pulleys, different gear ratios can be achieved. This is done by allowing the drive pulleys to expand and contract. In this way the drive belt is able to ride up and down within the pulleys, varying their size and therefore the gear ratio. As one pulley expands the other will contract equally. Since these pulleys do not rely on fixed gear sizes, a **stepless** gear ratio can be achieved, which maintains optimum efficiency for any engine speed or load.

Key terms

Continuously variable transmission (CVT) – a form of automatic transmission that uses drive belts and pulleys to provide a continuously variable transmission gear ratio.

Sprocket – type of gear used to drive a chain.

Stepless – when no gear change can be felt by the driver.

Find out

List the make and model of three cars that use a CVT gearbox.

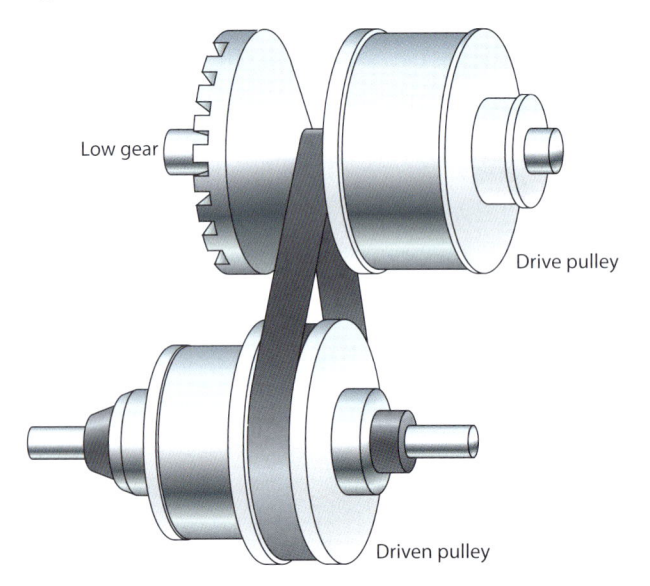

Figure 5.29 The pulleys and belts of a continuously variable transmission (CVT) system

Key engineering principles that are related to light vehicle manual gearbox systems

Torque multiplication

The turning effort provided by the engine's crankshaft is not strong enough to move the car from a standing start or to carry heavy loads, unless it is turning within a certain rev range. To overcome this, the transmission system uses gears of different sizes to multiply torque and allow the car to be driven at varying loads and speeds.

In order to multiply torque, a small input gear is engaged with a large output gear. The large output gear will turn slower than the input gear. As speed is reduced torque is increased. This is known as mechanical advantage.

The mechanical advantage gained through the rotational leverage is then transmitted to the road wheels.

Gear ratios

Gear ratios give an indication of how much torque is multiplied inside the gearbox. The gear ratio is a comparison of the number of teeth on the driving gear and on the driven gear (input and output).

To calculate a gear ratio, use this equation:

$$\text{Ratio} = \frac{\text{Driven}}{\text{Driver}}$$

Divide the number of teeth on the output (driven) gear by the number of teeth on the input (driver) gear.

Example:

The input gear has 16 teeth and the output gear has 32 teeth. So the input gear will need to turn two complete revolutions to make the output gear turn one complete revolution. This gives a gear ratio of 2:1.

- This 2:1 gear ratio means that speed is reduced by half (for example, an input turning at 200 rpm will give an output of 100 rpm).
- On the other hand, this gear ratio of 2:1 doubles the output torque or turning effort (for example, an input torque of 200 **Newton metres** will give an output torque of 400 Newton metres).

Gear ratios of compound gear sets

In practice, a gearbox uses a combination of gears to achieve torque multiplication. When drive is sent through a number of gears it is known as a compound gear set.

The equation will only give you the gear ratio for one set of gears. So, to calculate the overall gear ratio of a compound set, you need to calculate the ratio of each set of gears individually and then multiply them together.

> **Key term**
>
> **Newton metres** – unit of measurement for torque or turning effort.

$$\text{Ratio} = \frac{\text{Driven}}{\text{Driver}} \times \frac{\text{Driven}}{\text{Driver}}$$

Example:

Input gears: Driven gear has 48 teeth, Driver gear has 24 teeth (Ratio of 2:1)

Output gears: Driven gear has 36 teeth, Driver gear has 12 teeth (Ratio of 3:1)

Total gearbox ratio = Input gear ratio × Output gear ratio

For this example:

Total gearbox ratio = 2 × 3. This gives a total gearbox ratio of 6:1.

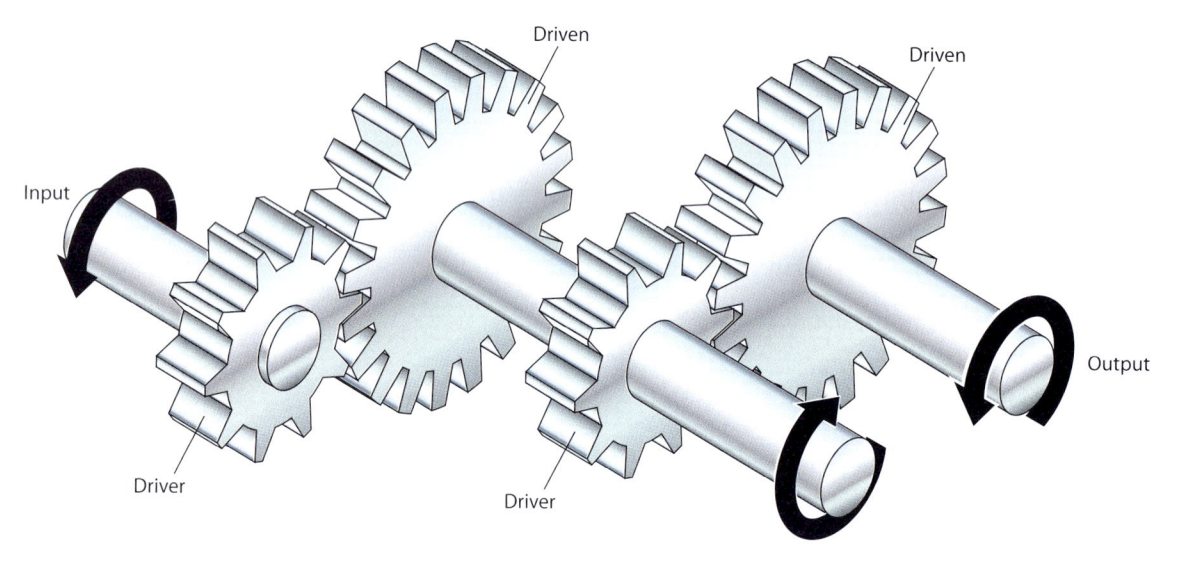

Figure 5.30 Gear ratios for a compound gear set

Gear ratios in the real world

In the examples given, an even number of teeth are used (16 and 32), but in reality gears can have odd numbers of teeth. When this is the case and you use the equation to calculate a gear ratio, you will get an answer containing a decimal point.

The reason for using odd numbers of teeth is that as the gears rotate against each other, it is less common for the same teeth on each gear to come into contact. This reduces overall wear and improves the lifespan of the individual gears.

By varying the number of teeth on the input and output gears, a set of ratios for all driving conditions can be achieved.

Torque multiplication

You can use gear ratios to calculate the final torque multiplication and output speed.

- Calculate the torque multiplication using the following formula:

  Input torque × Gear ratio = Output torque

- Calculate the output speed using the following formula:

$$\frac{\text{Input speed}}{\text{Gear ratio}} = \text{Output speed}$$

Don't forget that as torque is multiplied speed is reduced, and as speed increases torque is reduced.

> **CHECK YOUR PROGRESS**
>
> Calculate the following compound gear ratios, including the final output speed and torque multiplication.
>
> The drive from the input shaft of the gearbox turns a gear with 26 teeth. This gear is in constant mesh with another gear with 51 teeth. Drive is then transmitted along a countershaft to another gear with 17 teeth. This gear is in constant mesh with an output gear with 45 teeth. The engine crankshaft is turning at 3000 rpm, with an input torque of 100 Newton metres.
>
> 1 What is the overall gear ratio?
> 2 What is the output speed of the gearbox in rpm?
> 3 What is the output torque of the gearbox in Newton metres?

Light vehicle driveline systems

The light vehicle driveline system is one that takes the turning effort from the output shaft of the gearbox and transmits it all the way to the wheels.

The three main driveline layouts used on cars are:

- front-wheel drive
- rear-wheel drive
- four-wheel drive (also known as all wheel drive).

Depending on how the vehicle will be used, different manufacturers use different forms of layout.

> **Find out**
>
> Using sources of information available to you, find one light vehicle with each of the following drive layouts:
>
> - front-wheel drive
> - rear-wheel drive
> - four-wheel drive.
>
> Write down the make and model of each vehicle you find.

Front-wheel drive

Front-wheel drive layouts are useful for small passenger cars. Since there is no need to design a transmission tunnel in the floor of the passenger compartment, occupant space can be kept as large as possible.

Level 2 Light Vehicle Maintenance & Repair

> **Key terms**
>
> **Transaxle** – type of gearbox which also contains the final drive unit. It is often turned transversely (sideways) across the car and is commonly used with front-wheel drive vehicles.
>
> **Final drive** – transmission unit that houses the final fixed gear reduction. It is manufactured from the crown wheel, pinion and the differential unit.
>
> **Differential** – unit that allows one wheel to travel faster than the other when turning a corner.
>
> **Traction** – the friction between a body (vehicle) and the (road) surface on which it moves.

- If a **transaxle** is used, the gearbox will only have an input and output shaft.
- The output of the gearbox has a pinion gear, meshed with a crown wheel in the **final drive**.
- The crown wheel drives the **differential**, which splits the torque from the engine to each of the front wheels.
- This turning effort is transferred down drive shafts to the front wheels.
- To allow for steering and suspension movement, constant velocity (CV) joints are used at either end of the drive shafts.

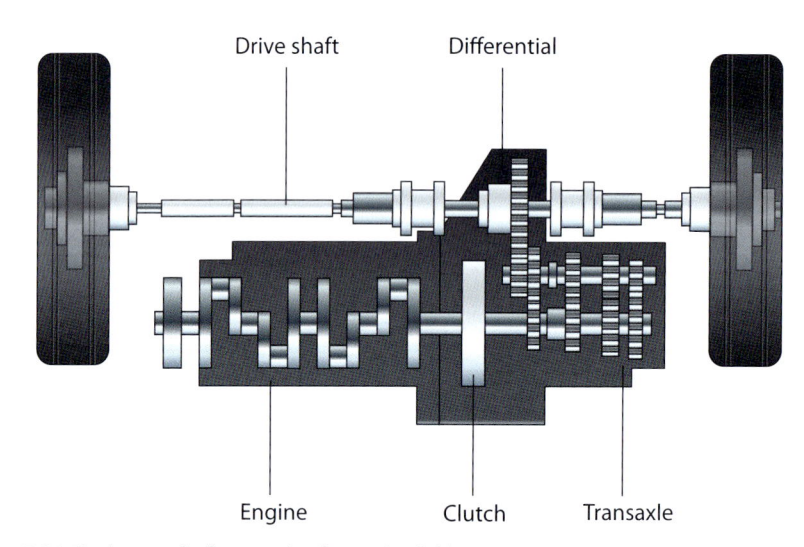

Figure 5.31 The layout of a front engine front-wheel drive car

Rear-wheel drive

Rear-wheel drive layouts are more common for slightly larger passenger cars, where the driveline travelling through the floor area is not so much of an issue.

- The engine and gearbox of many rear-wheel drive cars is mounted longitudinally (front to rear).
- The transmission usually contains an input shaft, a countershaft and an output shaft.
- As drive leaves the rear of the gearbox, it is transferred to the back of the car along a propeller shaft (usually called a prop shaft).
- The prop shaft normally has a universal joint at each end to allow for suspension movement.
- The prop shaft joins the rear axle at the final drive unit.
- Inside the final drive unit a pinion gear meshes with a crown wheel, turning the drive through 90 degrees.
- The crown wheel rotates the differential casing, which splits the turning effort from the engine and transfers it along drive shafts to the rear wheels.

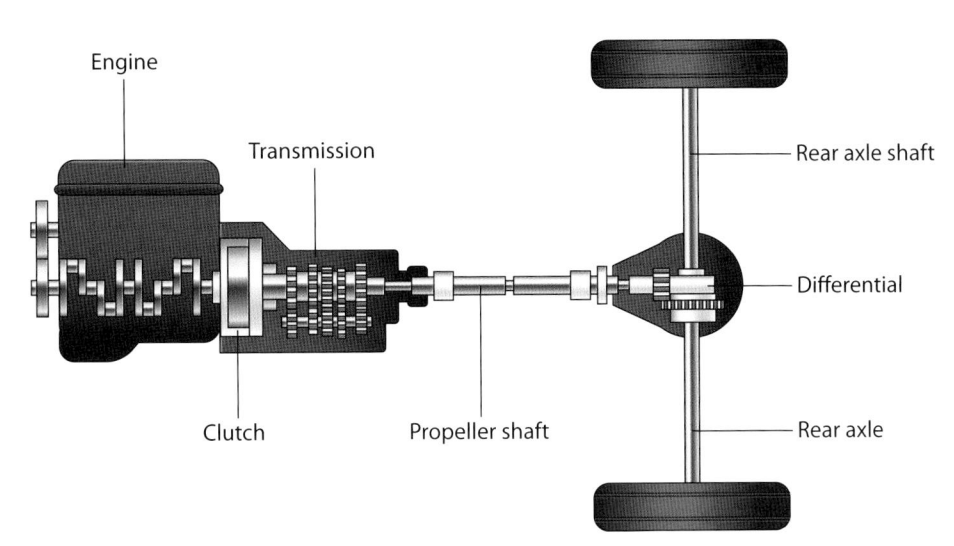

Figure 5.32 The layout of a front engine rear-wheel drive car

Four-wheel drive

Four-wheel drive layouts are used when large amounts of **traction** are required. Cars that are designed to be used off-road on loose surfaces and some high performance vehicles will require the largest possible amount of traction to be transferred to the road surface through all four wheels.

- A four-wheel drive vehicle can have the engine and gearbox mounted transversely (sideways across the car) or longitudinally (front to rear).
- When drive leaves the back of the gearbox, it normally enters an additional unit called a transfer box. This is another small gearbox that splits the drive to the front and rear axles.
- Two prop shafts are normally connected to the transfer box through universal joints, one leading to the front axle and one leading to the rear axle.
- At the front and rear axles, the turning effort is split through final drive and differential units and on to the wheels via drive shafts.

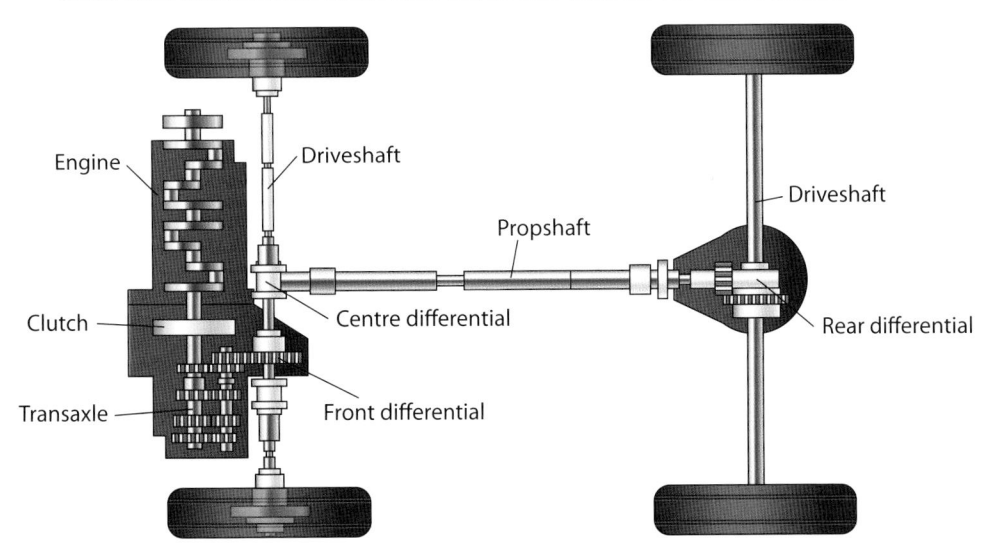

Figure 5.33 The layout of a front engine four-wheel drive car

Light vehicle driveline components

Once the gearbox has multiplied the torque from the engine, the turning effort must be transferred to the wheels. Depending on transmission layout (front-wheel drive, rear-wheel drive or four-wheel drive), drive shafts or prop shafts are used.

Propeller shaft

The propeller shaft is commonly known as a **prop shaft**. It is used on front engine rear-wheel drive vehicles. The prop shaft is simply a metal tube, strong enough to transmit the full power of the engine and the torque multiplied by the gearbox. The shaft is connected to the back of the gearbox and runs underneath the floor to join it to the back axle.

At each end of a prop shaft, a **universal joint (UJ)** is needed. As the suspension moves up and down a difference in height occurs between the rear axle and the gearbox. Using universal joints means that the drive is still transmitted without bending the prop shaft, as shown by Figure 5.34. (Some prop shafts are split into two, with an additional universal joint mounted in the middle.)

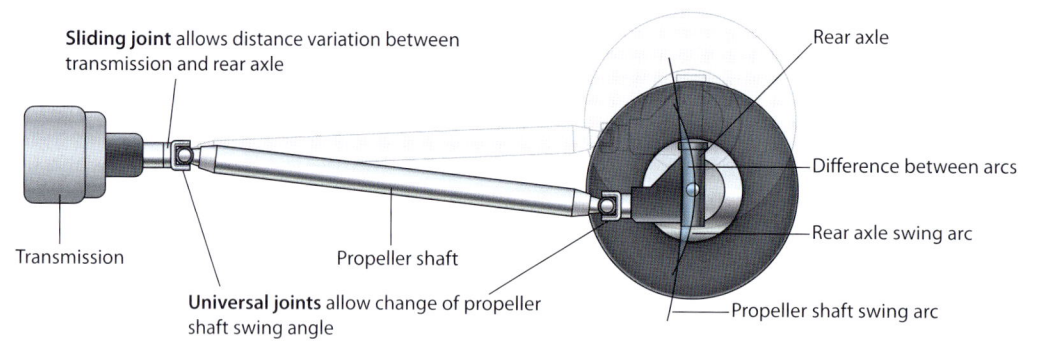

Figure 5.34 Why universal joints are needed

The most common type of universal joint is the Hooke's UJ (see Figure 5.35). This is made up of two yokes pivoted on a central crosspiece, sometimes called a spider. The spider is formed by two pins crossing over each other at right angles. The yokes, one on the input shaft and the other on the output shaft, are connected to the spider so that they are at right angles to each other. This arrangement allows the input and output shafts to rotate together even when their axes are at different angles.

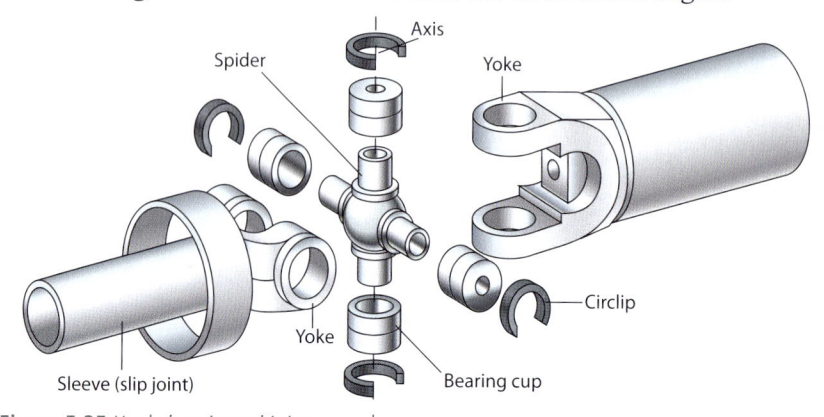

Figure 5.35 Hooke's universal joint

5 Light vehicle transmission & driveline units & components

As a universal joint turns through an angle, it will speed up and slow down. The universal joints need to be synchronised because they speed up and slow down as they travel through 90 degrees of revolution. The waveform shown in Figure 5.36 represents the speeding up and slowing down, which cancel each other out.

As the rear suspension moves up and down, this movement tries to stretch or compress the prop shaft. Because of this, a sliding joint is included to allow the prop shaft to get longer and shorter.

> **Working life**
>
> Matt has been replacing the universal joints on a vehicle's prop shaft.
>
> 1 What important procedure does he need to remember?
> 2 What are the consequences of not carrying out this procedure?

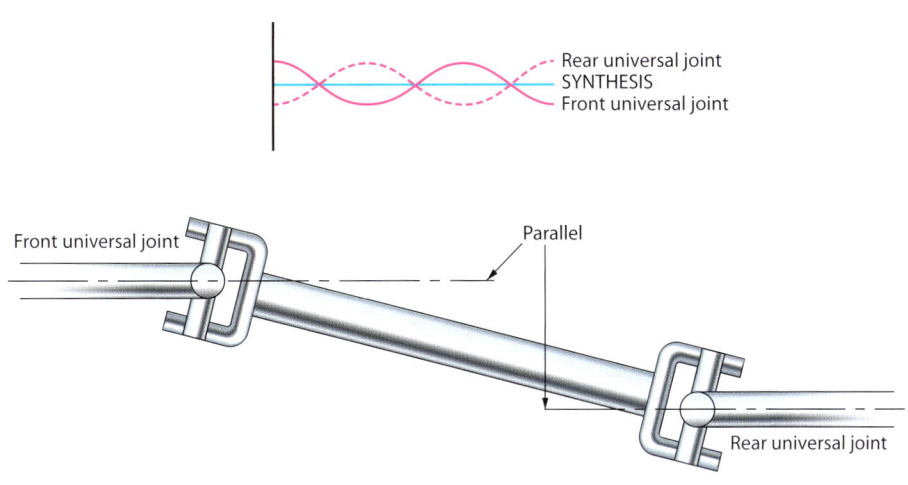

Figure 5.36 Universal joint synchronisation

Drive shafts

A drive shaft can be used with front engine front-wheel drive cars. Two shafts are placed across the width of the car and connect to the hub of the driving wheels. As with a prop shaft, these shafts must be able to cope with suspension movement and, in the case of front-wheel drive, steering movement as well.

CV joint

Since steering and suspension movement causes large angular movements, a universal joint would be unsuitable for use with front-wheel drive shafts. This is because the variations in speed would cause large amounts of vibration.

To replace the universal joint, a special type of coupling called a **constant velocity (CV) joint** is used. Different manufacturers use a number of different designs and styles of CV joint, but their job remains the same – to transmit drive at a constant speed regardless of steering and suspension movement.

The most common CV joint is the Birfield joint, which is based on a design patented in 1935 by the Ford engineer Alfred Hans Rzeppa. It transmits drive through a series of ball bearings housed in a cage (see Figure 5.37). Because ball bearings are spheres, it doesn't matter what angle or at which point the drive is transmitted through these balls – speed (velocity) is kept constant.

> **Key terms**
>
> **Prop shaft** – short form of the words propeller shaft.
>
> **Universal joint (UJ)** – mechanism that allows drive to be transmitted through an angle.
>
> **CV joint** – joint at the end of the drive shaft on a front-wheel drive vehicle that is able to transmit drive with no variation in speed.

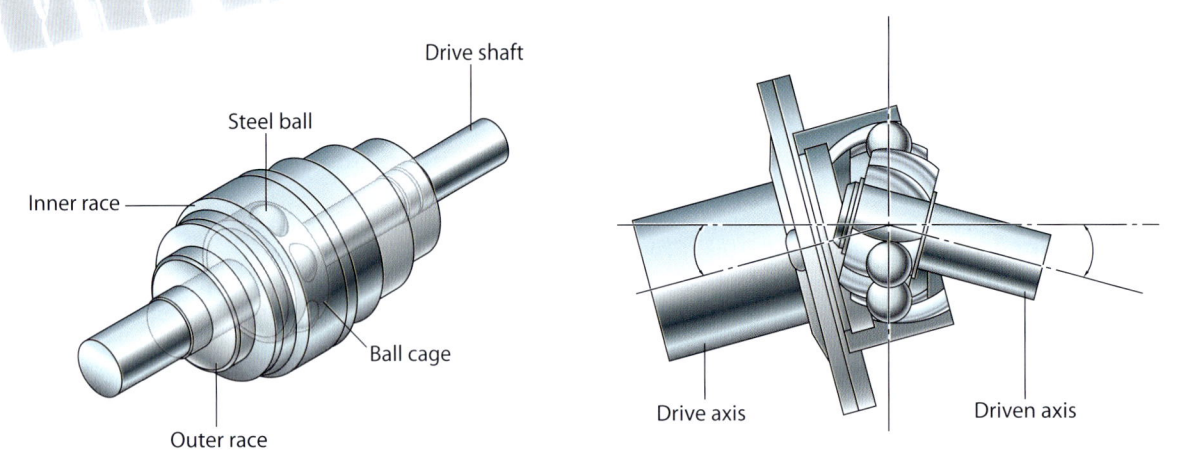

Figure 5.37 Constant velocity (CV) joint

CV joints are prone to wear because of the forces they undergo. They must be kept well lubricated with grease, so a rubber boot is used to keep grease in and water and dirt out. These rubber boots are often called 'drive shaft gaiters' or 'CV gaiters'.

Final drive

Having completed its drive through the gearbox, the turning effort is now transmitted to the **final drive** unit. The final drive unit contains a number of gears to further increase torque.

As with the gears in a gearbox, the final drive reduction depends on the number of teeth on the crown wheel and pinion. A typical figure for final gear reduction is approximately 4:1.

The final drive is the last stage in the transmission of power from the engine to the drive shafts that turn road wheels. A final drive unit normally consists of:

- **Crown wheel and pinion** – This provides a fixed **final gear reduction** and, in the case of a rear-wheel drive vehicle, will turn the drive through 90 degrees.
- **Differential unit** – This allows one wheel to travel faster than the other when turning a corner (see opposite).

Rear-wheel drive power flow

Figure 5.38 shows the flow of power in rear-wheel drive vehicles.

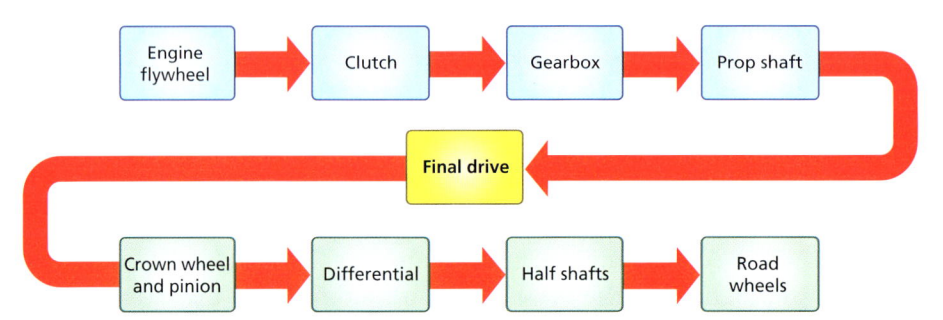

Figure 5.38 Rear-wheel drive power flow

> **Key terms**
>
> **Final drive** – the transmission unit which houses the crown wheel and pinion and the differential.
>
> **Crown wheel and pinion** – a large and small gear found in the final drive.
>
> **Final gear reduction** – the last torque multiplication before it is transmitted to the wheels.
>
> **Bevel gear** – gear with teeth that have a curved profile.
>
> **Hypoid** – in this type of design the final drive pinion is offset below the centre line of the crown wheel.

Front-wheel drive power flow

On front-wheel drive cars, the final drive is similar to that found on rear-wheel drive cars. The main difference is that there is no need for a prop shaft to take drive from the gearbox to the final drive. On cars with a transverse engine (where the engine is mounted across the car from left to right), the final drive doesn't need to turn the drive through a right angle. If the final drive and gearbox are manufactured into a single unit, this is often called a transaxle (see page 342). The crown wheel and pinion use helical-cut gears to provide the fixed final gear reduction (see Figure 5.39).

Figure 5.40 shows the flow of power in front-wheel drive vehicles.

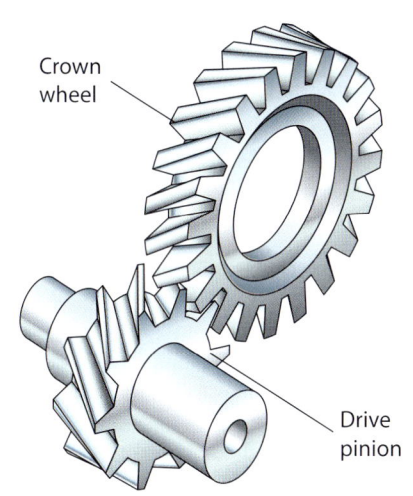

Figure 5.39 Crown wheel and pinion of a front engine front-wheel drive car

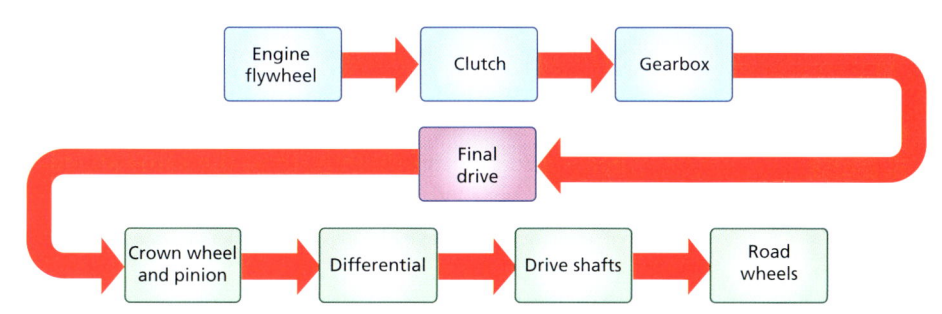

Figure 5.40 Front-wheel drive power flow

Bevel gears

If the crown wheel and pinion are used on a rear-wheel drive, a special type of gearing called **bevel gears** is needed. Most rear-wheel drive axles combine the bevel gear with a spiral **hypoid** design, as shown in Figure 5.41.

In this type of design, the bevel gear teeth are curved in a way that allows the pinion to be set below the centre line of the crown wheel. This means that the prop shaft can be set lower down.

The tunnel in the floor that houses the prop shaft can be made lower or may disappear altogether. This flatter floor area inside the car gives more space for passengers and lowers the centre of gravity of the vehicle, making it more stable on the road.

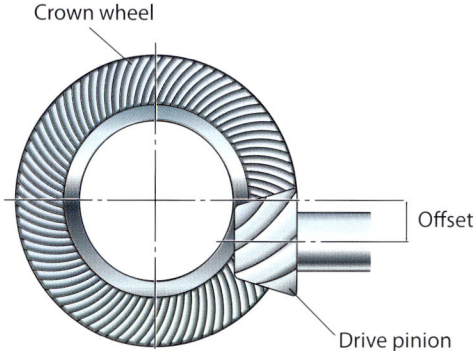

Figure 5.41 Hypoid rear-wheel drive crown wheel and pinion

Differential

When a vehicle is travelling along a straight piece of road, both of the driven wheels cover the same amount of ground at the same speed. When it comes to a bend, however, the inner driven wheel doesn't have to travel as far as the outer one, and so must travel more slowly. If both wheels rotated at the same speed when trying to turn the corner, the inner wheel would be forced into a skid.

On a bend, drive must be transmitted at different speeds. This is achieved by a differential unit, which is housed inside the final drive casing (see Figure 5.42).

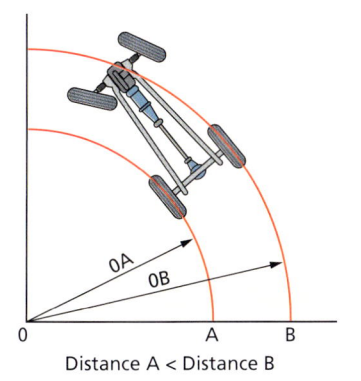

Distance A < Distance B

RPM of inside wheel < RPM of outside wheel

Figure 5.42 The need for a differential

347

Level 2 Light Vehicle Maintenance & Repair

When the car goes round a bend, the differential unit transfers some of the driving force from the inner wheel to the outer wheel, speeding the outer wheel up and slowing the inner wheel down. A differential unit achieves this by using a gearing system.

- The turning effort taken from the crown wheel is transmitted to the differential casing.
- A metal pin is fixed through the differential casing. As the differential casing turns, the drive pin moves end over end.
- On the drive pin are mounted two small gears (often called 'planet gears'). These are in constant mesh with two side gears (often called 'sun gears').

Travelling in a straight line

- When the vehicle is travelling in a straight ahead direction, the drive pin turns end over end, locking the planet gears directly to the side gears and driving them both at the same speed.
- If load is the same on both sides, all gears lock together, as shown in Figure 5.43.

Turning a corner

- As the car turns a corner, the extra load tries to slow down one wheel and reduces the speed at one sun gear.
- The drive pin still turns end over end, providing torque to the sun gears, but the planet gears now rotate on the pin.
- This allows more drive to be transmitted to one wheel than the other. The result is that one wheel travels faster than the other, but still transmits drive with the same amount torque.

Figure 5.44 shows that when travelling in a straight line, equal speed is delivered to both drive shafts, and when a left-hand corner is turned, the greater speed is delivered to drive shaft B.

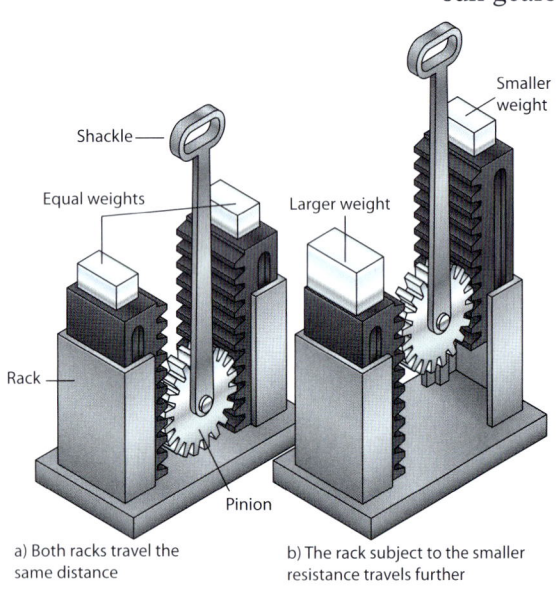

a) Both racks travel the same distance
b) The rack subject to the smaller resistance travels further

Figure 5.43 How loads affect the operation of a differential unit

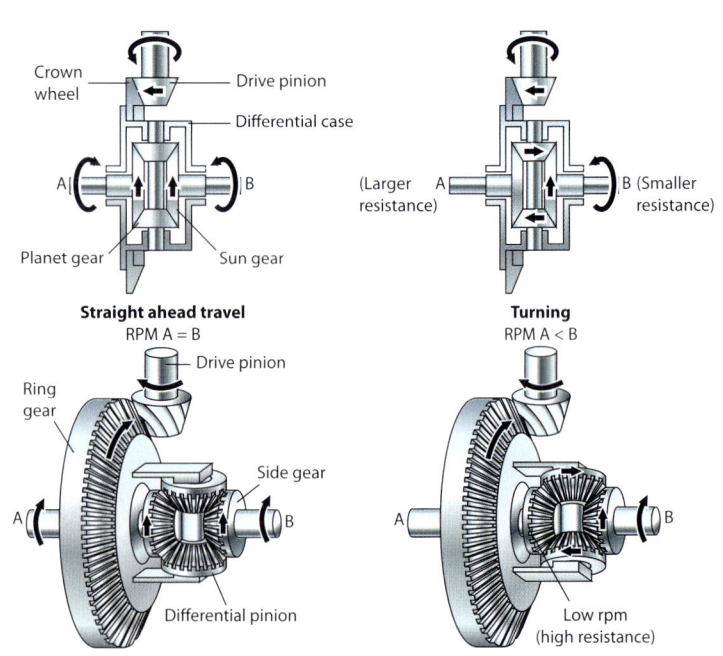

Straight ahead travel RPM A = B
Turning RPM A < B

Figure 5.44 Differential gears under different driving circumstances

Limited slip differentials (LSD)

A limited slip differential is designed to transmit an equal torque to both driving wheels when the car is travelling in a straight ahead direction, while still allowing differential action when going round a corner. Because of this, if one wheel loses traction, the other will still have some drive transmitted and this will give greater vehicle control.

Three main designs of limited slip differential are in common use:

1. **Clutch operation** – A series of multi-plate clutches are included in the design of this

differential unit. When travelling in a straight ahead direction, forces acting on the differential unit operate these clutches, clamping them together to create friction and transmit drive to both road wheels. As the car turns a bend, the load on the clutches is reduced, allowing them to slip. This lets normal differential action take place.

2. **Viscous coupling** – A chamber inside the differential is constructed that contains a viscous (thick) liquid and a series of rotor blades. When travelling in a straight ahead direction, the viscous liquid creates drag and reduces slip in the differential gears. As the car turns a bend, normal differential action can take place because the rotor blades are able to shear through the viscous liquid.

3. **Torsion wheel** – A series of **worm gears** are included in the design of this differential unit. In a worm and wheel gear set up, the worm can easily drive a gear, but the gear is unable to drive the worm (this will allow drive in one direction only). When travelling in a straight ahead direction, gear teeth lock against the worm drive, providing equal torque to both wheels at the same time. As the car turns a bend, one wheel will slow down while the other speeds up. When this happens, the worm turns the gear, allowing one wheel to travel faster than the other and provide normal differential action.

> **Key terms**
>
> **Worm gear** – gear consisting of a spiral threaded shaft and a wheel with teeth that mesh into it.
>
> **Transfer box** – mechanism used on four-wheel drive cars to split the drive between the front and rear axles.
>
> **Speedometer drive** – small gear or electrical pickup used to measure vehicle speed for the dashboard console.

Final drive lubrication

Lubrication of final drive systems is usually achieved using splash feed. To stop the oil being squashed out from between the gear teeth, a special type of oil must be used. These specialist transmission and final drive oils are often known as extreme pressure (EP) oils and they usually have a high viscosity grade. (For more on viscosity grades, see the section on *Classification of lubricants* in Chapter 3, pages 230–233.)

Transfer box

In a four-wheel drive vehicle, the engine can be mounted transversely (sideways) or longitudinally (in a straight line) depending on the manufacturer's design. As drive exits the gearbox, an extra unit called a **transfer box** is often used. This splits the drive so that it can be used by the front and rear axle. A four-wheel drive car has at least two differential assemblies, one for the front wheels and one for the back. In addition to this, cars with permanent four-wheel drive have a central differential that splits the drive from front to rear.

When a car is travelling with forward motion, the front axle reaches a bend first. Because of this the front axle needs to be travelling at a different speed from the rear axle. A central differential is sometimes used. This operates in the same way as a standard differential, but instead of allowing a difference in speed between the right-hand and left-hand wheels, it allows a difference in speed between the front and rear axle.

Speedometer drive

Some transmission and final drive units may include a **speedometer drive**. This can be mechanical or electronic.

Level 2 Light Vehicle Maintenance & Repair

> **Did you know?**
> Many modern cars use electronic information provided by the ABS wheel speed sensors or a dedicated vehicle speed sensor (VSS) to operate the speedometer on the dashboard.

- Mechanical systems normally use a worm gear drive mounted by the output of the gearbox or on the final drive unit. When the car is moving, this worm gear transmits the output speed through a speedometer cable to a gauge on the dashboard.
- Electronic systems normally use a sensor and pickup mounted in a similar position to the mechanical drive. When the car is moving, an electronic signal is produced which can be converted by an ECU to represent the vehicle speed on the dashboard.

The electrical and electronic components, including reverse lamp switch

In order that other road users are aware of the direction of travel, many transmission systems include a method for illuminating a reversing lamp. This is normally done by using a plunge-type switch, mounted in the side of the gearbox casing. When reverse gear is selected, the switch comes into contact with the selector mechanism. This closes the switch and completes an electric circuit, which will turn on a white light at the rear. (White light is used as it indicates that the vehicle is travelling towards you.)

Types of wheel bearing arrangements

In rear-wheel drive arrangements, there are three main methods of supporting the drive shaft, hub and wheels within the axle casing: semi-floating, three-quarter floating and fully floating.

- **Semi-floating axle** – Each drive shaft is supported at its inner end by a bearing that also carries the final drive unit. At the outer end there is a bearing between the shaft and the inside of the axle housing. In this arrangement the half shaft has to support the bending load created by the weight of the car, as well as transmitting the turning effort to the wheels.

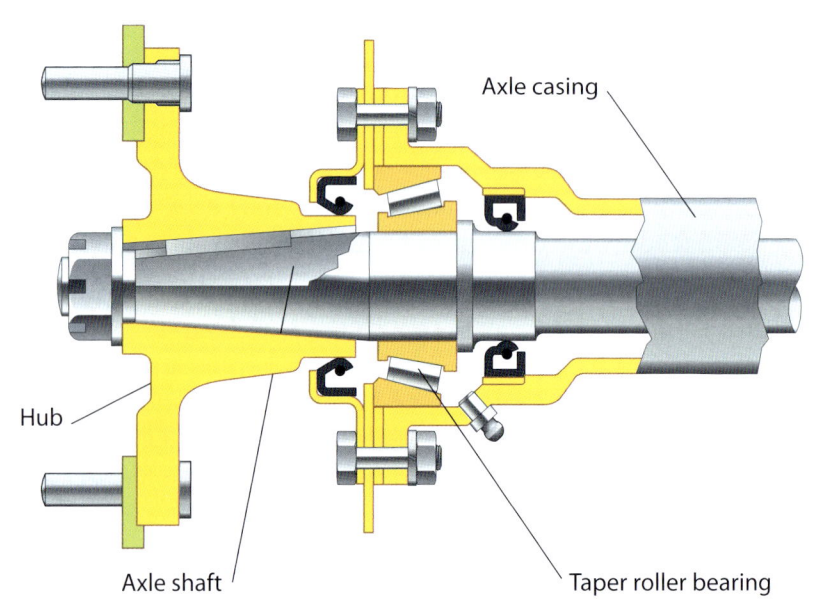

Figure 5.45 Semi-floating axle bearing arrangement

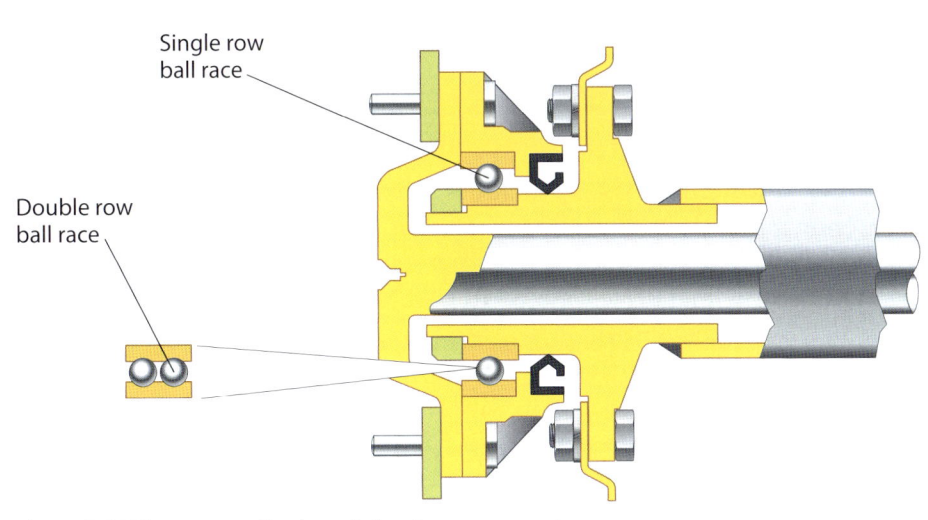

5 Light vehicle transmission & driveline units & components

Figure 5.46 Three-quarter floating axle bearing arrangement

- **Three-quarter floating axle** – The inner end is the same as on the semi-floating design, but the outer end is different. Instead of the bearing being between the drive shaft and the axle housing, it is mounted between the wheel hub and the axle housing. The advantage of this arrangement is that the bearing now supports the weight of the car instead of the drive shaft, so the drive shaft is now only subjected to bending force during hard cornering. Cars that have independent rear suspension usually have two bearings in the hub, instead of the one bearing found on cars with non-independent rear suspension.
- **Fully floating axle** – This is a more complex and expensive design. In the fully floating axle, there are two bearings between each hub and the axle housing. Because of the extra support provided by these two bearings, they will carry the weight of the car and any forces trying to bend the drive shaft during cornering.

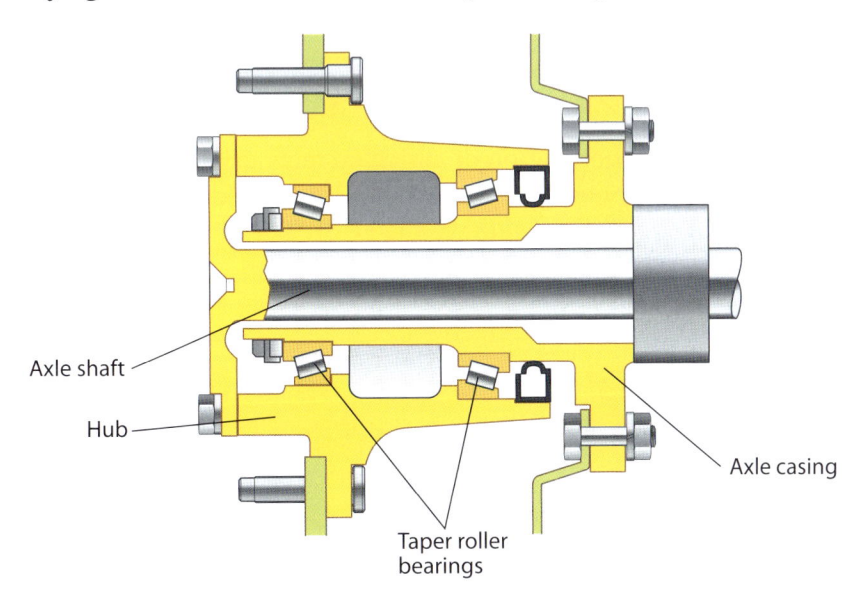

Figure 5.47 Fully floating axle bearing arrangement

Key engineering principles that are related to light vehicle driveline systems

Final drive and overall gear ratios

The final drive unit normally includes a fixed final gear reduction of approximately 4:1. To find the overall gear ratio of a car it is important to include its final gear reduction in the calculation. You therefore need to multiply the final drive ratio by the overall gearbox ratio:

Overall gear ratio = Gearbox ratio × Final drive ratio

Simple stresses

Three main types of stress exist in the driveline system:

1. torsional (twisting or turning)
2. bending
3. shear force (which tries to cut through the driveline components in a manner similar to a pair of scissors slicing through).

CHECK YOUR PROGRESS

1 Name three different driveline layouts.
2 Why are differentials required?
3 What does the term hypoid mean?

How to test and maintain vehicle transmission and driveline systems

Regular maintenance of transmission and driveline systems is very important for a long and fault-free service life. Even if these systems are maintained, the high levels of stress and strain involved with operation will eventually lead to wear and tear.

Testing and maintaining clutch systems

Testing for the correct operation of clutch systems usually involves putting the mechanism under strain. If the clutch needs replacement, it will normally be necessary to remove the gearbox.

Safe working

- If the engine is run in the workshop, make sure that you use exhaust extraction to remove harmful fumes.
- Do not test for clutch slip if anyone is likely to be in front or behind the car.
- Do not exceed the safe working loads of any mechanical lifting equipment.
- You must wear appropriate PPE to protect yourself from any clutch dust.
- The gearbox and components can be heavy, so make sure you use correct manual handling techniques.

Key term

Free play – a small amount of movement in order for components to adjust slightly.

Checking for slip and drag

Checklist			
PPE	**VPE**	**Tools and equipment**	**Source information**
• Steel toe-capped boots • Overalls • Latex gloves	• Wing covers • Steering wheel covers • Seat covers • Floor mat covers	• Wheel chocks	• Manufacturer's technical data

1. Start the engine.

Select a high gear. With the handbrake applied and the wheels chocked, try to take up drive.

2. After some practice you should become competent in making judgements about the condition from the 'feel' of the clutch as it begins to bite.

3. You can normally diagnose slip if the engine revs rise but drive to the rest of the transmission system is limited.

4. You can normally diagnose drag if the clutch is pushed fully to the floor but gear selection is difficult and the car tries to move even with the clutch disengaged.

5. Both slip and drag can sometimes be repaired with the correct adjustment of clutch **free play**.

6. If clutch free play is correct, it may require inspection and replacement of clutch components.

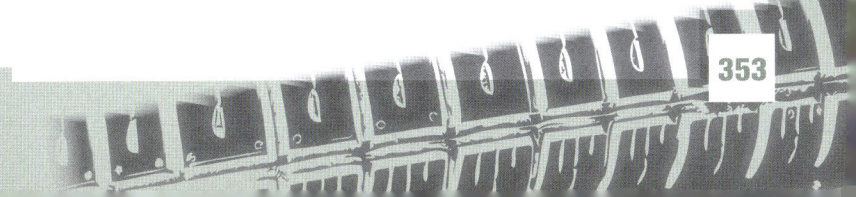

Replacing the clutch mechanism

Checklist			
PPE	**VPE**	**Tools and equipment**	**Source information**
• Steel toe-capped boots • Overalls • Latex gloves • Goggles • Particle mask	• Wing covers • Steering wheel covers • Seat covers • Floor mat covers	• Vehicle hoist/ramp • Spanners • Sockets • Screwdrivers • Hammers • Clutch aligner • Transmission jack • Torque wrench	• Vehicle workshop manual • Manufacturer's technical data

1. Raise the car so that you can gain access to the transmission components.

2. As you dismantle the components of the clutch, it is important to visually inspect each part as you remove it. This will give an indication of where the fault might lie.

3. It is recommended that if one component of the clutch requires replacement, you should change the three main parts at the same time. These are:

- the clutch cover
- the friction plate
- the clutch release bearing.

4. When you reassemble the clutch, the friction plate must be accurately aligned in the centre of the engine's flywheel. Use alignment tools to help you keep the clutch friction plate in position.

5. As you reattach the clutch cover to the flywheel, make sure you tighten the securing bolts in the correct order so that the pressure plate clamps evenly against the friction plate.

6. Now replace the old clutch release bearing and reassemble the gearbox to the back of the engine. Make sure that the input shaft is correctly located in the centre of the friction plate.

7. Once you have reassembled the gearbox and reconnected the clutch operating mechanism, adjust the clutch free play.

8. Road test the car to check for correct operation.

5 Light vehicle transmission & driveline units & components

Testing and maintaining gearboxes

To check the correct operation of a vehicle's gearbox, it is often necessary to road test the car under various driving conditions. This will allow you to check all gears for correct selection, transmission of drive and unusual noises.

The road test should help you identify the area in which the fault might lie. Any components that you find to be damaged or worn should be replaced.

If you find a fault, it might be necessary to remove the gearbox and strip out the internal components. As you dismantle the gearbox, make sure that you lay the removed parts out in order, under clean conditions, so that when you reassemble the gearbox you can be sure that everything is refitted correctly.

Once reassembled, it is important to use the correct grade and quantity of lubricating oil.

The oil level in manual and automatic transmissions, including final drives, will be affected by temperature. As the oil warms up, its level will increase. Transmission oil should be checked when cold, unless specified by the manufacturer. Some automatic transmission oil dipsticks may have different measurements marked on them for checking hot or cold.

Figure 5.48 A gearbox oil measurement being taken

Testing and maintaining the driveline

Checking for CV joint wear

Checklist			
PPE	VPE	Tools and equipment	Source information
• Steel toe-capped boots • Overalls • Latex gloves	• Wing covers • Steering wheel covers • Seat covers • Floor mat covers	• Vehicle hoist/ramp • Inspection lamp	• Vehicle workshop manual • Manufacturer's technical data

1. On a front-wheel drive car that uses drive shafts and constant velocity (CV) joints, carry out an inspection to make sure that the CV boots are in good condition and not split.

2. You can usually assess any wear in the CV joint by driving the car slowly with the steering on a full lock. As the car moves, you will normally hear any wear in the CV joint as a clicking noise, which might disappear when the steering is straightened up.

Safe working

When testing for CV joint wear, make sure you are in a suitable test area where you can drive safely in circles, for example in a closed car park.

When performing the check given in Step 3, make sure you do it in an area that doesn't pose a hazard to other road users or pedestrians.

3. Drive the car in both left-hand and right-hand circles, to ensure a correct diagnosis.

4. If the drive shaft gaiter is split or the CV joint is noisy, they should be replaced.

Drive shaft gaiter

To replace a split or damaged CV gaiter you might need to remove the drive shaft.

Replacing the drive shaft gaiter

Checklist			
PPE	**VPE**	**Tools and equipment**	**Source information**
• Steel toe-capped boots • Overalls • Latex gloves	• Wing covers • Steering wheel covers • Seat covers • Floor mat covers	• Oil collection pan • Sockets • Spanners • Screwdrivers • Side cutters • Workbench and vice • Torque wrench	• Vehicle workshop manual • Manufacturer's technical data

1. Gain access to and remove the drive shaft. (Place an oil collection pan beneath the gearbox to catch any gear oil lost during the drive shaft removal.)

2. Once you have removed the shaft, you can cut away the old gaiter and disconnect the CV joint (refer to the manufacturer's procedures). Replace the CV joint at this stage if necessary.

3. Replace the gaiter, clean and relubricate the CV joint with fresh grease and refit the assembly.

4. Make sure that all items are correctly aligned, and that any nuts and bolts are tightened to the manufacturer's recommended torque setting.

> **! Safe working**
>
> When the drive shaft is removed, gear oil can leak from the transmission. Clean up any spills immediately to reduce the risks of slips, trips and falls.
>
> Use barrier cream or latex gloves to protect your skin from CV joint grease.

5 Light vehicle transmission & driveline units & components

Prop shafts

Prop shafts will usually require little maintenance, but over a period of time the universal joints might develop wear and require replacement.

If the universal joints are out of sequence, or the prop shaft does not get reassembled to its original position, vibrations can occur in the transmission system.

> **Safe working**
> Do not exceed the safe working load of any mechanical lifting equipment.

Prop shaft universal joint replacement

Checklist			
PPE	**VPE**	**Tools and equipment**	**Source information**
• Steel toe-capped boots • Overalls • Latex gloves	• Wing covers • Steering wheel covers • Seat covers • Floor mat covers	• Vehicle hoist/ramp • Workbench and vice • Sockets • Spanners • Hammers • Torque wrench	• Vehicle workshop manual • Manufacturer's technical data

1. Remove the prop shaft. Make sure that you have marked its original position so that when you reassemble the component it can go back in the same place.

2. Remove the universal joints and replace them following the manufacturer's instructions.

3. Make sure that when you remove or refit universal joints, you keep the position of their yokes the same to maintain synchronisation.

Level 2 Light Vehicle Maintenance & Repair

Safe working

Always use correct manual handling techniques when removing and refitting heavy final drive components.

Final drive and differential unit

To assess the condition of the final drive and differential unit, it is often necessary to road test the car. The car should be driven under different conditions, especially acceleration and overrun, to make sure that no slack exists in the system which will create a whining noise. If you hear unusual noises, you will normally need to strip out the differential and final drive unit to check for wear.

Checking final drive units

Checklist			
PPE	VPE	Tools and equipment	Source information
• Steel toe-capped boots • Overalls • Latex gloves	• Wing covers • Steering wheel covers • Seat covers • Floor mat covers	• Vehicle hoist/ramp • Inspection lamp • Torque wrench	• Vehicle workshop manual • Manufacturer's technical data

1. Carry out a visual inspection to check for damage to any gear teeth.

2. Examine the components for wear and the contact position of gear teeth.

3. Apply a dye, such as 'engineer's blue' to the teeth. Then rotate them against each other to see their exact position. (They should align in the centre of the teeth.)

4. If the contact position of gear teeth is incorrect, you should follow the manufacturer's instructions for adjustment.

5. Once reassembled, make sure that you use the correct grade and quantity of final drive transmission oil to top up the system.

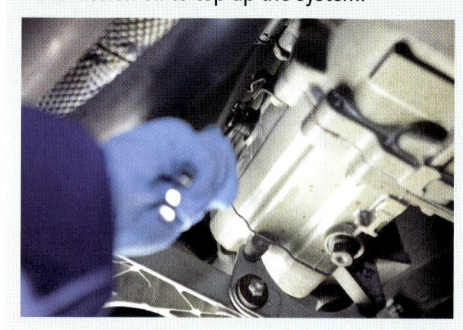

6. Road test the vehicle to check for correct function and operation.

358

FINAL CHECK

1 Which of the following is not a wheel bearing arrangement?
   a all floating
   b fully floating
   c three-quarter floating
   d semi-floating

2 Which of the following are components in a torque converter?
   a stator
   b impeller
   c turbine
   d all of the above

3 In a gearbox, a small input driver gear will turn a large output driven gear with:
   a reduced torque and reduced speed
   b higher torque and reduced speed
   c higher speed and higher torque
   d higher speed and reduced torque

4 Air bubbles in a hydraulic clutch system may cause:
   a leaks
   b corrosion
   c slip
   d drag

5 Limited slip differentials can be made using:
   a torsion wheels
   b clutches
   c viscous couplings
   d all of the above

6 Final drive lubrication is usually achieved by:
   a dry sump lubrication
   b total loss
   c splash feed
   d pressure feed

7 Torque converters create the greatest amount of torque multiplication:
   a when turbine and impeller are turning at the same speed
   b at coupling point
   c when the lock-up clutch is engaged
   d when turbine and impeller are turning at different speeds

8 Overall gear ratio can be calculated using:
   a Gearbox ratio − Final drive ratio
   b Gearbox ratio + Final drive ratio
   c Gearbox ratio × Final drive ratio
   d Gearbox ratio ÷ Final drive ratio

9 CVT stands for:
   a constant variation transmission
   b continuously variable torque
   c continuously variable transmission
   d constant variation torque

10 The component that helps keep gears in engagement is:
   a detent
   b interlock
   c baulk ring
   d selector fork

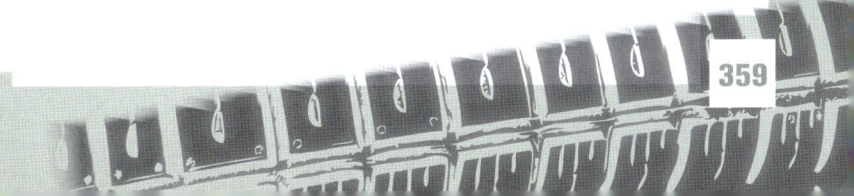

GETTING READY FOR ASSESSMENT

The information contained in this chapter, as well as continued practical assignments in your centre or workplace, will help you to prepare for both the end-of-unit tests and diploma multiple-choice tests. This chapter will also help you to prepare for working on and maintaining light vehicle transmission and driveline systems safely.

You will need to be familiar with:

- The operation and maintenance of clutch systems
- The operation and maintenance of manual transmission systems
- How to calculate compound gear ratios
- How to remove and refit light vehicle driveline components
- Common faults found in light vehicle transmission and driveline systems
- General maintenance procedures for light vehicle transmission and driveline systems

This chapter has given you an introduction and overview of the maintenance and repair of light vehicle transmission and driveline systems, providing the basic knowledge that will help you with both theory and practical assessments.

Before you try a theory end-of-unit test or multiple-choice test, make sure you have reviewed and revised any key terms that relate to the topics in that unit. You will need to read all the questions carefully. Take time to digest the information so that you are confident about what the question is asking you. With multiple-choice tests, it is very important that you read all of the answers carefully, as it is common for two of the answers to be very similar, which may lead to confusion.

For practical assessments, it is important that you have had enough practice and that you feel that you are capable of passing. It is best to have a plan of action and work method that will help you. Make sure that you have enough technical information, in the way of vehicle data, and appropriate tools and equipment. It is also wise to check your work at regular intervals. This will help you to be sure that you are working correctly and to avoid problems developing as you work.

When you are doing any practical assessment, always make sure that you are working safely throughout the test. Light vehicle driveline systems are dangerous and your precautions should include:

- Isolate electrical components located near the transmission.
- Where possible, use mechanical lifting equipment for moving heavy transmission components such as transmission jacks and hoists.
- Use correct manual handling techniques when lifting and carrying.
- Use appropriate cleaners to safely remove clutch dust and prevent it from becoming airborne.
- Observe all health and safety requirements and use the recommended PPE and VPE.
- Make sure that you use all tools correctly and safely.

Good luck!

6 Light vehicle electrical units & components

This chapter will help you to develop an understanding of the electrical and electronic systems used on light vehicles. It covers the principles behind these systems as well as their construction. You will also learn about the procedures you can use when removing, replacing and testing electrical systems for correct function and operation. This chapter provides you with knowledge that will help you with both theory and practical assessments. It will help you plan a systematic approach to inspection and maintenance by explaining the methods used with light vehicle electrics.

This chapter covers:

- Safe working on light vehicle electrical units and components
- Light vehicle electrical and electronic principles
- Light vehicle starting and charging systems operation
- Light vehicle auxiliary electrical systems operation
- How to check, remove, replace and test light vehicle electrical systems and components

WORKING PRACTICE

When you are working with light vehicle electrical and electronic systems, the main hazard is the possibility of electric shock. Although most light vehicle electrical systems operate with low voltages of around 12V, an accidental electrical discharge can be caused by incorrect circuit connection. This can be enough to cause severe burns. Some electrical components, such as car batteries, contain dangerous chemicals or acids. If you allow these to come into contact with your skin, they can cause injury. Your selection of personal protective equipment (PPE) and your manner of working on electrical systems should protect you from these hazards.

Personal Protective Equipment (PPE)

Safety helmet protects the head from bump injuries when working under cars.

Overalls provide protection from coming into contact with oils and chemicals.

Safety gloves provide protection from oils and chemicals. They also protect the hands when handling objects with sharp edges.

Barrier cream protects the skin from old engine oil, which can cause dermatitis and may be carcinogenic (a substance that can cause cancer).

Safety goggles reduce the risk of small objects or chemicals coming into contact with the eyes.

Safety boots protect the feet from a crush injury and often have oil- and chemical-resistant soles. Safety boots should have a steel toe-cap and steel mid-sole.

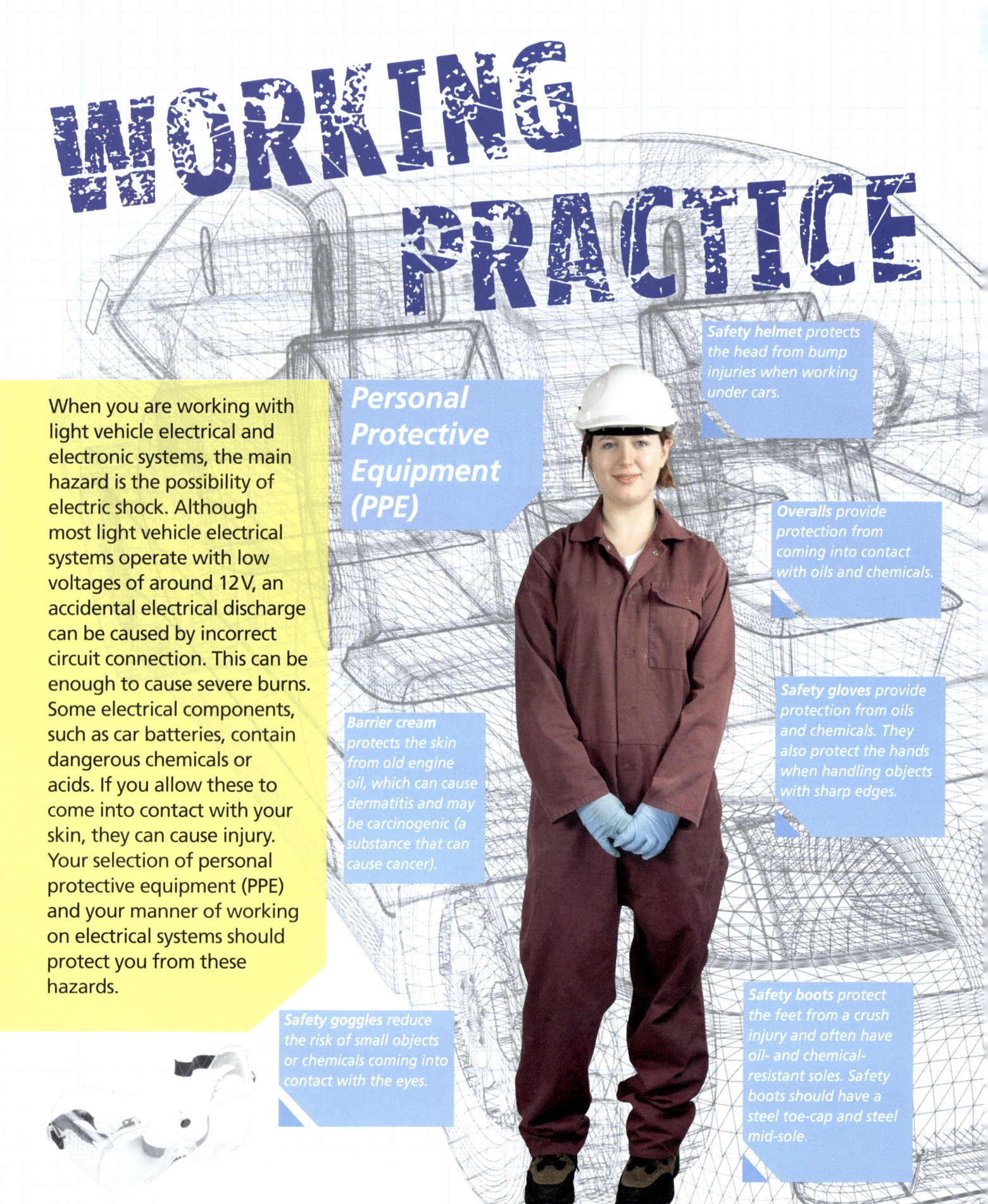

To reduce the possibility of damage to the car, always use the appropriate vehicle protection equipment (VPE):

Wing covers

Steering wheel covers

Seat covers

Floor mats

If appropriate, safely remove and store the owner's property before you work on the vehicle. Before you return the vehicle to the customer, reinstate the vehicle owner's property. Always check the interior and exterior to make sure that it hasn't become dirty or damaged during the repair operations. This will help promote good customer relations and maintain a professional company image.

Vehicle Protective Equipment (VPE)

Safe Environment

During the repair or maintenance of light vehicle electrical and electronic systems you may need to replace car batteries. Under the Environmental Protection Act 1990 (EPA), you must dispose of car batteries in the correct manner. They should be stored safely in a clearly marked area until they are collected by a licensed recycling company. This company should give you a waste transfer note as the receipt of collection.

To further reduce the risks involved with hazards, always use safe working practices, including:

1. Immobilise the vehicle by removing the ignition key. Where possible, allow the engine to cool before starting work.

2. Prevent the vehicle moving during maintenance by applying the handbrake or chocking the wheels.

3. Follow a logical sequence when working. This reduces the possibility of missing things out and of accidents occurring. Work safely at all times.

4. Always use the correct tools and equipment to avoid damage to components and tools or personal injury. Check tools and equipment before each use.
   - Inspect any mechanical lifting equipment for correct operation, damage and hydraulic leaks.
   - Never exceed safe working loads (SWL).
   - Check that measuring equipment is accurate and calibrated before you take any readings.

5. If you need to replace any electrical or electronic components, always check that the quality meets the original equipment manufacturer (OEM) specifications. (If the vehicle is under warranty, inferior parts or deliberate modification might make the warranty invalid. Also, if parts of an inferior quality are fitted, it might affect vehicle performance and safety.) Only replace electrical components if the new parts comply with the legal requirements for road use.

6. Following the replacement of any vehicle electrical components, it is important to test the vehicle thoroughly to ensure safe and correct operation. This may require road testing the car. Make sure that all work is correctly recorded on the job card and vehicle's service history, to ensure that any maintenance work can be tracked.

Preparing the car

Tools

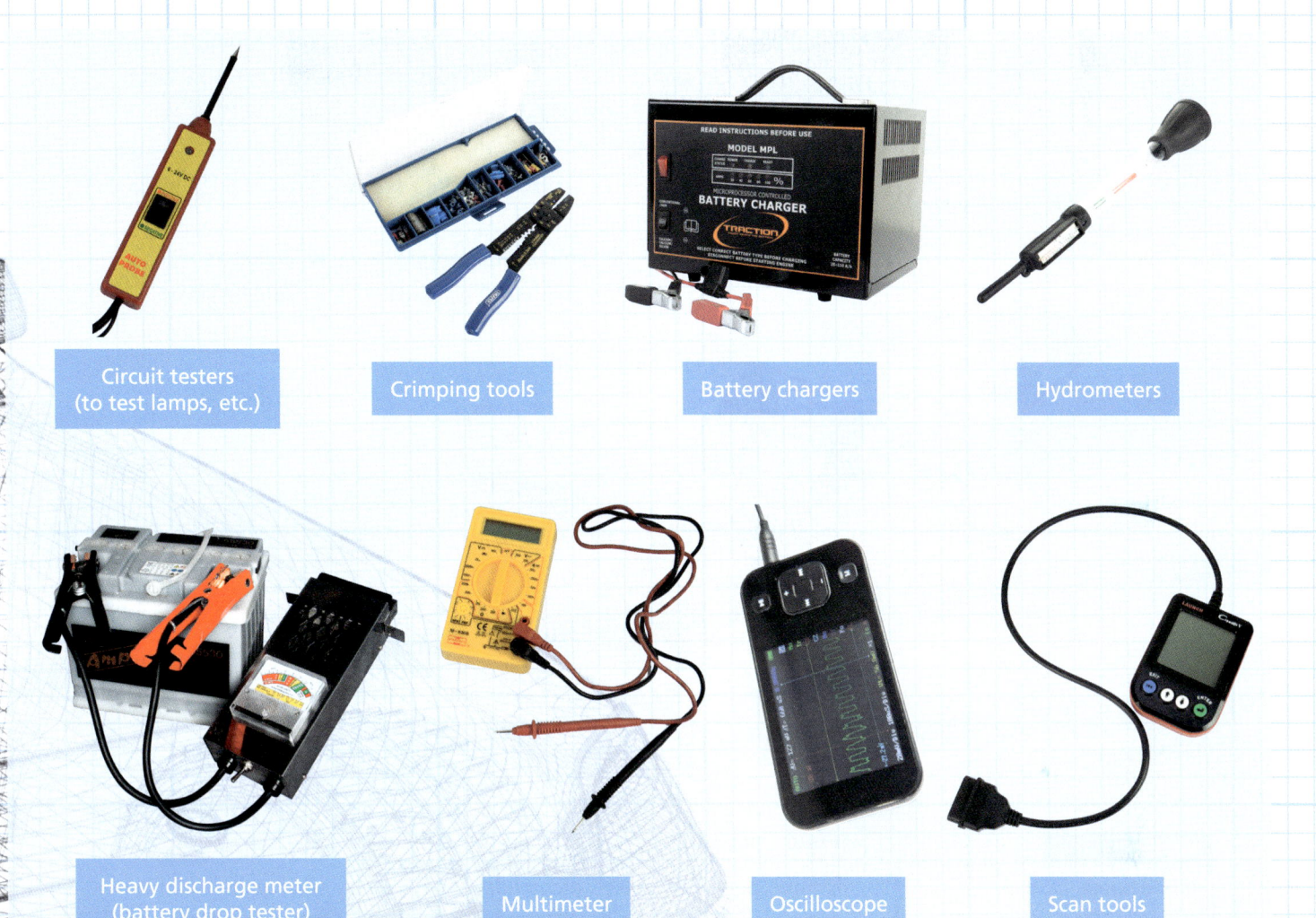

Circuit testers (to test lamps, etc.)   Crimping tools   Battery chargers   Hydrometers

Heavy discharge meter (battery drop tester)   Multimeter   Oscilloscope   Scan tools

Safe Working

- Always make sure the engine is cool before you carry out work on it.
- Always isolate (disconnect) the electrics before you replace electrical components.
- When you charge a battery, remove all sources of ignition and always work in a well-ventilated area.
- Always make sure that your tools and equipment are in good condition and working correctly.
- Always use the correct tool to avoid damage to the vehicle and its electrical systems.
- When you use a multimeter, always make sure that it is correctly fitted to the electric circuit for the type of measurement you are testing:
  - connected in parallel when testing electrical voltage
  - connected in series when testing electric current
  - system power switched off and disconnected from the circuit when testing for resistance.

Level 2 Light Vehicle Maintenance & Repair

Light vehicle electrical and electronic principles

Electricity is a source of energy that can be used to provide power for many car components. Lots of electrical parts have now been **miniaturised** and use very small amounts of electricity to operate – this is called electronics.

As the technology on cars develops, electricity is used to help control many vehicle systems such as:

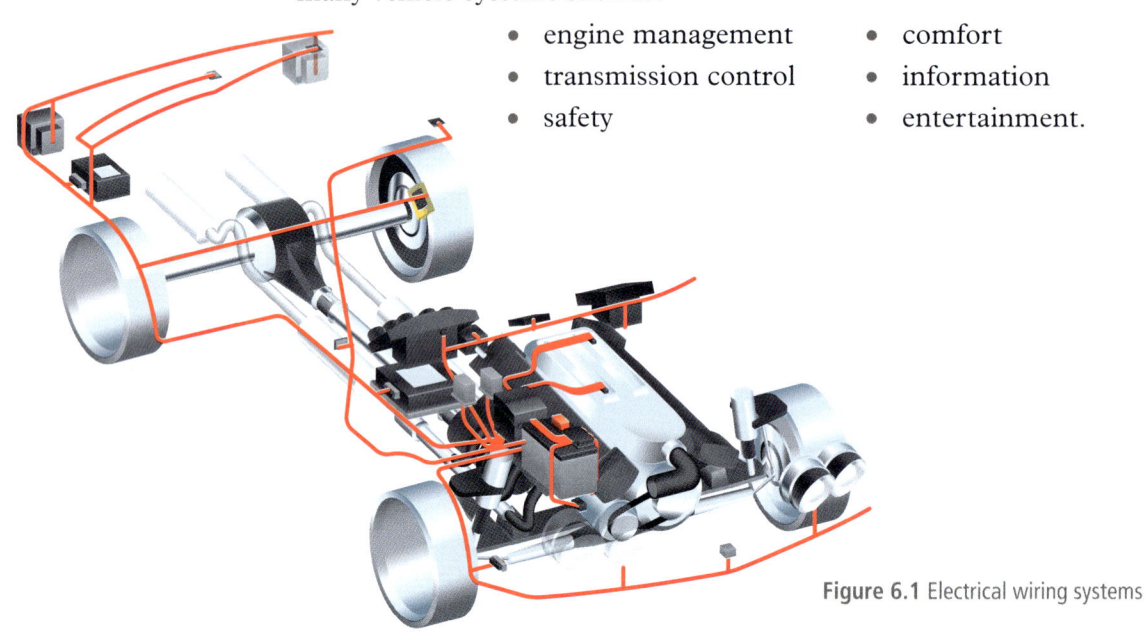

- engine management
- transmission control
- safety
- comfort
- information
- entertainment.

Figure 6.1 Electrical wiring systems

Figure 6.1 shows the electrical wiring systems in a car.

Unlike petrol or diesel, electricity is an energy source that can be generated by the car and stored in chemical form in a battery, ready for use at a moment's notice. When energy is stored in this way it has a 'potential'. This means that it is waiting to do some work.

If this **potential** electrical **energy** can be released and made to move, this movement can be converted into a source of heat, magnetism or chemical reaction:

- **Heat** – The heat produced can be used to make lighting, demisting and comfort systems used in cars.
- **Magnetism** – The magnetism produced can be used to create movement in electric motors used in starting, engine control (e.g. injectors, throttles) and body control (e.g. windows, central locking).
- **Chemical reaction** – A chemical reaction inside a vehicle's battery produces electrical energy that is stored in the battery. Once the battery is charged, the chemical reaction can be reversed to produce electricity for use in the car's electrical systems. Other types of chemical reaction are used to produce light, for example high-intensity gas discharge (HID) headlamp systems.

> **Key terms**
>
> **Miniaturise** – to make extremely small.
>
> **Potential energy** – energy that is stored and waiting to do some work.

Figure 6.2 Headlamps – an example of electricity used to create light from heat

How electricity works

Atoms

Every substance is made of molecules, and the molecules are made up from atoms. For example, in the substance water, the molecule is H_2O. This means that the molecule is made up of two hydrogen atoms joined to one oxygen atom, as shown in Figure 6.3.

The easiest way to imagine an atom is like a miniature solar system, with a sun in the middle and planets orbiting around the outside, as shown in Figure 6.4.

There is a nucleus in the centre of each atom. The nucleus is made up of:

- positively charged particles called protons
- particles with no charge called neutrons.

Orbiting around the nucleus (in a similar way to the planets) are:

- negatively charged particles called **electrons** (this is the electricity part).

The number of protons contained inside an atom determines what sort of atom it will be. Each type of atom makes a particular **element**. A list of chemical elements and the number of protons they contain can be found on a chart called the **periodic table**.

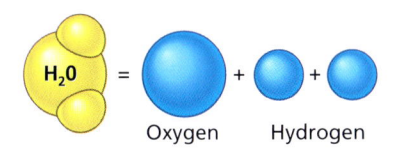

Figure 6.3 The atoms in water

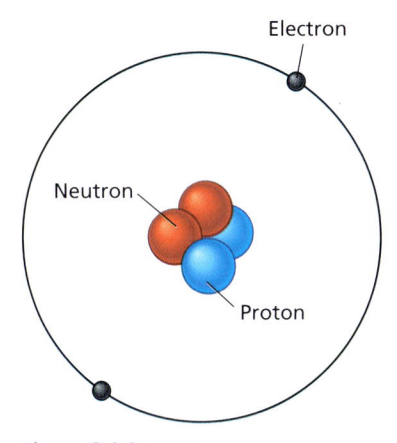

Figure 6.4 An atom

Movement of electrons

The simplest atom is the hydrogen atom (chemical element symbol H on the periodic table). It contains just one electron and one proton (see Figure 6.5).

The orbiting electron is held in place by an attraction force. This force is similar to the force of gravity which holds the planets in orbit around the Sun. Because the hydrogen atom is so simple, the bond between the nucleus and the electron is very strong.

A copper atom (chemical element symbol Cu on the periodic table) contains 29 electrons and 29 protons. The electrons in a copper atom orbit in ever-increasing circles called shells (see Figure 6.6).

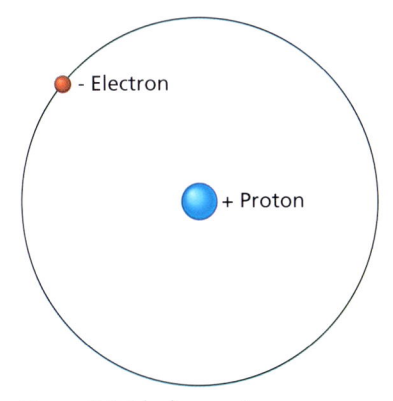

Figure 6.5 A hydrogen atom

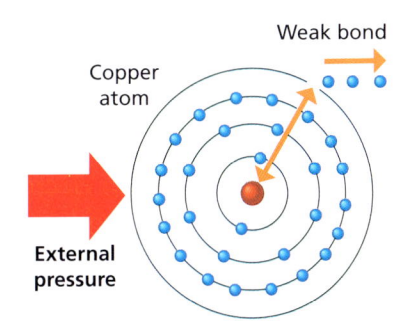

Figure 6.6 A copper atom

> **Key terms**
>
> **Electron** – a subatomic particle with a negative electric charge. It is the movement of electrons that creates electric current.
>
> **Element** – a substance which is made of just one type of atom, for example oxygen or copper.
>
> **Periodic table** – a list of elements arranged in order of the number of protons they contain.

Level 2 Light Vehicle Maintenance & Repair

> **Key terms**
>
> **Current** – moving electricity.
>
> **Alternator** – a component used in vehicles to generate electricity efficiently.
>
> **Conductor** – a substance that allows electricity to move through it easily.
>
> **Insulator** – a substance that doesn't allow the easy movement of electricity.
>
> **Polarity** – a term used to describe electrical connection to a circuit. It represents the positive and negative connections.
>
> **Generator** – a component that turns mechanical energy into electrical energy.
>
> **Circuit** – an unbroken path that an electric current can flow around.

The electrons in the furthest orbit (shell) have a far weaker bond to the nucleus than the electron in a hydrogen atom. These outer electrons are known as 'free electrons'. If the free electrons are placed under pressure, their attraction to the nucleus can be broken. This allows them to move from one atom to the next.

This movement of electrons is electric **current**. The pressure needed to move the free electrons can be provided by magnets, as used in an **alternator**, or by a chemical reaction, such as that in a battery.

- If a chemical substance or element allows electrons to move easily between atoms it is called a **conductor**. Copper is a conductor which is used in wiring.
- If a chemical substance or element restricts the movement of electrons from one atom to the next, it is called an **insulator**. For example, plastic is an insulator which is used to coat the outside of copper wire.

Electricity and magnetism are very closely linked. Both electricity and magnets have positive and negative (or North and South) poles, and this is where the word **polarity** comes from.

Creating electricity

- If a magnet is moved past a copper conductor, the magnetic attraction will draw electrons through that copper conductor and create electric current. By turning a magnet inside a copper winding, a **generator** like an alternator can be made.
- If an electric current is passed through a copper conductor it will generate an invisible magnetic field. This can be used to move other magnets to make a motor.
- If certain chemicals are mixed they react with each other and change from one substance to another. The chemical reaction releases or moves electrons, which creates electric current. This is the basic principle of how a battery works.

The moving electric current, created by chemical reaction, was first produced by an Italian scientist called Alessandro Volta. He gave his name to one of the common electrical units: the volt.

Electric circuit

When an atom loses an electron to the atom next to it, the lost electron must be replaced by another. Because of this, a conductor such as copper wire must be joined in a circle, called a **circuit** (see Figure 6.7). In this way, as one electron leaves an atom another electron can drop in from behind to take its place.

If the conductor is not connected in a circuit, no electrons can flow and there will be no current. This is because the last electron in the conductor has nowhere to go.

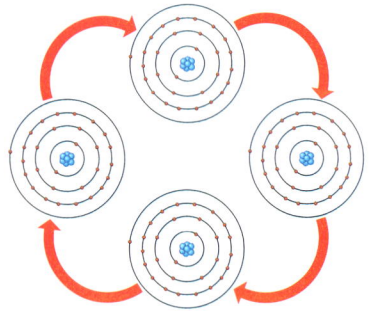

Figure 6.7 Atoms in a circle to show how electrons will move from one atom to the next

6 Light vehicle electrical units & components

Most electrical circuits on cars are **direct current** (**DC**). This means that electricity flows in one direction only.

Because it is very hard to imagine electrons moving from one atom to the other, a direct current electric circuit is often described by comparing it to the movement of water from a water tower, as shown in Figure 6.8:

- A water tower contains a reservoir of water at the top. (This represents the battery.)
- A pipe leads down from the bottom of the reservoir. (This represents the wire.)
- There is a tap on the end of the pipe. (This represents a switch.)
- There is a small water wheel at the end. (This represents a motor.)
- There is a pressure gauge on the pipe. (This represents voltage, as voltage is electrical pressure.)

The water tower works as follows:

- When the tap is turned on, gravity pushes water down through the pipe, under pressure, out through the tap, and on to turn the water wheel.

> **Key term**
>
> **Direct current (DC)** – electric current that only moves in one direction.

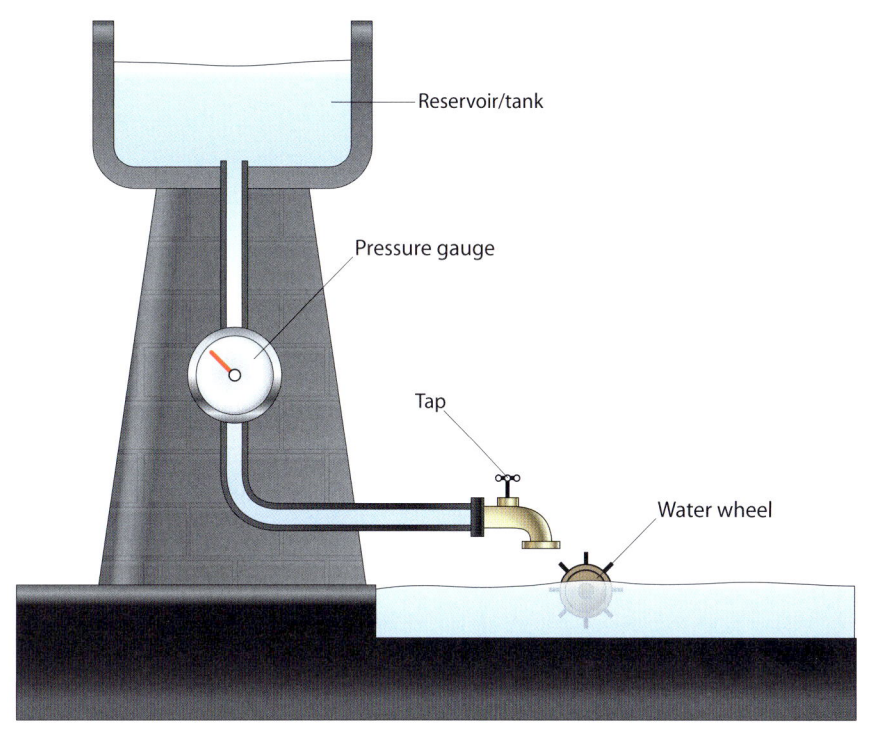

Figure 6.8 A water tower used to represent a direct current (DC) electric circuit

This can be compared with a starter motor circuit:

- When the ignition key (tap) is turned, pressure from the battery (water tank) pushes electricity through the wiring (pipe) to create magnetism in the motor (water wheel), making it turn.

If you are able to compare the movement of electric current in a circuit to water flowing through pipes, many electrical principles can be explained in this way. Once you have understood how electric current works, you will be able to understand the operation, faults and testing procedures of various vehicle components, circuits and systems.

Electrical and electronic principles

Electrical units

There are four main electrical units which are used to describe and measure electricity, as shown in Table 6.1. Each unit describes a different function or operation of electricity.

369

Level 2 Light Vehicle Maintenance & Repair

⚠ Safe working

Many systems on light vehicles use low voltage (12 V), which reduces the risk of electric shock. You must treat any system that uses high voltage (greater than 30 V AC or 60 V DC) with extreme caution, as these voltages can easily cause shock and injury.

High voltage systems include:

- ignition circuits
- hybrid vehicle drive circuits.

Table 6.1 The four main electrical units

Electrical unit	Meaning	Symbol
Volt	Electrical pressure	V or sometimes EMF
Amp	Electric current (quantity of moving electricity)	A or sometimes I **Alternating current** ∿ Direct current ⎓
Ohm	Electrical resistance (slows the flow of electricity)	Ω
Watt	Electrical power (the rate at which work is done)	W

Volt

The volt is used to measure the potential force or pressure in any part of an electrical circuit.

When everything is switched off and no electric current is flowing, the stored pressure is called **electromotive force** (**EMF**). This energy waiting to do some work is called a potential.

When an electric circuit is switched on and current starts to flow, the pressure (voltage) in the system will fall. You can see this happening by connecting a voltmeter to a car battery.

- When everything is switched off, you should get a reading of approximately 12 volts EMF.
- When the ignition key is turned and electric current is allowed to flow through the starter motor circuit, you will see the voltage reading on the meter fall to approximately 9 volts.

This voltage drop is caused by the movement of electricity in the circuit. The difference between the off and on pressures (3 volts) is called a **potential difference** (Pd).

This voltage drop is normal, but damage or wear to an electric circuit can cause a high resistance. The potential difference caused by this high resistance can mean that there might not be enough voltage to operate the starter motor properly.

Find out

- Set up a digital multimeter to measure 20 V direct current (DC).
- Connect it to a car battery in your workshop. With everything switched off, measure the electromotive force.
- Switch on various electric circuits and measure the voltage with current flowing.
- Calculate the potential difference (Pd) using the following formula:

  Potential difference = Electromotive force (EMF) – Voltage measured when current is flowing

Which electrical circuit that you switched on created the largest potential difference?

Figure 6.9 Voltmeter connected across a circuit in parallel to show potential difference

Ampere (electrical current)

The ampere (usually called amp) was named after a Frenchman, André-Marie Ampère. Amps are the unit used to measure the amount of electric current (moving electricity) flowing in any part of an electric circuit.

The quantity of stored electricity is measured in amp hours (Ah). For example, a car battery such as that shown in Figure 6.10 can be rated as 100 Ah. This battery should contain enough electricity to provide:

- 100 amps for 1 hour
- 1 amp for 100 hours
- 10 amps for 10 hours
- any other combination that multiply together to make 100 (for example, 25 amps for 4 hours).

The battery rating gives the theoretical value, but the actual value will be slightly less in reality. This is normally due to the demands placed on the battery by electrical components.

Figure 6.10 A 100 Ah car battery

Ohm

The ohm is the unit used to measure resistance to electric flow. It was named after a German mathematician called Georg Ohm.

As current passes through any conductor or **consumer**, the electricity is slowed down. This slowing down is called resistance.

The flow of electricity can be compared to water flowing inside a pipe, as shown in Table 6.2. Pipe size and friction will create a difference in flow.

Table 6.2 Resistance to flow in water pipes and electrical wire

Water pipe	Resistance to flow	Electrical wire
Large diameter pipe: water flows easily	Low	Large diameter wire: current flows easily
Small diameter pipe: water flow is slowed	High	Small diameter wire: current is reduced
Short pipe: water flows easily	Low	Short wire: current flows easily
Long pipe: water flow is slowed	High	Long wire: current is reduced

> **Key terms**
>
> **Alternating current (AC)** – electric current that continuously moves backwards and forwards in a circuit.
>
> **Electromotive force (EMF)** – the electrical voltage when no current is flowing.
>
> **Potential difference (Pd)** – the difference between the EMF and the electrical voltage when current is flowing.
>
> **Consumer** – component in an electrical circuit which does useful work, for example a light, motor, heater, radio, engine management sensor or actuator.

Level 2 Light Vehicle Maintenance & Repair

> **Find out**
>
> A mathematical calculation called Ohm's law is used to show the relationship between voltage, current and resistance.
>
> Ohm's law states that: 'The current flowing in an electrical circuit is in proportion to the voltage applied and inversely proportionate to the resistance.' This means that:
>
> - Current goes up as voltage goes up.
> - Current goes down as resistance goes up.
>
> 1 Find out how to draw the Ohm's law triangle.
> 2 Explain how the Ohm's law triangle is used.

Variable resistance

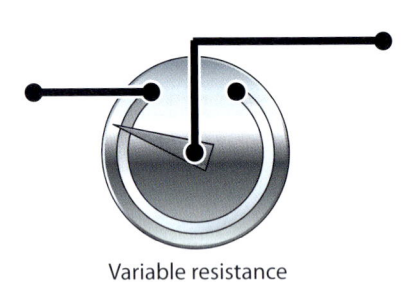

Figure 6.11 Variable resistor or potentiometer

If you are replacing wires on a car, make sure that electricity doesn't have to travel too far. The longer the wire, the greater the volt drop (potential difference). Not all of the electricity will reach the component.

Resistance can be used to help you control the flow of electricity in a circuit. A **variable resistor** (also called a **potentiometer**) can be used to vary the amount and pressure of electric current, and change the operation of a component. (It works like a tap for electricity.) For example, a dimmer switch for the control of dashboard lighting.

Watt (power)

Power is the rate at which work is done.

The watt is the unit used to measure electrical power that is used or made. It is named after the Scottish engineer James Watt.

- The amount of energy needed to operate an electric component (e.g. to light a bulb or turn a motor) is the power measured in watts.
- Electricity is generated by an alternator, its power can also be measured in watts (although its output is normally shown in volts and amps).

To make a car electrical circuit you need the components shown in Figure 6.12.

> **Did you know?**
>
> James Watt developed the steam engine and his surname is used as the international unit of power (not just electrical power).

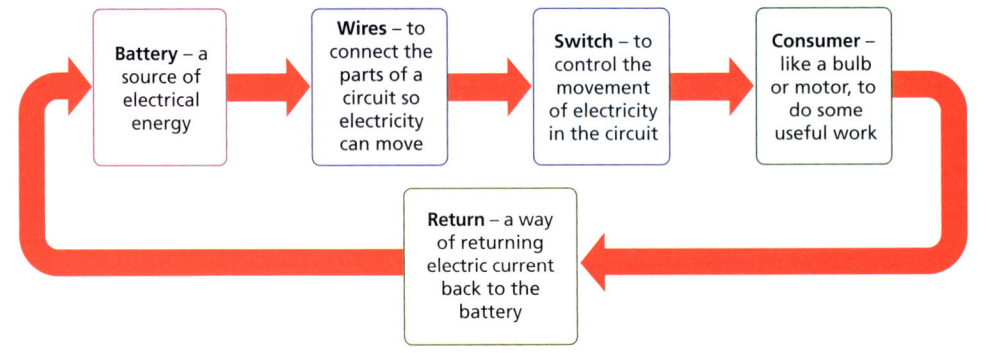

Battery – a source of electrical energy → Wires – to connect the parts of a circuit so electricity can move → Switch – to control the movement of electricity in the circuit → Consumer – like a bulb or motor, to do some useful work → Return – a way of returning electric current back to the battery

Figure 6.12 The components needed to make a light vehicle's electrical circuit

The battery

Because vehicles are mobile, the energy source must be portable. The electrical energy needed for a light vehicle can be carried in a chemical container called a battery.

> **Did you know?**
>
> Lead acid batteries come in different types and constructions. You should check the manufacturer's specifications to ensure you know which type you are working on.

6 Light vehicle electrical units & components

How the battery works

A standard lead-acid battery, the type often found in cars, contains a number of sections called cells (see Figure 6.13). Each cell is capable of producing approximately 2.1 V. A standard battery contains six cells linked together, creating a battery with a voltage of 12.6 V. This is rounded down to 12, so we say that the battery has a voltage of 12 V.

Each cell contains a number of lead plates, which are chemically different:

- The negative plate is made of lead.
- The positive plate is made of lead peroxide.

To prevent the plates from touching each other and causing a **short circuit,** thin sheets of material called separators are inserted between them.

Lead peroxide contains extra oxygen compared with normal lead. This means that the positive plate is chemically different from the negative plate, and it would like to share its electrons with the negative plate. If connected in a circuit, the electrons are allowed to move from the positive plate to the negative plate. This creates electric current, which provides the energy to power components in the vehicle.

The first half of the circuit is made with an **electrolyte** which consists of sulphuric acid and **deionised water**. The electrolyte covers the plates and allows electrons to move from one plate to another (just like swimming). The top of each plate is then connected to the rest of the circuit. The circuit must contain a consumer to use up the electrical potential energy.

The chemical reactions taking place inside the battery are shown in Figure 6.14.

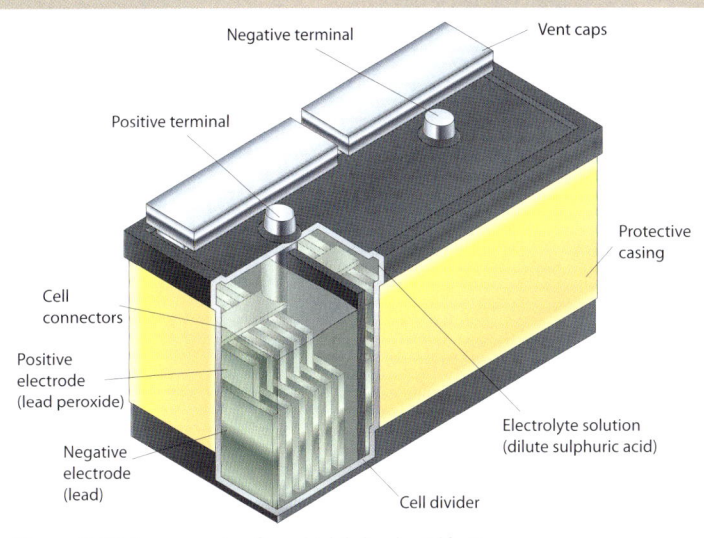

Figure 6.13 Components of a vehicle's lead-acid battery

Key terms

Variable resistor/potentiometer – an adjustable resistor to control the flow of electricity.

Short circuit – electricity taking a short cut and not travelling the full length of the circuit.

Electrolyte – the liquid inside a battery cell, usually made from deionised water and sulphuric acid.

Deionised water – water that has been treated to remove any electrical charge.

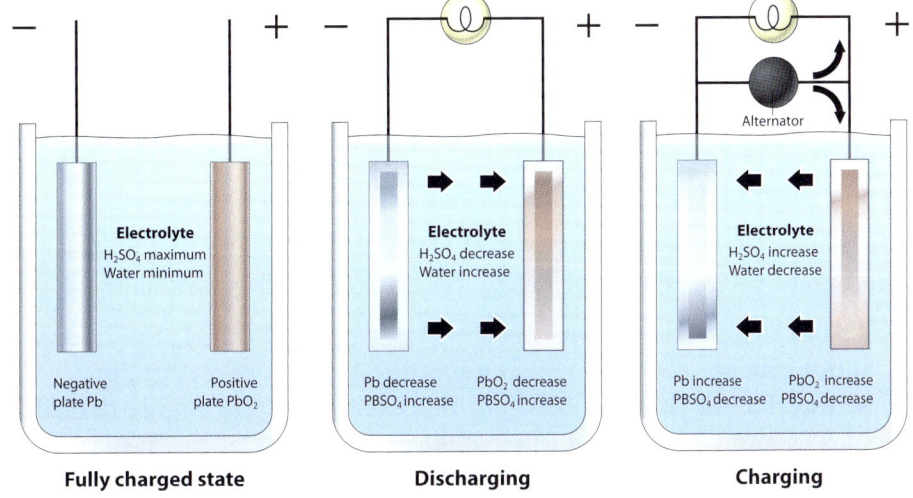

Figure 6.14 The chemical reactions taking place inside a lead-acid battery

Safe working

The electrolyte in a lead-acid battery is made up from deionised water and sulphuric acid. Sulphuric acid is highly corrosive and can cause severe burns if splashed on your skin or eyes. Always wear the appropriate PPE when handling electrolyte, including: steel toe-capped boots, overalls, goggles and chemical-resistant gloves.

When the circuit is complete, the electrons combine with the electrolyte and move from one plate to another as a chemical reaction, creating current.

Why batteries need to be recharged

Batteries work with electricity flowing in one direction only (direct current). This means that eventually both plates will become chemically the same (they change to a substance called lead sulphate) and current will stop. When this happens, no more electricity flows through the circuit, which means that the battery is flat.

When the battery is flat, it needs to be recharged. To do this, an electricity pump or generator is connected to the engine. Then, during normal operation, an electric voltage is produced in the battery circuit with a pressure that is higher than the EMF (approximately 12.6 volts with all electrical consumers switched off). This reverses the chemical reaction – it forces electrons back through the electrolyte to their original positions in the lead and lead peroxide plates, which recharges the battery.

An alternator is the most common type of generator. It supplies a voltage of around 14 volts to recharge the battery.

The direction of current flow and electron flow

Electric current will always move from an area of high pressure to an area of low pressure.

In conventional or standard electrics, the positive (+) side of the circuit has high pressure, and the negative (−) side of the circuit has low pressure. This mean that electricity leaves the battery at the positive terminal, flows through the circuit and then re-enters the battery at the negative terminal. Whenever you test an electric circuit on a vehicle you should use conventional electrics theory (follow the circuit from positive to negative).

Electrolyte liquid level

The chemical processes taking place during **charging** and **discharging** create heat, which gradually causes the electrolyte liquid to evaporate away. As this happens, the level of electrolyte inside the cell slowly drops. When carrying out regular battery maintenance procedures, you should check and top up the electrolyte with deionised water.

The level of electrolyte must be kept above the top of the lead plates (see Figure 6.16). If it drops below this level, two things will happen:

1. The exposed parts of the plates will not be used during the charge and discharge process. This reduces the amount of electricity within the cell that can be used. For example, if the engine doesn't start straight away when it is cranked, the energy in the battery will be quickly used up. The starter motor will slow down and eventually stop.

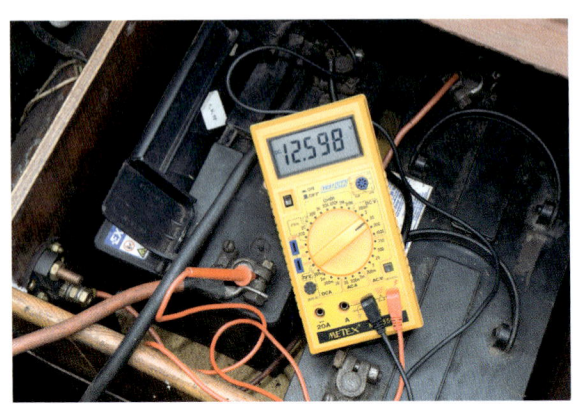

Figure 6.15 The voltage in a battery

Safe working

During the charging process of a lead-acid battery, hydrogen sulphide gas is given off. This gas is extremely flammable – if it is ignited, it could cause the battery to explode. When charging a battery, remove all sources of ignition and work in a well-ventilated area.

Did you know?

True electron flow is in the opposite direction to standard or conventional current (negative to positive). Remember this if the words 'electron flow' are ever used in a description or test.

Key terms

Charging – generating electrical energy.

Discharging – releasing electrical energy.

2. Because the plates are not submerged in a cooling electrolyte, the exposed parts might become overheated and can warp or bend. If the plates bend too much they can eventually touch, creating a short circuit.

All the cells in the battery are linked in series (one after the other). If a short circuit occurs, it will damage the cell and stop all the others working, so that the entire battery is dead.

Maintenance-free batteries

Some batteries are classified as maintenance-free (MF) and are sealed for life. This means that you can't open the cells and top up with deionised water. The plastic casing of the cells is designed so that as the electrolyte evaporates it is captured, condensed back into liquid and returned to the cell.

Do not attempt to open a maintenance-free battery. If you do it will damage the internal system designed to recover electrolyte during its normal life cycle.

Checking a flat battery

If a battery is flat and needs jumpstarting but will still not start the car after it has been run for a while, you should check for a fault in the charging circuit (from the alternator to the battery).

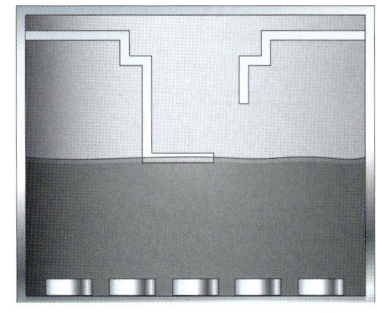

Figure 6.16 The electrolyte level of a car battery

Did you know?

A car battery can only be charged and discharged so many times. Eventually the chemical reactions inside the battery will fail and it will need to be replaced.

The average lifespan of a lead-acid battery is around 3 to 5 years.

Checking for a charging circuit fault

Checklist			
PPE	VPE	Tools and equipment	Source information
• Steel toe-capped boots • Overalls • Latex gloves	• Wing covers • Steering wheel covers • Seat covers • Floor mat covers	• Multimeter capable of reading volts and amps • Spanners • Screwdrivers	• Manufacturer's technical data

1. To carry out a quick check, connect a voltmeter to the battery terminals and run the engine. You can then see if the voltage is higher than the EMF. If the alternator is trying to charge the battery, a typical voltage will be somewhere between 13 and 14 volts.

2. Also check the current output of an alternator.

3. To carry out a current check, partially drain the battery by leaving the lights on for a few minutes.

4. Connect a suitable ammeter in series. (Be careful, as many ammeters can only read up to 10 amps.)

5. Start the engine and measure the current output.

375

Nickel cadmium battery

An alternative to the lead-acid battery is the nickel cadmium (nicad) type, which is found in some cars. It works in a similar way to a standard battery but requires less maintenance and cannot be overcharged. This type of battery is made of the following materials:

- positive plates – nickel hydrate
- negative plates – cadmium
- electrolyte – potassium hydroxide and water.

These batteries tend to be larger and more expensive than normal lead-acid batteries. However, they are better at coping with the extreme loads placed on them by modern electrical systems, expecially in hybrid or battery electric vehicles.

Other electrical components

Cables

Insulated copper wiring is used to transport electricity around the vehicle to where it is needed. Thin strands of copper (which is a conductor) are bundled together and coated with a plastic shield (which is an insulator) to help prevent electricity conducting to any other metal components. If this happened, it would cause a short circuit.

Because a large number of wires are used in motor vehicle construction, the external plastic coating is usually colour-coded (see Figure 6.17). When diagnosing an electrical circuit fault you can use these colours to help trace cable routing or identify them on a wiring diagram.

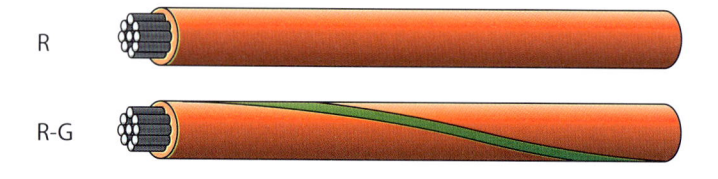

R

R-G

Wiring colour code
Wire colours are indicated by an alphabetical code.

B = black	L = light blue	R = red
BR = brown	LG = light green	V = violet
G = green	O = orange	W = white
GR = grey	P = pink	Y = yellow

The first letter indicates the basic wire colour and the second letter indicates the colour of the stripe.

Figure 6.17 Electric wire with common electrical colour codes

Electrical wires come in different sizes. The copper strands are bundled together, which means that if one or more strands are damaged electricity can still flow. Automotive wires are normally labelled with the number of strands they contain and the diameter of each strand in millimetres. This gives an indication of the amount of current the wire is able to carry.

Some typical wire size designations and uses are shown in Table 6.3.

> **Did you know?**
>
> Maintenance free batteries are sealed so they must be charged using a constant current to prevent the expansion of gases inside becoming too great.
> A suitable battery charger should be chosen when recharging a maintenance free battery to help prevent damage or explosion. See the *Checking for a charging circuit fault* procedure on page 375.

> **Did you know?**
>
> To reduce the amount of wiring used in the construction of some vehicle electrical circuits, computers can be used to transmit commands between switches and consumers along a single communication wire. This is known as networking or multiplexing.

> **Did you know?**
>
> The thicker the wire, the more electricity it can carry and the less internal resistance it has. This means that more voltage and current will be available for the component.
>
> The longer the wire, the higher the resistance. This means that less voltage and current will be available by the time it reaches the component.
>
> - Double the length of the wire and you double the resistance.
> - Double the diameter and you halve the resistance.

6 Light vehicle electrical units & components

Table 6.3 Wire sizes and uses

Number of strands / Wire diameter	Continuous current rating	Uses of the wire
9 / 0.30 mm	5.75 amps	Side lamps, tail lamps, reversing lamps, horns
14 / 0.30 mm	8.75 amps	Side lamps, tail lamps, reversing lamps, horns, general wiring
28 / 0.30 mm	17.5 amps	Headlamps, fog/driving lamps, windscreen wiper motor
44 / 0.30 mm	27.5 amps	Charging cable, battery feed
65 / 0.30 mm	35 amps	Charging cable for alternator, dynamo
84 / 0.30 mm	42 amps	Charging cable for alternator, dynamo
97 / 0.30 mm	50 amps	Heavy-duty alternator
120 / 0.30 mm	60 amps	Heavy-duty alternator
80 / 0.40 mm	70 amps	Heavy-duty alternator
37 / 0.71 mm	105 amps	Starter/battery cable
37 / 0.90 mm	170 amps	Starter/battery cable
61 / 0.90 mm	300 amps	Starter/battery cable

> **Find out**
>
> You have to change a headlamp wire.
>
> - What might happen if you use a wire marked 9 / 0.30 mm?
> - What size wire should you choose?

Terminals, connectors and continuity

When a manufacturer designs a car, the electrical wiring used to create the circuits can be bundled together as insulated sections called wiring looms. When the car is assembled, these looms can be routed around the vehicle in the most efficient way and hidden from view behind panel work, carpets and trims. The wiring looms are made in sections and joined together by **connectors**. At the ends of the looms, **terminals** are used to connect the wires to electrical components.

For electricity to operate the components correctly, the circuits must be continuous and unbroken – this is called **electrical continuity**.

Earth return systems

Light vehicle designers and manufacturers try to keep the amount of wiring used to a minimum. This will save on materials, improve efficiency and reduce costs. Because many vehicles are manufactured mainly from metals, which are good conductors of electricity, it is not always necessary to complete an electrical circuit back to the battery using wire alone.

- The negative end of electrical circuit wiring can be connected to the vehicle body or chassis. This is called an **earthing** point, as shown in Figure 6.18.
- The negative terminal of the battery can also be connected to the vehicle body or chassis to complete the circuit. This is called **earth return**.

> **Key terms**
>
> **Connector** – a component that joins two parts of a circuit together.
>
> **Terminal** – where the circuit ends (terminates).
>
> **Electrical continuity** – when an electrical circuit conducts current easily and is unbroken (i.e. continuous).
>
> **Earth** – an electrical ground connection, designed to complete a circuit.
>
> **Earth return** – using the metal chassis to complete the car's electrical circuit.

Figure 6.18 Battery earthing point

Disconnecting and connecting the battery

When you connect or disconnect the battery from a vehicle, as shown in Figure 6.19, you should remove and refit the terminals in a certain order, as this will help reduce the possibility of a short circuit.

- **Order for disconnecting** – Always remove the negative lead first when you are disconnecting the battery. Once you have disconnected the negative terminal, the vehicle's electrical system is now **open circuit**. Therefore, if the spanner you are using accidentally touches the car's bodywork, they both have the same electrical potential or pressure. If the pressure in an electric circuit is equal, no current can flow (because a difference in pressure creates flow).

- **Order for connecting** – When reconnecting a battery, always connect the positive terminal first and the negative terminal last, for the same reasons.

> ### ⚠️ Safe working
>
> Because the metal of the vehicle body now forms part of the car's electrical system, it is possible to cause damage or injury by mistakenly allowing the positive side of the electrical circuit to come into contact with this earth return. The most common time for this to happen is when you are replacing a car battery.
>
> To reduce the possibility of injury or damage, you must follow the procedure described opposite when connecting or disconnecting a car battery.

> ### Working life
>
> Paul is attending a breakdown at the side of the road. A car with a flat battery needs jump starting. Using a set of jump leads and the battery from the recovery vehicle, he must connect up to the flat battery to start the car.
>
> 1. What safety precautions should Paul take?
> 2. How should the jump leads be connected?
> 3. How could Paul connect the negative lead to reduce the possibility of sparks or a short circuit?

Insulated earth return

There are certain safety issues that you must consider when you work on a vehicle with an earth return system. If the vehicle body is used as part of the vehicle's electrical circuit, it has the potential to ignite fuel or damage sensitive electrical equipment if a short circuit occurs.

Vehicles such as fuel tankers, ambulances and emergency vehicles might have an insulated earth return system. These vehicles are designed so that the wiring of the various electric circuits terminates at an electrical connector called a **bus bar** (see Figure 6.20). The bus bar is insulated from the body or chassis. A single insulated earth return wire can now connect the bus bar back to the negative side of the battery.

Figure 6.19 Battery terminal being disconnected

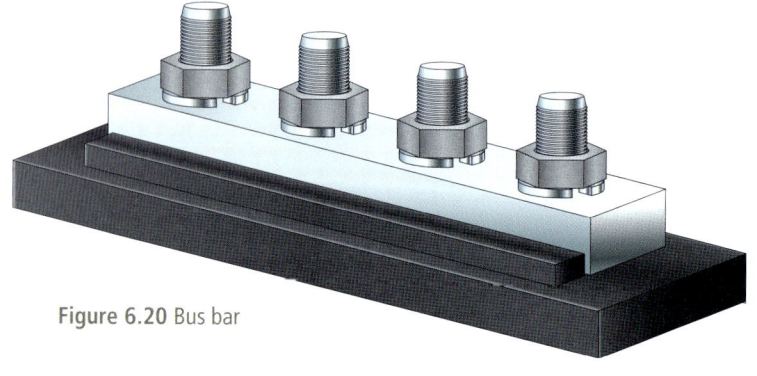

Figure 6.20 Bus bar

Switch

A switch is an electrical component that is designed to control the circuit. This can be as simple as making or breaking the electrical circuits so that current can be turned on and off. Alternatively, the switch could be a variable resistor (potentiometer) that can regulate the flow of electricity (as with a dimmer switch).

Circuit relays

A **relay** is an electromagnetic switch. It is designed to allow a small current to switch a much larger current. A coil of wire is wrapped around a soft iron core. When a small electric current is passed through it, this creates an **electromagnet**. The electromagnet is then able to open or close a switch inside the relay unit.

Uses of relays

Relays can be useful in circuits that use large amounts of electricity.

Consider if a heavy-duty consumer, such as a powerful headlamp bulb, was attached directly to a normal on and off switch. The wiring, switch and connectors would also have to be heavy duty so that they could cope with the high current flow. There are several problems that would arise from this:

- Large-diameter wires would have to be routed all over the car, increasing cost and overall weight.
- Any electrical switches would be under excessive strain and likely to fail early.
- The length of wire used in the circuit would increase overall resistance. This would reduce the amount of electricity available by the time it reached the consumer, reducing electrical performance.

To help overcome these problems relays are used.

- Wiring carrying the heavy-duty current can be designed to take the shortest route to the consumer, thereby reducing overall circuit resistance, but will be controlled by the mechanical switch part of the relay, mounted nearby.
- A lightweight switch can now be used by the driver to control the electromagnet (which only consumes a small amount of current) within the relay.
- When the driver operates the switch to control the circuit, a magnetic field is created inside the relay, causing the heavy duty switch to open or close.

Types of relay

There are three main types of relay:

1. **M4** – This type of relay has four electrical terminals. When operated, it switches on (makes) the circuit.

> **Key terms**
>
> **Open circuit** – a broken electrical circuit where no electricity can flow.
>
> **Bus bar** – an electrical connector that makes a single connection between several circuits.
>
> **Relay** – an electromagnetic switch used to control large currents.
>
> **Electromagnet** – a metal core made into a magnet by passing electric current through a surrounding coil.

Did you know?
When testing a relay, don't disconnect it, otherwise you'll break the circuit and stop it from working.

2. **B4** – This type of relay has four electrical terminals. When operated, it switches off (breaks) the circuit.
3. **Double throw** – This type of relay normally has five electrical terminals. When operated, it switches between two electrical circuits.

The three different types of electric relay symbol are shown in Table 6.4 on page 384.

Current-consuming devices

The scientific law of conservation states:

> 'Energy cannot be created or destroyed but only released and converted into some other form of energy.'

If electrical energy in a circuit is not used up, it will be changed into heat and a fire could result. So something is needed in an electrical circuit to do some useful work, and this is called a consumer. Some examples of consumers are:

- lights
- heaters
- motors
- radios
- engine management sensors and actuators.

Electrical consumers are designed to work when the system voltage is switched on. If working correctly, these consumers should use all of the electrical pressure available. For example, if 12 V are supplied to the consumer, there should be 0 V after the consumer. This is called **volt drop**.

Bulbs

Automotive electrical lightbulbs are usually made from a thin tungsten wire (called a filament), through which an electric current is forced.

- The resistance of the wire causes the electrical energy to be released as heat, making the filament glow. When the filament reaches a temperature of 2300 °C, it gives off white light.
- If this filament is exposed to oxygen, the extreme heat will be enough to burn the wire away. To stop this, the filament is normally enclosed in a glass bulb with all of the air removed. Even though there is no oxygen, if the filament gets too hot, it will melt and turn into a vapour.
- The bulb normally has two electrical terminals so that it can be connected to the lighting circuit. It is shaped so that it fits securely in a socket.

Types of bulb

Bulbs come in different shapes, sizes and power ratings depending on their usage.

Headlamp bulbs can contain an **inert gas** such as argon. The gas is put inside the bulb under pressure. This will allow it to run hotter than a standard bulb and produce a brighter light.

Quartz-halogen headlight bulbs run even hotter. Instead of glass they are normally enclosed in a bulb made from the rock crystal quartz. Quartz can withstand much higher temperatures than glass.

Figure 6.21 Quartz halogen headlamp bulb

6 Light vehicle electrical units & components

Quartz-halogen bulbs are pressurised with a gas made from **halogens**. This gas helps repair the tungsten filament as it starts to vaporise. It works almost like a non-stick coating – as tungsten is released from the filament due to extreme heat, it cannot stick to the inside of the quartz globe and instead reattaches itself to the filament as it cools.

Fan and heater

A simple motor, such as that shown in Figure 6.22, can be made by passing an electric current through a coiled wire that is wound around a central shaft called the **armature**. This creates an electromagnet.

> **Key terms**
>
> **Volt drop** – the fall in electrical voltage when current flows through a consumer.
>
> **Inert gas** – gas that doesn't support combustion, unlike oxygen.
>
> **Halogens** – non-metallic elements found on the periodic table, such as iodine or bromine.
>
> **Armature** – the central shaft of an electric motor.
>
> **Commutator** – a segmented electrical contact mounted on the end of a motor armature. It is designed to change electrical polarity as the motor turns.
>
> **Brushes** – spring-loaded electrical contacts that transfer current to the rotating armature.

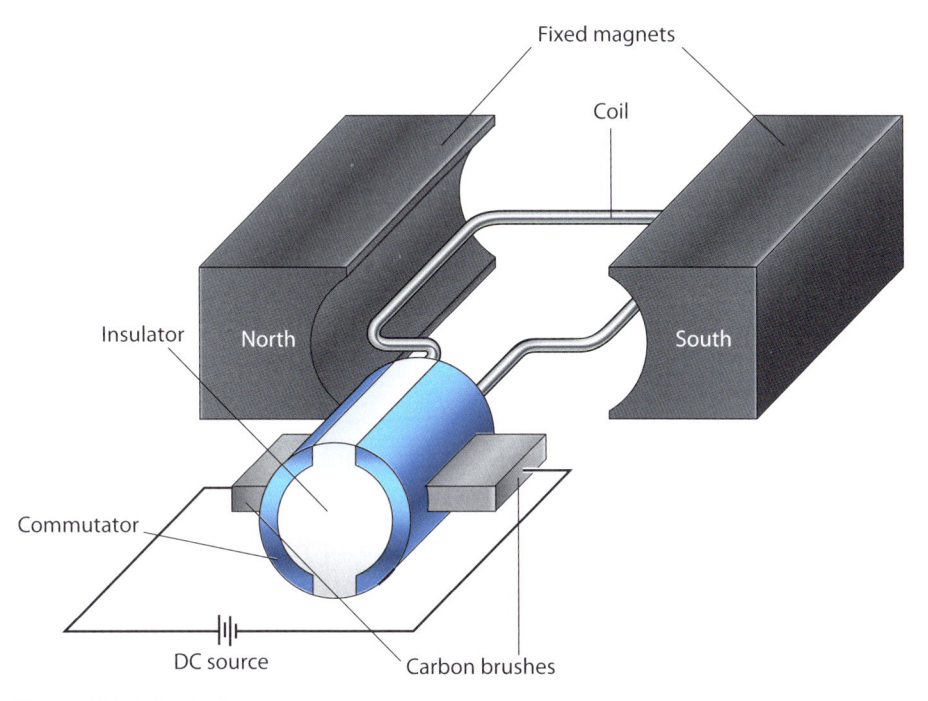

Figure 6.22 A simple electric motor

- The electric current produces an invisible magnetic field, which is repelled (pushed away) or attracted (pulled towards) by the permanent magnets surrounding it. This causes the armature to turn.
- Once the armature has turned out of the magnetic field, it would normally stop.
- To keep it rotating, the polarity of the electricity passing through the electromagnet mounted on the armature must be changed. This is done by a component called a **commutator** (see Figures 6.22 and 6.23).
- Two spring-loaded electrical contacts called **brushes** are mounted on the end of the armature to maintain electrical connection with the commutator as the shaft rotates.
- When electric current is switched off, the motor will stop.

The end of the armature shaft on a simple electric motor can be attached to a set of fan blades, used to direct air through the radiator to help with cooling, or a heater matrix to warm the inside of the car.

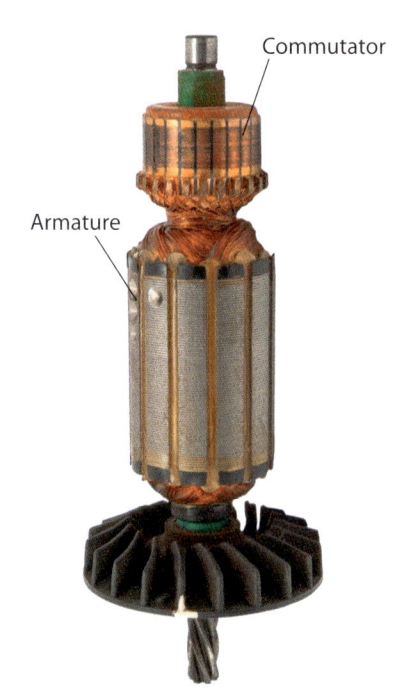

Figure 6.23 Armature and commutator

Fuses

If you allow electrical current to flow in a circuit without passing through a consumer, then the energy will be converted into heat. This can rapidly cause the circuit wiring and components to burn out, which will be expensive to repair.

To protect the circuit, **fuses** or circuit breakers are commonly used. A fuse is a weak link placed in the circuit in series. It is a thin piece of wire with a current rating just above that of the current intended to flow in the system.

This relatively inexpensive component is designed to burn out if a rapid increase in current flow occurs, to prevent any further damage to the rest of the circuit. Once the fuse has burnt out, an open circuit exists and no further current can flow (electricity stops), hopefully saving more expensive parts.

Fuses come in different sizes, shapes and types, including glass, ceramic and blade. Blade fuses (see Figure 6.24) are the most common type found on light vehicles today.

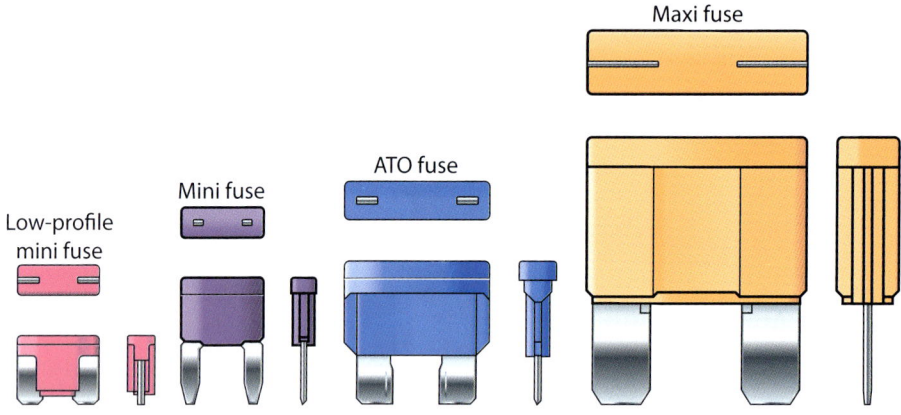

Figure 6.24 Blade fuses

Series and parallel circuits

Series circuit

In a **series circuit** there is only one path from the power source through all of the components and back to the source. All the consumers share the electricity, so if more than one consumer is fitted it will only get part of the voltage available.

Voltage drop will occur across each consumer in the circuit, until all available voltage has been used up. This means that if you connect a voltmeter to a series circuit, you will see the reading on the display fall lower after each consumer. This will continue until you see 0 volts after the last component.

If any one of the consumers fails, the circuit is broken and no electricity can flow. The rest of the consumers stop working. This makes series circuits unsuitable for many systems on cars. For example, if you wired a lighting circuit in series, not only would the bulbs glow dimly, but if one bulb broke all of the others would go out.

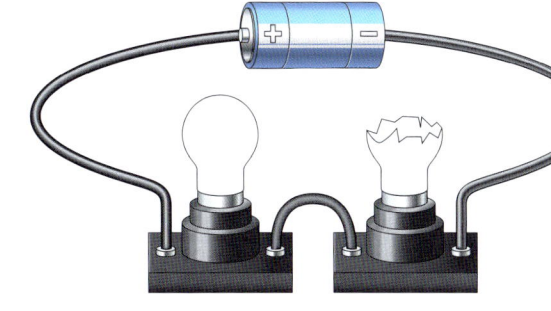

Figure 6.25 A simple series circuit

Figure 6.26 When a bulb in a series circuit is damaged the others will no longer work.

Parallel circuit

In a **parallel circuit** there are multiple parallel paths for the electricity to flow through. Each consumer has its own power supply and earth return back to the battery.

Because each consumer has its own power supply and earth, all the consumers receive the full voltage available and work at full power.

So if one consumer in a parallel circuit fails, the others keep working. For example, in a headlight circuit each bulb has its own 12-volt supply and earth return to the battery. If one bulb breaks, the other bulbs will keep working.

Voltage drop also occurs across the consumers of a parallel circuit. However, unlike in a series circuit, if you connect a voltmeter to a parallel circuit, you will see full supply voltage before each component and 0 volts after it.

> **Key terms**
>
> **Fuse** – a weak link in an electrical circuit, designed to burn out if current flow is too high.
>
> **Series circuit** – circuit where there is only one path for the electricity to take. The consumers are connected one after another and share the voltage available.
>
> **Parallel circuit** – circuit where there is more than one path for the electricity to take and the consumers are connected side by side. Each path and consumer receives the full voltage available.

Figure 6.27 Bulbs in a parallel circuit

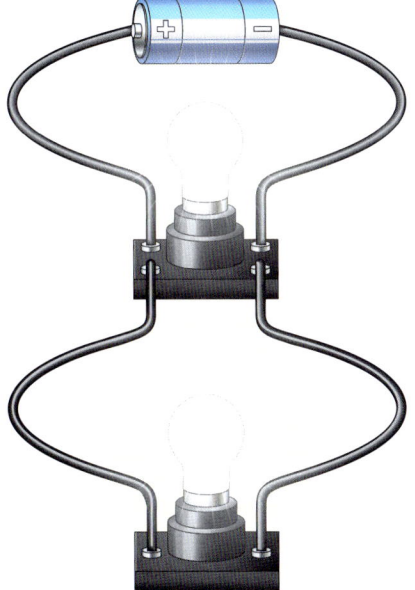

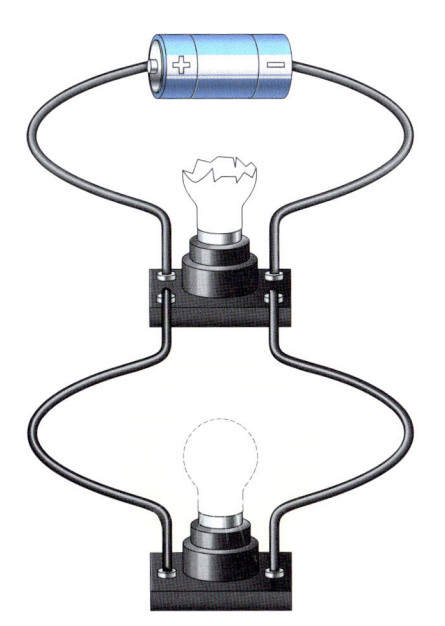

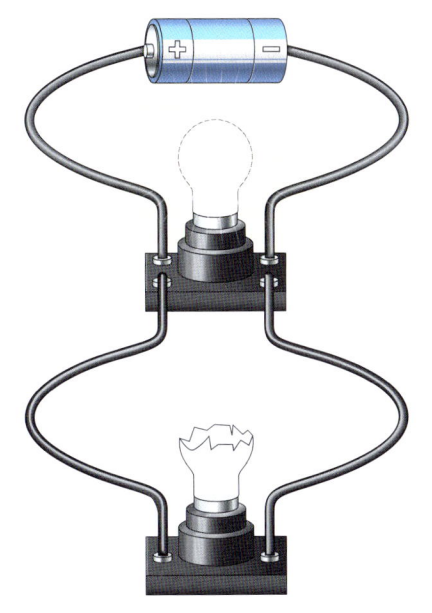

Figure 6.28 When one or more bulbs in a parallel circuit are damaged the others will still work

383

Electric symbols and wiring diagrams

Many electric circuits are designed using wiring diagrams. To help with the design and reading of wiring diagrams, manufacturers use graphics to represent electrical system components.

Common electrical and electronic symbols

Examples of some common electrical and electronic symbols are shown in Table 6.4.

Table 6.4 Symbols for common electrical and electronic system components

Symbol	Component
	Battery – source of portable electrical power
	Switch – component that controls electrical current in a circuit
	Motor – component that uses electromagnetism to provide mechanical movement
	Fuse – component that protects an electric circuit from excess current
	Bulb – component that converts electrical heat into light
	Earth – electrical ground connection designed to complete a circuit
	Diode – component that acts as a one-way valve for electricity
	Transistor – component that acts as a switch with no moving parts
	Relay – remote, electromagnetic switch (see pages 379–380). The three different types of electric relay symbol are M4, B4 and double throw.

6 Light vehicle electrical units & components

Wiring diagrams

Some simple wiring diagrams are shown in Figures 6.29 to 6.31. Study these to help you get used to the layouts and symbols.

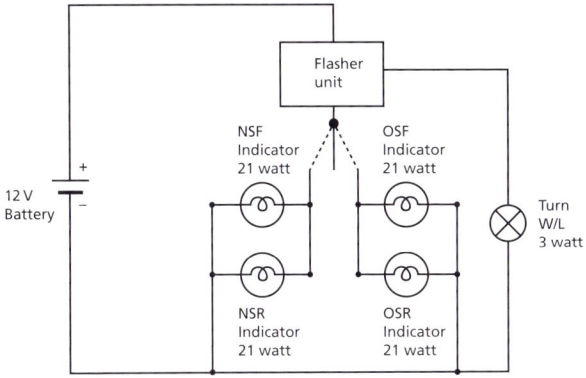

Figure 6.29 Indicator wiring circuit diagram

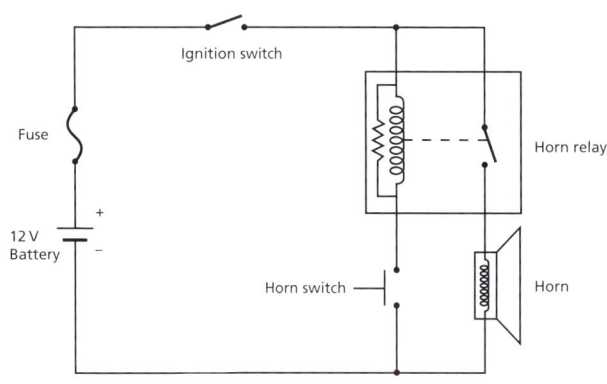

Figure 6.30 Horn wiring circuit diagram

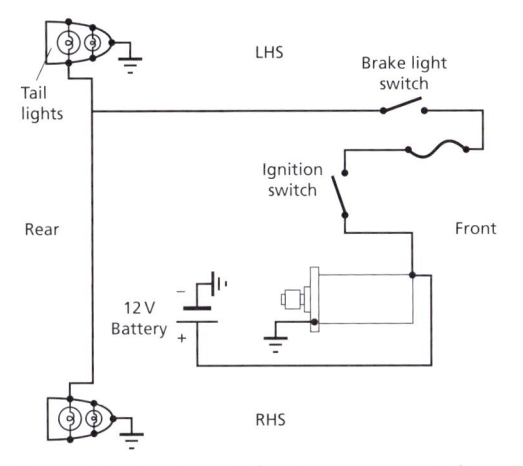

Figure 6.31 Starting and charging wiring circuit diagram

Multimeters

The multimeter is a piece of electrical test equipment designed to measure a number of different units within an electrical circuit. There are two types of multimeter: analogue and digital.

Analogue multimeter

Analogue multimeters use a needle that moves across a graduated scale to record electrical readings within a circuit. The old-fashioned name for this type of unit was 'AVO meter', which stands for amps, volts and ohms.

The problem with analogue meters is that they are only as good as the operator. The graduated scale can be difficult to read, so an inexperienced operator might take inaccurate measurements. Depending on the range of the scale provided by the manufacturer, when a needle is lying between two units marked on the scale, you will have to make your best guess as to what the exact reading is.

If the needle flicks all the way to the end of this scale, it means that the scale doesn't go high enough for the measurement you are taking. This is called full-scale deflection (FSD) and it means that the reading is invalid.

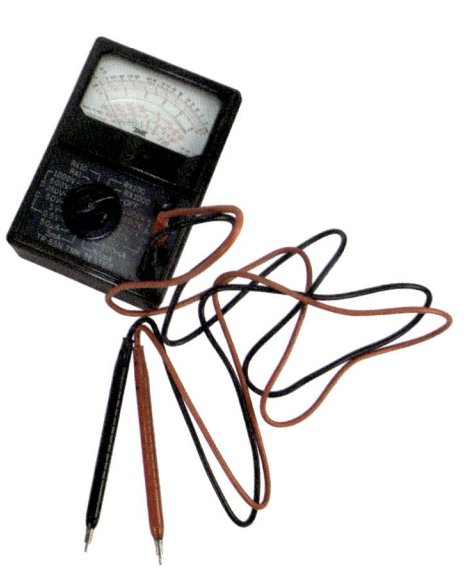

Figure 6.32 Analogue multimeter

385

Digital multimeter

Digital multimeters display digits or numbers on a liquid crystal display (LCD) screen. These numbers are clearly displayed and are easy to read accurately, unlike when using an analogue multimeter.

It is quite normal for the last digit (on the far right end of the screen) to continuously change. As high accuracy is not often required, you can ignore this.

Types of digital multimeter

Two types of digital multimeter are common: manually operated and autoranging.

- **Manual multimeter** – With this type of multimeter, you select the unit (e.g. volts, amps or ohms) and the scale to be measured, normally by turning a dial on the front of the unit.
- **Autoranging multimeter** – With this type of multimeter, you choose the unit but the scale is automatically selected by the multimeter. The scale is known as the unit multipliers. The unit multipliers show the size of the readings being taken. This is indicated by the position of a decimal point or the number of zeros that follow the reading shown on the screen.

Unit multipliers include:

- **Micro (μ)** – one millionth, rarely used in automotive applications.
- **Milli (m)** – one thousandth, e.g. millivolts used in electronic components, such as electronic control modules (ECMs).
- **Kilo (K)** – one thousand, e.g. kilovolts used to measure the high voltages that create sparks at the spark plug.
- **Mega (M)** – one million, rarely used in automotive applications.

When using an autoranging multimeter, you should pay close attention to any scale reading, for example millivolts, volts, kilovolts and megavolts, shown in the corner of the screen. If you ignore these scales you could misinterpret the results, which will lead to incorrect diagnosis and wasted time and effort.

Using a manual multimeter

When using a manual multimeter, if you don't know the scale to be measured use the following procedure:

- To test volts and amps, turn the dial on the front of the multimeter to the highest scale setting. Then slowly rotate the dial down through the scales until an accurate reading is shown.
- To test ohms, turn the dial on the front of the multimeter to the lowest scale setting. Then slowly rotate the dial up until an accurate reading is shown.

Figure 6.33 Digital multimeter

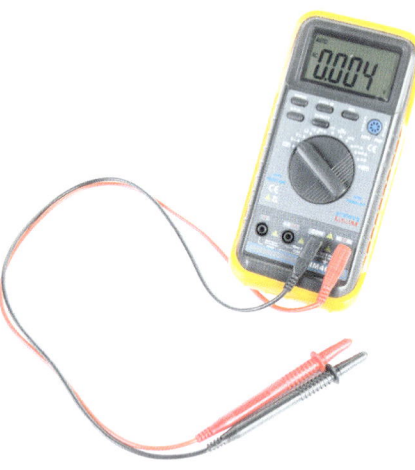

Figure 6.34 Autoranging digital multimeter

Safe working

When using a multimeter, always make sure that it is correctly fitted to the electric circuit for the type of measurement to be tested:

- connected in parallel when testing electrical voltage
- connected in series when testing electric current
- system power switched off and disconnected from the circuit when testing for resistance.

What a multimeter can measure

You can measure a number of electrical units using a digital multimeter, including volts, amps and ohms. Other measurements can also be taken. Extra facilities on a digital multimeter can include:

- temperature
- frequency
- audible continuity testing.
- diode testing
- transistor tests

The electrical units of volts and amps are often broken down into two further areas, direct current and alternating current:

- The direct current scale is normally shown on the meter as a straight line with a number of dots underneath it. ⎓

    This symbol is designed to help prevent confusion. If just a single line was used it might be mistaken for a minus sign and if two lines were used it might be mistaken for an equals sign.

- The alternating current scale is normally shown on the meter as a wavy line. ∼

The ohms scale on a multimeter is normally represented by the Greek letter omega (Ω) as this will help prevent it being confused with the letter 'O' or the number zero.

Connecting the multimeter

When connecting the test leads to a digital multimeter, make sure that they are plugged into the correct sockets for the type of measurement you will be taking. There are normally three sockets:

- Socket 1, marked 'common' or 'ground' – used with the black test probe.
- Socket 2, marked 'volts, ohms and milliamps' – used with the red probe.
- Socket 3, marked '10 amps' – also used with the red test probe, but only when measuring amperage. It is separate from the others to help protect the multimeter from damage if you connected it wrongly to a circuit. (When measuring amps the multimeter must be connected in series, otherwise serious damage can result.)

Health and safety information

There are a number of safety issues that you need to consider when operating multimeters. These include:

- Never use a multimeter if the leads or unit appear to be damaged.
- Never touch the test leads to a voltage source when they are connected to either the 10-amp or 300-milliamp sockets, as major damage to the multimeter can occur.
- Never measure electrical resistance (ohms) with a component still connected to an electrical circuit. You must always switch off the power and disconnect the component first. If you don't do this, your readings will be inaccurate, as the ohmmeter will try to measure the resistance of the entire circuit, not just the component that you are trying to test.

> **Emergency**
> If the low battery symbol appears on a digital multimeter screen, replace the battery straightaway. Otherwise, you might get inaccurate readings that could lead to electric shock or personal injury.

> **Safe working**
> Be aware of the voltage you are measuring. Voltages over 60 V DC and 30 V AC RMS may cause electric shock due to the electrical pressures involved. (RMS stands for root mean square and is roughly the average of an AC voltage.)

> **Safe working**
> When taking electrical measurements, always make sure that your fingers are behind the finger guards or are holding on to the insulated plastic part of the test probes.

Using a multimeter to check voltage

When you check voltage, you are using the multimeter as a voltmeter. A voltmeter measures the pressure difference in an electric circuit between where you put the black probe and where you put the red probe. For example, if you are testing a battery, you will normally put the black probe on the negative terminal and the red probe on the positive terminal. The difference in electrical pressure between these two points, for example 12.6 V will be measured and displayed on the digital screen.

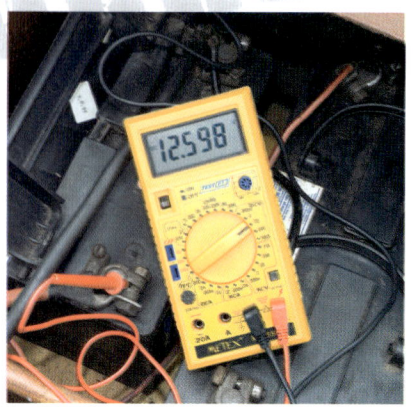

Figure 6.35 Digital multimeter connected to a battery showing approximately 12.6 volts

Checking voltage

Checklist			
PPE	**VPE**	**Tools and equipment**	**Source information**
• Steel toe-capped boots • Overalls • Latex gloves	• Wing covers • Steering wheel cover • Seat covers • Floor mat covers	• Multimeter capable of measuring volts, amps and ohms	• Manufacturer's technical data

1. Connect the appropriate probes to the correct sockets on the front of the multimeter.
   - Connect the black probe and test lead to the common socket.
   - Connect the red probe and test lead to the voltage socket.

2. Most voltage that you will measure on a light vehicle is direct current (DC), so select the scale with the straight and dotted lines (‒‒‒). Remember that if you do not know which scale to select when using your voltmeter, always start at the highest setting and work down until a sensible reading is shown.

3. Remember to connect the voltmeter in parallel.

4. Make sure you connect the tip of the black lead to a good source of earth, such as the battery terminal, metal bodywork or engine.

5. Now use the tip of the red lead to probe the electrical circuit that you want to test.

Safe working

Do not measure the voltages of a hybrid drive system unless you have been specifically trained. Hybrid drives operate with high voltages that can cause electric shock and death.

Did you know?

A voltage test can be done with the circuit switched on or off, but remember that two different readings will be shown:

- electromotive force (EMF), which is the electrical pressure in the circuit with everything switched off
- potential difference (Pd), which is the electrical pressure in the circuit when it is switched on.

Using a multimeter to check amperage

When measuring the electric current in a circuit, you are using the multimeter as an ammeter. You need to take care because if the ammeter is connected incorrectly it can be damaged.

6 Light vehicle electrical units & components

- You must connect an ammeter in series to the electrical circuit being tested. This means breaking into and using the ammeter as part of the circuit. You can do this by disconnecting a wire, then putting the black probe on one end of the connector and the red probe on the other. This will complete the circuit and allow current to flow.
- A good place to connect an ammeter is at the fuse box. You will need to remove the fuse completely and replace it with the ammeter.
- Be careful that the expected current is below the maximum stated on the ammeter, as many multimeters are only rated to 10 amps.
- If the ammeter is connected correctly, when switched on, current will flow and the circuit will operate normally. You can now take the reading from the screen.

> **Safe working**
>
> Never connect an ammeter in parallel (across a circuit). A good ammeter has a very low internal resistance, and if you connect it in parallel too much current will flow. This will create a short circuit that could damage the ammeter or even cause a fire.

Checking amperage

Checklist as for *Checking voltage* on page 388.

1. Connect the appropriate probes to the correct sockets on the front of the multimeter. • Connect the black probe and test lead to the common or ground socket. • Connect the red probe and test lead to the socket used for measuring amps. (This socket is normally separate from the one used to measure volts or ohms.)	2. Turn the selector dial to amps measurement.	3. You now need to break into the circuit being tested, being careful to avoid short circuits.	4. Connect the ammeter in series, turn on the circuit and measure the current.

Non-contact inductive clamp ammeters

Non-contact inductive ammeters are available that can test the current flow without breaking into the circuit. You place a clamp around the wire of the circuit you want to test and then switch the ammeter on to take the measurement. An advantage of this type of meter is that it is less likely to be damaged and can normally take much higher current readings than a standard ammeter.

Using a multimeter to check resistance (ohms)

When measuring the resistance in a circuit, you are using the multimeter as an ohmmeter. An ohmmeter is different from the ammeter and voltmeter functions of a multimeter – instead of using the electrical energy from the circuit for testing, it uses its own internal battery.

When checking for electrical resistance, you need to switch off the power and disconnect the component to be tested from the circuit. If you try to test the component while it's in the circuit, you may be getting readings from the entire system.

> **Safe working**
>
> Do not use an ohmmeter on wiring that has yellow colour-coded insulation or shielding. This may be part of the supplementary restraint system (SRS) and can cause an airbag to deploy.

Figure 6.36 Inductive clamp ammeter

Checking resistance

Checklist as for *Checking voltage* on page 388.

1. Connect the appropriate probes to the correct sockets on the front of the multimeter:
   - Connect the black probe and test lead to the common or ground socket.
   - Connect the red probe and test lead to the socket marked with the omega symbol (Ω).

2. Before you take any measurements, you need to calibrate the ohmmeter to check that it is accurate.

3. Turn the selector dial to the lowest ohms setting and join the tips of the two probes together.

4. When the leads are connected, the readout should show zero or very nearly zero. (If any figures are shown on the screen you will need to add these to or subtract them from your final results.)

5. When the leads are disconnected you should see OL (meaning off limits) or the number 1, which is used to represent the letter 'I' (meaning infinity).

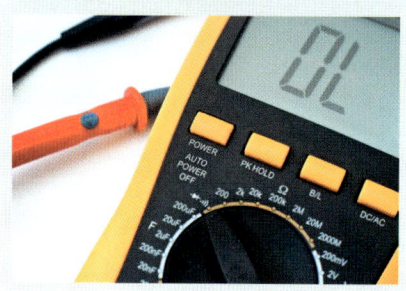

6. Now connect the ohmmeter in parallel across the component so that you can measure the resistance. (Remember to start on the lowest setting and work up.)

Checking for continuity

The ohmmeter can also be used to check for continuity (a continuous or unbroken conductor). You need to set it up and calibrate the ohmmeter in the same way as for resistance. Then connect it across the component to be tested. If a reading is shown then continuity exists. If infinity or off limits is shown then the component may be broken.

Some multimeters have an audible continuity tester. You need to connect the probes in the same way and calibrate the meter by joining the wires together. Once you've done this you should hear a beeping sound or tone. The device can then be used like an ohmmeter, but you don't need to be looking at the screen to see if the component is functioning.

Common electrical faults

Trying to diagnose an electrical fault can be confusing, as the symptoms can be very wide and varied. If you follow a simple approach, the diagnosis can be reduced to four main electrical faults:

1. open circuit
2. high resistance
3. short circuit
4. parasitic drain.

Open circuit

In an open circuit, electricity cannot flow. This is normally because there is a physical break in the system. This is like a blocked pipe which prevents water from flowing – when you turn on the tap, no water comes out.

To diagnose an open circuit, you can use the multimeter as a voltmeter. This acts like a pressure gauge for electricity.

If the circuit is working properly, you should see full voltage all the way up to the consumer, at which point the electrical pressure should be used up. In an open circuit, the voltage will disappear. You can use a voltmeter to see at what point in the circuit this happens.

For example, Figure 6.37 shows how a voltmeter is used to check an open circuit in which the bulb does not light up. The voltmeter is connected at various points of the circuit to find out the voltage at these points. Where the voltage is different (between 12 V and 0 V), this shows the position of the open circuit. In this example, the open circuit is between points B and C.

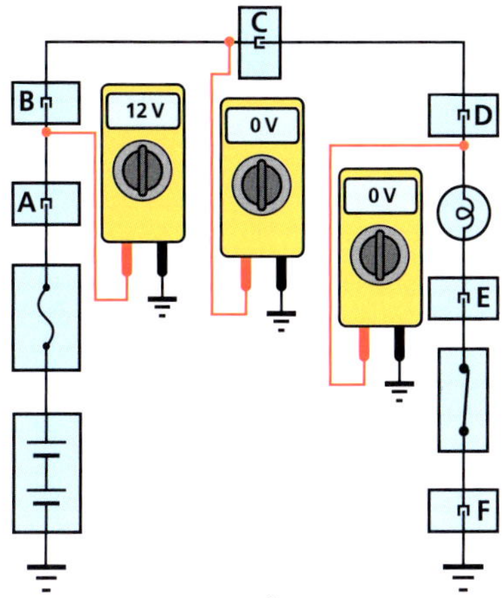

Figure 6.37 Using a voltmeter to check an open electric circuit (12V battery)

Checking an open circuit

Checklist as for *Checking voltage* on page 388.

1. Connect a voltmeter at the feed to the consumer.

2. If there is no voltage, the open circuit lies before this point.

3. If there is voltage, the open circuit lies after this point.

4. Use the voltmeter to probe the circuit at different points, to see where voltage disappears. The position of the open circuit will be between the last point where you got a voltage reading and the point where the voltage disappears.

Working life

Pete is trying to find the cause of an electric window motor that has stopped working. He thinks it might be a broken wire in the hinge.

1. How should Pete set up the multimeter?
2. How should the meter be connected to the circuit?
3. What readings should Pete get if the wire is OK?
4. What readings should Pete get if the wire is broken?

Level 2 Light Vehicle Maintenance & Repair

High resistance

In a high resistance circuit, the electricity slows down. This is normally because of a partial restriction in the system. Many high resistance faults are caused by poor, corroded or loose connections. (This is like squeezing a pipe so that the water inside slows down.)

The symptoms of high resistance are that the component does not work properly, for example a bulb glows dimly. This is because the pressure (voltage) is shared with the resistance.

To diagnose this fault, you can use the multimeter as a voltmeter. This acts like a pressure gauge for electricity. If the circuit is working properly, you should see full voltage all the way up to the consumer, at which point all of the electrical pressure should be used up.

Checking for high resistance

Checklist as for *Checking voltage* on page 388.

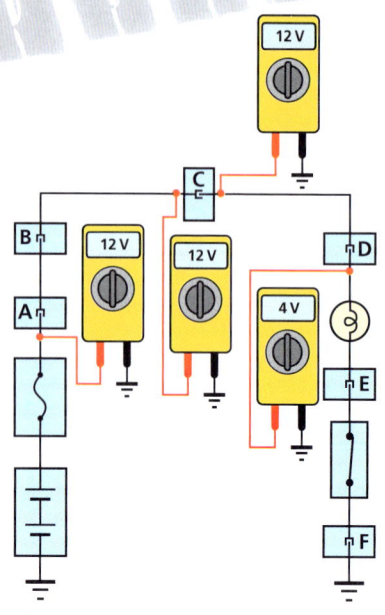

Figure 6.38 Using a voltmeter to check a high resistance electric circuit (12V battery)

1. Connect a voltmeter at the feed to the consumer.

2. If there is low voltage, the high resistance lies before this point.

3. If there is full voltage, the high resistance lies after this point.

4. Use the voltmeter to probe the circuit at different points to see where voltage changes (goes from high to low or disappears). This will locate the position of the high resistance.

> **Key terms**
>
> **Dead short to earth** – direct connection between the positive and negative sides of an electric circuit.
>
> **Chafed** – rubbed against something so that the surface has been damaged.

> **Safe working**
>
> When checking for an electrical short circuit, only bypass the fuse with an electrical consumer like a bulb. Using other electrical components (such as wire) could cause a sudden discharge of electricity that may burn you.

Bad earth

A bad earth is a high resistance after the consumer. If this exists the symptoms will be the same, for example the light on the car is dim. Sometimes a bad earth can cause the electrical energy to find an alternative path to the negative side of the battery. If this happens you may see symptoms such as all of the lights in the same unit operating at the same time, for example brake lights flashing with the indicators.

To diagnose a bad earth use the same procedure as for high resistance.

Short circuit

Electricity is lazy, and will always take the path of least resistance. (Why travel the full length of the circuit when it can take a shortcut?)

In a short circuit, the electricity doesn't make it all the way to the end. This is like a burst water pipe – the water leaks out before it gets to its destination.

The sudden discharge of current in a short circuit can cause a lot of damage, so the fuse that is used to protect the system should blow. If this happens the symptoms can make you think that the problem is an open circuit. Indeed, as the blown fuse has broken the circuit no current can flow.

In this situation, you can test the system with a voltmeter as explained on page 388. But once you have discovered the blown fuse, you should change your diagnostic routine to look for a short circuit. Any heat damage, including blown fuses, is a good indication that a short circuit might exist.

If a **dead short to earth** exists, for example because the insulation of a wire has **chafed** against the metal bodywork of the vehicle, you can use a test lamp to help diagnose this fault (see Figure 6.39). It is important to use a test lamp containing a bulb and not an LED, as this could lead to system damage.

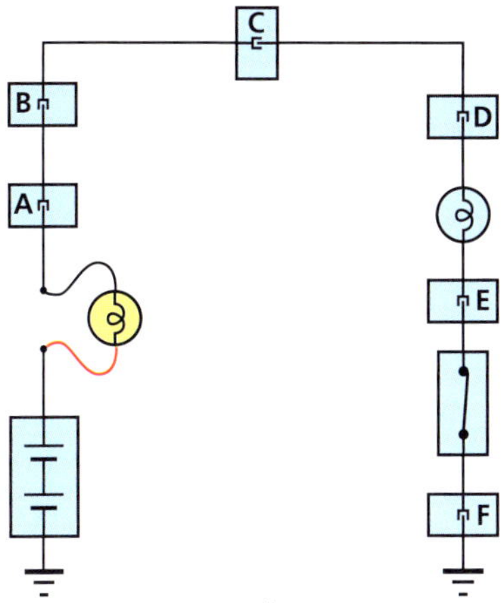

Figure 6.39 An electric circuit with a test lamp bulb connected in the place of the fuse to check for an electrical short circuit (12V battery)

Checking a short circuit

Checklist			
PPE	VPE	Tools and equipment	Source information
• Steel toe-capped boots • Overalls • Latex gloves	• Wing covers • Steering wheel cover • Seat covers • Floor mat covers	• Test lamp • Spare fuses	• Manufacturer's technical data

| 1. Remove the blown fuse and replace it with a test lamp. | 2. Switch the circuit on. | 3. If you see the bulb light up, the system might have a dead short to earth. | 4. Put the bulb where it can be seen and start systematically disconnecting and moving wires and components. | 5. If the bulb goes out, you have broken the earth connection and found the source of the short circuit. |

Parasitic drain

A parasitic drain is similar to a short circuit – electricity will continue to flow even if the system is switched off. (This is like a dripping tap, which allows the water to leak away.) The symptom reported is normally that the battery goes flat if left for a period of time.

To help diagnose this fault, you can use the multimeter as an ammeter. This acts like a flow gauge to measure the amount of electric current moving in a circuit.

Checking a parasitic drain

Checklist

PPE	VPE	Tools and equipment	Source information
• Steel toe-capped boots • Overalls • Latex gloves	• Wing covers • Steering wheel cover • Seat covers • Floor mat covers	• Multimeter or inductive clamp ammeter	• Manufacturer's technical data

1. Switch off all electric systems and connect the ammeter.

2. The ammeter must be inserted into the circuit (connected in series) so that it isn't damaged. To do this you may need to disconnect one lead from the battery and use the ammeter to bridge the gap, so that current flows through it.

3. With everything switched off, there should be no current on the display of the meter. If any current (measured in amps) is shown, then a parasitic drain exists.

4. To help find the parasitic drain, remove the fuses one at a time until **amps draw** falls to zero. This will help you locate the circuit containing the drain.

5. Once you have identified the circuit, disconnect the components in that system until the current draw falls once again. This will allow you to locate the parasitic drain.

6. You can now replace the faulty component.

Key term

Amps draw – the amount of current being used.

Did you know?

If you don't know what voltage or timescale to use on an oscilloscope, find out in the same way as you would with a multimeter. Start with the highest setting available and work downwards until you can see an image on the screen.

Working life

A customer complains that their battery keeps going flat when the car is left overnight. John thinks that one of the electric components might not be switching off properly, and says that this is called 'parasitic drain'.

1. What sort of things might cause a parasitic drain?
2. How could John find out what's wrong?

Oscilloscopes

The problem with a multimeter is that the measurement readout can't change fast enough to deal with modern electronic systems on motor vehicles – the numbers on the screen can't keep up. The answer to this is to use an oscilloscope.

Many people are put off using oscilloscopes because they think they're complicated, when really they are only a different type of voltmeter. Unlike a voltmeter, oscilloscopes not only show volts but also time. Instead of a digital readout, the results are shown as a graph of volts against time on a screen (as shown in Figure 6.40).

- The graph normally shows voltage on the vertical y-axis, and this axis is often labelled 'amplitude'. A scale setting switch can be used in a similar way to the dial on a manual multimeter to choose the amount of volts shown on the screen, e.g. 20 volts.
- The graph normally shows time across the horizontal x-axis. This axis is often labelled 'frequency'. A timescale switch can be used in a similar way to the dial that is used to choose the amount of volts, e.g. 5 milliseconds.

Many motor vehicle workshops own an oscilloscope but do not use them to their full potential, if at all. This can be due to lack of time, knowledge or ability. Lots of vehicle technicians are put off by the large box containing many wires and connectors. They feel that it will be complicated and time-consuming to set up, so they don't bother. However, to use an oscilloscope for simple electrical testing, only two probes are needed – a common and voltage wire – just like with a multimeter. Most of the diagnostic sockets on oscilloscopes are colour-coded, so after a quick check of the manufacturer's instructions it should be fairly easy to know where to plug these probes in.

If correctly connected or used with an inductive pickup, oscilloscopes can also be used to measure amps.

Using an oscilloscope for electrical testing

Note: The oscilloscope probes may come in different colours, but for the sake of simplicity we will call them red and black here.

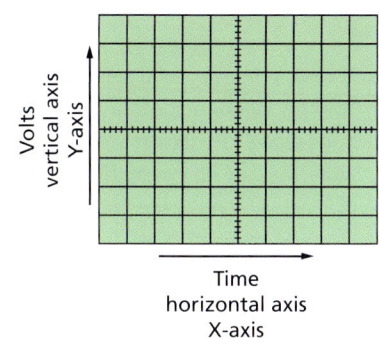

Figure 6.40 An oscilloscope screen

Safe working

The nice thing about oscilloscopes is that in theory they should not put out any power down the red probe. It should, therefore, mostly be safe to probe around until you find the signal wire from the components that you are testing.

However, you need to look out for wiring that is shielded in yellow insulation. You should never test this wiring, as it could be connected to the vehicle's supplementary restraint system (SRS) and can cause an airbag to deploy.

Checklist			
PPE	**VPE**	**Tools and equipment**	**Source information**
• Steel toe-capped boots • Overalls • Latex gloves	• Wing covers • Steering wheel cover • Seat covers • Floor mat covers	• Oscilloscope	• Manufacturer's technical data

1. Connect the black probe to a good source of earth or ground (e.g. the battery negative terminal). This will then only leave you with the red wire to connect.

2. Now connect the red probe to the circuit to be tested.

3. Adjust the scales until you see an image on the screen.

After some practice, you will become familiar with the patterns and waveforms created by different vehicle systems.

395

Level 2 Light Vehicle Maintenance & Repair

> **Did you know?**
>
> It's a common misunderstanding to think that plugging a fault code reader into the vehicle's OBD system will tell you what the fault is. It actually only points you in the direction of the fault. You must test the system and components to find the fault.

> **Find out**
>
> Describe some symptoms and faults associated with electrical and electronic systems, including:
>
> - high resistance
> - loose and corroded connections
> - short circuit
> - excessive current consumption
> - open circuit
> - battery faults, including:
>   - flat battery
>   - failure to hold charge
>   - low state of charge
>   - overheating
>   - poor starting.

Manufacturer dedicated equipment and scan tools

As vehicles have developed, electrical systems and control have become extremely sophisticated. This means that even with good diagnostic routines and equipment, faults can be extremely difficult to find, especially when they are intermittent (don't happen all the time).

Faults with many modern vehicle systems would be difficult to diagnose without the aid of a scan tool. The electronic processes mean that these systems are being controlled many thousands of times a second, and faults can happen so quickly that you could miss them.

Since the 1980s, manufacturers have been including on-board diagnostic (OBD) systems as part of their vehicle design. The computers that control the vehicle's electrical systems have a self-diagnosis feature. This allows them to detect certain faults and store a code number. Because these computers – called electronic control units (ECUs) – are monitoring functions, they are able to record intermittent faults and store them in a 'keep alive memory' for retrieval by a diagnostic trouble code (DTC) reader.

Figure 6.41 A technician connecting and using a scan tool

Fault code reading is the most common use of scan tools by automotive technicians, although many other functions exist on modern scan tools. Typical features include:

- retrieval of electronic control module (ECM) fault codes
- erasing of system ECM fault codes
- serial data/live data
- readiness monitors
- resetting of ECU adaptions
- freeze frame data
- coding of new components, such as fuel injectors
- access to information on various vehicle electronic systems
- resetting of service reminder lights
- vehicle key coding.

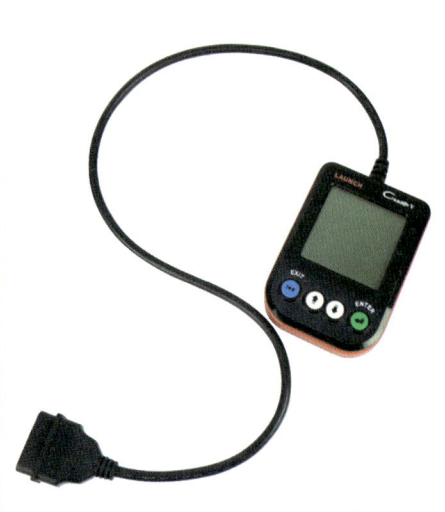

Figure 6.42 Diagnostic scan tool

6 Light vehicle electrical units & components

> **CHECK YOUR PROGRESS**
>
> 1 List the four main electrical units.
> 2 Why are fuses used in an electrical circuit?
> 3 Why are relays sometimes used in an electrical circuit?
> 4 What is the difference between a parallel and a series circuit?

Light vehicle starting and charging systems operation

An electric motor is used to drive the engine crankshaft. The motor is provided with electricity from the battery. Over a period of time the electrical energy in the battery will be used up, so a system is required to keep the battery charged. This is normally done with a generator called an alternator.

Charging the battery

As the energy in the battery is used up by the vehicle's electrical system, chemical reactions take place which reduce the strength of the sulphuric acid in the electrolyte. Once most of this electrical energy is used up, the battery becomes flat.

The reason why the lead-acid battery is so popular in light vehicles is that it is rechargeable. If an electrical voltage (pressure) is produced by the vehicle's charging system which is higher than the voltage coming out of the battery, then the chemical process can be reversed. The pressure forces electrons back in the opposite direction and increases the strength of the sulphuric acid.

Hydrometer

You can test the strength of the sulphuric acid in the electrolyte with a hydrometer (see Figure 6.43). This uses a floating indicator, which shows the **specific gravity** (or density) of the liquid being tested. If you take a sample of electrolyte from each battery cell, its specific gravity can be compared with that of the other cells. As all the cells are joined together to form one complete circuit, if you find a large difference in specific gravity between one cell and the others, the battery condition is poor and it might need replacing.

State of charge

On the scale of specific gravity, water is used as the base and is given a figure of 1. If the liquid is denser than water, its specific gravity will be higher than 1.

- A discharged battery cell will have a specific gravity of around 1.06 SG, which is almost water.

> **Did you know?**
>
> Electronic car battery testers are now available which, when connected to a battery and programmed with details found on the battery casing, will run through a series of checks and display the result of the analysis on a screen.

Figure 6.43 Hydrometer

> **Key term**
>
> **Specific gravity** – the density of a liquid when compared to water.

Level 2 Light Vehicle Maintenance & Repair

- A fully charged battery cell will have a specific gravity of around 1.28 SG, which is denser than water.

Table 6.5 shows the specific gravity of battery cells at different states of charge.

Table 6.5 Approximate battery cell state of charge and specific gravity

State of charge	Specific gravity
100%	1.280 SG
75%	1.230 SG
50%	1.190 SG
25%	1.145 SG
0%	1.060 SG

Recharging the battery

Electricity is needed to recharge the battery. To generate electrical voltage, an engine-driven magnet can be rotated inside a coil of copper wire. As the poles of the magnet turn, the invisible magnetic field moves past the copper windings, creating an electric current. The amount of electric voltage produced will depend on the strength of the magnets used and the speed with which they are turned. Early cars used a **dynamo** (see Figure 6.44) to generate this electricity, but they were inefficient in operation.

In many light vehicles, the generation of electrical energy to recharge the battery is now done by an alternator, which is a much more efficient generator than a dynamo.

Construction of an alternator

Many car alternators are bolted to the side of the engine with brackets. They are turned by rubber drive belts, which sit in pulleys and are normally driven by the engine crankshaft. The tension on the rubber drive belt can usually be adjusted, so that it is able to operate the alternator without slipping.

Rotor

The component that spins (rotates) in an alternator is usually referred to by the name rotor. In many cases it is a combined electromagnet with a number of North and South poles, as shown in Figure 6.45. For an electromagnet to operate, it must be supplied with electric current from the battery. Because the rotor moves, a way of connecting it to the battery circuit is needed – this is provided by a **slip ring** and **brush** assembly.

Figure 6.44 A car dynamo

> **Key terms**
>
> **Dynamo** – old-fashioned and inefficient type of electrical generator.
>
> **Slip rings** – pair of copper tracks connected to the alternator's rotor, which help provide a movable electrical connection.
>
> **Brush** – piece of carbon or metal which ends in wires or strips. These are designed to make contact with rotating slip rings or generator/motor commutators.

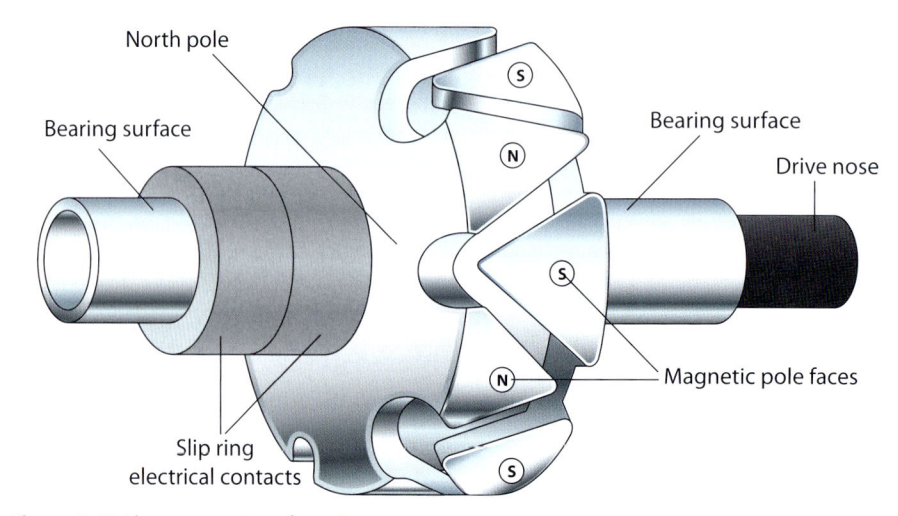

Figure 6.45 The rotor section of an alternator

6 Light vehicle electrical units & components

Stator

The electromagnet rotor needs to spin inside a coil of copper wire. This wire coil is normally stationary and is called the stator (see Figure 6.46). As the electromagnetic rotor spins inside this coil of wire, an electrical current is created which has a high enough voltage to charge the battery.

> **Did you know?**
>
> The generation of electrical energy creates heat, which means that alternators need to be cooled. This is done by attaching a fan to the alternator's rotating shaft, which draws air through the alternator to help keep it cool. To improve cooling efficiency, some manufacturers now produce liquid-cooled alternators, which are plumbed into the engine's cooling system.

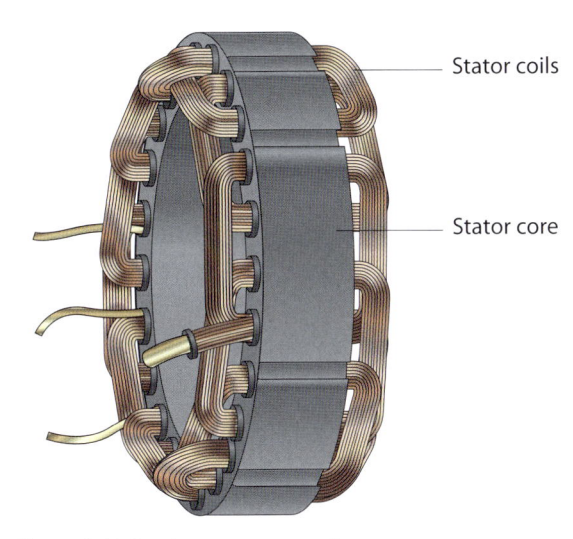

Figure 6.46 An alternator stator unit

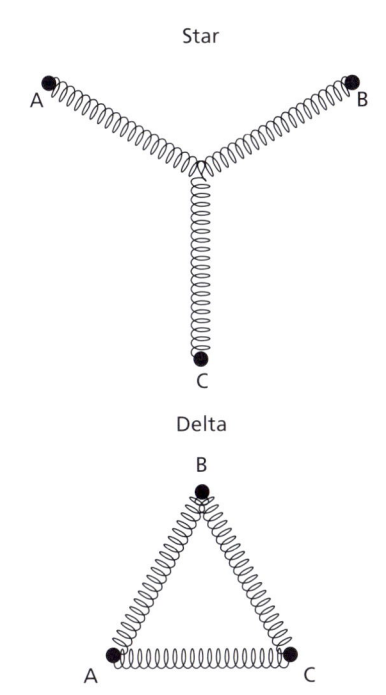

Figure 6.47 Stator phase wiring

To increase the output of the alternator, more than one coil of copper wire is used – in fact, many systems use three. Each coil of copper wire is called a **phase** (which leads to the term **three-phase**). These three phases can be joined together in either a 'star' or 'delta' formation, as shown in Figure 6.47.

If the current from a three-phase alternator is represented on an oscilloscope graph, it will look like the graph shown in Figure 6.49.

When three phases are used together, they give three times the output of a single phase generator. The three phases also overlap each other, which helps to smooth out surges in electrical current and gives a much more even delivery of electric charge – see Figure 6.49.

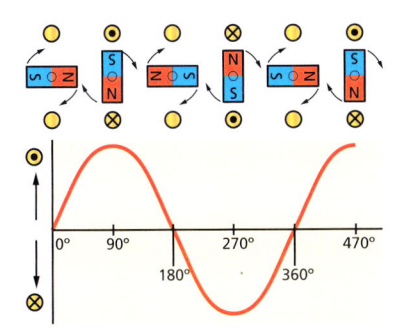

Figure 6.48 The generation of electrical energy by rotating a magnet past a copper wiring

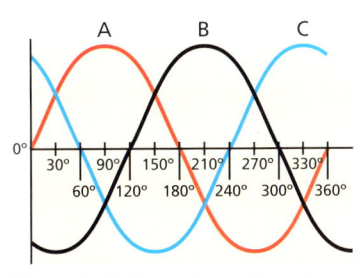

Figure 6.49 Three-phase electricity generation

> **Key terms**
>
> **Phase** – generator winding in which an alternating current (AC) wave is created.
>
> **Three-phase** – three individual generator phases connected, to triple the output of an alternator.

399

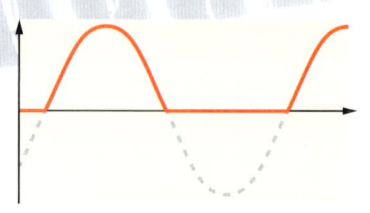

Figure 6.50 Half wave rectification (where half the wave has been used to create direct current)

Rectifier

As the name suggests, an alternator produces alternating current (AC). AC is electricity that moves in one direction then the other. This is because a magnetic field has North and South poles, and as the electromagnet spins, electricity is created first in one direction and then the other.

A battery cannot accept alternating current – if the current was left unchanged, only half of the generated electricity could be used. To solve this problem, a component called a **rectifier** is used to convert alternating current to direct current (AC to DC), as shown in Figure 6.50.

A series of **diodes** are connected to the output of the alternator phases. Because the diodes only allow electricity to flow in one direction, they can be arranged so that by the time the electricity reaches the battery it has all been converted into direct current. This is called full wave rectification and is very efficient.

Regulator

With mechanically generated electricity, the faster the magnet spins, the more voltage (or pressure) is produced. Left unrestricted, the engine could run so fast that the voltage output from the alternator increases until it damages the battery. To prevent this from happening, an electrical pressure relief valve called a **regulator** is used.

Some of the electric current produced by the three-phase windings in the alternator is directed back into the electromagnet of the rotor. This means that as the engine spins faster and faster, the electromagnet in the rotor becomes stronger and stronger.

- When a preset output voltage is reached, a **semiconductor** component called a **Zener diode** in the regulator shuts off electric current to the rotor so that it no longer acts as an electromagnet.
- As electric output voltage falls, the Zener diode switches the electromagnet in the rotor back on, which starts the alternator charging again.

This switching on and off of the electromagnet rotor regulates the output voltage while allowing a large amount of current (amps) to be produced.

Starter motor

For the engine to start, the crankshaft must be turned at speed (normally faster than 180 rpm). On early cars this was done with a starting handle attached to the front of the engine. The driver had to turn this by hand. This took a great deal of effort and, if the engine started suddenly, the handle could be thrown around with such force that it could break the driver's arm. An easier and safer method was needed, and this was achieved using an electric starter motor.

Did you know?

One way of understanding full wave rectification is to imagine a traffic policeman standing at a T-junction with cars coming at him from the left and from the right. The traffic policeman is able to redirect them so that they all travel down the road that is directly in front of him. It doesn't matter which way the cars arrive from, they all leave down the same road.

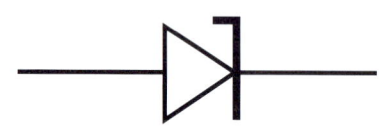

Figure 6.51 Zener diode symbol

Did you know?

Since the 1990s, some manufacturers began producing smart charging systems where the regulated output of the alternator is controlled by an ECU. This gives accurate control of the current used when recharging. The lifespan of the battery is increased due to better temperature control and prevents overcharging. Smaller loads are placed on the engine from the alternator so performance and fuel economy are increased and exhaust emissions are reduced.

6 Light vehicle electrical units & components

A simple starter motor is made using the same principles as the motor described in the section on *Fan and heater* on page 381, although the starter motor has a much more powerful design (see Figure 6.52).

The stronger the magnetic forces inside the starter motor, the more power it will produce. The magnets inside the starter casing that surround the armature can produce stronger magnetic fields if they are wrapped in wire coils and electric current is passed through them. These external magnet wires are called **field coils**.

> **Key terms**
>
> **Rectifier** – component that converts AC to DC.
>
> **Diode** – electronic component designed to act as a one-way valve for electricity.
>
> **Regulator** – electronic device for controlling the output voltage from a generator.
>
> **Semiconductor** – substance that has the ability to be both an electrical conductor and an insulator.
>
> **Zener diode** – electronic component designed to act as a one-way valve for electricity, until a certain voltage is reached, when it allows electricity to move in the opposite direction.
>
> **Field coil** – copper wiring wrapped around magnets, which can increase the magnetic field produced when supplied with electricity.

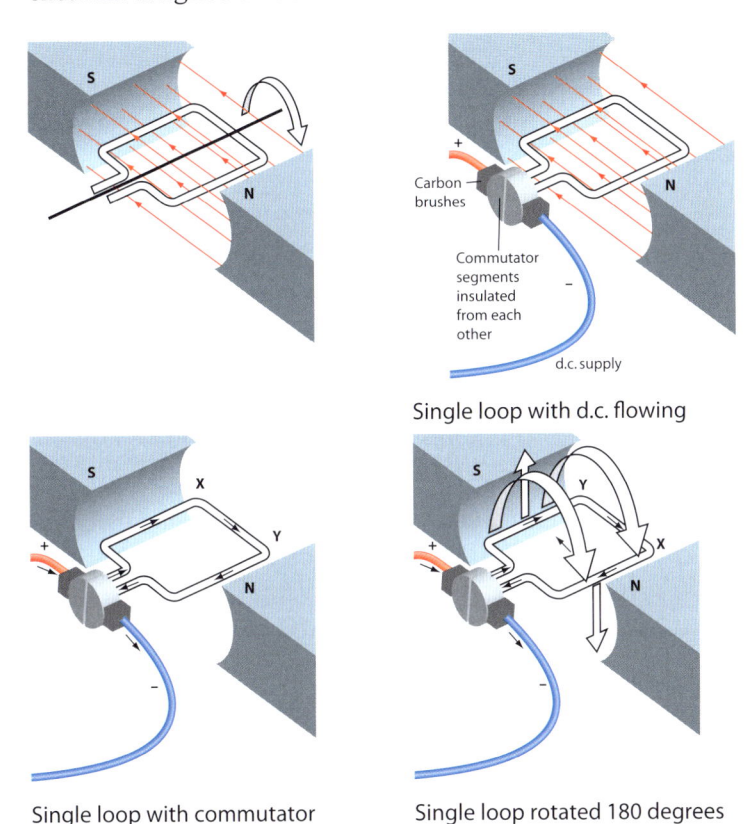

Single loop with d.c. flowing

Single loop with commutator

Single loop rotated 180 degrees

Figure 6.52 How an electric motor uses magnetism to produce rotation

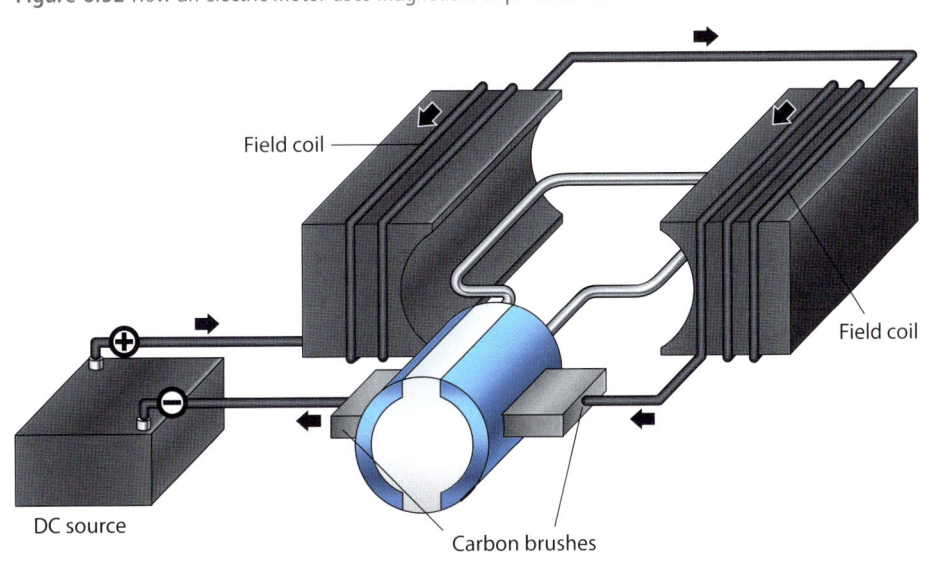

Figure 6.53 An electric motor which uses an electromagnet at the edge to create more power

401

Torque multiplication in the starter motor

Even with a very strong starter motor, a large amount of torque (turning effort) is required.

To multiply the torque, gears can be used in a way similar to their use in the gearbox (see the section on *Torque multiplication* on pages 339–340 of Chapter 5). If a small gear on the end of the armature shaft (called the pinion) is engaged with a large gear mounted around the edge of the engine flywheel (called the **ring gear**), the engine crankshaft can be turned.

You can calculate the torque multiplication by using the formula:

$$\text{Torque multiplication} = \frac{\text{Driven}}{\text{Driver}}$$

Figure 6.54 Starter pinion and flywheel ring gear

> **Example:**
>
> If there are 100 teeth on the flywheel ring gear and 10 teeth on the starter **pinion gear**, you need to do the calculation 100 divided by 10. This gives a gear ratio of 10:1. Therefore, if the starter motor is able to produce a turning effort of 80 Newton metres, the crankshaft will receive a torque of 800 Newton metres.
>
> As torque is multiplied, speed is reduced. So in this example, if the crankshaft needs to turn at 180 rpm, the starter motor armature must spin at 1800 rpm.

A big disadvantage of this gearing system is that when the engine starts, the gearing is reversed, with the flywheel ring gear becoming the driver and the starter motor pinion becoming the driven gear. This will turn the pinion gear with a reduced torque but a multiplied speed. If an engine is left to tick over at 1000 rpm, then the pinion gear of the starter motor will be rotating the armature at a speed of 10,000 rpm. So a method is needed to prevent this from happening. Depending on starter motor design, this can be achieved with the use of a Bendix or one-way clutch (see page 404).

Starter motor electrical feed

A large amount of electrical energy (current) is needed to operate a starter motor. The cable that runs from the battery to the starter motor is much larger than standard vehicle wiring. This large-diameter cable allows easy movement of electricity from the battery to the starter motor with very little resistance.

The rest of the starting system controls are also subjected to these large electrical loads. The small, lightweight switches (such as the starter/ignition switch) that are found in the controls of a car cannot cope with the amperage required. Because of this, heavy-duty switches are needed to connect the starter motor to the battery.

6 Light vehicle electrical units & components

A **solenoid** can be used to act as a heavy-duty switch.

- A metal rod surrounded by a coil of copper wire is supplied with a small current from the starter switching circuit inside the car.
- As this current is applied, an invisible magnetic field is generated in the coil of copper wire, forcing the metal rod to one side in a linear motion.
- If two large electrical contacts are mounted on the end of the metal rod of the armature, these can connect the large current-carrying wires attached to the starter motor.
- Because of this a small electric current is able to switch the large electric current required by the starter.

Starter motor designs

A motor is designed to turn electrical energy into mechanical or movement energy, but if the starter motor is turned by the engine it will become a generator. The generation of electrical energy within the starter motor will create a large amount of heat, and as the starter motor is not cooled, it would quickly be damaged or burn out.

To avoid the possibility of this happening, the pinion gear must not stay in mesh with the ring gear once the engine has started. A number of starter motor designs produce methods of engaging and disengaging the pinion gear during starter motor operation. The two most common designs used on cars are the inertia and pre-engaged types of starter motor.

Figure 6.55 Inertia starter motor

Inertia starter motors

An inertia starter motor is an old-fashioned form of starter motor. It uses the difference in speed of the armature and the pinion gear to spin it along a **helix** called a 'Bendix'.

As voltage is applied to the armature of the starter motor, it turns on so fast that it 'outruns' the speed of the pinion gear. This moves it along the helix, allowing it to engage with the ring gear on the flywheel. When the engine starts, the speed of the flywheel is far quicker than the pinion, and the gear is thrown back down the helix towards the armature of the starter motor (out of engagement).

Pre-engaged starter motors

Another type of starter motor, which is more commonly used, is the pre-engaged type (see Figure 6.56). It is easy to recognise a pre-engaged starter motor as it has a solenoid mounted on top. In this type of starter motor the solenoid performs two functions:

1. It acts as the heavy-duty switch to connect the power to the motor.
2. One end of the solenoid shaft is connected to a pivoting rod that pushes the pinion gear along the armature shaft and into engagement with the ring gear on the flywheel. When the starter button or key is released, a large spring pushes the solenoid back in the opposite

> **Key terms**
>
> **Ring gear** – large-diameter gear wheel mounted on the engine flywheel. It is used to engage with the starter motor drive pinion to start the engine.
>
> **Pinion gear** – small gear wheel, which is the driver of a meshing pair of gear wheels.
>
> **Solenoid** – electric motor that moves in a straight line (linear motion) instead of rotating.
>
> **Helix** – a spiral, similar to a screw thread.

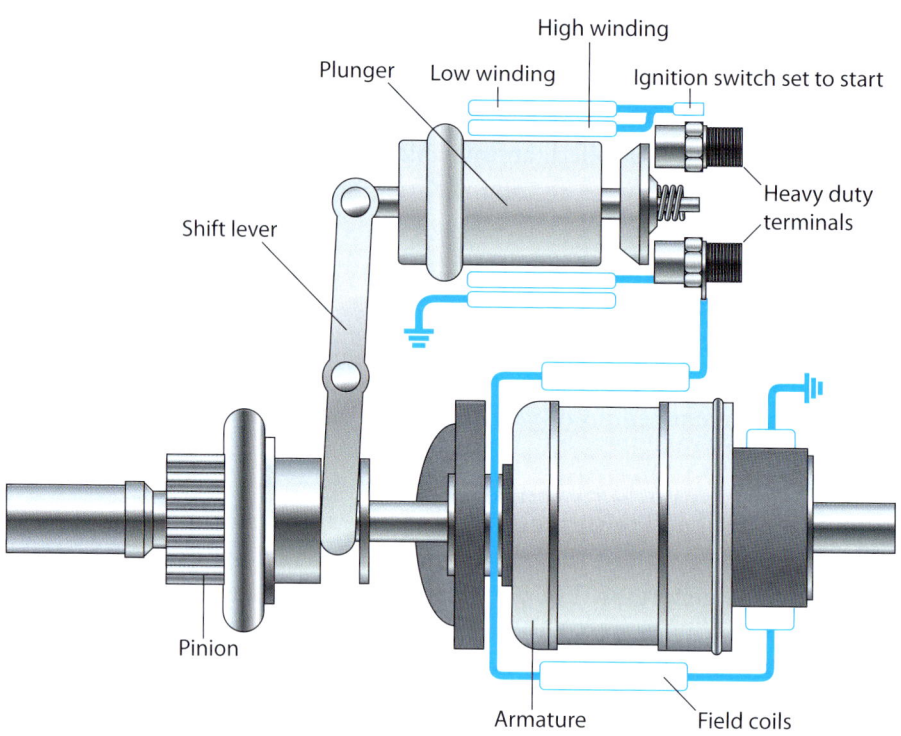

Figure 6.56 A pre-engaged starter motor

direction. This disconnects the heavy-duty power to the starter motor and pulls the pinion gear out of mesh from the flywheel.

One-way clutch

Even though starter motors are not left engaged when the engine is running, there is a short period of time at start-up when the engine speed will outrun the speed of the starter motor and can cause damage. To help prevent this, pre-engaged starter motors use a one-way clutch. This allows the motor to turn the engine but does not allow the engine to turn the motor.

A one-way clutch is normally made using metal rollers. When the armature shaft spins, the centrifugal force produced throws the rollers outwards. The rollers wedge themselves into a groove, locking the pinion gear to the starter motor shaft.

When the engine starts, force created from the spinning ring gear spins the starter clutch back in the opposite direction. This moves the rollers back out of the wedge, allowing the pinion gear to freewheel.

To check the operation of the one-way clutch with the starter motor removed, simply turn the pinion gear on the end with your fingers:

- In one direction the gear should lock to the armature and turn the starter motor shaft.
- In the other direction, the pinion should move freely.

6 Light vehicle electrical units & components

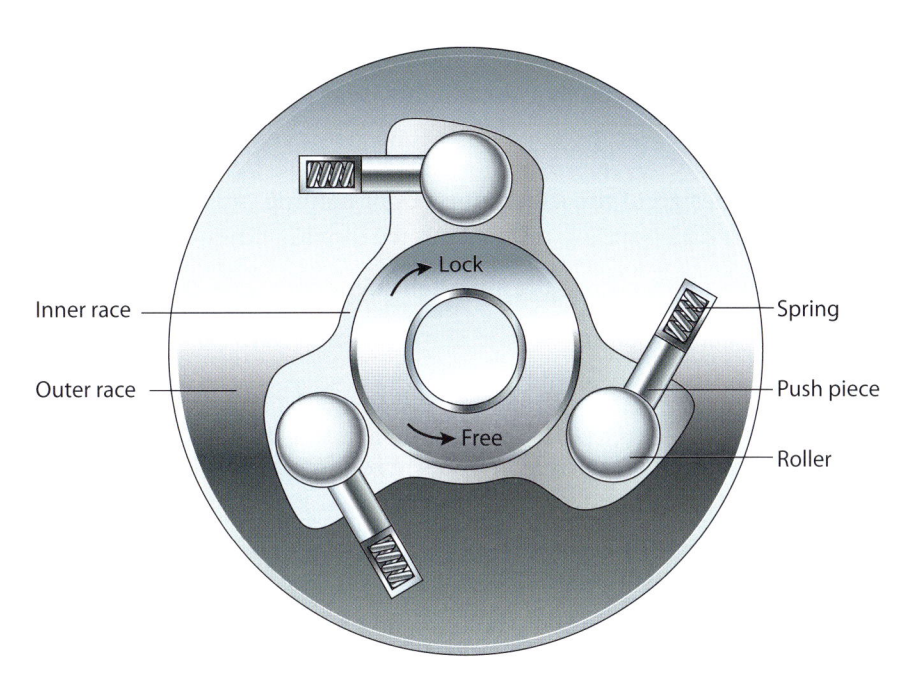

Figure 6.57 A starter motor one-way clutch

CHECK YOUR PROGRESS

1 Name two types of engine starter motors.
2 Describe how an alternator generates electrical energy.
3 Explain the safety procedures you should take when removing and refitting a battery.
4 What is the purpose of a rectifier?
5 What is the purpose of a regulator?

Light vehicle auxiliary electrical systems operation

Light vehicle auxiliary electrical systems are all the electrical systems that are not directly related to starting, charging and ignition. These can include:

- lighting
- comfort and convenience
- entertainment.
- anti-theft
- safety

Auxiliary systems also cover new technology such as: keyless entry, external view cameras, lane change safety, self parking, anti collision and pedestrian safety.

Find out
Choose a vehicle in your workshop and list all of the electrical auxiliary systems that you can find.

Lighting

Most standard vehicle lighting is created by forcing electric current through a thin piece of tungsten wire called a filament. The electric current causes the filament to get very hot – when it reaches a temperature of 2300 °C, it produces white light. By placing the bulb

405

inside a housing and using reflectors and lenses, different types of lighting can be produced such as headlights, rear lights and indicators.

Lighting colours

The colour and design of the different types of lighting must meet certain legal requirements, depending on the country in which they are used.

In the United Kingdom, the following lighting colours are required:

- All lighting facing forward should show a white light
- All lighting to the rear should be red.

This colour-coding helps other road users work out which direction the car is travelling. For example, if you see red lights, you know that the vehicle is moving away from you, but if you see white lights, this tells you that the vehicle is heading towards you. (This is why reversing lamps show white light to the rear.)

Direction indicators are normally colour-coded amber and are designed to flash on and off at a speed of between 60 and 120 flashes per minute. This differentiates them from the other lamps and helps other road users anticipate a direction change.

Lighting design

There are two main types of lighting design:

- **Lighting to help you see where you're going** – This type of lighting focuses a beam of light to assist with vision. It is used for headlamps, spot lamps and fog lamps.
- **Lighting to show the vehicle's position or direction** – This type of lighting uses diffused light. This is where a lens spreads out the light emitted, making it glow. It is used for tail lights, indicators, rear fog lights and reversing lamps.

Headlamps

Headlamps use a type of bulb called **pre-focus** (see Figure 6.58). Pre-focus bulbs allow vision in the dark by using a curved reflecting lens to direct light forwards in a concentrated beam. The bulb illuminates where you point it, just like a torch. Often shields are used on the bulb which direct the light backwards into the reflector (see Figure 6.59).

The position of the filament in the bulb will determine the direction of the light:

- When the filament is placed above the centre line of the reflector, most of the light hits the upper part of the reflector and is then pointed downwards in a dip beam.
- When the filament is placed on the right-hand side, the beam will be directed to the left.
- When the filament is placed on the left-hand side, the beam will be directed to the right.

Figure 6.58 Headlamp unit with a pre-focus bulb

Figure 6.59 Pre-focus bulb and reflector directing light

The beam is then directed through a shaped lens covering the front of the lamp unit. **Prisms** moulded into this lens further bend the light to achieve a precisely directed headlamp beam, as shown in Figure 6.60.

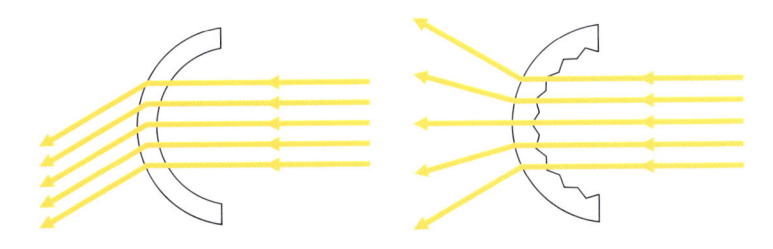

Figure 6.60 Prisms used in headlamp lenses

Did you know?
Some modern headlamp systems use a spark inside a bulb to create a chemical reaction with xenon gas, which will produce a very bright blue/white light. These systems are known as HID or high intensity gas discharge lighting.

Key terms
Pre-focus – bulb mounted in front of a reflector to create a beam of light. It is usually found in headlamps.

Prism – shaped transparent glass designed to bend light.

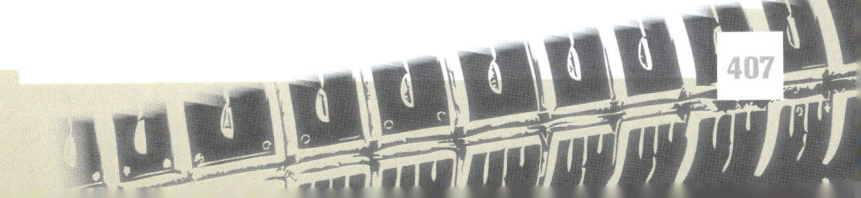

Dip beam and full beam

In the United Kingdom headlamps are designed to have a dip beam that points down to the left. Some lenses also allow the beam to be projected upwards on the left-hand side to help illuminate pavements and road signs. The beam of light is allowed to project upwards slightly on the left-hand side only, since this helps to avoid dazzling oncoming traffic.

If there is no oncoming traffic, the driver can switch the headlamps to a main beam filament which sits right in the middle of the reflector. This allows a direct beam of light to be projected in front of the car to illuminate all areas of the road ahead.

As part of the legal requirements for using vehicles, headlamp aim must be regularly checked and adjusted to ensure correct alignment.

Fog and spot lamps

Front fog and spot lamps work on the same principle as the pre-focus bulbs used for headlamps. The pre-focus bulbs cast a light that can dazzle oncoming drivers. Spot lamps are used for extra illumination in addition to the main beam headlamps.

- If mounted on the front of the car higher than 1200 mm from the ground, they are classed as spot lamps. These lamps should be connected so that they can only be switched on when the main beam headlights are illuminated.
- If mounted below 1200 mm from the ground, they are classed as fog lamps and should only be used in foggy conditions.

Fog lamps can sometimes be fitted with a yellow lens. If white light is used in foggy or snowy conditions, glare is produced which makes it more difficult to see. If yellow light is used the light will penetrate better and improve vision in these conditions.

Direction indicators

Direction indicators are used to help other road users know which way the car is going to move. They are normally amber in colour. They switch on and off and are controlled by the **flasher unit**.

> **Key term**
>
> **Flasher unit** – electrical component used to control the flash rate of direction indicators.

Bimetallic strip flasher unit

In a bimetallic strip flasher unit, the electrical contacts are mounted on two strips of different metal.

- When current is passed through the electric contacts, the indicator bulbs light up.
- When the bulbs illuminate, current is also passed through a small heating element, which warms up the bimetallic strip.
- As the bimetallic strip heats up, the metals expand at different rates and begin to bend. At this point the circuit is broken and no current flows to either the bulb or the heater element.
- As soon as the circuit is broken, the bulb goes out and the heater element begins to cool down. The bimetallic strip cools, contracts and reconnects the circuit, starting the whole process over again.

If a bulb fails, the change in circuit resistance can cause the indicator to speed up on that side and flash faster.

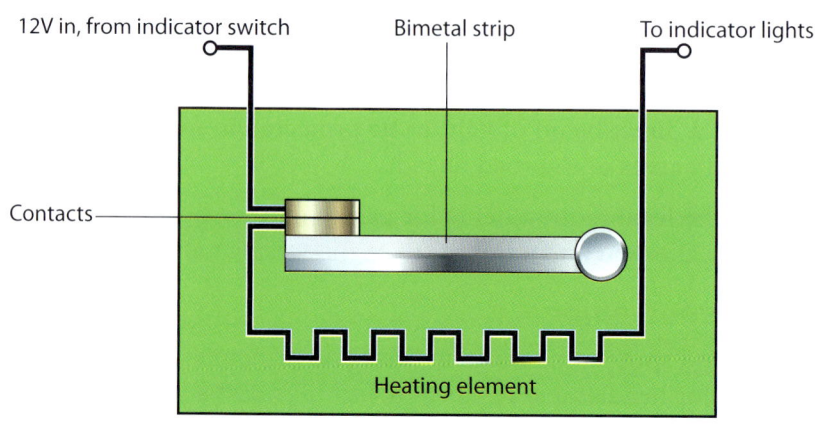

Figure 6.61 A bimetallic strip indicator flasher unit

6 Light vehicle electrical units & components

Electronic relay flasher unit

In an electronic relay flasher unit, a small processing chip acts as a timer to open and close a relay connection, which makes the indicator lights flash on and off. With this type of flasher unit, changes in circuit resistance do not affect indicator speed.

Circuit diagrams for vehicle lights

Study the circuit diagrams shown in Figures 6.63 to 6.67 to familiarise yourself with the layouts and operation of vehicle lights.

Side and tail lamps

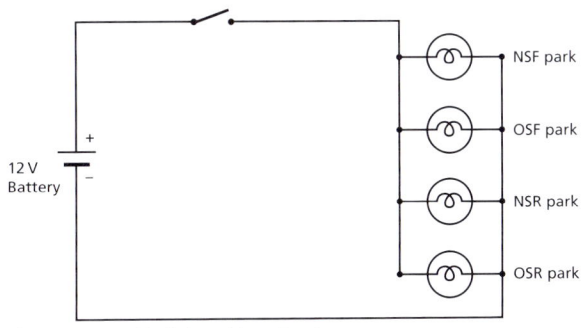

Figure 6.63 Side light wiring circuit

Figure 6.62 An electronic relay indicator flasher unit

Headlamps

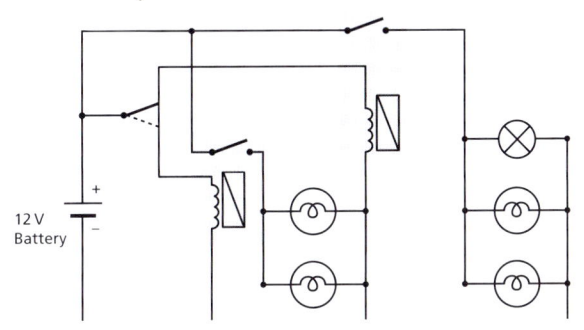

Figure 6.64 Headlamp wiring circuit

Interior lamps

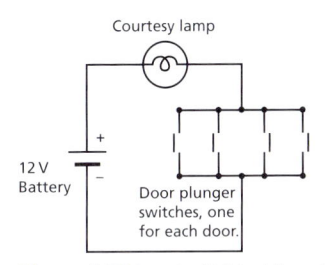

Figure 6.65 Interior light wiring circuit

Fog and spot lamps

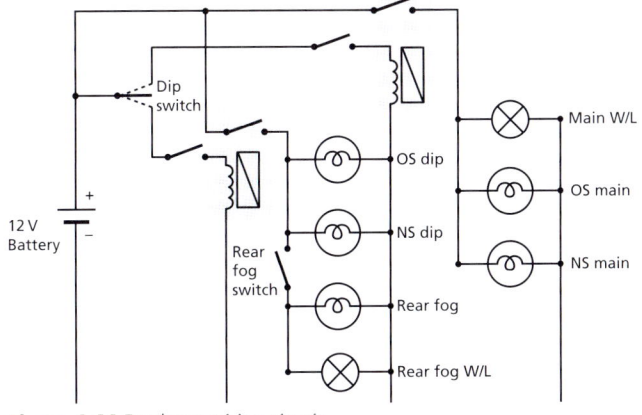

Figure 6.66 Fog lamp wiring circuit

Direction indicators

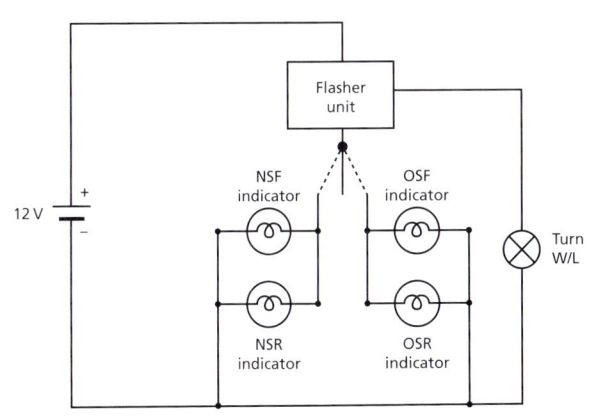

Figure 6.67 Direction indicator lamp wiring circuit

409

Auxiliary systems

As the use of electricity has developed, many electrical convenience devices have been produced to make the operation of vehicle systems easier.

Central door locking

Central locking uses solenoids or actuators in each door (see Figure 6.68). These are operated from a master control, which is normally the driver's door, but can be both front doors. When operated, they automatically lock all of the other doors, including the boot and petrol flap. As the key is turned in the master door lock, a **microswitch** sends a signal to the control unit, which then operates the solenoids or actuators to open all of the other doors.

- Many solenoids or actuators reverse their direction (i.e. lock or unlock) by changing the polarity of the voltage applied.
- All central locking solenoids or actuators are connected in parallel, so if one fails the others will still continue to operate.

Lots of modern central locking systems operate by remote control, which can be infrared or radio wave controlled, normally from the key fob. An advancement of this system is called 'lazy lock'. With lazy lock there is a main control unit, which not only operates the central locking but can also close electric windows, sunroofs and in some cases convertible hoods.

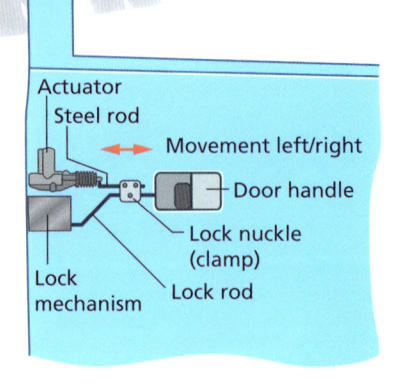

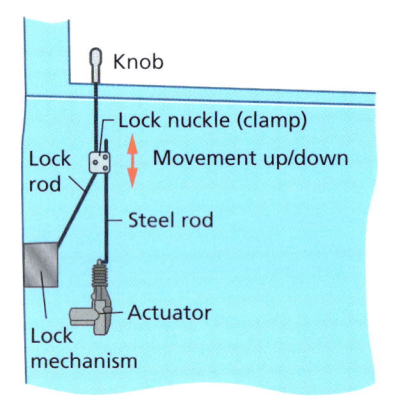

Figure 6.68 Central door locking mechanisms

Anti-theft devices

Theft of and theft from cars makes up a large proportion of all reported crime. To combat this, many manufacturers now include anti-theft systems in their vehicles, which are constructed as part of the vehicle manufacture process.

The two main types of anti-theft system used are hardware and software.

Hardware

Hardware systems are physical systems that help prevent theft of vehicles. Examples include:

- **Deadlocks** – When the key is turned in the door-locking mechanism, it operates a mechanical system that prevents the lock or latching mechanism being operated unless the correct key is used.
- **Steering locks** – When the key is removed from the ignition barrel, a mechanical latch operates against the steering column, preventing it from being used.

A number of aftermarket hardware devices are also available that can attach to the steering wheel (see Figure 6.69), the handbrake and gear lever mechanism, or around the actual road wheel to help prevent theft.

Figure 6.69 Steering wheel lock

Software

Anti-theft software normally involves an alarm or immobiliser system to reduce the possibility of theft. When actuated, either manually by a switch or automatically (a short time after the ignition is switched off and all doors are locked), alarms or sirens can be used to deter a potential thief.

When forced entry is detected, the alarm should sound for a short period of time before resetting.

Different methods that can be used to trigger an alarm system include:

- **Switches** – This could be, for example, a courtesy light switch mounted on the door, so that opening the door triggers the alarm system.
- **Volumetric sensing** – A radar or sonar system is used inside the car. If movement is detected within a certain area the system is triggered.
- **Voltage-sensed systems** – If any electrical circuit is switched on, the potential difference or voltage drop is sensed by the alarm system and the alarm is triggered.

Care should be taken if a voltage-sensed system is being fitted to a vehicle. Some electrical systems, such as electric cooling fans which are designed to continue running after the vehicle has been switched off, may inadvertently trigger the alarm system.

Immobilisers

Immobilisers are devices used to prevent the car from being started or driven. Many immobilisers are 'passive'. This means that the immobiliser doesn't stop the car being broken into or damage being done, but it does help prevent theft.

A number of systems on a vehicle can be immobilised, including:

- fuel supply
- spark ignition
- starter motor system
- engine management ECU operation.

These systems are connected so that if theft is attempted, an open circuit is created. This makes it difficult for a potential thief to get the vehicle started.

When fitting an immobiliser system, it is recommended that two or more systems are installed. Having more systems will increase the time required to bypass them and get the vehicle started, which will make it more difficult for someone to steal the vehicle.

Electric windows

Instead of a manual winder handle connected to the door window mechanism, an electric motor can be used. This is operated by a **rocker switch** and controls the movement of the window up or down.

It is common for an electric window system to include one touch operation. This means that when the switch is pressed once either up or down, the window will go either fully open or fully closed.

> **Key terms**
>
> **Microswitch** – very small electrical switch.
>
> **Rocker switch** – switch that goes both ways.

> **Did you know?**
>
> For a one-touch system to operate correctly, it is usually necessary to program the fully open and fully closed positions. If battery power is disconnected to the one-touch system, it will normally not work after reconnection and will have to be reprogrammed.
>
> A common method of reprogramming one-touch systems is to hold the rocker switch in the fully open position for at least 3 seconds after the window has reached the end of its travel. You then repeat this process for the fully closed position. (Different manufacturers may use different methods of programming.)

One-touch operation can pose a significant health and safety risk. If operated incorrectly, it is possible for someone to become trapped in the window mechanism and even be strangled. To help prevent this, many systems operate a safety stop or 'inch back'. If the window becomes jammed or does not fully close, the motor can be stopped and some systems can reverse the direction of the motor to open it slightly. To allow the system to operate, a motor position sensor is needed (such as a Hall effect sensor) or a current-sensing device that shows that the motor is struggling to operate.

Front and rear windscreen demisting

Screen demisters (see Figure 6.70) allow the driver to have clear vision at both the front and the rear of the car. During the windscreen manufacture process, a thin filament of electric wire is laid between the layers of glass. This filament is connected to a source of electrical power and current flow, and the energy is converted into heat.

One main difference between front and rear windscreen demisters is that front screen elements tend to have a much thinner wire so that forward vision is not affected.

The heating elements of both the front and rear windscreen can draw large quantities of current, so a timer circuit is included, which stops the flow of current after a preset amount of time. This reduces the possibility of damage caused by overheating or excessive load on the electrical system.

Electric door mirror operation

Many door mirrors can be adjusted by the driver from inside the car without moving from the driving seat. A control switch mechanism is used to operate two direct current motors behind the mirror glass – one motor goes up and down and one motor goes side to side. The combination of the two allows the driver to adjust the mirror glass exactly as he or she needs it for clear vision.

Figure 6.70 Heated window element

Some door mirror glasses include a heating element similar to that found in a windscreen demister. This can be operated when the vehicle is first started for a short period of time, to help demist the external door mirrors.

Interior lights

Interior lights, sometimes called courtesy lights, are usually switched on when a door is opened. This is to assist the driver and passengers getting in and out of the car in the dark. Most courtesy light systems use switches on the door shuts – when a door is opened the switch closes, earthing the circuit and switching on the lights.

The courtesy light circuit can be connected to a timer or a charging **capacitor**, to keep the lights on for a short period of time after the doors have been closed. This will help the occupants correctly adjust seats, put on seat belts and find the ignition key slot.

Electric sunroof operation

As with electric windows, sliding sunroofs can be operated by an electric motor and a rocker switch. A direct current motor is used to drive the sunroof open or closed by reversing the polarity of electricity flowing and changing the direction of the motor.

Many sunroofs use a **limit switch** so that the system knows it has reached the fully open or fully closed position. Some systems use safety measures similar to those found on electric windows (see opposite) to reduce the possibility of injury due to being trapped in the sunroof as it closes.

Comfort and convenience systems

Heated seats and electrically adjusted seats

Some vehicle manufacturers produce heated seat systems. These use elements and time and control circuits similar to those found in window demisting.

Seat positioning can also be controlled by electric motors. These allow the seat to be moved forwards, backwards, up, down or tilted to suit the driver or passenger. This system can be monitored by an ECU and programmed so that driver preferences for seating position are stored in a memory and set with one-switch control. This means that if the car has a number of drivers, each individual can have their own seat position setting, which is quickly replicated when needed.

Heating and ventilation

To help with occupant comfort inside the cabin of the car, a heating and ventilation system is provided.

Ventilation

Air travelling through the cabin can be controlled by a series of flaps and vents. The speed at which air travels through the cabin can be boosted by the use of an electric fan.

Heating

The temperature inside the vehicle can be raised using a heating system. A small internal radiator, called a **heater matrix**, is connected to air ducting behind the dashboard. Liquid coolant from the engine is passed through the heater matrix. As air is forced over the heater matrix by an electric fan, the air is warmed and is directed out of a series of ventilation flaps inside the vehicle. It is usually possible to direct the air vents at the feet, face or windscreen.

Air conditioning

Some manufacturers now include air conditioning or climate control as part of their vehicle design.

> **Key terms**
>
> **Capacitor** – device used to store electricity temporarily.
>
> **Limit switch** – switch that operates when a component has opened or closed fully.
>
> **Heater matrix** – the heater radiator for the passenger compartment.

Key term

Compressor – a mechanical device that compresses gases.

Find out

Find circuit diagrams for the following electrical auxiliary systems:

- central door locking
- anti-theft devices
- electric windows
- demisting systems
- door mirror operation mechanisms
- sunroof operation.

Study the circuit diagrams to familiarise yourself with layouts and operation.

An engine-driven **compressor** circulates a refrigerant gas around a sealed system containing two radiators:

- a radiator mounted at the front of the car under the bonnet, called the condenser
- a radiator mounted inside the car just behind the heater matrix, called the evaporator.

During operation, the evaporator becomes cold. As the air from inside the cabin passes over it, heat is absorbed into the refrigerant gas. This gas continues to circulate and, when it reaches the condenser, the heat is radiated to the surrounding air. This system works as a heat exchanger, taking the heat out of the air from inside the car and getting rid of it outside.

The system is called air conditioning because it not only cools the air, it also helps remove moisture, dirt and dust.

With a standard air-conditioning system, the driver must manually select the required temperature and electric fan speed. But with climate control the system is automatically controlled by a series of sensors and an ECU.

CHECK YOUR PROGRESS

1. Are alarm systems 'active' or 'passive' security?
2. How many motors are normally used to operate adjustable door mirrors?
3. Why is a timer circuit used in a heated rear windscreen?

How to check, remove, replace and test light vehicle electrical systems and components

Whenever you replace an electrical component on a light vehicle, you must make sure that:

- the component is switched off
- the electrical supply is isolated (disconnected) where possible
- the component is cool.

Remember that when an electrical component is being used, some of the energy will be turned into heat, so the component can get hot.

Starting and charging

Before replacing any starting or charging components, such as alternators or starter motors, it is important that you test them thoroughly.

These components are expensive and are normally replaced on a service exchange basis. This means that when you replace the starter motor or alternator of a car, the old unit will be sent back to the supplier for overhaul. Once it has been repaired or remanufactured it will then be sold on to another customer. This is a form of recycling.

Information on the operating values of starter motors (power) and alternators (current output) can be found in the manufacturer's technical data, as shown in the example in Figure 6.71.

Technical data on the vehicle

Make	Volkswagen	Date	15-12-2010
Model	Golf IV 1,6i	Owner	
Year	2000-2006	Registration No.	
Engine	AUS	VIN	
Variant	Hatchback	1. Reg. Date	

Technical item	Data
Electrical system	
Battery	12 V 44 Ah
Starter motor current (cranking), A	12 V - 0,9 kW
Voltage relay, Volt at/amp.	12,5 - 14,5 V/
Terminal definitions DIN 72 552	
Alternator max, A	70

Remarks

GDS Diagnostics Order No.:
Address Mechanic

Figure 6.71 Manufacturer's technical data

Correct method to replace a halogen headlamp bulb

When you replace a quartz-halogen headlamp bulb, take care not to touch the quartz globe. The natural oils found on the skin of your hands can boil at the extremely high temperatures produced, leading to damage and premature failure of the bulb.

When handling this type of bulb, always hold it by the metal base. When fitting it into the headlamp unit, it is important to mount the locating lugs (small metal legs at the base of the bulb) in the correct position.

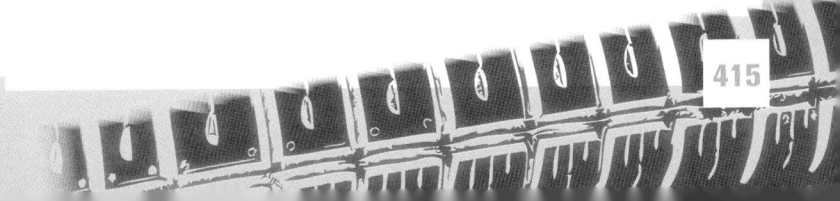

Level 2 Light Vehicle Maintenance & Repair

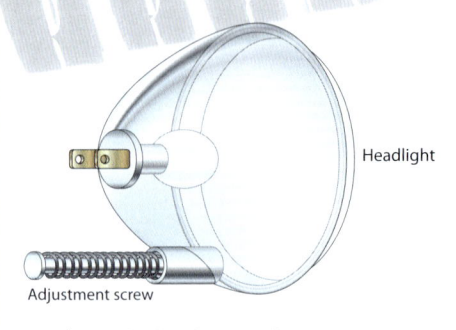

Figure 6.72 Using an adjuster screw to change the angle of a headlamp unit

Key terms

Tolerance – an allowable difference from the required measurement.

Obligatory – required.

Correct method to replace a headlamp unit

If the headlamp unit needs to be replaced, it is important to follow certain procedures.

- Make sure the headlamp unit is switched off and allow it to cool.
- Follow the manufacturer's instructions for the correct removal, as it is easy to damage clips and mounting positions.
- Following safe removal, always use the correct type of headlamp to replace the original, otherwise the type of light emitted, beam pattern and direction of beam may be affected. For example, replacing a headlamp unit with one designed for a left-hand drive vehicle will direct the headlamp beam to the right-hand side of the vehicle and dazzle oncoming drivers.
- The alignment of the headlamp beam can usually be adjusted by turning screws at the back of the headlamp unit (see Figure 6.72).

Once fitted, you must check the headlamp for alignment to make sure that it complies with the UK regulations. Before you make any headlamp alignment checks, certain criteria must be met:

- The car should be on level ground.
- Tyre pressures should be correct.
- Suspension and ride height should be within the manufacturer's specifications, so make sure that any wear in suspension will not affect the alignment of the headlights.
- The car should be correctly loaded, so that any weight in the boot, for example, will not affect the headlamp alignment.

You will need to set up the headlamp alignment tool following the manufacturer's instructions for the type of headlamp to be tested. Depending on the type of headlamp and its operation, when it is switched on a beam pattern will be cast on a scale in the alignment tool.

The beam pattern must be appropriate for the headlamp type, and it must fall within scale **tolerances** laid down by lighting regulations. If the headlamp alignment is incorrect, adjusters are normally included in the design of the headlamp unit so that corrections can be made. Examples of headlamp patterns are shown in Figure 6.73.

Statutory requirements for vehicle lighting when using a vehicle on the road

Lights that are required for use on roads in the UK are known as **obligatory** lights. They are covered by the Road Vehicle Lighting Regulations 1989 and include those shown in Table 6.6.

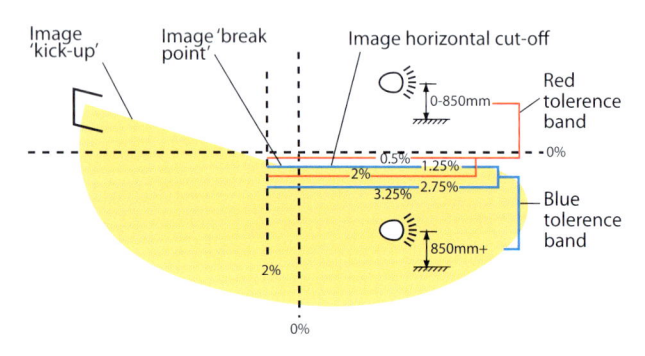

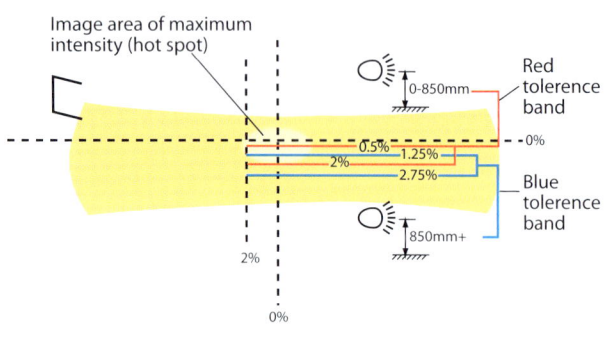

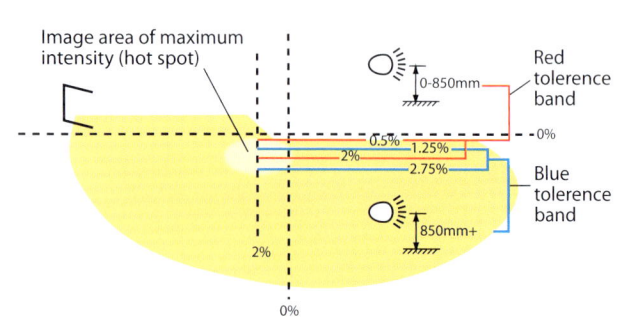

Figure 6.73 Different headlamp beam patterns as seen in a headlamp alignment tool

6 Light vehicle electrical units & components

Table 6.6 Obligatory vehicle lights in the UK

Lights	Regulations
Obligatory front position lamps (normally called side lights)	• These lamps must be fitted to a vehicle to indicate its presence and width when viewed from the front. • They must show a steady white light to the front, or a yellow light if included in a headlamp which emits yellow light. • The light must be visible from a reasonable distance. • Direction indicators can be combined in the same lamp housing as front position lamps.
Obligatory rear position lamps (rear lamps)	• These lamps must be fitted to a vehicle to indicate its presence and width when viewed from the rear. • They must show a steady red light to the rear that is visible from a reasonable distance. • Stop lamps or direction indicators can be combined in the same lamp housing as rear position lamps.
Obligatory registration plate lamps	• Registration plate lamps are the lamps that illuminate the rear registration plate. • They are white in colour but shielded so that none of the light can be seen from behind.
Obligatory rear fog lamps	• Cars only require one rear fog lamp, which must be fitted to the centre or offside of vehicles first used on or after 1 April 1980. • A rear fog lamp is allowed to operate independently of headlamp, position lamp or ignition systems.

Obligatory headlamps	• Headlamps are used to illuminate the road ahead of the vehicle, but are not fog lamps.
	• Headlamps must be a matched pair, which means that they must both:
	  – emit light of substantially the same colour and intensity
	  – be the same size and shape
	  – be symmetrical to one another.

Obligatory stop lamps (also called brake lamps)	• Vehicles first used before 1st January 1936 do not need to be fitted with a stop lamp. • Cars first used before 1st January 1971 must be fitted with one stop lamp, either on or to the offside of the vehicle centre line. • Vehicles first used on or after 1st January 1971 must be fitted with at least two stop lamps. • High level stop lamps are normally fitted in the rear window or boot spoiler of a car. They can consist of a number of light sources, and at least 50 per cent of the light sources must illuminate.
Obligatory direction indicators and hazard warning lights	• Direction indicators are not required on cars first used before 1st January 1936. • Cars first used after 1st April 1986 must be fitted with one side repeater indicator on each side. Instead of a separate lamp, the side repeater can be part of the front direction indicator if it includes a wraparound lens. • Cars first used on or after 1st April 1986 must have a hazard warning device fitted.

Inspection

You will usually conduct an inspection of a vehicle's electrical systems during routine service and maintenance. You should record any faults or recommendations on the service inspection sheet and report them to the customer so that they can give their authorisation for any repairs required.

An example of a service inspection sheet is shown in Figure 6.74.

Level 2 Light Vehicle Maintenance & Repair

Service Inspection Checklist

Repair Order Number: 12345	**Service Date:** 22/05/2011
Customer Name: Mr Bloggs	**Vehicle Make:** Sangsong
Telephone Number: 07978 564504	**Vehicle Model:** Coupe
Vehicle Registration: SJ58 SWJ	**VIN:** SCBGA1114XHC37189
Current Mileage: 40012	**Engine Size:** 1600cc

FLUID LEVELS & REPLENISHMENTS	A	B	C	Recommendation
1. Engine Oil/Filter Change		✓		
2. Coolant, Washer Brake Fluid Level		✓		
3. Power steering/Auto Box Level		✓		
4. Air filter/Pollen Filter Change		✓		
5. Coolant Change				
6. Brake Fluid Change				
7. Timing Belt/Auxiliary Belt Change				

LIGHTS				
8. Indicators / hazard warning lights		✓		
9. Side lights / number plate		✓		
10. Dipped / main beam headlights		✓		
11. Reversing Lights / Stop lights		✓		
12. Front / rear fog lights		✓		
13. Wiper blade wear and operation of front / rear wash / wipe		✓		

STEERING, BRAKES & SUSPENSION				
14. Tyre wear / condition		✓		N/S/F 3 mm O/S/F 3 mm N/S/R 4 mm O/S/R 6 mm Spare 7 mm
15. Adjusted tyre pressures		✓		Front 1.8 bar Rear 1.8 bar Spare 1.8 bar
16. Exhausts (leaks, mounting)		✓		
17. Condition of front /rear dampers		✓		
18. Condition of brake system		✓		
19. Leaks in brake system (external)		✓		
20. Wheel Bearing/Steering Joints		✓		
21. Condition of drive shaft bellows		✓		

BODY & ROAD TEST				
22. Horn		✓		
23. Mounting of bumpers		✓		
24. Windscreen		✓		
25. Rear view mirrors		✓		
26. Operation of doors and windows – Lubricate Locks/Hinges		✓		
27. Road Test		✓		

A – 10,000 Miles/1 Year Interval   ✓ Inspected – Good Condition
B – 20,000 Miles/2 Years Interval   R Inspected – Adjustment Made
C – 60,000 Miles/6 Years Interval   X Inspected – Fault Found

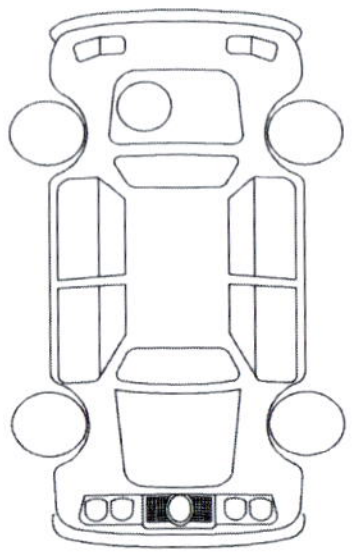

* Consumables in addition, a full estimate may be requested
** Price estimate only provided without dismantling work. This price can only illustrate approximate cost. You may request a full estimate detailing the exact nature and estimated cost of repairs.

Note Body Damage

D = Dents ☐   S = Scratches ☐
C = Chips X   R = Rust ☐
O = Other ☐

Estimate Given   Y   N

Additional Comments

Inspected by: JBarnes   Customer authorisation: CBloggs

Figure 6.74 Service inspection sheet

Find out

Carry out a lighting inspection of a car in your workshop.

- Are all obligatory lights present and working?
- What is the indicator flash rate of the car you inspected?

CHECK YOUR PROGRESS

Using sources of information available to you or a car from your workshop, list the power rating of the following bulbs (marked on the bulbs in watts):

1. headlamp bulb
2. indicator light bulb
3. stop lamp/brake lamp bulb
4. front side light bulb
5. number plate light bulb.

FINAL CHECK

1. What is the best method of isolating the electrical system before starting work?
   a. remove the ignition key
   b. disconnect the negative battery terminal
   c. disconnect the positive battery terminal
   d. remove the ignition fuse

2. An autoranging multimeter automatically selects:
   a. ohms
   b. multipliers
   c. volts
   d. amps

3. When testing resistance:
   a. the meter should be in series
   b. the component should be connected
   c. the power should be on
   d. the power should be off

4. Voltage is checked:
   a. in parallel
   b. in series
   c. hot
   d. cold

5. Oscilloscopes read:
   a. volts and amps
   b. amps and watts
   c. watts and volts
   d. ohms and amps

6. Amperage can be described as:
   a. electrical pressure
   b. electrical resistance
   c. electrical current
   d. electrical power

7. To measure electrical resistance, you should set a digital multimeter to:
   a. voltage AC
   b. ohms
   c. voltage DC
   d. audible continuity

8. To measure electrical current, you must connect a multimeter:
   a. in series
   b. in parallel
   c. across the battery
   d. back to front

9. On an oscilloscope the x-axis represents:
   a. voltage
   b. time
   c. space
   d. resistance

10. Hall effect sensors might be used as a safety device in:
    a. heated rear windows
    b. electric windows
    c. alarm systems
    d. door mirrors

GETTING READY FOR ASSESSMENT

The information contained in this chapter, as well as continued practical assignments in your centre or workplace, will help you to prepare for both the end-of-unit tests and diploma multiple-choice tests. This chapter will also help you to prepare for working on and maintaining light vehicle electrical systems safely.

You will need to be familiar with:

- Electrical safety precautions
- The electrical units – volts, amps, ohms and watts
- How to calculate Ohm's law
- How to select appropriate power supply, wiring, consumer devices, switching and circuit protection
- The basic operating principles of light vehicle electrical systems, such as starting charging and auxiliary systems
- How to maintain a vehicle's electrical system
- Common electrical symbols
- Electrical wiring diagrams
- The basic use of multimeters and oscilloscopes
- How to test and replace electrical components

This chapter has given you an introduction and overview of light vehicle electrical systems, providing the basic knowledge that will help you with both theory and practical assessments.

Before you try a theory end-of-unit test or multiple-choice test, make sure you have reviewed and revised any key terms that relate to the topics in that unit. You will need to read all the questions carefully. Take time to digest the information so that you are confident about what the question is asking you. With multiple-choice tests, it is very important that you read all of the answers carefully, as it is common for two of the answers to be very similar, which may lead to confusion.

For practical assessments, it is important that you have had enough practice and that you feel that you are capable of passing. It is best to have a plan of action and work method that will help you. Make sure that you have enough technical information, in the way of vehicle data, and appropriate tools and equipment. It is also wise to check your work at regular intervals. This will help you to be sure that you are working correctly and to avoid problems developing as you work.

When you are doing any practical assessment, always make sure that you are working safely throughout the test. Light vehicle electrical systems are dangerous, so your precautions should include:

- Isolate electrical circuits by disconnecting the battery where possible.
- Make sure you disconnect and reconnect the battery leads in the correct order.
- Where possible, use hand tools that are specifically designed to insulate you against electric shock.
- Make sure that electrical test equipment is set to the appropriate scale/unit and correctly connected to the circuit that is tested.
- Only replace electrical components with parts of the same quality and electrical rating.
- Observe all health and safety requirements and use the recommended PPE and VPE.
- Make sure that you use all tools correctly and safely.

Good luck!

7 Light vehicle inspection & servicing

This chapter will give you an overview of light vehicle serviceable systems, areas and components. It will give you the knowledge and skills you need to be able to inspect a vehicle and recommend appropriate repairs, using a variety of prescribed testing and inspection methods. You will find out about the relevant information you need to use to help you carry out light vehicle inspection and servicing.

The procedures involved in carrying out a light vehicle service are covered in detail – for each of these you will learn about the appropriate PPE, tools and equipment you need to use. You will also learn how to produce the required records of your work. Safe working features throughout the chapter provide reminders of how to work safely when carrying out light vehicle inspection and servicing.

This chapter covers:

- Safe working when carrying out light vehicle inspection and servicing
- The importance of carrying out light vehicle inspection and servicing
- Using relevant information to carry out light vehicle inspection and servicing
- Carrying out light vehicle inspection and servicing using the appropriate tools and equipment
- Recording information and make suitable recommendations

WORKING PRACTICE

While carrying out vehicle inspection and servicing work you will be exposed to hazards from oils and liquids, equipment and machinery. You should always use appropriate personal protective equipment (PPE) when you work on these systems. Make sure that your selection of PPE will protect you from these hazards.

It is your responsibility to:

- protect yourself
- protect the safety of others in the workshop
- comply with all safety requirements
- behave appropriately at all times.

Personal Protective Equipment (PPE)

Safety goggles/glasses reduce the risk of small objects or chemicals coming into contact with the eyes.

Safety mask protects against brake dust inhalation.

Overalls provide protection from coming into contact with oils and chemicals.

Safety gloves provide protection from oils and chemicals. They also protect the hands when handling objects with sharp edges.

Barrier cream protects the skin from old engine oil, which can cause dermatitis and may be carcinogenic (a substance that can cause cancer).

Safety helmet protects the head from bump injuries when working under cars.

Safety boots protect the feet from a crush injury and often have oil- and chemical-resistant soles. Safety boots should have a steel toe-cap and steel mid-sole.

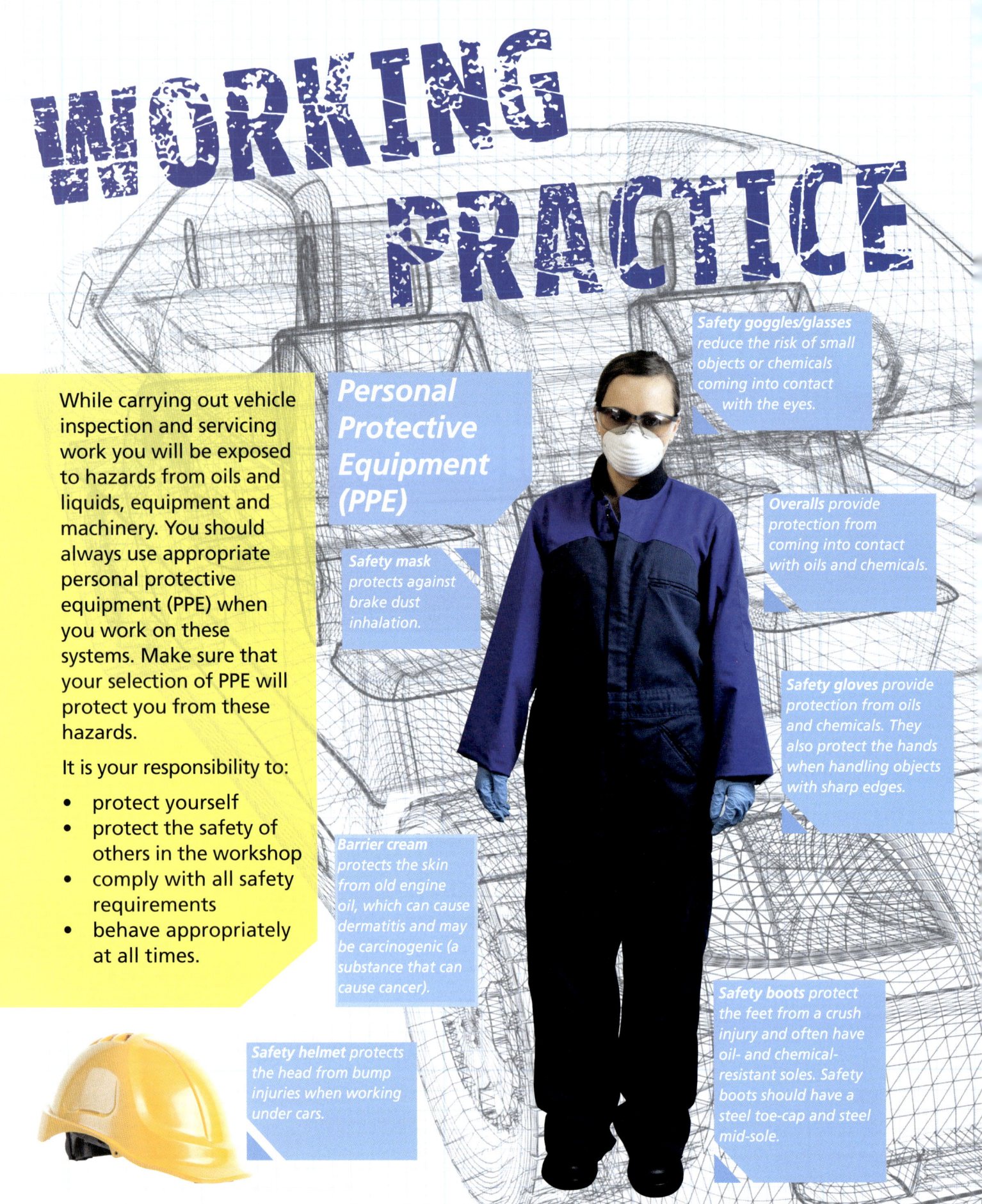

To reduce the possibility of damage to the car, always use the appropriate vehicle protection equipment (VPE):

Wing covers

Steering wheel covers

Seat covers

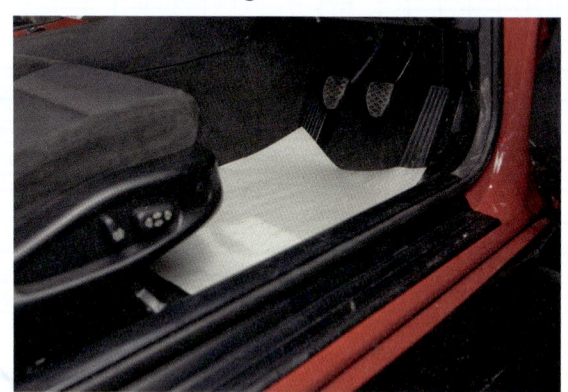

Floor mats

If appropriate, safely remove and store the owner's property before you work on the vehicle. Before returning the vehicle to the customer, reinstate the vehicle owner's property. Always check the interior and exterior to make sure that it hasn't become dirty or damaged during the repair operations. This will help promote good customer relations and maintain a professional company image.

Vehicle Protective Equipment (VPE)

Safe Environment

During servicing and inspection you may be required to dispose of waste such as oil filters, brake fluid, coolant and used or damaged PPE, for example gloves. Under the Environmental Protection Act 1990 (EPA), you must dispose of all waste in the correct manner. You should store all waste safely in a clearly marked container until it is collected by a licensed recycling company. This company should give you a waste transfer note as the receipt of collection.

To further reduce the risks involved with hazards, always use safe working practices, including:

1. Immobilise the vehicle by removing the ignition key.

2. Prevent the vehicle moving during maintenance by applying the handbrake or chocking the wheels.

3. Follow a logical sequence when working. This reduces the possibility of missing things out and of accidents occurring. Work safely at all times.

4. Always use the correct tools and equipment to avoid damage to components, tools or personal injury. Check tools and equipment before each use.
   - Inspect any mechanical lifting equipment for correct operation, damage and hydraulic leaks.
   - Never exceed safe working loads (SWL).
   - Check that measuring equipment is accurate and calibrated before you take any readings.

5. If components need replacing, always check that the quality meets the original equipment manufacturer (OEM) specifications. (If the vehicle is under warranty, inferior parts or deliberate modification might make the warranty invalid. Also, if parts of an inferior quality are fitted, it might affect vehicle performance and safety.)

6. Following the replacement of any vehicle components, road test the vehicle thoroughly to ensure safe and correct operation. Make sure that all work is correctly recorded on the job card and vehicle's service history, to ensure that any maintenance work can be tracked.

Preparing the car

Tools

You will use a wide range of tools when you are inspecting and servicing light vehicles. You will find out which tools you need for different tasks throughout the chapter.

Dial gauge

Engine oil and filter replacement tools

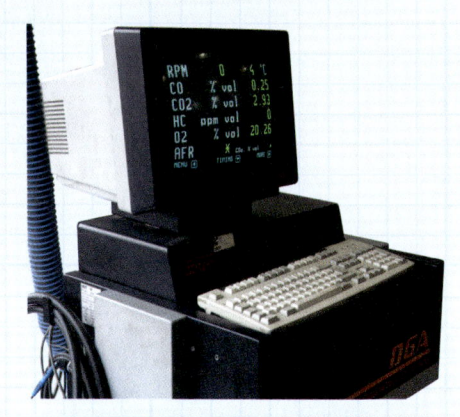

Exhaust gas analyser

Brake tester

Tyre pressure gauge

Tread depth gauge

Safe Working

It is essential that you follow all safety procedures when carrying out inspection and servicing activities. Make sure you know the specific safety procedures for any system that you are servicing. General safety rules include:

- Always clean up any fluid spills immediately to avoid slips, trips and falls.
- Always disconnect the ignition when you are working under the bonnet of a vehicle, for example when inspecting and servicing spark plugs.
- Always use correct manual handling techniques when you remove and refit heavy components.
- Always remove all sources of ignition (e.g. smoking) from the area and have a suitable fire extinguisher to hand when you work on petrol or diesel fuel systems.
- Always work in a well-ventilated area.
- Always use exhaust extraction when you run engines in the workshop.
- Always wear safety goggles and a particle mask when you are working on components such as brake linings, fuel and exhaust systems.

The importance of carrying out light vehicle inspection and servicing

The servicing and inspection of a vehicle is vital. Regular servicing and inspection will ensure that the vehicle is:

- safe
- efficient on fuel
- kind to the environment
- **roadworthy**
- compliant with warranty requirements.

It will also be safer for other road users in the event of an accident if the vehicle is in good repair. The severity of an accident involving the vehicle could be reduced, as the performance of important systems such as braking and suspension will be improved by regular maintenance.

Warranty requirements

Every new vehicle comes with a vehicle warranty. This is the manufacturer's guarantee for the replacement of major components. Depending on the manufacturer and vehicle model, the warranty can last for different lengths of time. For the warranty to be valid, the vehicle must be serviced by an approved repairer at the proper intervals, using the correct parts agreed by the manufacturer.

Road test

A vehicle must be fit to be driven on the road. A road test before and after inspection and servicing activities will determine the performance and condition of the vehicle.

If you have been asked to complete a road test on a vehicle, you have been trusted to complete the test thoroughly. It is your responsibility to check the vehicle over carefully to ensure it is roadworthy. Always record your findings on the appropriate pre- and post-work check sheets (see pages 432–433).

Table 7.1 highlights all the vehicle systems that you need to check before road testing a vehicle.

Carrying out the road test

During the road test, you will need to check brake operation, steering and suspension under varying road conditions. You also need to check performance during acceleration and any abnormal noises. It is vital that you document all your findings.

When taking a vehicle for a road test you need to make sure that:

- it has a valid tax disc
- you are insured by your employer to drive the vehicle
- the vehicle has an up-to-date MOT certificate.

> **Key term**
>
> **Roadworthy** – the term used to describe a vehicle that complies with the Road Vehicles (Construction and Use) Regulations 1986.

Table 7.1 Vehicle system checks prior to road testing a vehicle

Vehicle system	Checks to be performed
Tyres	Is tyre tread a minimum of 1.6 mm? Are the tyres damaged in any way? Are they inflated to the correct pressures?
General overview of the vehicle	Does the vehicle have anything that could come loose and fall off? Is there anything that could injure pedestrians if they came into contact with the vehicle?
Oil and fluid levels	Are levels correct for: • engine oil • brake fluid • washer fluid • coolant?
Lights and windscreen wipers	Do lights and windscreen wipers operate correctly?
Fuel	Is there enough fuel for the test?
Brakes	Does the pedal feel firm?
Steering	Is there any excess movement in the steering?
Seat belts	Do the seat belts operate correctly and smoothly? Are they damaged in any way?

Working life

Over the workshop radio Tom hears that there has been a serious accident involving a van and passenger vehicle. The next day Tom is summoned to the manager's office. His manager asks him to confirm that he carried out a complete test of the roadworthiness of the van booked in at the beginning of the week. It turns out that this was the van in the accident mentioned on the radio.

Tom can confirm that he carried out the check and he is able to show his manager the documentation that he completed when carrying it out. His manager gives this information to the police. After police investigation, the cause of the accident is established: the driver was not paying attention to other traffic.

1 What could have happened if Tom had not carried out the road test thoroughly?

Did you know?

It is a legal requirement to have all lights working correctly on a vehicle at all times. It is also an offence to drive a vehicle if the washer bottle is empty or ice has frozen the washer jets. These requirements are part of the Road Vehicles (Construction and Use) Regulations 1986, which also covers other rules regarding the construction and use of vehicles on the road.

The law and vehicle safety

There are several laws regarding the use of vehicles on the UK's road. These laws are designed to ensure that all vehicles on the roads are in a safe condition and that drivers hold the correct documentation.

The Road Traffic Regulation Act

This law provides powers to regulate or restrict traffic on UK roads in the interests of safety. This Act is regularly updated – for example, since 1988 this law requires that drivers hold vehicle insurance to cover their liability for any damage to other people's property or injury to others (including passengers) which is caused by their use of a vehicle on a public road or in other public places.

The Road Safety Act

This law also covers the use of vehicles on the roads and was last updated in 2006. The Act covers:

- driving offences (such as drink driving and speeding)
- driving standards (such as driving tests)
- registration plates
- the central registration of all vehicles.

Road Fund Licence

The Road Fund Licence is more commonly known as car tax. This states that:

- Every vehicle registered in the UK must be taxed if it is used or kept on a public road.
- If the vehicle is kept off-road it must either be taxed or have a SORN (Statutory Off Road Notification).

If a vehicle has neither of these, it could be wheel clamped or removed. Exceptions to this rule are where **trade plates** (see Figure 7.1) are used by a garage for a road test either before or after repair of the vehicle or if it is in transit to an MOT testing station.

Figure 7.1 Trade plates

> **Did you know?**
> Vehicles constructed before 1 January 1973 are exempt from paying car tax, but must still display a tax disc in the bottom right of the windscreen.

> **Key term**
> **Trade plates** – vehicle licences issued to motor traders and repairers. They can only be used on a vehicle that is in the temporary possession of the holder for business purposes. Trade plates have a white background and red digits. They must be placed on the vehicle so that they can be easily read from the front and the rear.

> **Working life**
> Farid takes delivery of a customer's car for a routine service. He notices that there is no road tax disc in the window.
>
> 1 What should Farid do?

7 Light vehicle inspection & servicing

Ministry of Transport (MOT) certificate

If a vehicle is 3 years or older it must be MOT tested every 12 months. This is to ensure that it complies with minimum vehicle safety requirements, is roadworthy and that the exhaust emissions fall within the environmental standards. An MOT is a legal requirement which is monitored by the Vehicle and Operator Services Agency (VOSA).

If a garage has been approved to complete MOT testing, at least one technician must have successfully completed an MOT inspector training course run by the Department of Transport. The MOT logo (see Figure 7.2) will be displayed on the premises.

The MOT certificate is not a guarantee of the general mechanical condition of a vehicle. Later in this chapter you will find out about the specific items contained within the MOT test (see page 435).

Figure 7.2 MOT logo

Did you know?
Ambulances and taxis have to be MOT tested when they are a year old.

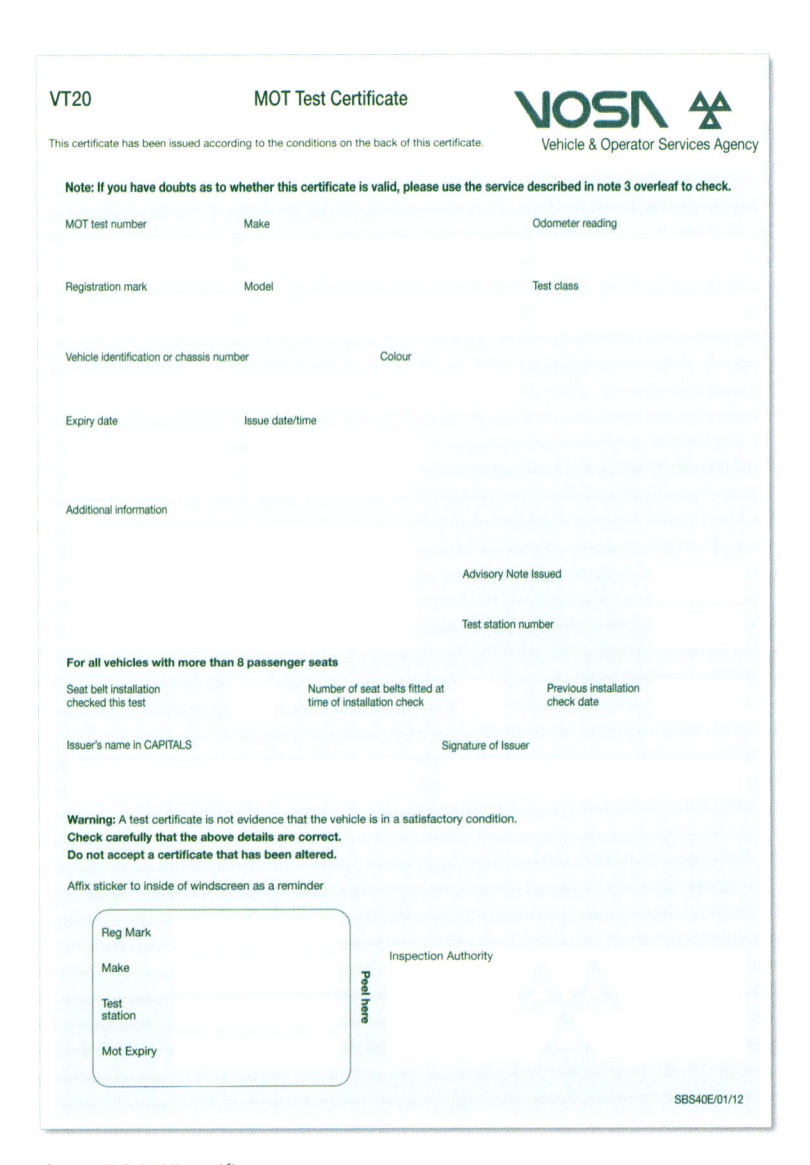

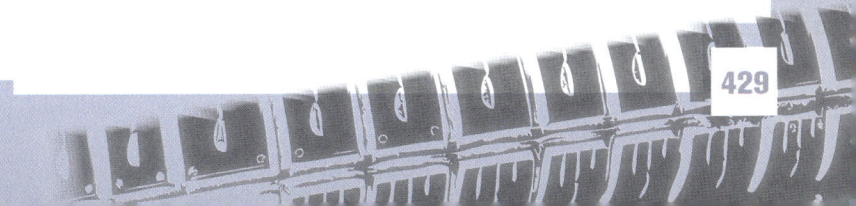

Figure 7.3 MOT certificate

429

Using relevant information to carry out light vehicle inspection and servicing

When you carry out inspection and servicing procedures, you need to:

- use a range of information to help you do the job
- record information on the condition of the vehicle and any servicing activities that you carry out.

To do this, you will use the documents listed in this section

Job card

During the inspection and servicing of a light vehicle there is a flow of information. First the customer gives you or the service adviser their contact details and basic information about their vehicle. From this information, the service receptionist (or whoever is responsible for administration tasks in the service department) will create a job card. This sets out all the information required for the inspection or service, including:

- job number
- date
- customer details
- vehicle details
- customer's instructions
- description of work carried out
- labour – hours and costs

- parts and consumables used and costs
- total cost
- customer authorisation (signature)
- technician's name and signature.

In-vehicle service record

Every new vehicle is supplied with a service book, which is kept by the owner. Every time you service a vehicle, you should stamp the service book with the vehicle's mileage, the date and the type of service carried out. This will form part of the vehicle's service history.

Vehicle handbook

The vehicle handbook is an instruction manual for the driver which is produced by the vehicle's manufacturer. It also gives information needed for inspection and service operations, including:

- oil and fluid capacities
- tyre pressures
- bulb sizes
- vehicle dimensions.

It is essential to refer to the vehicle handbook when servicing a vehicle.

Manufacturer's workshop manuals

Every manufacturer produces a detailed instruction manual on how to repair, inspect and service their vehicles. These are now more commonly produced in electronic format. They include:

- step-by-step guides for the diagnosis and repair of each vehicle system
- all **specifications** relating to the vehicle for inspection and service.

> **Key term**
>
> **Specification** – the technical information provided by the manufacturer, which gives details of how the vehicle works to its optimum performance.

Generic manufacturer data

To assist smaller garages who do not work on just one make or model of vehicle, automotive data information for vehicles is available from various companies in paper form, on CD or online.

One such company is Autodata, which publishes technical manuals, charts and diagnostic CDs for around 11,000 vehicles made by 80 different manufacturers. These publications are particularly useful for the independent sector of the motor industry, as one annual subscription for this product prevents them having to buy every workshop manual for the full range of vehicles they repair. Examples of data covered include:

- technical data such as torque settings
- wiring diagrams
- fault code information
- inspection and service data.

To find out more about these publications, go to hotlinks to visit the Autodata website.

> **Did you know?**
> Information produced in paper form is often referred to as a 'hard copy'.

Vehicle Identification Number

The Vehicle Identification Number (VIN) consists of 17 characters that make the vehicle unique. The VIN tells you when the vehicle was built and how it is constructed. It also indicates the correct spare parts required during inspection and servicing.

The VIN is printed on the chassis plate and can be found in any of the following locations:

- under the bonnet near to the bonnet catch (as shown in Figure 7.4)
- on the bulkhead
- on the inner wing
- stamped into the floor next to one of the front seats.

In newer vehicles the chassis plate may also be at the foot of the windscreen. You need to check the vehicle handbook or manufacturer's repair specifications to confirm the location.

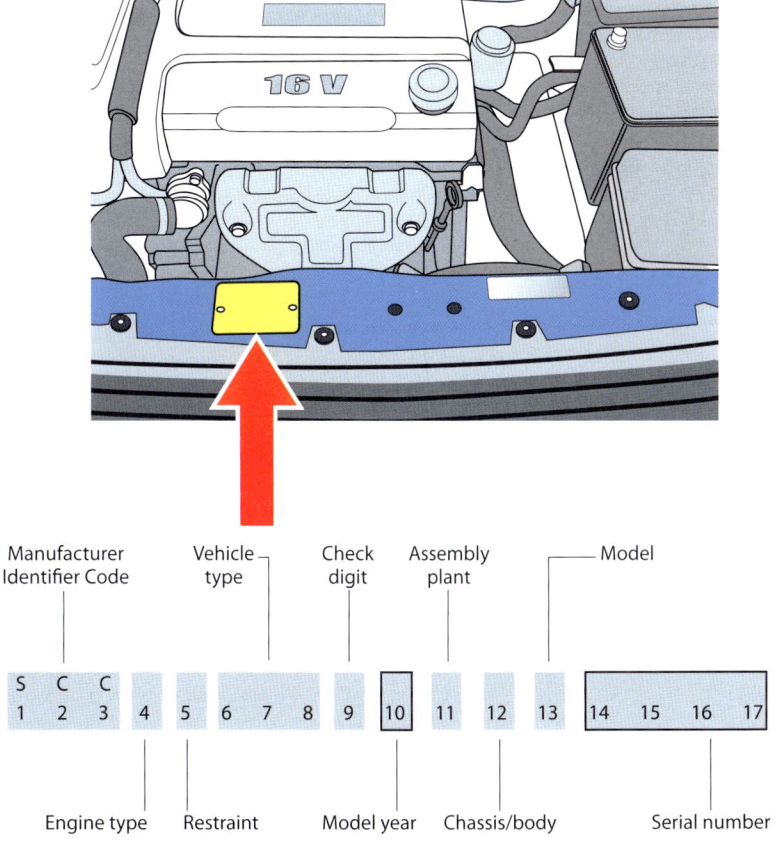

Figure 7.4 Vehicle Identification Number

Pre-work check sheet

It is good practice to perform pre-checks before you carry out any service. These checks usually include a visual inspection of the vehicle's exterior and interior for any damage or cosmetic faults. Many garages use pre-check sheets to identify and record any damage.

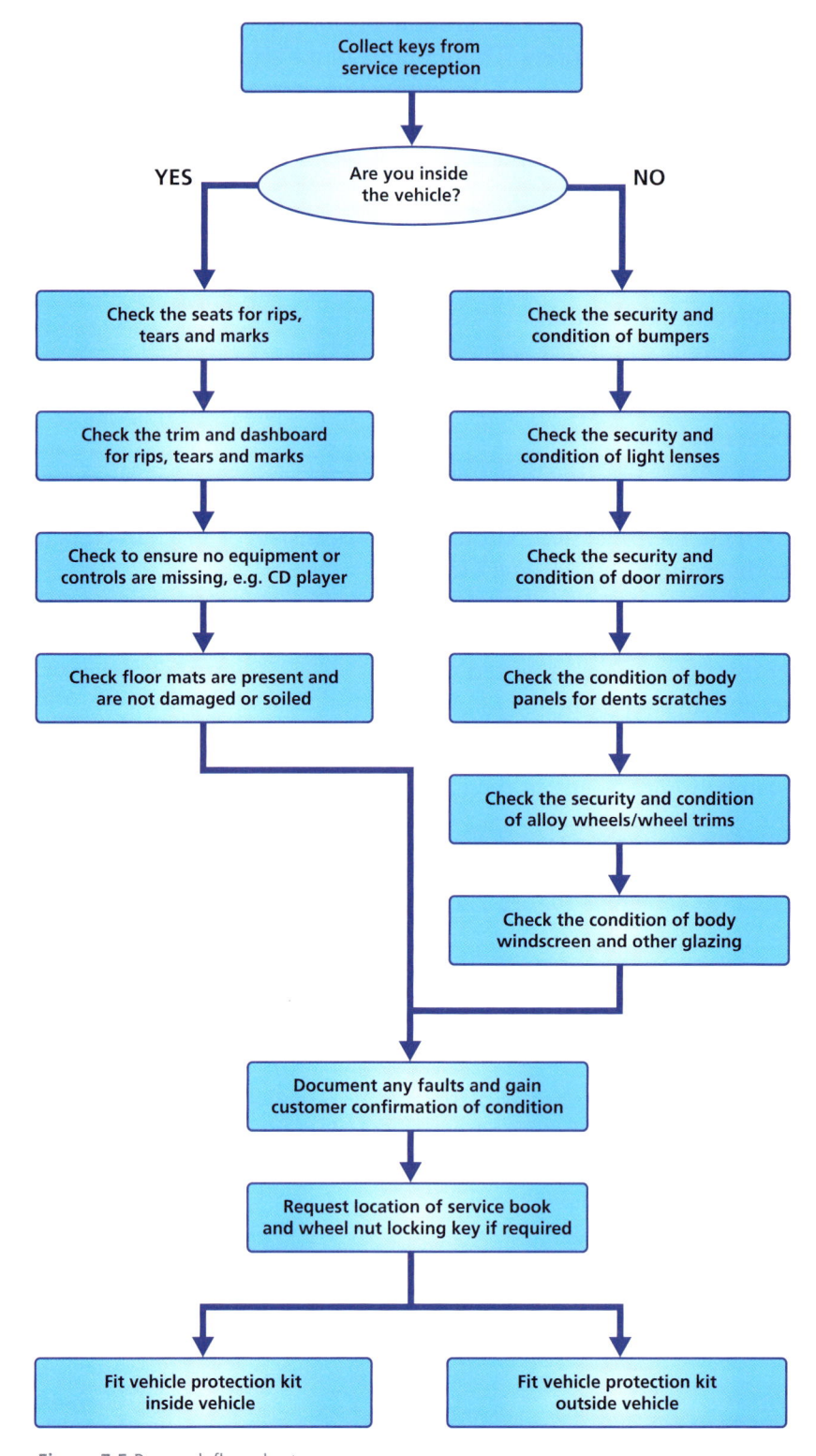

```
                    Collect keys from
                    service reception
                            │
                            ▼
        YES         Are you inside         NO
        ┌───────────  the vehicle?  ───────────┐
        ▼                                      ▼
  Check the seats for rips,            Check the security and
      tears and marks                   condition of bumpers
        │                                      │
        ▼                                      ▼
  Check the trim and dashboard          Check the security and
   for rips, tears and marks           condition of light lenses
        │                                      │
        ▼                                      ▼
  Check to ensure no equipment or       Check the security and
  controls are missing, e.g. CD player  condition of door mirrors
        │                                      │
        ▼                                      ▼
  Check floor mats are present and      Check the condition of body
   are not damaged or soiled           panels for dents scratches
        │                                      │
        │                                      ▼
        │                            Check the security and condition
        │                             of alloy wheels/wheel trims
        │                                      │
        │                                      ▼
        │                              Check the condition of body
        │                             windscreen and other glazing
        │                                      │
        └──────────────────┬───────────────────┘
                           ▼
                Document any faults and gain
              customer confirmation of condition
                           │
                           ▼
              Request location of service book
            and wheel nut locking key if required
                           │
              ┌────────────┴────────────┐
              ▼                         ▼
      Fit vehicle protection kit   Fit vehicle protection kit
          inside vehicle                outside vehicle
```

Figure 7.5 Pre-work flow chart

By completing this inspection before you begin work, you can get the customer to sign the check sheet and confirm the condition of the vehicle before the repair. This will help to avoid any potential conflict after the repair if the customer claims they didn't know about any damage. Ideally the customer should be present while the check is completed, so that any queries can be cleared up immediately.

The flow chart in Figure 7.5 shows what needs to be included in a pre-work inspection.

Post-work check sheet

To meet the customer's expectations, you should carry out all service and repairs so that they conform to a high standard. You also need to return the vehicle in a clean and acceptable condition following any work carried out on the vehicle.

You will need to complete a post-work check sheet by referring to the notes written on the original pre-work check sheet and the road test checklist. This will make sure that you haven't missed anything and that you send the vehicle back to the customer in a roadworthy condition.

> **Working life**
>
> A customer has dropped off his newly purchased family car for its first full service. It was Emma's birthday yesterday and she really doesn't feel like being at work. She has completed many services before and decides to retune the radio. She doesn't put any VPE in the car. She shows off to her workmates by revving the engine and driving at speed into the workshop. She misjudges the entrance and catches the wing mirror. She jumps out of the car and adds the damage to the pre-check sheet (which has already been signed by the customer).
>
> The customer returns to pay for this expensive service only to find oily footprints on the carpet, handprints on the pillars and steering wheel and scratches to the front wings.
>
> 1. How do you think the customer will react?
> 2. What could be the legal implications of Emma's actions?

> **Did you know?**
>
> Buying a new car is often the second most expensive purchase in your life – after buying a house.

Pre-delivery inspection sheet

Before delivering a new vehicle to a customer you are required to conduct a pre-delivery inspection (PDI). This is to make sure that the vehicle is in the condition agreed and expected by the customer. When you complete the PDI you need to check that:

- everything is correctly assembled
- all fluids are at the required levels
- any in-car entertainment, clock and radio stations are in working order and have been set correctly.

Level 2 Light Vehicle Maintenance & Repair

Record any additional information for future servicing on the inspection sheet (see Figure 7.6). You should complete your PDI by carrying out a vehicle road test (see pages 426–427). If the inspection highlights problems such as tyre pressures being set too high, then they must be adjusted and rechecked before the customer takes delivery of their vehicle.

Pre-delivery Inspection Checklist

Equipment Make/Model/Year: _____ V.I.N. #: _____ New ☐   Inspection Completed ☐
Used ☐   _____
Vehicle Inspector/Date

Under Bonnet – Engine Off
- ☐ Engine Oil Level
- ☐ Steering gear housing fluid level
- ☐ Brake master cylinder fluid level
- ☐ Windscreen washer reservoir fluid level
- ☐ Radiator coolant level
- ☐ Battery fluid level
- ☐ Drive belt tensions
- ☐ Battery voltage and load-voltage drop

Under Bonnet – Engine Operating
- ☐ Automatic transmission fluid level

Under Bonnet – Engine Operating & Hot
- ☐ Alternator and voltage regulator operation
- ☐ Spark plug & hi-tension wire condition

Other Operations
- ☐ Idle speed check
- ☐ Hydraulic lines, fittings, connections/components for leaks
- ☐ Adjust mechanical valve tappet clearance

On Outside
- ☐ Latches, keys, and locks – operation
- ☐ Bonnet, deck and door panels for fit and alignment
- ☐ Bumpers and moldings – alignment
- ☐ Weather strips – adhesion and fit
- ☐ Wheel nut torques

Miscellaneous
- ☐ Obtain vehicle/equipment Operation/maintenance manual
- ☐ Check and operate auxiliary equipment
- ☐ Maintenance records (used vehicle)

On Inside
- ☐ Operation of lights, turn signals, stop signals, courtesy lights
- ☐ Operation of oil pressure & alternator warning lights
- ☐ Front seat control – operation (manual or power)
- ☐ Rear seat and floor operation

- ☐ Inside locks & door handles – operation
- ☐ Windows and vents – operation & fit
- ☐ Glass condition
- ☐ Cigarette lighter – install & test

On Hoist – Or Underside
- ☐ Axle fluid level
- ☐ Hydraulic lines, fittings, connections/components for leaks
- ☐ Tyre pressures
- ☐ Steering linkage and connections

Alignment Stall
- ☐ Front/rear wheel toe-in

Road Test
- ☐ Neutral switch – operation (Automatic transmission)
- ☐ Parking brake – operations
- ☐ Horn, windscreen washer/wipers – operation
- ☐ Heater & air vents – operation
- ☐ Air conditioning – operation
- ☐ Brakes – operation
- ☐ Transmission shift lever – operation
- ☐ Automatic transmission shift timing & quality
- ☐ Accelerator pedal – operation
- ☐ Engine performance
- ☐ Drive belts
- ☐ Steering control
- ☐ Squeaks, rattles, and wind noise
- ☐ Speedometer, odometer, fuel & temperature gauge – operation
- ☐ Other optional equipment – operation

After Road Test
- ☐ Wash car and check for leaks
- ☐ Inspect for interior & exterior metal & paint damage
- ☐ Check soft trim for soilage & excess sealer

Other Items specific to vehicle
- ☐

The vehicle should not be accepted until the campus vehicle inspector has performed the pre-delivery inspection on the vehicle in accordance with the above recommended vehicle pre-acceptance checklist and has authorised the acceptance.

☐ Acceptance Authorised   _____(Signature)
☐ Acceptance Not Authorised   _____(Date)

6/11/09   1

Figure 7.6 Pre-delivery inspection sheet

7 Light vehicle inspection & servicing

> **Working life**
>
> John is completing an interim service. He notices on the job card that the customer has commented that the steering is pulling to the right and there is a knocking noise. John knows what this problem is – he goes ahead and rectifies it by replacing the nearside spring and damper, as he wants to impress his manager and the customer. John notes the work on the job card and informs the service desk to invoice the customer for £135.40. The customer refuses to pay for the extra work and the manager agrees with the customer.
>
> 1 John can't understand what he has done wrong. Can you?

> **Find out**
>
> Research car retailers in your local area. Ask them about the types of inspection they carry out on both new and used cars.

A pre-delivery inspection can be used for new or used vehicles. On a used vehicle it may be called a 'used car inspection' rather than PDI. Some vehicle retailers incorporate a service into this type of inspection.

Pre-MOT inspection sheet

A pre-MOT test covers all the items carried out on an MOT. This test is done before the actual MOT takes place, to ensure the vehicle is roadworthy and that it will pass the actual MOT test.

A pre-MOT inspection is not widely offered today, as re-test fees are rarely charged if the vehicle owner or garage representative returns the repaired vehicle to the testing station within two weeks of the initial test. The pre-test does not cover the general appearance or care of the body and drive train.

Table 7.2 shows some of the checks required on an MOT and those checks that are part of a routine inspection.

Table 7.2 MOT and routine inspection checks

System, components or assembly	Inspection and MOT item	Inspection and non-MOT item
Paint and body appearance	If edges are corroded and may cause injury to pedestrians	General appearance
Lubricant leaks	If leaking on to brake linings	Small leaks to ground are advisory items
Lights	Side lights, headlights, indicators, number plate lights, brake lights and rear fog lights	Reverse lights, front fog lights, interior lights
Tyre sizes	If different on the same axle If wider than wheel arch	If tyre sizes are increased and profile reduced
Mirrors	If the number of mirrors is reduced to less than two	Passenger side mirror or interior mirror broken
Seat belts	Operation of the seat buckle Condition of seat belt webbing Operation of inertia mechanism	Seat belt indicator light

Types of service offered

Vehicles will need different types of service depending on the manufacturer's specifications and seasonal changes. You should refer to the manufacturer's service sheet and other general service specifications found in the vehicle's service manual.

Winter service

Some vehicle garages offer a winter service. This consists of a cooling system pressure check, coolant change, winter screen wash addition and a general check of vehicle systems such as lights and tyres.

Interim service

An interim service is a service interval between two full services. It requires fewer **replenishments** and less in-depth inspection of vehicle systems such as brakes.

Full service

A full service or main service is the largest service possible. It requires the replacement of all filters and fluids and an in-depth inspection of brakes.

A, B and C services

Some manufacturers use letters to denote the depth of service. An A service is equivalent to an interim service and a B service is equivalent to a full service. The C service will be required after a longer period of time where items such as brake fluid, timing belt and coolant changes are necessary.

Service checklist

It is important that you complete a service or repair in a logical order. This will help you make sure that:

- you complete all items of the service
- you carry out servicing to the correct standard
- you use your time productively
- the quality of service is of the highest level.

To make sure that you carry out the service in a logical order, you can use a service checklist.

> **Key term**
>
> **Replenishments** – new vehicle parts and fluids used during inspecting and servicing.

> **Did you know?**
>
> - A **time-based service** is where the customer has not covered the mileage in the time recommended between service intervals. However, the vehicle will still need to be inspected, as age can have an effect on some of the vehicle components.
> - A **mileage-based service** is where the customer drives long distances and will need the vehicle servicing more frequently than the service intervals specified.

CHECK YOUR PROGRESS

1. What does the abbreviation VIN mean?
2. When would an interim service take place?
3. The job card sets out all the information required for the inspection or service. Write down five pieces of information that you will find on a job card.

Carrying out light vehicle inspection and servicing using the appropriate tools and equipment

Inspection overview

An inspection may involve measuring, testing and looking for general wear and tear. The tests are carried out on the ground, when the vehicle is in the air and then with the vehicle at chest height. When you complete an inspection you will identify whether the components fit the specified requirements. Some faults may be obvious to you, such as an unusual knocking sound in the engine, whereas others may require several testing methods to diagnose.

If you are investigating a problem which the customer has highlighted, consult the note on the job card before you complete the inspection. Always add information to the job card if you find additional faults. The job card can then be used when contacting the customer for authorisation if extra work is needed.

Inspection sheets

An inspection sheet lists the checks that should be made and the order in which they should be completed. The checks will be laid out in a specific way by the manufacturer, to make sure that inspection processes are carried out efficiently and consistently. Some examples are shown in Figures 7.7 to 7.9. See also Chapter 1, page 75, Figure 1.47.

> **Find out**
> Obtain inspection sheets for your local vehicle retailer and check on the order of the tasks to be carried out.

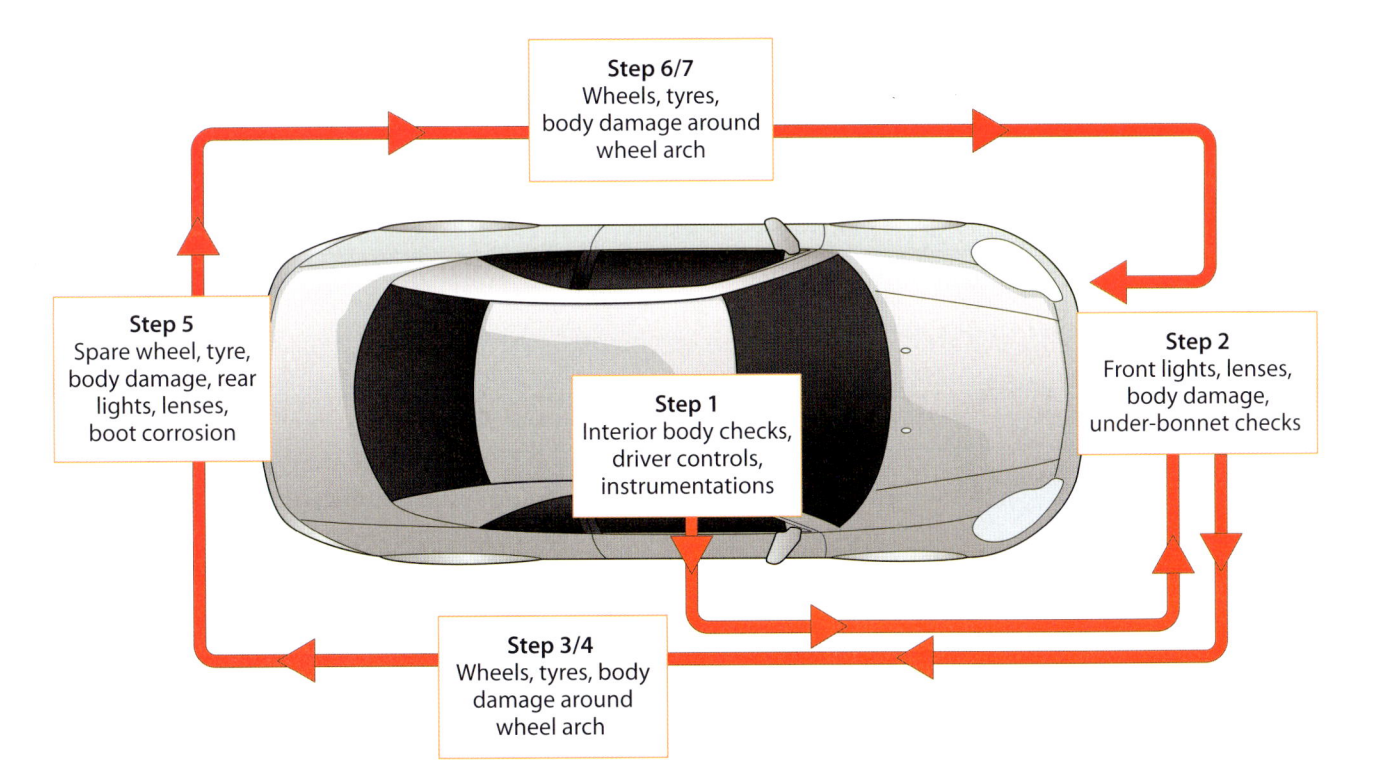

Step 6/7 Wheels, tyres, body damage around wheel arch

Step 5 Spare wheel, tyre, body damage, rear lights, lenses, boot corrosion

Step 1 Interior body checks, driver controls, instrumentations

Step 2 Front lights, lenses, body damage, under-bonnet checks

Step 3/4 Wheels, tyres, body damage around wheel arch

Figure 7.7 Inspection sheet: vehicle with wheels on the ground

437

Level 2 Light Vehicle Maintenance & Repair

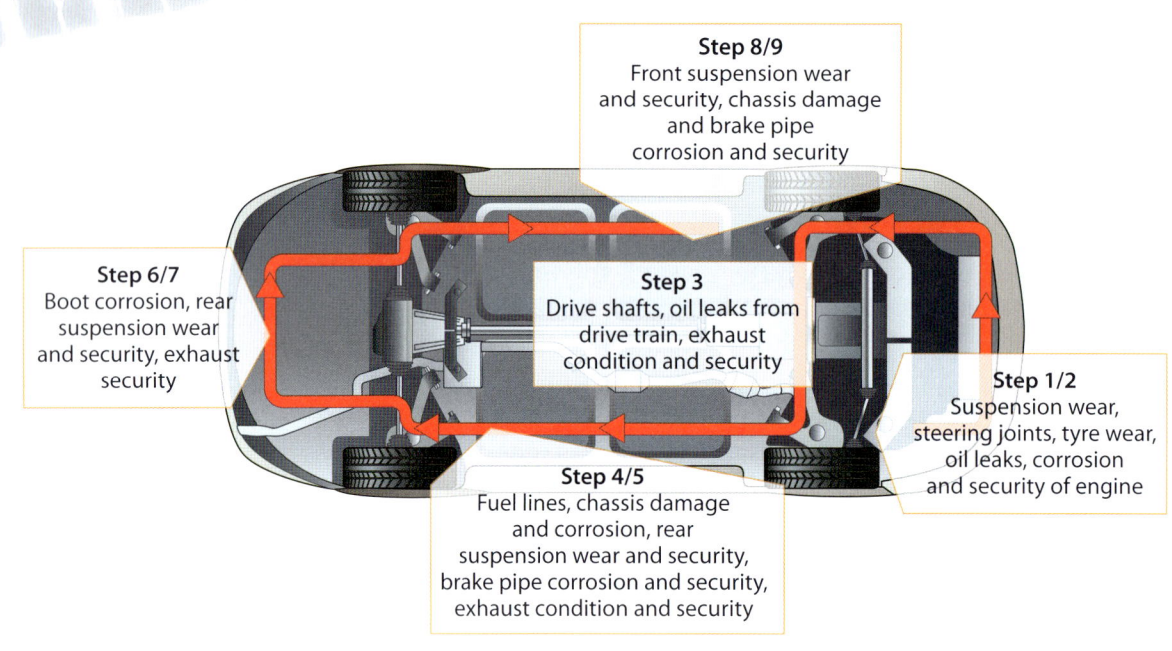

Step 8/9
Front suspension wear and security, chassis damage and brake pipe corrosion and security

Step 6/7
Boot corrosion, rear suspension wear and security, exhaust security

Step 3
Drive shafts, oil leaks from drive train, exhaust condition and security

Step 1/2
Suspension wear, steering joints, tyre wear, oil leaks, corrosion and security of engine

Step 4/5
Fuel lines, chassis damage and corrosion, rear suspension wear and security, brake pipe corrosion and security, exhaust condition and security

Figure 7.8 Inspection sheet: vehicle in the air

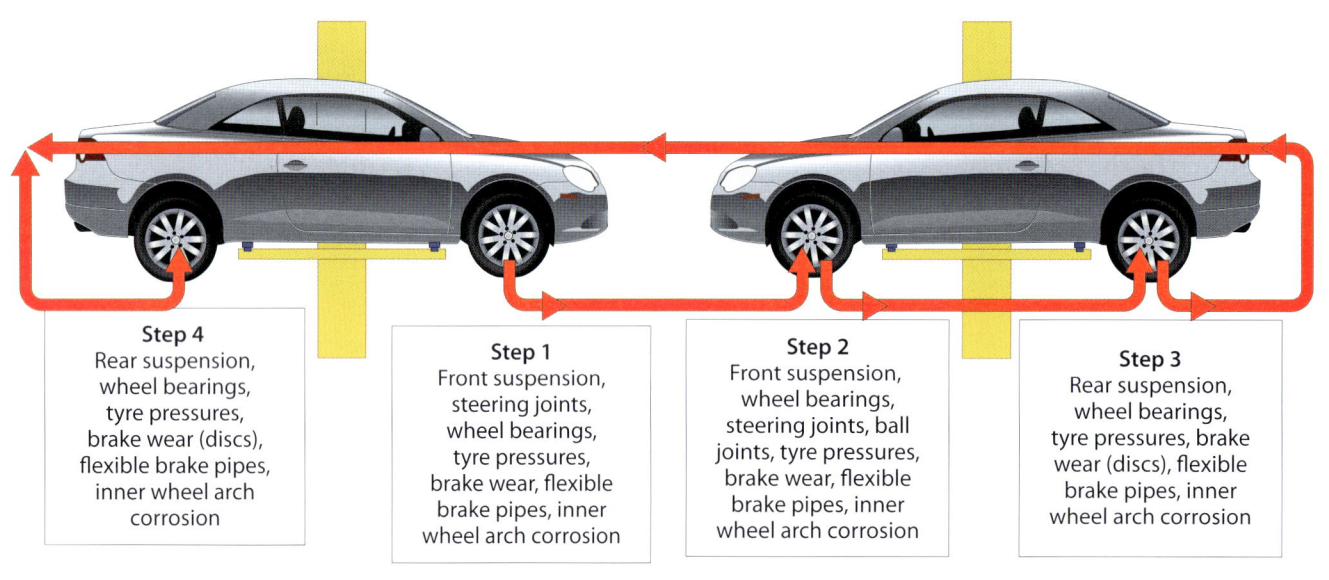

Step 4
Rear suspension, wheel bearings, tyre pressures, brake wear (discs), flexible brake pipes, inner wheel arch corrosion

Step 1
Front suspension, steering joints, wheel bearings, tyre pressures, brake wear, flexible brake pipes, inner wheel arch corrosion

Step 2
Front suspension, wheel bearings, steering joints, ball joints, tyre pressures, brake wear, flexible brake pipes, inner wheel arch corrosion

Step 3
Rear suspension, wheel bearings, tyre pressures, brake wear (discs), flexible brake pipes, inner wheel arch corrosion

Figure 7.9 Inspection sheet: vehicle at chest height

Inspection and service good practice

Even though you will be using VPE, it very likely that you will leave some handprints or marks inside and outside the vehicle you are working on. Always make sure that you fully inspect the vehicle for marks before you hand it back to the customer. You should do this by checking body panels, interior carpets and controls, seats and paint surfaces. This will ensure that the customer is confident in the high quality of service provided. If the customer is delighted with a job well done, this is likely to lead to repeat business.

For a checklist of how to meet your customer's expectations, see Chapter 1, page 22.

Assessment methods

Using the senses

During an inspection and service you will need to use your senses to determine the condition of the vehicle systems. Some examples are shown in Table 7.3.

Table 7.3 Using your senses to assess the condition of the vehicle

Sense	Example of service check carried out
Visual (seeing)	• Look for lights not operating correctly • Look for wiper blades split
Aural (hearing)	• Listen for knocks on suspension during road test • Listen for exhaust gas leakage
Olfactory (smelling)	• Smell brakes binding during road test • Smell coolant leaking on to exhaust manifold
Touch (feeling)	• Feel for wear on a disc brake surface • Feel for vibrations through steering wheel for wheel imbalance during road test

Functional assessment

Functional assessment is where you carry out a **simulation** of system operation. You can do this by connecting a test instrument to the vehicle system to be checked. Examples of functional testing include:

- brake roller testing
- coolant pressure testing
- wheel balancing on a machine.

> **Key term**
>
> **Simulation** – recreating an inspection activity to mirror the actual operation.

CHECK YOUR PROGRESS

1. What does an aural vehicle inspection involve?
2. List the order of an inspection overview. Name two areas to be inspected at each point.
3. What is an inspection sheet?

Engine and engine systems inspection and servicing

The engine provides the source of power for the vehicle. Servicing the engine will ensure that it lasts longer and provides optimum performance. (For more detail on the engine and its systems, see Chapters 3 and 4.)

Level 2 Light Vehicle Maintenance & Repair

An engine service will cover:

- the lubrication system
- the cooling system
- spark plugs for SI engines
- valve assemblies
- drive belts
- the fuel and air system.

Start by carrying out an initial engine inspection as shown in the flow chart in Figure 7.10.

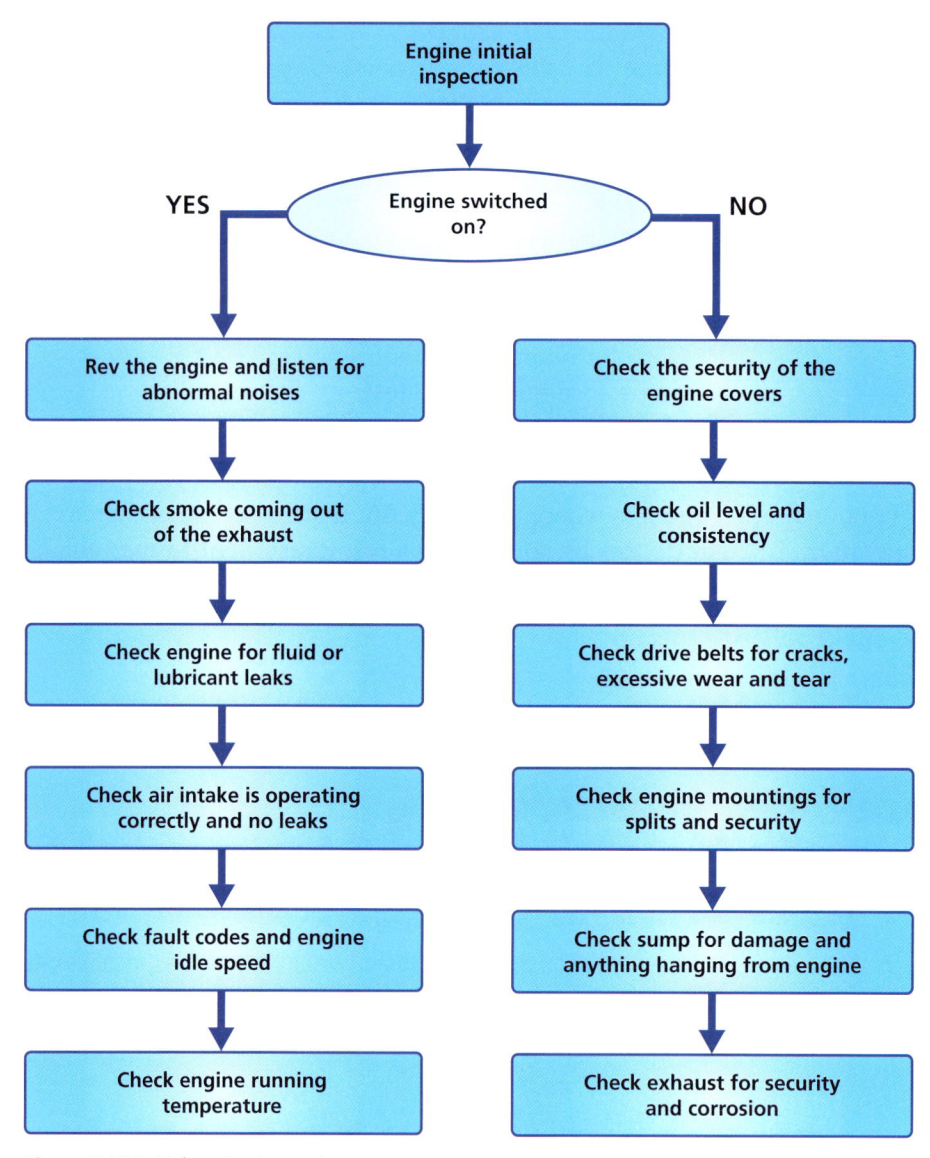

```
                    Engine initial
                     inspection
                          |
                          v
         YES      Engine switched      NO
        /              on?               \
       v                                  v
Rev the engine and listen for    Check the security of the
    abnormal noises                   engine covers
       |                                  |
       v                                  v
Check smoke coming out           Check oil level and
    of the exhaust                   consistency
       |                                  |
       v                                  v
Check engine for fluid or        Check drive belts for cracks,
    lubricant leaks              excessive wear and tear
       |                                  |
       v                                  v
Check air intake is operating    Check engine mountings for
correctly and no leaks               splits and security
       |                                  |
       v                                  v
Check fault codes and engine     Check sump for damage and
    idle speed                   anything hanging from engine
       |                                  |
       v                                  v
Check engine running             Check exhaust for security
    temperature                      and corrosion
```

Figure 7.10 Initial engine inspection

Did you know?

Drivers should check their vehicle's engine oil level once a week.

Engine oil level check

It is essential to check the level of oil in an engine. If a vehicle has incorrect levels of oil the engine will be subjected to unnecessary stress. This can result in severe damage.

7 Light vehicle inspection & servicing

Before you check and adjust the oil level:

- Check that the vehicle is on a flat surface.
- Make sure that the engine is cool.
- Make sure that there is no build-up of dirt or **sludge** around the oil cap/component.
- Have all the tools and the correct oil to hand.
- Check the oil grade you need to use by looking up this information in the vehicle handbook.

How to use a dipstick to check engine oil level is shown in Figure 7.11, and the procedure to follow is given in the flow chart in Figure 7.12. You must correct the oil level if it is above the maximum limit by letting oil out of the sump (see *Replacing the engine oil and filter* on page 443).

> **Key term**
>
> **Sludge** – used oil that has dried and mixed with dirt to form a black paste.

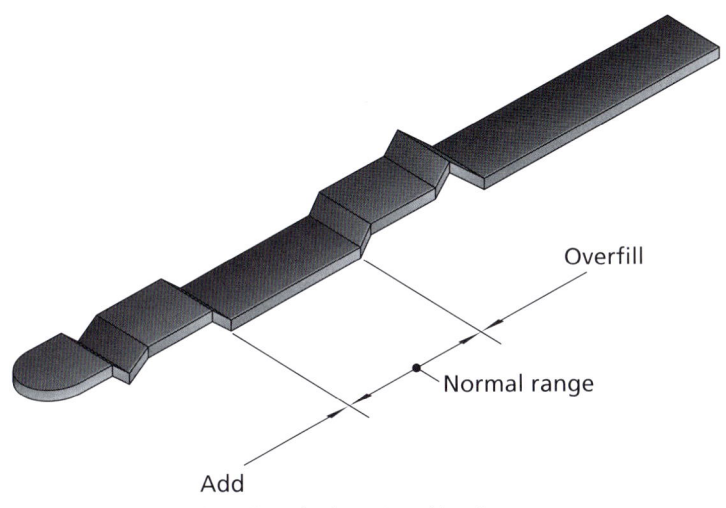

Figure 7.11 Using a dipstick to check engine oil level

Checking the oil level

1. With the bonnet open, remove dipstick and wipe it. 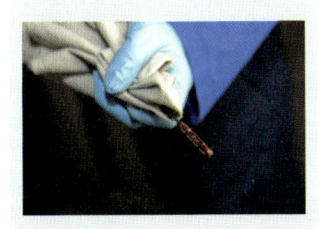
2. Insert the dipstick. 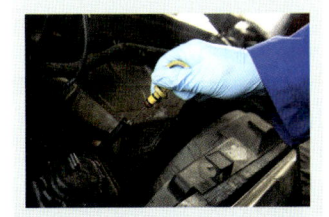
3. Remove the dipstick and check the oil level (see Figure 7.11). 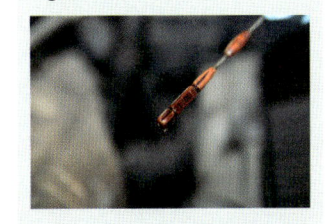
4. Replace the dipstick and top-up with oil if necessary.

Once you have topped up the engine oil you must check the new level. To do this, run the engine to allow the oil to move around the engine. Then switch off the engine and allow the engine to stand for five minutes before rechecking the oil level.

441

Level 2 Light Vehicle Maintenance & Repair

> **Safe working**
> Never put oil-soaked rags in your overall pockets. Oil will soak through them on to your skin and it may cause dermatitis.

> **Safe working**
> You must dispose of any oil-soaked rags and old engine oil filters following the local government environment regulations.

> **Key term**
> **Viscosity** – resistance to flow. This can be described as the 'thickness' of the lubricant.

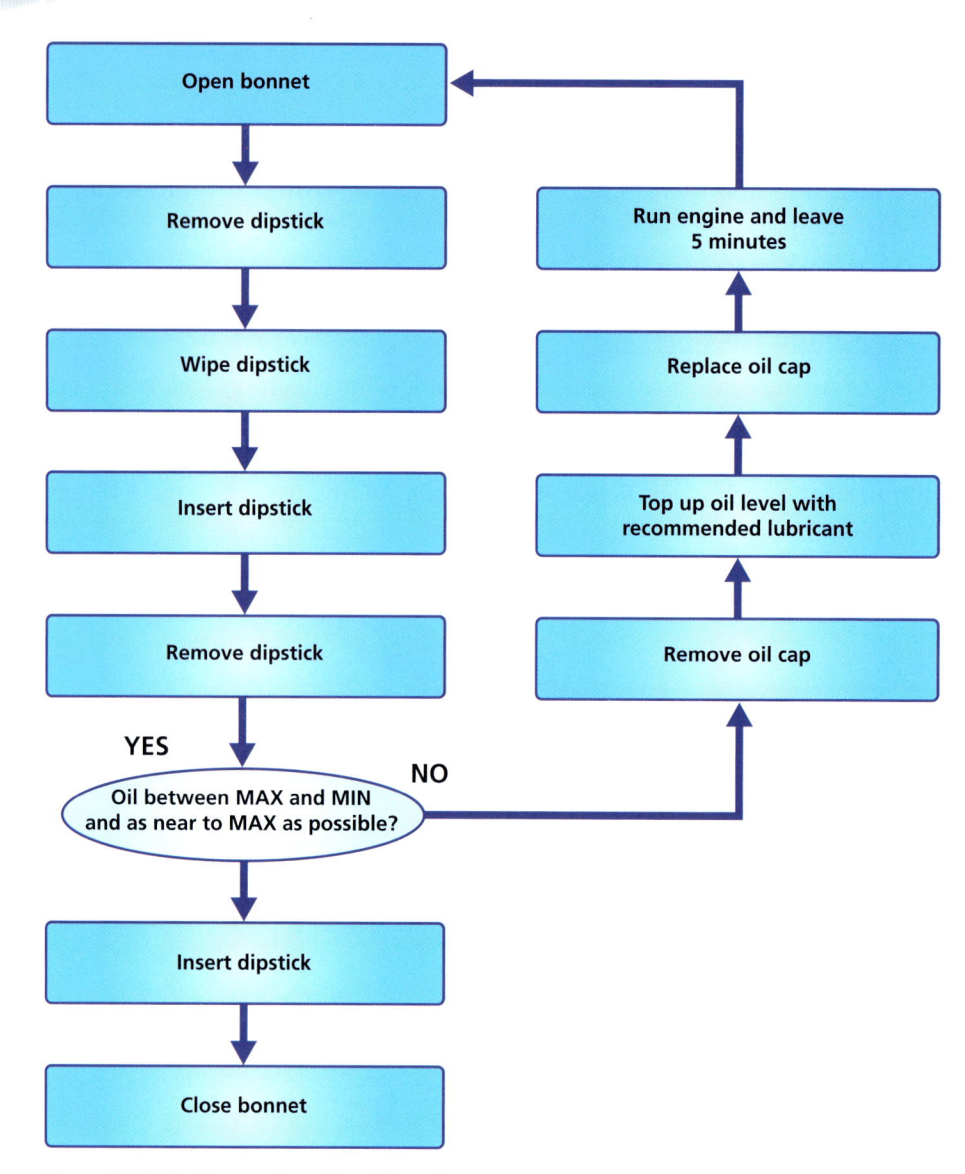

Open bonnet → Remove dipstick → Wipe dipstick → Insert dipstick → Remove dipstick → Oil between MAX and MIN and as near to MAX as possible?
- YES → Insert dipstick → Close bonnet
- NO → Remove oil cap → Top up oil level with recommended lubricant → Replace oil cap → Run engine and leave 5 minutes → (back to Open bonnet)

Figure 7.12 Checking engine oil level flow chart

Outside temperature range anticipated before next oil change

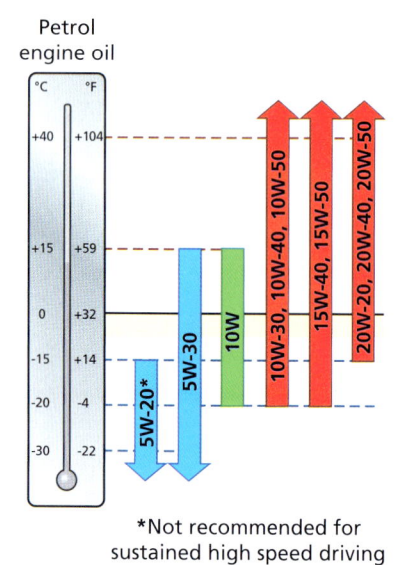

Petrol engine oil

(Thermometer showing °C and °F scales with oil grade ranges: 5W-20* (not recommended for sustained high speed driving), 5W-30, 10W, 10W-30, 10W-40, 10W-50, 15W-40, 15W-50, 20W-20, 20W-40, 20W-50)

*Not recommended for sustained high speed driving

Figure 7.13 Oil grades

Oil grades

Engine oil is available in different **viscosity** grades. Additives in the oil ensure that as the oil gets warmer, it does not get too thin, as this would damage the engine. It is important to use the same grade of engine oil when topping up the engine. The correct grade to use can be found in the manufacturer's workshop manuals and the vehicle handbook. (For more on engine oil grades, see Chapter 3, pages 230–233.)

Engine oil and filter replacement

Engine oil cools and lubricates the components of a vehicle's engine. Over a period of time, the oil will gather particles of dirt and debris from engine wear. These particles can get between moving parts and cause more wear and even engine failure.

7 Light vehicle inspection & servicing

An oil pump forces oil to the vehicle's crankshaft bearings, valve assemblies and other components such as turbochargers. Between the pump and system components is an oil filter. This collects all the dirt and debris to protect the engine from damage. (For more on the oil filter, see Chapter 3, page 227.)

At the same time, oil loses its viscosity because of heat changes and because it becomes diluted with unburnt fuel. For these reasons, the engine oil and oil filter must be changed regularly following the manufacturers' specifications, to ensure the life of the engine is optimised.

Figure 7.14 Tools and equipment for engine oil and filter replacement

Replacing the engine oil and filter

Checklist				
PPE	**VPE**	**Tools and equipment**	**Consumables**	**Source information**
• Steel toe-capped boots • Overalls • Latex gloves	• Wing covers • Steering wheel cover • Seat covers • Floor mat covers	• Hoist • Oil filter • Wrench • Oil drainer • Exhaust extraction	• Oil • Oil filter • Sump washer	• Oil capacity

1. Open the bonnet. Remove the oil filler cap.

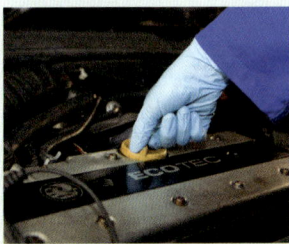

2. Lift the vehicle safely. Position the oil drainer under the sump plug and filter.

3. Remove the sump plug using the correct socket.

4. Remove the filter using the oil filter wrench. Allow the oil to drain fully.

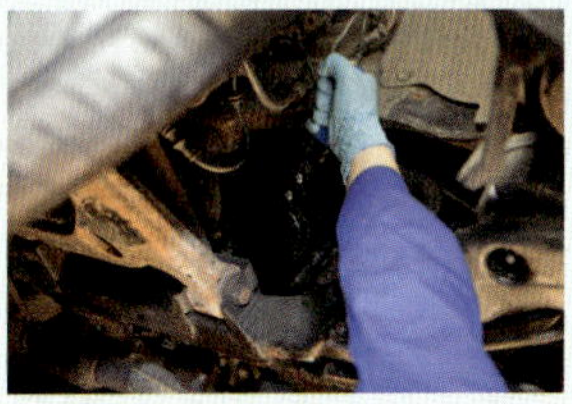

5. Wipe clean the sump bung and mating surface. Replace the bung with a new washer.

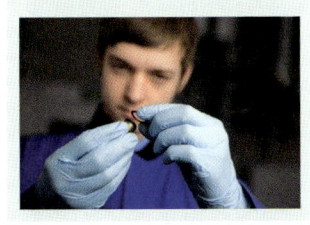

6. Wipe clean the filter mating surface which has the rubber seal attached. Apply a thin layer of new engine oil to the engine oil filter seal before installing the engine oil filter.

7. Tighten the oil filter by hand until it touches the mating surface. Then tighten the filter by two-thirds of a turn.

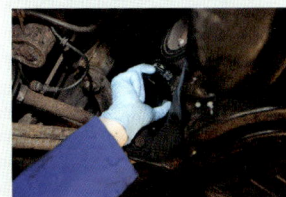

8. Lower the vehicle safely ensuring the drainer is clear from underneath the vehicle.

9. Fill the engine with new oil, using only the amount stated by the manufacturer.

10. Replace the oil filler cap and dispose of any oil-soaked gloves used for PPE.

11. Run the engine for five minutes using the exhaust extraction equipment. (Do not rev the engine as the oil needs to circulate in the system.)

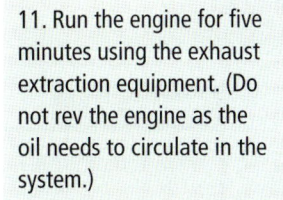

12. Check the oil level using the procedure described on page 441.

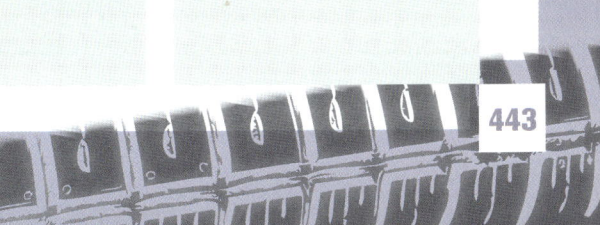

Level 2 Light Vehicle Maintenance & Repair

> ⚠ **Safe working**
>
> Always allow an engine to cool down before adjusting coolant levels, as at higher temperatures the coolant is under pressure. If the pressure cap is opened, engine coolant will expand and force boiling liquid over the technician.

Coolant level check

The coolant level should be checked regularly because it enables the engine to maintain a constant temperature, which keeps the oil at the correct viscosity.

- During hot weather, engine coolant removes heat from a vehicle and prevents it from overheating.
- When the weather is very cold, coolant prevents the engine freezing.

Coolant is a mixture of antifreeze and water. (For more on coolants, see Chapter 3, pages 235–236.)

Before you check and adjust the coolant level:

- Refer to the specifications given in the vehicle manual and service manual. This information will tell you at what mileage to change the coolant and if it has been changed recently.
- Park the vehicle on a level surface.
- Make sure the vehicle has been at rest for a considerable time and is cool.

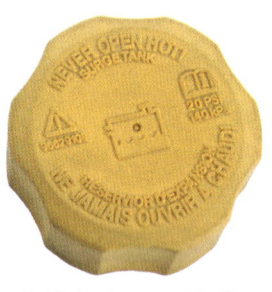

Figure 7.15 Coolant cap labelling

Checking the coolant level

Checklist				
PPE	VPE	Tools and equipment	Consumables	Source information
• Steel toe-capped boots • Overalls • Latex gloves	• Wing covers • Steering wheel cover • Seat covers • Floor mat covers	• Suitable cloth • Measuring jug	• Coolant (water and antifreeze mix)	• Coolant capacity • Ratio of water to antifreeze mix

1. Open the bonnet and locate the coolant reservoir. Make a note of the coolant 'cold' or 'cool' level. If the level is below cold/cool, you will need to add coolant.

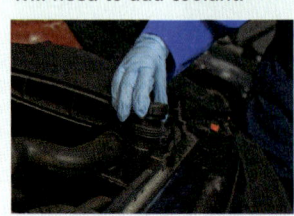

2. Use a suitable cloth to wipe around the filler cap, to prevent dirt entering the system. Remove the coolant pressure cap.

3. Add the coolant and watch to see if the level increases. If it does not there may be a leak. If there is a leak repair this before continuing.

4. Replace the cap securely.

> **Did you know?**
>
> Over a period of time, antifreeze loses its strength. This means that every two years the coolant should be drained and replaced.

> **Working life**
>
> Diana forgot to change the coolant on the vehicle she was servicing. One month later there was a really hard frost. The customer contacted the garage to say the engine was leaking.
>
> 1 What do you think happened to the engine?

Testing the cooling system

If there is a large drop in the amount of coolant in the coolant reservoir, you will need to inspect the system for leaks:

1. Do a visual check under the vehicle and around cooling system components.
2. Run the cool engine without the pressure cap on. If there are large bubbles of air in the reservoir, there may be an internal engine leak.
3. Carry out a pressure test using a cooling system pressure tester to determine the source of the leak (see Chapter 3, pages 237–238).

> **Safe working**
>
> Antifreeze is toxic so take care to avoid spills and skin contact. It must be disposed of properly at a fluid recycling centre.

Figure 7.16 Testing the cooling system

Replacing the coolant

Checklist				
PPE	**VPE**	**Tools and equipment**	**Consumables**	**Source information**
• Steel toe-capped boots • Overalls • Latex gloves • Safety goggles	• Wing covers • Steering wheel cover • Seat covers • Floor mat covers	• Hoist • Screwdriver • Socket and wrench • Catch basin	• Coolant (antifreeze plus water)	• Coolant capacity • Ratio of water to antifreeze mix

1. When the system is cool, remove the coolant pressure cap.

2. Lift the vehicle safely off the ground. Place a catch basin under the area you will be draining the fluid from.

3. If the system is provided with a drain tap, use this to drain the system. If not, undo a pipe at the lowest part of the cooling system, usually at the bottom of the radiator.

4. When the coolant is drained, tighten all drain plugs or replace the pipes.

5. Fill the system at the pressure cap with the amount of coolant specified by the manufacturer. You may need to run the vehicle for a few minutes so the coolant can circulate throughout the system. This is known as bleeding the system.

6. When the correct amount of fluid is in the system and the cooling system reservoir is at the correct level, replace the pressure cap.

Level 2 Light Vehicle Maintenance & Repair

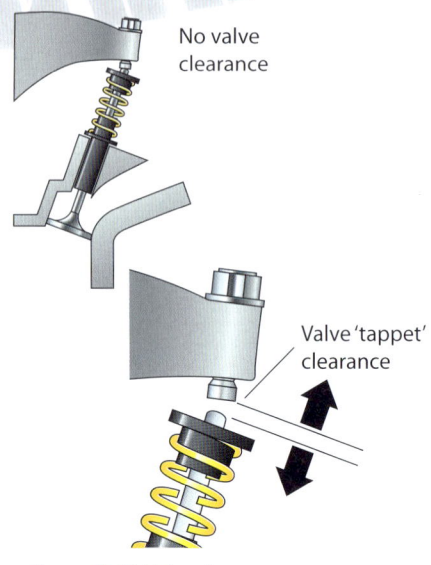

Figure 7.17 Valve clearances

Did you know?
New vehicles are fitted with automatic hydraulic valve lifters (see Figure 7.18), which do not require adjustment or servicing.

Figure 7.18 Hydraulic valve lifter

Valve clearances

Valve clearance is the small gap between a valve lifter (also called a valve rocker arm) and the top of the valve. The clearance is similar to the gap set with the spark plug.

The valve clearance ensures that that valve closes fully when the rocker arm has lifted from the valve stem.

- If the gap is too big the rocker arm and valve will knock or tap loudly whenever they make initial contact.
- If the gap it too small the valve will not close fully when the engine warms up, because metal expands at high temperatures.

You should always adjust valve clearances following the specific manufacturer's instructions.

Timing belt

The timing belt (also called the cam belt) is a rubber belt connected to the crankshaft to help drive the camshaft. Its purpose is to control the timing of the engine's valves. (For more on valve timing, see Chapter 3, pages 218–219.)

The belt often drives other auxiliaries such as diesel **injection pumps** and **ignition distributors**. All of these need to be timed exactly to allow the cam to open valves and diesel pumps to inject fuel at the correct time (hence the name timing belt). If the belt were to fail, major internal engine damage could occur, resulting in very expensive engine repair or even an engine replacement.

Different types of timing systems are shown in Figure 7.19. Some engines use chains and do not require inspection or servicing. (Always refer to the manufacturer's servicing and inspection manuals or information systems before replacing timing belts or chains.)

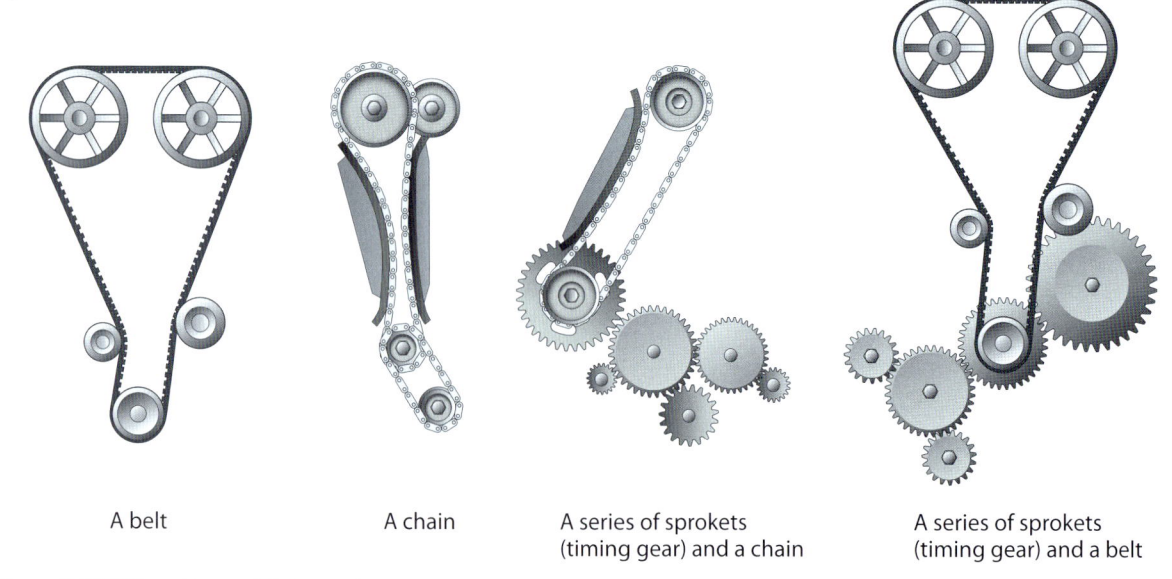

A belt │ A chain │ A series of sprokets (timing gear) and a chain │ A series of sprokets (timing gear) and a belt

Figure 7.19 Different types of timing systems

7 Light vehicle inspection & servicing

Timing components

Taking the time to understand how the basic components in the timing system work will give you valuable knowledge to help you carry out routine service inspections.

The main components of the timing system are shown in Figure 7.20.

Inspection methods

It may not be possible to carry out a visual inspection of the timing system because of protective belt covers and engine mounts. However, you can still carry out an aural inspection by listening for abnormal noise from the pulleys and belt bearings. You should record such noises on the service sheet.

Timing belt replacement

Always refer to manufacturer's manuals and information systems when carrying out a timing belt replacement.

- Most new vehicles have self-adjusting tensioners, which automatically set the tension of the belt.
- With older vehicles, you may need to adjust the timing belt manually. To do this you must use the correct tensioner tool, such as a **frequency meter** (see Figure 7.21) and set this according to the manufacturer's instructions. Check the tension on the longest portion of the belt where the **deflection** is at its greatest.

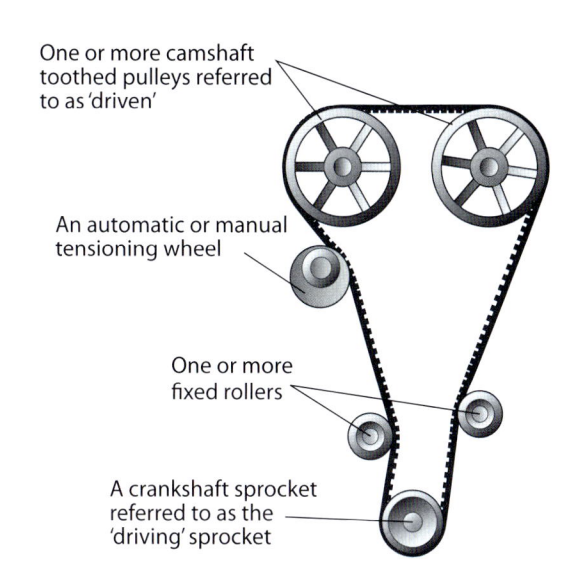

Figure 7.20 Timing system components

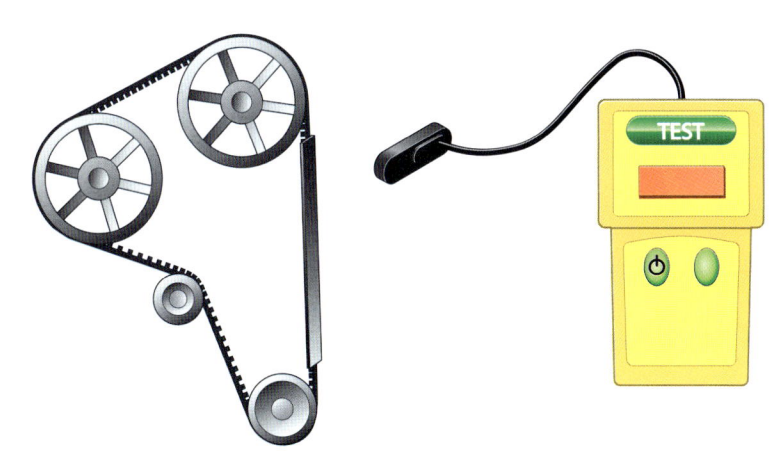

One or more camshaft toothed pulleys referred to as 'driven'

An automatic or manual tensioning wheel

One or more fixed rollers

A crankshaft sprocket referred to as the 'driving' sprocket

Figure 7.21 Measuring the tension of a timing belt using a frequency meter

Key terms

Injection pump – distributes fuel to each injector in a compression ignition (diesel) engine.

Ignition distributor – distributes high-tension voltage to each spark plug in a spark ignition (petrol) engine.

Frequency meter – electrical device which uses vibrations to measure the correct tension of a belt.

Deflection – the amount that the belt can be bent or moved.

Fan belts and auxiliary belts

Fan belts are used to drive cooling fans which keep the coolant radiator and coolant cool. Belt-driven fans are not needed in modern vehicles, as electric fans are used instead. However, the engine still needs to drive other auxiliary equipment such as alternators, air conditioning pumps and power steering pumps. So you are more likely to hear the term auxiliary belt than fan belt when working with modern vehicles. Figure 7.22 shows two different types of belts.

Level 2 Light Vehicle Maintenance & Repair

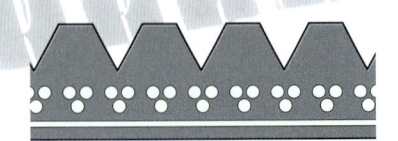

The flat ribbed belt

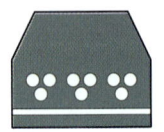

The V-belt

Figure 7.22 Types of belts

Checks to carry out on the fan/auxiliary belt include:

- **Visually inspecting** the belt for signs of ageing, cracking in the rubber, etc.
- **Listening** to confirm if the belt is slipping. Slipping belts will often make a loud squealing noise, which is more notable on initial start up when cold.
- **A functional inspection** to check the tightness of the belt. If the belt is too loose it will slip and not drive the fan or auxiliary equipment correctly, which could lead to the engine overheating.

Adjusting the fan/auxiliary belt

You can usually adjust the fan belt manually by following the manufacturer's information systems. As with the timing belt, the fan belt can be adjusted correctly using a frequency meter, as shown in Figure 7.23.

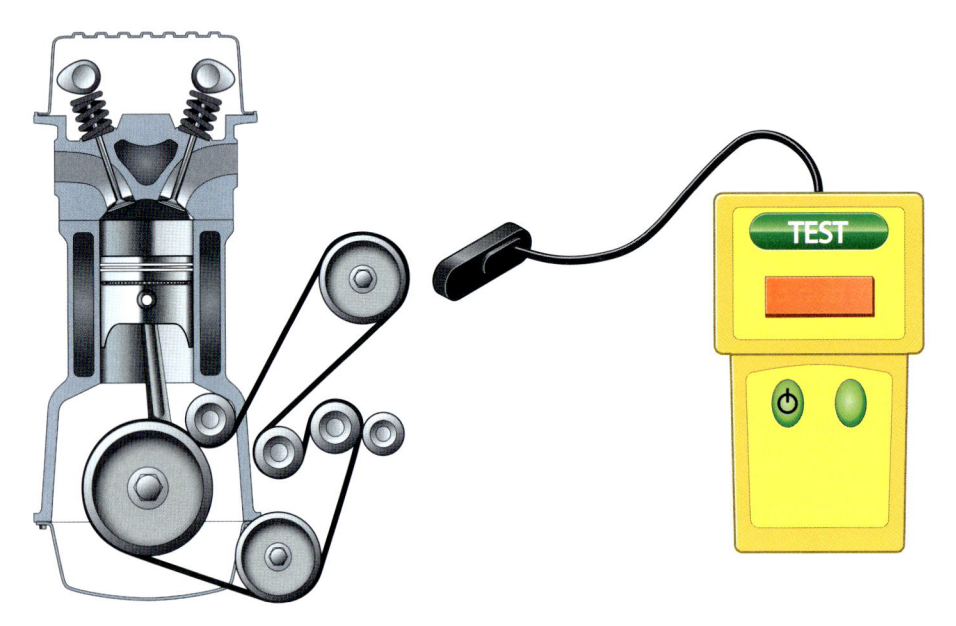

Figure 7.23 Measuring tension

The tension must be checked on the longest portion of the belt where the deflection is at its greatest. The effects of incorrect tension are shown in Table 7.4.

Table 7.4 Effects of incorrect tension in the fan belt

Belt too slack	Belt too tight
• Excessive squealing noise during acceleration • Reduced output of the alternator due to slippage of the belt	• Extra tension, causing premature failure of alternator bearings

Spark plugs

Spark plugs are used to ignite the air/fuel mixture within the cylinder in a spark ignition engine. (For detailed information about spark plugs see Chapter 4, pages 281–293.)

When replacing spark plugs, always refer to the manufacturer's information systems, as some spark plugs will require the **electrode gap** to be adjusted. If the gap is incorrect, the performance of **combustion** and the overall power output of the engine will be severely affected. An illustration of the spark plug gap is found in Figure 4.40 on page 291.

> **Key terms**
>
> **Electrode (spark plug) gap** – the distance the spark has to travel to complete the spark plug circuit to earth. This spark will ignite the air/fuel mixture.
>
> **Combustion** – the process of burning. This occurs when the petrol and air mixture is ignited by the spark.

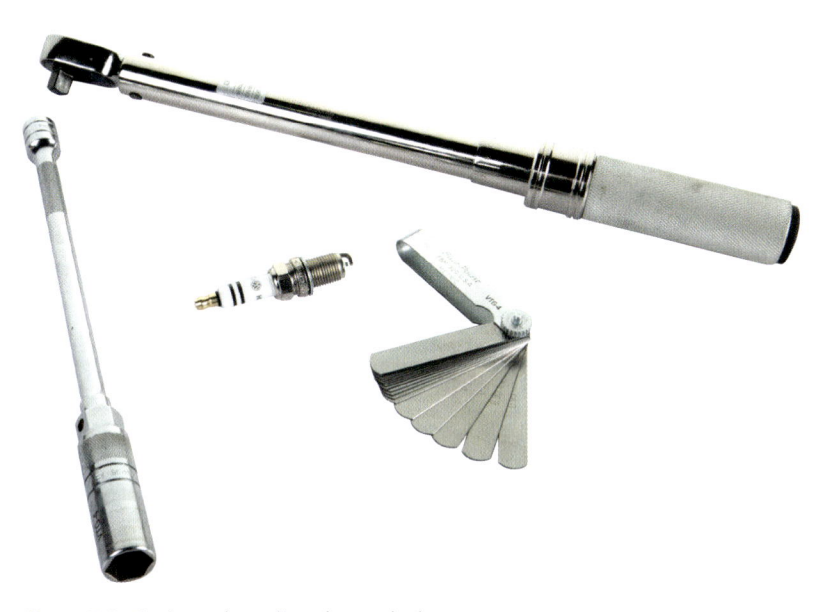

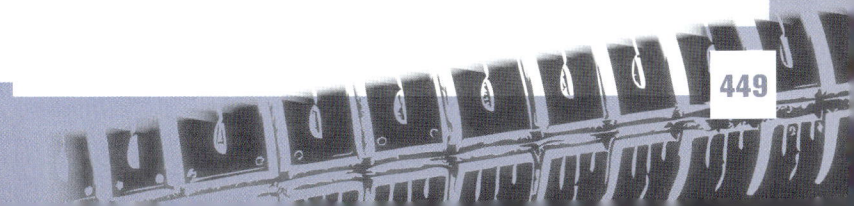

Figure 7.24 Tools used to adjust the spark plug gap

The main issues with spark plug gaps are shown in Table 7.5.

Table 7.5 Issues when the spark plug gap is too narrow or too wide

Gap	Risk	Benefit
Too narrow	Spark may be too weak to ignite fuel	Plug always fires
Too wide	Plug may not fire or may miss at high speed	Spark has a clean burn (is strong)

> **Did you know?**
>
> If you use a normal socket tool to tighten a new spark plug there is a risk that the outer porcelain will crack. Therefore, you need to use an extended socket with a rubber washer insert.

Level 2 Light Vehicle Maintenance & Repair

Remove, adjust and replace spark plugs

Checklist				
PPE	VPE	Tools and equipment	Consumables	Source information
• Steel toe-capped boots • Overalls • Latex gloves	• Wing covers • Steering wheel cover • Seat covers • Floor mat covers	• Spark plug socket • Wrench • Extension • Feeler gauges • Torque wrench • Masking tape • Pen	• Spark plugs	• Spark plug gap data • Spark plug torque setting

1. Open the bonnet. Remove any engine covers to access the spark plugs. Remove the spark plug lead from each plug. Mark which cylinder it came from using masking tape and a pen.

2. Use the correct socket and wrench to remove each spark plug.

3. Inspect the spark plug. This will give you an indication of how the engine has been running.

4. Check the new spark plug gap using feeler gauges. The gap should be the size specified by the manufacturer.

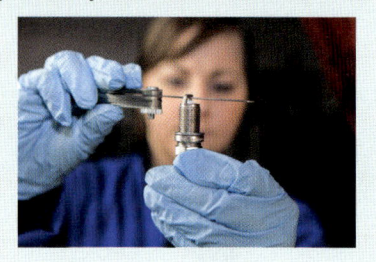

5. Place the spark plug into the socket and insert the plug into the cylinder head. Take care to make sure it goes in straight. Tighten the plug by hand to make sure that the threads on the plug are not damaged by **cross-threading**.

6. Use a torque wrench to complete the tightening. Take care not to overtighten the plug.

7. Replace the plug leads and engine cover. Make sure that the vehicle starts and close the bonnet.

Key term

Cross-threading – where one set of threads is out of line with the other. The damaged thread may lead to failure of the fixing. A thread can be corrected using a tap and die set.

Working life

Connor is about to change some spark plugs but has not got the correct plug socket.

1   What will happen if Connor uses the wrong socket?

7 Light vehicle inspection & servicing

Fuel system

The purpose of the fuel system is to supply the correct amount of fuel at the right time to ensure that the engine can perform powerfully and economically. (For more on the fuel system, see Chapter 4, pages 262–290.)

When inspecting the fuel system, you need to check that:

- there are no leaks
- the pipes are secure and damage free
- the gases that come out of the exhaust are within legal limits.

When you are servicing the fuel system you will also need to replace the fuel filter (see Figure 7.25). This needs to be done at the manufacturer's specified intervals.

Figure 7.25 Fuel filter

Changing the fuel filter

Checklist				
PPE	**VPE**	**Tools and equipment**	**Consumables**	**Source information**
• Steel toe-capped boots • Overalls • Latex gloves • Safety goggles • Safety helmet	• Wing covers • Steering wheel cover • Seat covers • Floor mat covers	• Hoist • Filter pipe release tool • Oil drainer • Socket and wrench	• Fuel filter	• None required

1. Raise the vehicle off the ground. Locate the fuel filter near the fuel tank and position the drainer under the filter.

2. Remove the filter from the mounting using the socket and wrench.

3. Wipe around the filter pipes with a rag to prevent dirt getting into the pipes. Remove the filter pipes with the pipe release tool.

4. Refit the new filter in the reverse order of removal.

5. Lower the vehicle. Run the engine and check for leaks.

> ⚠️ **Safe working**
>
> Fuel is highly flammable. Make sure that you change the fuel filter in a well-ventilated area and that there are no ignition sources in the vicinity.

451

Level 2 Light Vehicle Maintenance & Repair

Figure 7.26 Air intake filter

Air intake system

The purpose of the air intake system is to supply enough clean air to burn the fuel in the engine for maximum power and efficiency. (For more information on the air intake system, see Chapter 4.)

When inspecting the air intake system, you need to check for:

- the cleanliness of the air filter element and element housing
- air leaks in the pipes supplying clean air to the engine.

The vehicle manufacturer will specify when to replace the air filter. You may need to replace the filter more regularly if the vehicle is being used in dusty conditions, such as at or near a quarry.

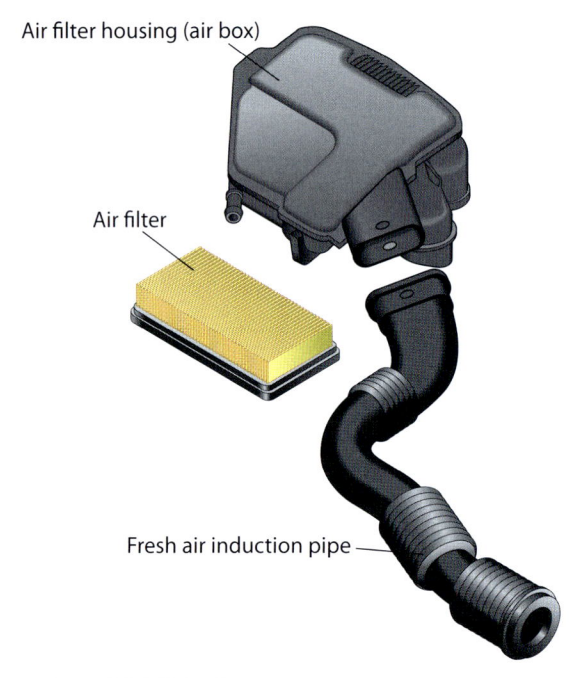

Figure 7.27 Air intake system components

Replacing the air filter

Checklist				
PPE	**VPE**	**Tools and equipment**	**Consumables**	**Source information**
• Steel toe-capped boots • Overalls • Latex gloves	• Wing covers • Steering wheel cover • Seat covers • Floor mat covers	• Hand tools	• Air filter	• None required

1. Open the bonnet. Locate the filter housing. Remove the fixings retaining the filter housing and lift off the filter housing.

2. Remove the filter. Remove any debris from the housing. Inspect the filter.

3. Replace the filter if it is dirty or at the service interval recommended by the manufacturer. Replace the filter housing.

4. Inspect the pipes (there may be more than one pipe, depending on the system) to check that they are secure and undamaged.

452

Exhaust system

The purpose of the exhaust system is:

- to silence the noise from the engine
- to cool and direct gases to the rear of the vehicle to prevent gases entering the passenger compartment.

It is very important that you visually check the exhaust for leaks and functionally check the exhaust for security, making sure that rubbers and mountings are in good order.

The main components of a petrol and diesel exhaust system are shown in Figures 7.28 and 7.29. (For more detail on the exhaust system, see Chapter 4, pages 301–305.)

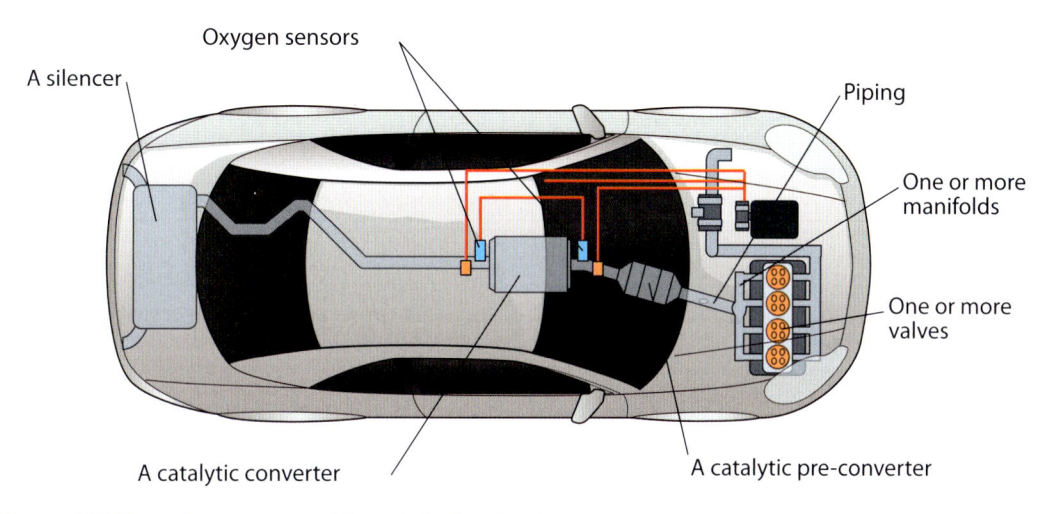

Figure 7.28 The main components of the petrol exhaust system

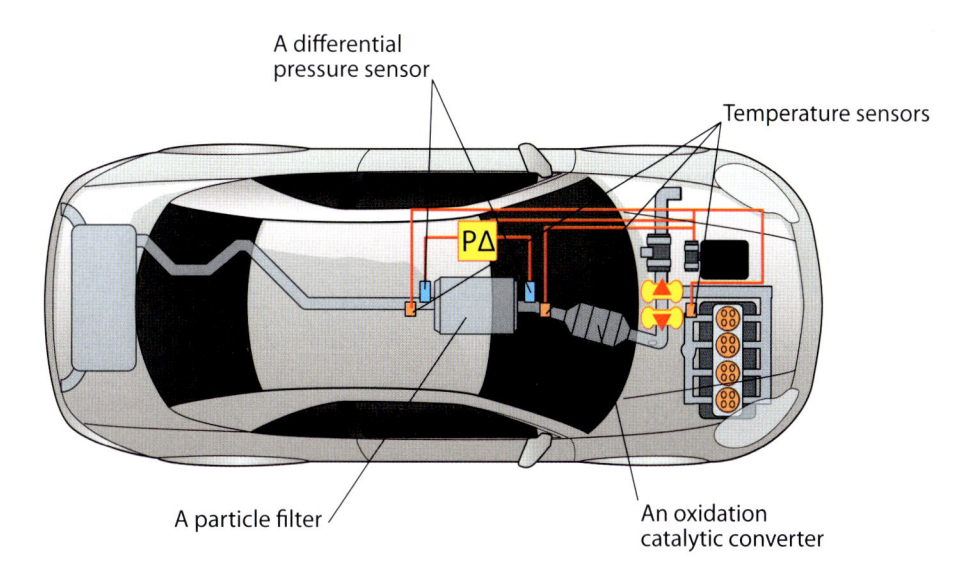

Figure 7.29 The main components of the diesel exhaust system

Figure 7.30 Oxygen sensor

Inspecting the exhaust system

You will need to carry out the following checks in your inspection of the exhaust system:

- **Aural check** – Check the mounting position for knocking noises and for exhaust gas leaks.
- **Functional check** – Grasp the exhaust (when cold) and move it side to side to make sure the system doesn't knock against the vehicle body.
- **Visual check** – Inspect the electrical connectors to the oxygen sensor and temperature sensors for corrosion at terminals. Confirm they are connected securely. Check pipes for corrosion and security.

Exhaust emissions testing

As the air/fuel mixture is burnt in the engine, gases are produced which are harmful to the environment. To reduce pollution, limits have been set on the emission levels that are allowed.

During some servicing activities and as a compulsory item on an MOT inspection, exhaust gases need to be checked to make sure they are within the legal limits set.

Pre-checks

Before you test the exhaust emissions, there are some checks you need to make. Otherwise, the vehicle may fail the test prematurely or lasting damage to the engine may occur.

You need to check that:

1. The condition of the engine is good so that the engine will not be damaged when revved up.
2. The air filters are clean.
3. The exhaust is in good condition and not leaking.
4. The engine is running at normal operating temperature.
5. The engine is running at the correct constant idle speed and not **hunting and surging**.
6. All electrical consumers are off, including the cooling fan.

You also need to input all the vehicle details (such as make, model and engine size) into the emissions testing machine.

> **Safe working**
>
> Exhaust gases are poisonous. Make sure that:
> - the exhaust extraction is fitted to the vehicle during warm-up
> - the area is well-ventilated during exhaust emissions testing.

> **Key term**
>
> **Hunting and surging** – where the engine revs rise and fall erratically during idle.

7 Light vehicle inspection & servicing

Testing exhaust emissions

Checklist			
PPE	VPE	Tools and equipment	Source information
• Boots • Overalls • Latex gloves • Particle mask	• Wing covers • Steering wheel cover • Seat covers • Floor mat covers	• Emissions testing machine • Vehicle exhaust extraction equipment	• Emissions settings

1. Following all pre-checks, raise the engine speed to about half the maximum engine speed. Hold this speed steady for 30 seconds to ensure that the inlet and exhaust system is fully cleared of the extra fuel needed on idle.

2. Allow the engine to return to idle and the emissions to stabilise.

Assess the smoke emitted from the tailpipe using the emissions tester.

3. Rapidly increase the engine speed to around half the maximum engine speed. Assess the smoke emitted from the tailpipe during acceleration.

The machine will give readings that indicate whether the vehicle has passed or failed the test.

If a vehicle hunts and surges during idle, there may be air leaks in the inlet system which are causing extra air and fuel to enter into the engine.

CHECK YOUR PROGRESS

1. What are you checking when you measure the deflection of either the auxiliary or timing belt at their longest point?
2. What initial type of inspection would you use to detect possible sources of exhaust gas leak?
3. What is the first thing you must do before testing the exhaust emissions of an engine?

Level 2 Light Vehicle Maintenance & Repair

The braking system

The braking system slows the vehicle down, stops the vehicle and keeps it stationary. The main components of the braking system are shown in Figure 7.31. (For more detail on the braking system, see Chapter 2, pages 164–186.)

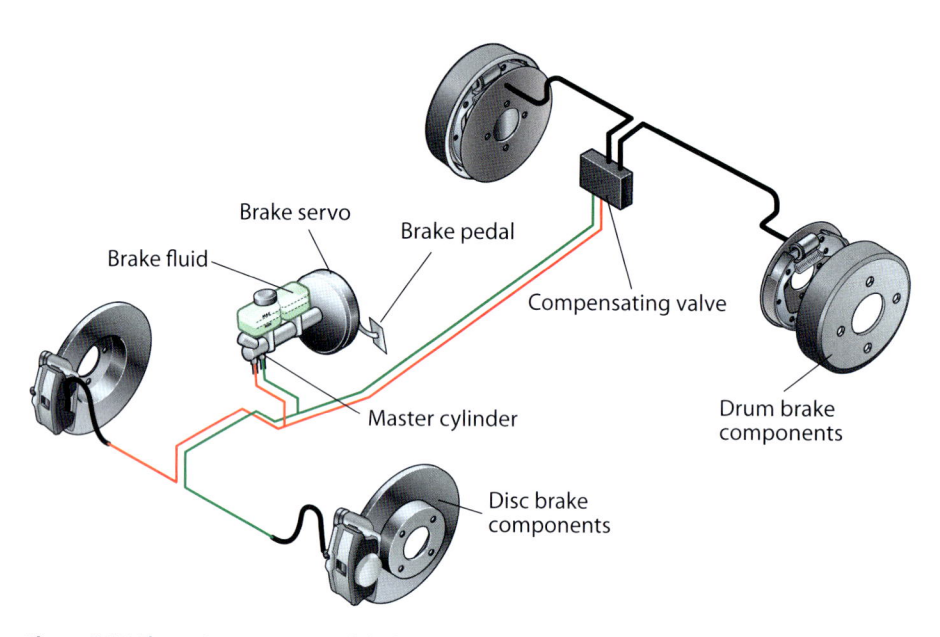

Figure 7.31 The main components of the braking system

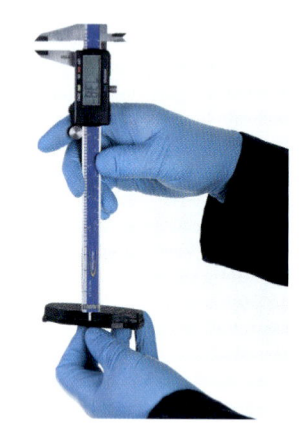

Figure 7.32 Measuring the thickness of brake linings

Disc brake servicing and inspection

Servicing disc brakes involves checking the condition of the pads, discs and hoses. Servicing is particularly important if the brake fluid level falls below the maximum mark, as this is an indication that there may be a fluid leak or that the brake linings are worn.

Checking the brake pads for wear involves:

- visually checking that the brake pads are not greasy
- measuring the thickness of the brake linings to ensure there will be enough material to last through to the next service.

If you find a brake pad fault on one side of vehicle, you must replace all the pads on the same axle (see opposite).

Checking the brake discs

To inspect the brake disc:

- check the surface condition
- measure the disc's thickness using a micrometer (see Figure 7.33)
- measure the **run-out** using a dial test indicator.

Run-out can be caused by heat generated by the friction surfaces during the brake operation.

Figure 7.33 Measuring the thickness of the brake disc

Key term

Run-out – the amount of variation measured between rotating and fixed parts.

7 Light vehicle inspection & servicing

Replacing brake pads

Checklist				
PPE	**VPE**	**Tools and equipment**	**Consumables**	**Source information**
• Steel toe-capped boots • Overalls • Latex gloves • Dust mask • Safety goggles	• Wing covers • Steering wheel cover • Seat covers • Floor mat covers	• Hoist • Torque wrench • Lever • Bleed tube and bottle • Piston retraction tool	• Brake pads • Brake fluid • Copper grease • Brake cleaner • Squeal-reduction shims (if fitted)	• Minimum brake pad thickness • Brake fluid type

1. Remove the brake fluid reservoir cap.

2. Raise the vehicle. Remove the wheel and position the steering to allow access to the disc assembly.

3. Spray brake cleaner on the assembly to remove any dust.

4. Remove the fixings holding the pads in place, then remove and inspect the pads.

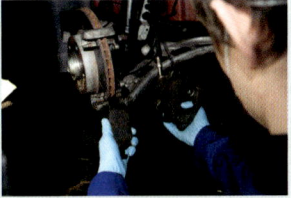

5. Fix the bleed pipe and bottle to the bleed nipple on the caliper and slacken the bleed nipple.

6. Use the piston retraction tool to push back the piston and tighten the bleed nut.

7. Smear a small amount of copper grease on the back of the new pads.

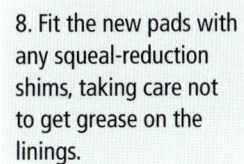

8. Fit the new pads with any squeal-reduction shims, taking care not to get grease on the linings.

9. Replace the fixings holding the pads in place.

10. Remove the bleed pipe and refit the wheel.

11. Pump the brake pedal until it goes firm.

12. Check the brake fluid level and top up if necessary to the maximum level.

13. Carry out the same procedure on the other side of the vehicle on the same axle.

14. Refit the brake fluid reservoir cap.

! Safe working

When replacing brake discs or pads, you must wear a particle mask to protect you from inhaling brake dust.

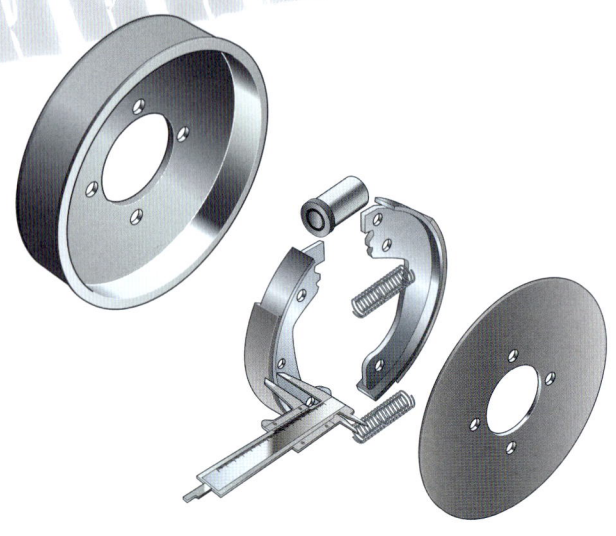

Figure 7.34 Checking the brake shoes

Drum brake servicing and inspection

Over time, the drum and brake shoes become worn. Inspecting the drum brake involves checking the condition and wear of the brake shoes and drums.

Run-out can be caused by heat generated by the friction surfaces during the brake operation. You should inspect the drum by:

- checking its surface condition for wear and grease
- measuring its **ovality** and diameter.

If the drums are not in good order, braking efficiency can be reduced.

A quick way for you to determine if the drum is oval is:

- Lift the vehicle off the ground.
- With the hand brake off, spin the wheel.
- If the wheel starts to stick in at the same position during rotation, the drum is oval.
- Then take two measurements across the drum lining surface at right angles to each other (see Figure 7.35). The two measurements should be the same. If they are different, the drum is oval and needs to be replaced.

The automatic wear compensation mechanism pushes the shoes closer to the drum as the brake linings wear. You should check the operation of the automatic wear compensation mechanism and the wheel cylinders at the same time as the brake drum.

Figure 7.35 Measuring the drum for ovality

When inspecting the automatic wear take-up system, you need to check that:

- it is positioned correctly
- it is clean
- when the brakes are applied, the ratchet mechanism holds the shoes in an outward position.

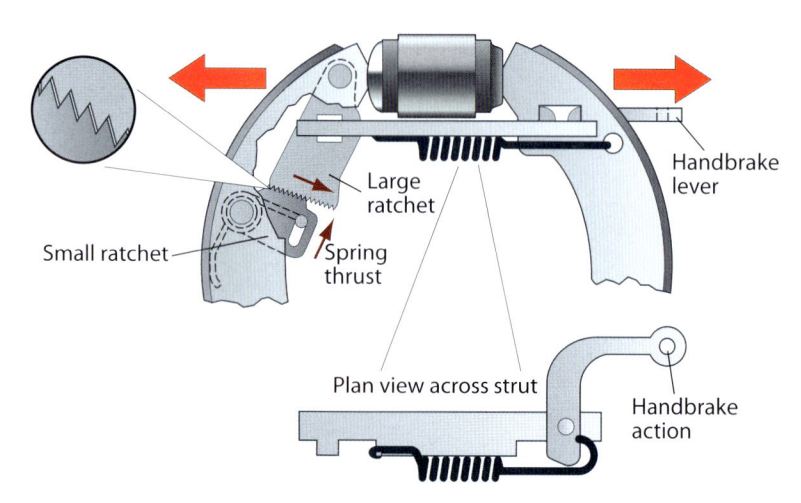

Figure 7.36 Checking the automatic wear take-up system

You carry out this check by spinning the drum and observing the drum's speed. If the drum does not slow down, this shows that there is no resistance from the brake shoes. This indicates that the brakes are not correctly adjusted.

If you find a brake shoe fault on one side of the vehicle, you should replace all shoes on the same axle.

Wheel cylinders

Wheel cylinders allow the brake fluid pressure to force the brake shoes towards the brake drum. Inspecting the wheel cylinders involves:

- checking for leaks
- checking the sliding action of the pistons
- peeling back the dust cover on the cylinder – if there is any damage to the seal, fluid will be visible.

Inspecting the parking brake

To inspect the parking brake (also called the handbrake) you need to check:

- the condition of the cables and sheaths
- the amount of handbrake travel
- that the brake releases fully when the handbrake is off.

To inspect the sheathed cables you need to check that:

- the cable slides in its sheath
- there is no fraying
- the protective sheath is not cracked or torn.

Excessive **handbrake travel** may be due to worn linings and/or the automatic adjuster on a drum brake system not working. If the handbrake does not release fully, mechanical linkages may have seized, due to corrosion, or there may be a fault with a handbrake cable.

The hydraulic circuit

Inspect the hydraulic circuit by checking for leaks and for any corrosion, cracking or damage to pipes or mountings.

You must check the brake fluid level during every inspection and the brake fluid must be replaced every two years. This is because brake fluid is a hygroscopic fluid – in other words, the fluid will absorb water from the air over a period of time. The water can dilute the fluid and reduce the boiling point. This may cause brake failure. For this reason, manufacturers state that brake fluid must be changed on a regular basis (every two years).

Figure 7.37 Wheel cylinder

Figure 7.38 The handbrake cable

> **Key terms**
>
> **Ovality** – where the drum wears in an oval shape. Over a period of time, the combination of heat and pressure on the drum causes the drum to deform and become oval.
>
> **Handbrake travel** – the amount the lever can be lifted by the driver. The normal rule is that it should not go above four 'clicks' when the handbrake button is not pressed in during lifting.

Level 2 Light Vehicle Maintenance & Repair

Brake hoses

When inspecting the brake hoses you should visually check that they have not perished or been fitted incorrectly. You will need to check:

- surface condition
- positioning
- in the case of flexible pipes, the amount the pipe has swelled.

The positioning is important if the pipe is on a steering axle, as wear can occur during vehicle cornering.

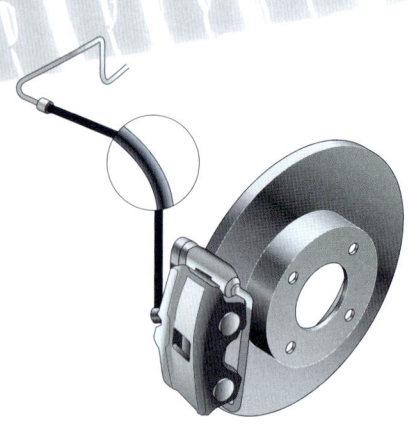

Figure 7.39 The correct alignment of the brake hose

Did you know?

- It is important not to twist the flexible pipe when working on the braking system, as this can weaken it.
- If the water content in the brake fluid reaches the 3% mark, the boiling point of the fluid will be reduced and air bubbles will be produced.

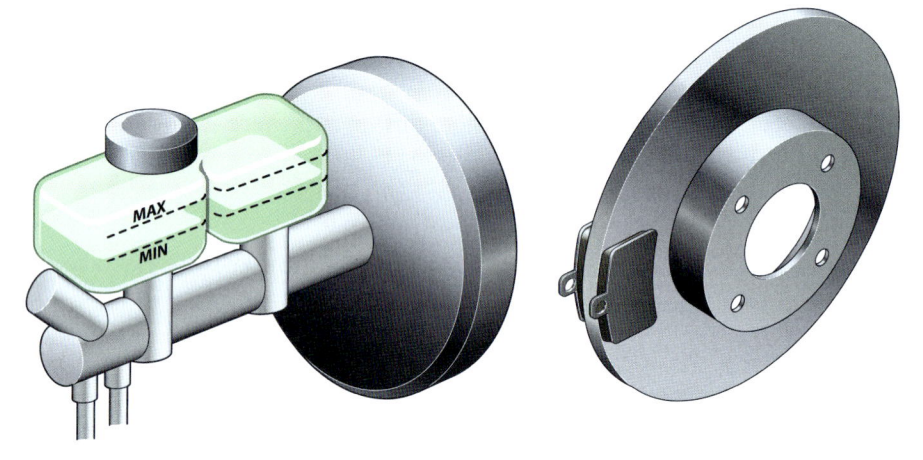

Figure 7.40 Brake fluid level depends on the wear of the pads

Figure 7.41 Brake fluid level warning indicator light

Brake fluid level check

The brake fluid level must always be between the maximum and minimum marks marked on the reservoir (see Figure 7.40).

An instrument panel warning light (see Figure 7.41) illuminates if the level is below the minimum mark. This is activated by a switch in the brake fluid reservoir.

If brake fluid is not changed at the intervals recommended by the manufacturer, the driver will only become aware that there is a problem when the brakes feel 'spongy' and the car fails to stop.

Pre-checks

Before you check and adjust the brake fluid level:

- Refer to the vehicle manual to check the type of brake fluid used.
- Make sure you follow all specifications given.
- Use brake fluid from a unopened container.

	Brake fluid is highly corrosive.
	Brake fluid is toxic if ingested or if fumes are inhaled.
	Brake fluid irritates the eyes.
	Used brake fluid should never be reused.
	Brake fluid should never be mixed with other products.
	Brake fluid should never be stored without sufficient identification.

Figure 7.42 Brake fluid safety warning symbols

Safe working

Avoid spilling or splashing brake fluid on to painted surfaces as a chemical reaction may occur which will spoil the surface. If there is a spill, wash it off straight away with water.

Checking and adjusting brake fluid level

Checklist				
PPE	**VPE**	**Tools and equipment**	**Consumables**	**Source information**
• Steel toe-capped boots • Overalls • Latex gloves	• Wing covers • Steering wheel cover • Seat covers • Floor mat covers	• Suitable cloth	• Brake fluid	• Brake fluid type

1. Locate the master cylinder.

2. Check the level of the brake fluid. (You should be able to see this through the reservoir.) The brake fluid should be at the 'full' level.

3. If the brake fluid is below this level, unscrew the cap and add the specified brake fluid until it reaches the full mark.

4. Replace the brake fluid cap and check that it is secure.

Brake bleeding

You must bleed the braking system if any component within the hydraulic line is changed or when changing the brake fluid during a service.

If air gets into the hydraulic system, pressure from the master cylinder will not be transmitted to the hydraulic wheel actuators. The reason for this is that air may be compressed and, when it is under light compression, it will not transfer force.

The procedure for brake bleeding is given in Chapter 2, page 180.

Figure 7.43 Different types of brake fluid

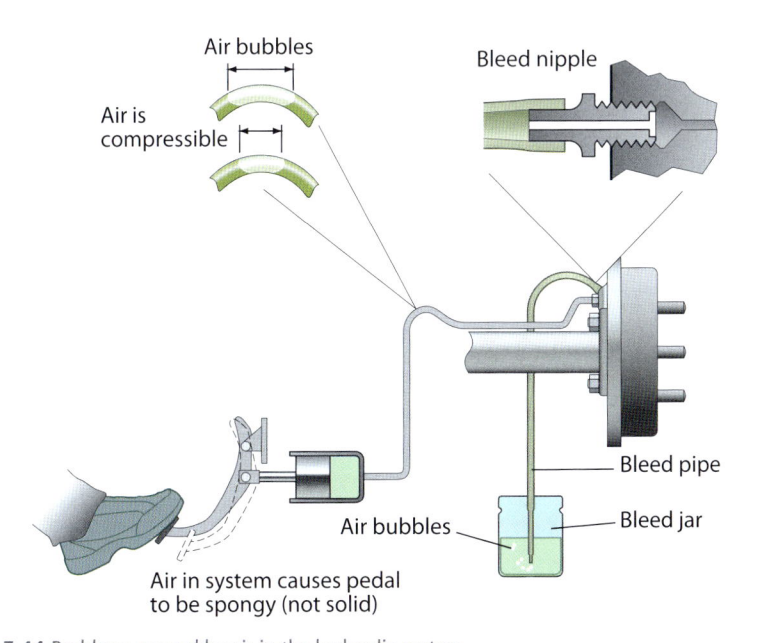

Figure 7.44 Problems caused by air in the hydraulic system

Did you know?

Most light vehicles use DOT 3 or 4 brake fluid. DOT stands for Department of Transport. (For more on DOT ratings, see Chapter 2, page 180.)

Suspension system

The suspension system provides a smooth ride for the driver, passengers and the load. (For more detail on the suspension system, see Chapter 2, pages 148–163.)

The main areas that you need to inspect for wear and damage are:

- swivelling and pivoting joints
- springs
- dampers
- linkages.

Inspecting the suspension system

1. Check the vehicle ride (trim) height by placing a tape measure from the hub centre to the base of the wheel arch. The measurements need to be the same on each axle and within the manufacturer's recommendations.

2. Check the dampers while the vehicle is on the ground by bouncing each corner. The vehicle should return to its natural position after bouncing one and a half times.

3. Check the dampers for leaks and corrosion when the vehicle is at chest height with the wheels removed.

4. Check that the springs are not broken and are located correctly in their housings.

5. Check the suspension bushes for wear, damage and splits. You may need a lever or pry bar to check some suspension bushes and joints.

6. Check the joints for excessive wear.

7. Check the wheel bearings for excessive movement and rumbling.

> **Find out**
>
> During routine servicing and inspection you discover hydraulic fluid leaking from a front brake.
>
> - What is the first thing you should do?

The procedure for checking wheel bearings is given in Chapter 2, page 146.

Steering system

The steering system of a vehicle allows the driver to control the direction of the vehicle. (For more detail on the steering system, see Chapter 2, pages 129–147.)

7 Light vehicle inspection & servicing

Inspecting the steering system

Checklist				
PPE	**VPE**	**Tools and equipment**	**Consumables**	**Source information**
• Steel toe-capped boots • Overalls • Latex gloves • Safety helmet	• Wing covers • Steering wheel cover • Seat covers • Floor mat covers	• Hoist • Lever bar • Inspection lamp	• PAS fluid	• None required

1. Push the steering wheel at right angles to the column, while applying light pressure downwards and upwards. This will indicate any excessive wear in the column or that the column is loose.

2. Push and pull the steering wheel in line with the column. Any movement at the centre of the steering wheel will indicate column wear or insecurity.

3. Rotate the steering wheel from lock to lock with the engine running. Listen for abnormal noises and check for smooth operation of the power assisted steering (PAS) system.

4. With the vehicle at chest height, check for excessive movement in the steering arms.

5. Check that the steering rack is securely mounted to the body or subframe.

6. Check the steering rack gaiters for splits.

7. Squeeze the steering rack gaiters to check if they have any power steering fluid in them – this will indicate that the rack is leaking.

8. Check that locking devices and fixings are present and secure.

9. Check for PAS system fluid leaks. Check that the PAS hoses are secure, not damaged or chafing and are positioned correctly.

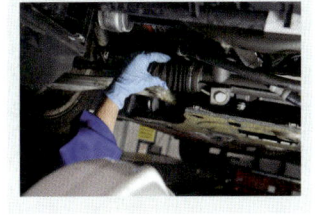

10. Check the PAS fluid level.

463

Power assisted steering (PAS) fluid level check

PAS is taken for granted but it is very difficult to drive a vehicle which has failed power steering. For this reason, it is important to check the PAS fluid level at the required service intervals.

The PAS service process is shown in Chapter 2, in the flow chart on page 136.

Pre-checks

Before you check and adjust the PAS fluid level, make sure:

- the engine is cool
- the hand brake is on
- the gears/transmission is in neutral.

Checking and adjusting the PAS fluid levels

Checklist				
PPE	VPE	Tools and equipment	Consumables	Source information
• Steel toe-capped boots • Overalls • Latex gloves	• Wing covers • Steering wheel cover • Seat covers • Floor mat covers	• Suitable cloth	• PAS fluid	• Vehicle handbook or manual

1. Unscrew and remove the cap to the power steering reservoir.

2. The cap normally has a dipstick attached. Wipe off the dipstick and replace the cap.

3. Remove the cap and inspect the level of the fluid on the dipstick.

4. Make sure you read the correct marking on the dipstick as specified in the vehicle handbook or manual. Top up the system if fluid has dropped below the marking on the dipstick.

Overfilling will result in fluid spraying out the top of the reservoir and onto the engine and other components.

Wheels and tyres

The role of the wheel is to bear the vehicle load, provide a sealed mounting for the tyre and make the vehicle look attractive. Tyres must support the vehicle load, provide a smooth ride for the occupants and transmit drive and braking forces to the road surface in all weather conditions.

For more detail on wheels and tyres, see Chapter 2, pages 186–198.

> **Did you know?**
>
> The reason for taking a reading of the fluid when the engine is cold is because the fluid level will rise on the dipstick as the PAS fluid gets warmer.

7 Light vehicle inspection & servicing

Checking wheel alignment

Wheel alignment must be correct to prevent unnecessary tyre wear and to ensure that the vehicle handles correctly. If wheel alignment is incorrect, the tyre will be worn on the inner or outer portion of the tread. This is sometimes called feathering.

As the vehicle is propelled forward, forces act on the front wheels. These forces, together with the movement in the steering and suspension joints, contribute to the wheels being pushed outwards or inwards depending on the drive train layout. To compensate for this and ensure that the wheels run true:

- front-wheel drive vehicles are normally set to toe-out
- rear-wheel drive vehicles are set to toe-in.

For more on toe-out and toe-in, see Chapter 2, pages 140–141.

Pre-checks

Before you check wheel alignment, you need to make the following pre-checks to ensure that your wheel alignment readings are accurate:

- Tyres must be inflated to the correct pressures.
- Wheels and tyres must be the correct type and size for the vehicle.
- Tyres must not have any uneven tread wear patterns.
- Wheels must not be buckled.
- There must be no worn or damaged steering or suspension components.
- The vehicle must be loaded correctly.
- The vehicle must be on a flat, level surface.
- The suspension must have been allowed to settle following any lifting or servicing and inspection.
- The steering wheel must be centralised and held in the straight ahead position.
- The alignment tools must be calibrated.

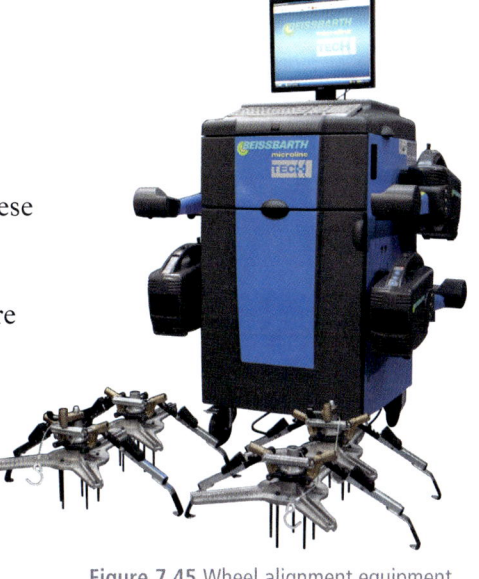

Figure 7.45 Wheel alignment equipment

Wheel alignment procedure

As there are many different systems, you must always follow the manufacturer's instructions when carrying out any wheel alignment. If the alignment is incorrect, you will need to adjust both track arms by the same amount to ensure that the steering wheel is positioned correctly. This will also prevent excessive tyre wear during cornering.

Inspecting wheels

1. Check for rim damage and any signs of cracks on the inside and outside of the wheel.

2. Spin the wheels by hand to check they are not buckled. Make sure that the wheels are of the same size on the same axle.

465

Level 2 Light Vehicle Maintenance & Repair

Inspecting tyres

Checklist				
PPE	**VPE**	**Tools and equipment**	**Consumables**	**Source information**
• Steel toe-capped boots • Overalls • Latex gloves	• Wing covers • Steering wheel cover • Seat covers • Floor mat covers	• Tread depth gauge • Tyre pressure gauge • Compressor	• Air from compressor	• Tyre pressures

1. Check that there are no bulges, cracks or tears in the walls or distortion of the tyre carcass.

2. Check that there are no objects embedded in the tread.

3. Check the tyre wear, using a tread depth gauge. Tread depth must be a minimum of 1.6 mm across the centre three-quarters of the width of the tyre around the whole circumference.

4. Check the direction of rotation is correct if the tyre is directional.

5. Check that the tyre valves are not damaged and dust caps are fitted.

6. Check that the tyre pressures are in line with the manufacturer's specifications. Inflate the tyres if necessary. Don't forget the spare tyre if the vehicle has one.

Find out

Inspect the wheels of a vehicle in your workshop. Copy the table below and use it to record your findings.

Wheel check – observations		
Wheel	**Tyre**	**Wheel rim**
Example	Blistering on the tyre wall	Minor deformation of the wheel rim edge
Front left		
Front right		
Rear left		
Rear right		
Spare wheel		

Did you know?

A chemical-filled aerosol is used on vehicles without an emergency spare wheel. In the event of a puncture, this chemical forms a seal around the puncture area so the driver can re-inflate the tyre and get to the nearest tyre specialist for a lasting repair.

The emergency aerosol kit has an expiry date which you need to check during an inspection.

Checking tyre pressures

Check tyre pressures using a pressure gauge on cold tyres.

The pressure varies according to the vehicle's load and speed. A vehicle driven at speed with a full load needs a greater tyre pressure than a vehicle that is driven at low speed and with a light load. You will find the required pressure values for the inflated tyres in the vehicle handbook.

The emergency spare wheel should be set to the highest recommended pressure for the vehicle – if it needs to be used, air can be let out by the person replacing the wheel to suit the vehicle use. Some emergency spare wheels have the pressure setting marked on the wheel itself.

Repairing a tyre

A puncture in a tyre can be repaired if:

- the vehicle has not been driven while the tyre was flat
- the puncture is within the tread area
- the tyre is not damaged in any other way.

Replacing a tyre

You must use the correct equipment to replace a tyre (see below). If a tyre is going to be repaired, take care when removing it not to damage it in any way.

Wheel balancing

The two methods of balancing a wheel (on and off the vehicle) are explained in Chapter 2, page 197.

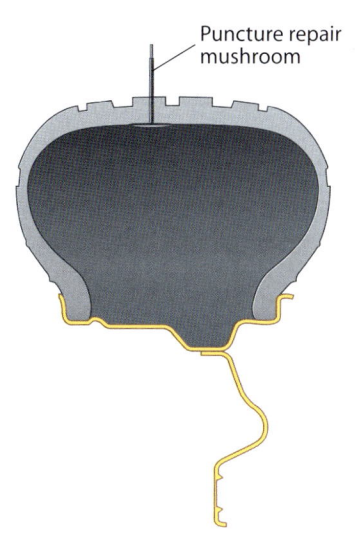

Figure 7.46 A puncture repair

Replacing a tyre

Checklist				
PPE	VPE	Tools and equipment	Consumables	Source information
• Steel toe-capped boots • Overalls • Latex gloves	• Wing covers • Steering wheel cover • Seat covers • Floor mat covers	• Tyre machine • Balance machine • Torque wrench • Valve replacement tool • Tyre lever • Wire brush	• Tyre • Balance weights • Valve • Compressed air • Tyre sealant	• Tyre pressures

1. Remove the wheel from the vehicle and place the wheel on the tyre machine.

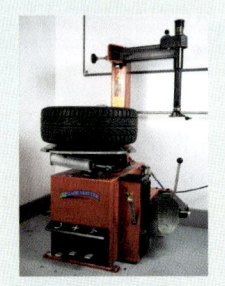

2. Remove the valve centre to deflate the tyre.

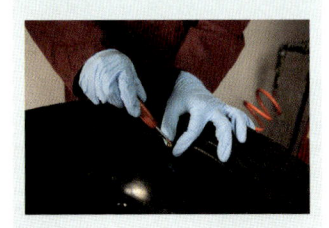

3. Break the tyre seal using the machine.

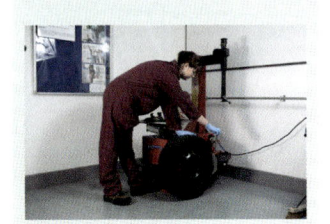

4. Lubricate the rim of the tyre. Use the tyre machine to peel the tyre away from the rim.

5. Clean the rim of the wheel.

6. Lubricate the tyre rim.

7. Replace the tyre in the reverse order of removal, taking care not to damage the tyre.

8. Inflate the tyre to the correct pressure.

Level 2 Light Vehicle Maintenance & Repair

CHECK YOUR PROGRESS

1. List two different tools which are required when:
   a inspecting a tyre
   b replacing a tyre.
2. During a vehicle inspection you discover that the front tyre treads are 'feathering' on the outer edges. What does this indicate?
3. What is the legal minimum tyre tread depth for passenger cars and trailers up to 3500 kg GVW?

Transmission system

The transmission system transfers the energy produced by the engine to the vehicle's wheels. (There is a full explanation of the transmission system in Chapter 5.)

Inspecting the transmission system

1. Check the transmission and axle casings for fluid leaks, damage and security.

2. Check the drive and prop shafts for excessive play and the gaiters for splits.

3. Check clutch fluid, gearbox and final drive levels.

4. Check the wheel bearings for play and noise.

Clutch fluid level check

Clutch fluid is used to lubricate the clutch assembly, reduce friction and assist in smoother gear changing by the use of hydraulic pressure. It must be changed every two years as this is essentially brake fluid, which is hygroscopic.

7 Light vehicle inspection & servicing

Checking the clutch fluid level

Checklist

PPE	VPE	Tools and equipment	Consumables	Source information
• Steel toe-capped boots • Overalls • Latex gloves	• Wing covers • Steering wheel cover • Seat covers • Floor mat covers	• Suitable rag	• Clutch fluid (brake fluid)	• Clutch fluid (brake fluid) type

1. Turn the engine off.

2. Locate the clutch fluid reservoir and remove the cap.

3. Check the fluid level. If the fluid is not up to the maximum mark, add more fluid.

4. Replace the clutch fluid cap and check that it is secure.

Vehicle body exterior and interior inspection

An inspection of the vehicle body exterior and interior should be completed at regular intervals in line with the manufacturer's service recommendations. This is because the exterior body is subject to corrosion if there is any damage to paintwork and the interior is subject to wear and tear. (See Chapter 2, pages 124–129, for further information on body and chassis construction.)

Body exterior inspection

A vehicle body exterior inspection involves checking the registration plate, vehicle identification number, doors, locks, hinges, bonnet, body panels and windows.

Inspecting the vehicle body exterior

Checklist

PPE	VPE	Source information	Tools and equipment
• Steel toe-capped boots • Overalls • Latex gloves	• Wing covers • Steering wheel cover • Seat covers • Floor mat covers	• Job card	• Lubricant • Clipboard and pen

1. Check that the registration number and the vehicle identification number (VIN) match the job card.

2. Check that the doors open and close securely from inside and outside. Lubricate the locks and hinges.

3. Check the door hinges for excessive movement.

4. Check that the bonnet release operates easily and smoothly. Lubricate the bonnet release mechanism.

5. Check the condition of all the body panels for scratches, marks and dents.

6. Check the windows for cracks and stone chips.

Chassis inspection

Checklist		
PPE	**VPE**	**Tools and equipment**
• Steel toe-capped boots • Overalls • Latex gloves • Safety helmet	• Wing covers • Steering wheel cover • Seat covers • Floor mat covers	• Inspection lamp

1. Check for damaged or cracked underseal resulting from incorrect jacking or grounding of vehicle.

2. Check that the chassis rails are straight and smooth. Look for signs that the vehicle has been lifted or supported incorrectly.

3. Check the floor pan for any damage and corrosion.

If you notice that the vehicle has been repaired following an accident, make sure you inform the customer. They may have just bought the vehicle and might not be aware of the repair. Signs of accident damage can include:

- damaged chassis members
- incorrect alignment of panels
- (in the worst cases) re-welded frames.

Vehicle interior inspection

Checklist	
PPE	**VPE**
• Steel toe-capped boots • Overalls • Latex gloves • Safety helmet	• Steering wheel cover • Seat covers • Floor mat covers

1. Inspect the condition of all seat belt webbing for cuts or obvious signs of fraying. Belts wear most around the anchorages, buckles and loops.

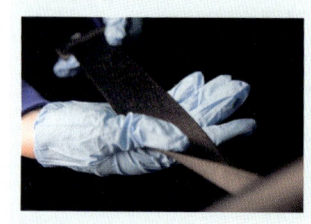

2. Check that the seat belts are smooth when pulled and lock when pulled sharply.

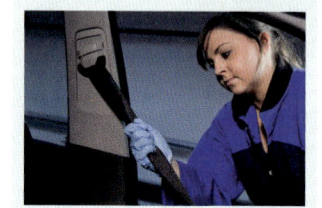

3. Check that all mirrors are secure and the mirror glass is not cracked.

4. Check that seats are secure and that the hinges lock in place.

7 Light vehicle inspection & servicing

Electrical and electronic systems

The electrical and electronic systems on a vehicle consist of the lighting system, wipers, audio, security and comfort systems. These systems are covered in detail in Chapter 6.

Battery

The battery is the main source of power for all the vehicle's electrical and electronic systems. There are two main types of battery:

- maintenance-free (MF) batteries
- batteries which require servicing and inspection.

Check the label on the top of the battery before any servicing and inspection. It will state 'Maintenance Free' if it is a maintenance-free battery.

Light vehicle batteries are covered in detail in Chapter 6, pages 372–376.

> **Safe working**
>
> Never add an electrolyte mixture or acid to the battery when topping up. This will cause a chemical reaction which will make acid spurt out of the battery.

Checking the battery

Checklist				
PPE	**VPE**	**Tools and equipment**	**Consumables**	**Source information**
• Steel toe-capped boots • Overalls • Latex gloves	• Wing covers • Steering wheel cover • Seat covers • Floor mat covers	• Sockets, ratchet and extension bar • Battery tester • Wire brush	• Suitable cloth	• Battery voltage

1. Check that there are no cracks or breakages of the casing (bottom of battery) and the cover (top of battery). The battery will need to be replaced if cracks or breakages appear.

2. Check that the top of the battery (cover) is clean and has no water or dirt on it. Wipe clean any deposits from the battery cover.

3. Check that the battery hold-down mounting is secure. Use sockets to tighten any loose mountings.

4. Check that the terminals are well connected and secure and that the cables are not frayed. Replace any frayed cables.

5. Check that there is no corrosion around the terminals. Remove any corrosion with a wire brush.

6. Check the charge of the battery by attaching a multimeter (see Chapter 6, page 388).

Figure 7.47 shows all the things you should be looking for when checking a battery.

A battery which requires servicing and inspection needs regular checking of the electrolyte level, in line with the service intervals given in the vehicle handbook and inspection sheets. The electrolyte level must be above the top of the plates inside the battery. If the level is low, you need to add deionised water (see Chapter 6, page 373).

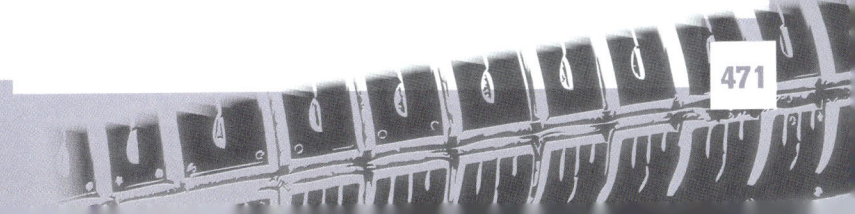

Level 2 Light Vehicle Maintenance & Repair

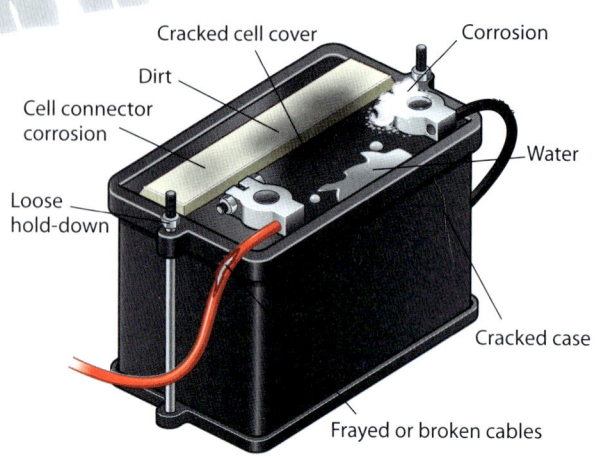

Figure 7.47 Items to check on a battery

Labels: Cracked cell cover, Dirt, Cell connector corrosion, Loose hold-down, Corrosion, Water, Cracked case, Frayed or broken cables

Checking the lighting system

The lighting system allows the driver to see and the vehicle to be seen. It is covered in detail in Chapter 6, pages 405–409.

You should check that the front, rear and interior lights work correctly. Figure 7.48 shows the rear lights on a light vehicle and Figure 7.49 shows the front lights.

- All forward-facing continuous use lights must show white light only.
- All rear-facing continuous use lights must show red light only.

The reverse light is white because it indicates that the vehicle is moving towards you. The flashing indicator lights and hazard warning lights are amber to differentiate them from the front and rear lights.

Did you know?

If the brake or tail lamp bulb has failed, take care to fit the new bulb correctly. Otherwise, the brake light will come on during tail lamp use and the tail lamp will come on during brake light use.

Figure 7.48 Rear lights

Labels: Brake light, Reversing light(s), Brake lights/side lights, Indicator lights/hazard warning lights, Registration plate light(s), Fog light(s)

Did you know?

All rear lamps use the same **earth** (electrical ground connection, designed to complete a circuit). If this is a bad connection, other lamps will earth through each other to find their easiest route to earth. For this reason, you must carry out tests on all the rear lights at the same time rather than testing them individually.

Figure 7.49 Front lights

Labels: Indicators/hazard warning lights, Side mounted repeater indicators, Dipped beam or dipped headlights, Main beam or full beam headlights, Side lights or parking lights, Fog lights

Figure 7.50 shows the interior lights in a light vehicle.

While you are conducting the vehicle light checks, you must also check that the warning lights on the interior instrument panel are operating correctly (see Figure 7.51).

472

7 Light vehicle inspection & servicing

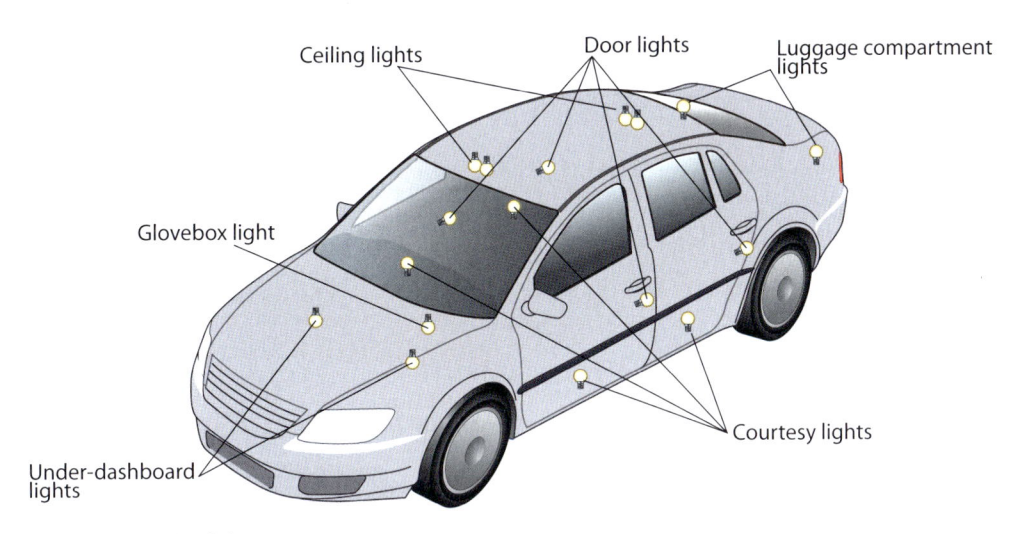

Figure 7.50 Interior lights

Figure 7.51 Example of warning lights on the interior instrument panel

Other checks on the lights

You also need to check that:

- the headlight and rear light units are secure
- water has not entered the system
- the reflectors are in good condition
- the plastic or glass is not broken, cracked or obscured
- the rear reflector lenses are not cracked
- the contacts are not corroded following the replacement of a bulb.

Checking and adjusting headlamp alignment

You need to adjust the headlamp aim if:

- the vehicle has been involved in a collision to the front
- a headlamp unit is replaced
- a suspension component has been replaced
- new wheels/tyres have been fitted
- the headlamp bulb has been replaced.

> **Did you know?**
> Most vehicles have a self-test procedure on the instrument panel which enables you to check all the warning and indicator lights.

> **Did you know?**
> When adjusting xenon lighting, there must be no faults in the vehicle's computer. This is because the automatic dip beam set by the vehicle computer may position the lights incorrectly and give an inaccurate beam setting.

- A halogen or xenon bulb should never be touched with bare hands.
- Never power the bulb when it is outside its lamp unit.
- The brightness can cause burns to the eye, and the voltage can cause an electric shock.

Figure 7.52 Safety symbols for the lighting system

473

Level 2 Light Vehicle Maintenance & Repair

> **Did you know?**
>
> Lightbulbs contain a gas around the metal filament inside the sealed glass bowl to prevent the filament burning out. Most lightbulbs contain a gas made from halogens, but xenon is another type of gas that can be used. A xenon lightbulb gives a bluer, brighter light.

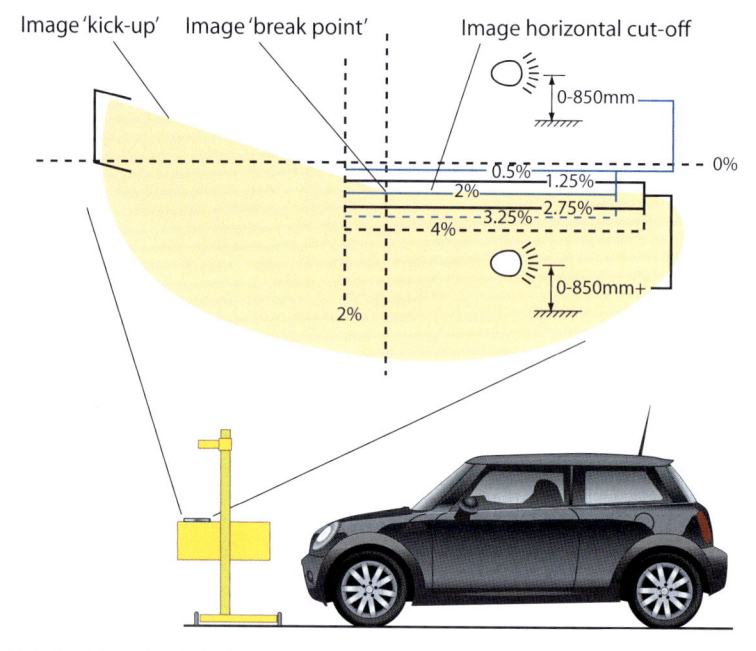

Figure 7.53 Checking aim of the headlamp beam using a headlamp aligner

Pre-checks

Before you check the headlamp alignment, there are some checks you need to make. Otherwise, the readings you take might be inaccurate.

- Check that the tyres are inflated to the correct pressures.
- Make sure the vehicle is on a flat, level surface.
- Check that the suspension dampers are not faulty.
- Check that the vehicle is correctly loaded.
- Settle the vehicle suspension by bouncing the wheels.
- Calibrate the headlamp alignment tool.
- For vehicles with manual vertical aim control the switch must be set at zero.
- For vehicles with automatic aim control headlights, check there are no fault codes stored on the system ECU prior to alignment checking and always use the manufacturer's specifications.

Headlamp beam setter

You can adjust the headlamp beam accurately by using a headlamp beam setter.

> **Did you know?**
>
> The law which covers light vehicle lighting is the Road Vehicle Lighting Regulations 1989 (see Chapter 6, pages 416–417). Failure to comply with the regulations can result in a fixed penalty fine, endorsement of the offender's driving licence or prosecution.

Adjusting the headlamp beam

1. Place the headlamp alignment equipment squarely in front of the vehicle headlamp to be checked.

2. Set the height of the aligner following the manufacturer's instructions.

3. Turn on the headlamps. You should see the pattern shown in Figure 7.53.
   - **The break point cut-off** is where the beam rises so that the left-hand side of the road can be viewed. This must be within 2%. If it is not, you will need to adjust the beam left or right using adjusting screws at the back of the light unit.
   - **The horizontal cut-off** needs to be at about 1.3%. If it is not, you will need to adjust the beam up or down, again using the screws behind the headlamp unit.

4. Carry out the same procedure on the other headlamp.

The wiper system

The wiper system uses a rubber blade to move dirt particles and water that land on the windscreen away from the swept area. There are two types of blade, as shown in Figure 7.54.

Servicing the wiper system

During the service, you are required to complete the following wiper checks:

- The wiper blades must not be split or cracked.
- Top up the washer fluid bottle with washer fluid.
- The wipers need to operate at all speeds.
- The wipers must not travel past the screen or touch each other.
- The wipers must clear the screen completely and not leave a thin film of water on the screen.

Washer fluid is a mixture of an alcohol-based solvent and water. This prevents freezing in cold weather.

Conventional blade – incorporates a metal frame

Flexible blade – incorporates a stiffener which adapts to the different types of windscreen

Figure 7.54 Types of wiper blades

Did you know?

The flexible blade has a longer service life than the conventional blade.

The different types of wiper blade are not interchangeable.

CHECK YOUR PROGRESS

1. The table shows data collected from a tyre inspection. All readings are given in mm. Which of these tyres are legal and which are illegal?

Front nearside tread depth				Front offside tread depth			
Left	Middle left	Middle right	Right	Left	Middle left	Middle right	Right
1.6	1.6	1.0	1.0	1.0	1.0	2	3

Rear nearside tread depth				Rear offside tread depth			
Left	Middle left	Middle right	Right	Left	Middle left	Middle right	Right
2	2	2	1.6	3	3	3	3

2. The rear lamps and stop lamps are often checked by first switching on the rear lamps and then operating the stop lamps. What does this sequence check?
3. What are the checks made on the windscreen wiper operation during routine vehicle inspection?

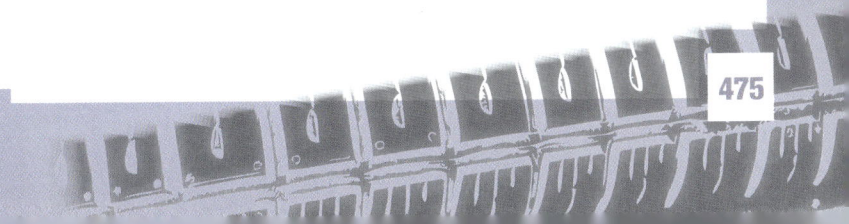

Level 2 Light Vehicle Maintenance & Repair

> **Emergency**
>
> If a child seat is being used in the passenger seat, the airbag located in front of the seat will need to be deactivated. This is to prevent the airbag crushing the child if the vehicle is involved in a collision and the airbag is activated. 
>
> *Do not put a child seat in front seat*

Other service operations

Airbag labels

It is the manufacturer's responsibility to make the driver aware that the vehicle has airbags. If they are not aware, then they may be injured or unnecessarily shocked in the event of a collision where the airbags are deployed. Airbag labels need to be displayed where the airbags are situated. Figure 7.55 shows various airbag safety labels.

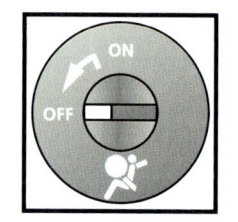

| Airbag warning light | Airbag warning notice | Airbag switch | Airbag warning sticker |

Figure 7.55 Airbag safety labels

Engine compartment safety labels

Various labels under the bonnet warn the driver of hazards. They will also highlight hazards to the technician during routine servicing and inspection. For example, the cooling system pressure cap has a warning label indicating that it should not be removed if the system is hot (see Figure 7.15 and opposite).

Serviceable items have a label with yellow background or are made from a yellow material. These include:

- oil filler cap
- dipstick
- brake fluid reservoir cap
- washer bottle cap
- cooling system reservoir cap.

Figure 7.56 Engine compartment safety label – high voltage

476

Items on the vehicle to be checked following inspection and servicing

Following the inspection or service of a vehicle, you should check that you have completed the following.

1. You have carried out the customer's instructions in full.
2. You have removed all grease and oil marks from the interior and exterior of the vehicle.
3. If the battery has been disconnected, make sure that the clock is set and radio stations are restored.
4. Following a service, make sure that the **service indicator(s)** have been reset correctly.
5. Wash the vehicle and remove all VPE.
6. Carry out a post-work check (see page 433).
7. Complete all paperwork (see Table 7.7).

> **Key term**
>
> **Service indicator** – light on the dashboard which illuminates to inform the driver when the next service is due.

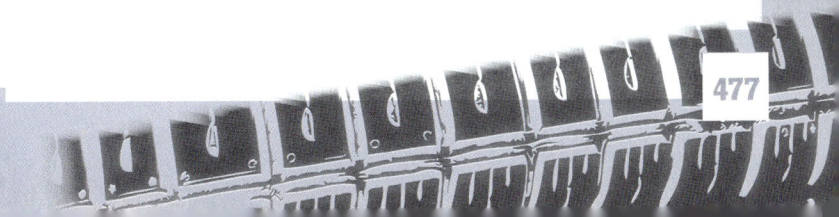

Record information and make suitable recommendations

Before, during and after inspection and servicing activities you need to record faults and activities carried out on the vehicle. This will keep the customer informed of the work you have carried out and be added to the vehicle's service history. Table 7.6 lists the paperwork that you need to complete as part of your inspection and service procedures.

Table 7.6 Items of paperwork to be checked or completed

By you (the technician)	By your supervisor or accounts department
• Service, PDI or inspection sheets • In-vehicle service book • Job card	• Database service record • Staff rota • Customer invoice

Customer documentation

- A copy of the service or inspection checklist must always be given to the customer, as this will confirm the extent of work you have carried out.
- Following an MOT, the customer must receive a copy of the MOT certificate and any recorded emission data.
- Copies of all other documentation should be available on request by the customer.

Customer satisfaction

If there is additional work to be carried out, you need to make the customer aware of this. Discuss a convenient time that suits the customer to return the vehicle for the work to be completed. It is good practice to:

- Provide service and MOT date stickers informing the customer when the next MOT or service is due.
- Ask the customer if they would like to pre-book the service or MOT.
- Ask the customer to complete a customer satisfaction survey, as this will help you to gain feedback and improve on your level of service.

> **Working life**
>
> It is a busy weekday. Mr Banks had his car serviced last week. He enters the workshop stating that he wants to make a complaint. He is very angry and says loudly, so everyone can hear, that he has found dirty fingerprints on his upholstery after the service last week.
>
> 1   What would you do to resolve the problem?

7 Light vehicle inspection & servicing

FINAL CHECK

1. The Road Vehicles (Construction and Use) Regulations 1986 is a set of rules concerning:
   a speed limits and road construction
   b how vehicles are constructed
   c the construction of buildings and safety procedures
   d control of substances hazardous to health

2. After rectifying an engine fault you need to road test the customer's vehicle. Before you start the test you must:
   a check that you hold the correct driver's licence for the vehicle
   b check the vehicle is clean and the VPE has been removed
   c make sure your work bay is tidy and all tools have been put away
   e complete the job card with all details concerning the vehicle repair

3. Which of the following is **not** part of the MOT inspection for light vehicles?
   a measuring the brake balance
   b operation of number plate lamp
   c condition of the spare wheel
   d operation of the brake servo

4. Why is it important to report any existing vehicle body damage before routine servicing?
   a It is not necessary to report existing vehicle body damage
   b It is a legal requirement under the Road Vehicles (Construction and Use) Regulations
   c To avoid any potential conflict between the company and the vehicle owner
   d It is a legal requirement under the Supply of Goods and Services Act

5. Which of the following will affect the accuracy of the headlamp alignment?
   a condition of the headlamp reflector
   b badly worn front shock absorbers
   c any tyres with very low pressures
   d all of the above

6. Which of the following is **not** true when checking and adjusting the front wheel alignment?
   a You should remove any very heavy loads from the boot
   b You should check the tyre pressures are correct.
   c Adjustments must be made equally to both track rods
   d Buckled wheels will not affect the wheel alignment

7. You can check the effectiveness of dampers by:
   a bouncing the vehicle to check for damping action
   b inspecting the bump stops to check for impact
   c testing the efficiency using a roller tester
   d inspecting the front tyres for abnormal wear

8. During a pre-MOT check you notice the front position (side) lamps are blue. You would advise the customer:
   a that blue lights can only be fitted to the rear
   b a blue side lamp is legal providing it is incorporated in a blue headlamp
   d that the vehicle will fail the MOT test
   e you would not advise the customer as the lights are legal

9. What information is not usually shown on a job card?
   a the vehicle's specifications and settings
   b the make and model of the vehicle
   c the customer's name and contact details
   d a description of the work to be done

10. When carrying out a routine vehicle service, you should report any defects that you find to your supervisor. This is so that:
    a the additional parts required can be ordered without any further delay
    b you can show your supervisor that you are observant and knowledgeable
    c the customer can be contacted to get their authority to do the extra work
    d you can't be blamed if the vehicle goes wrong after you have done the service

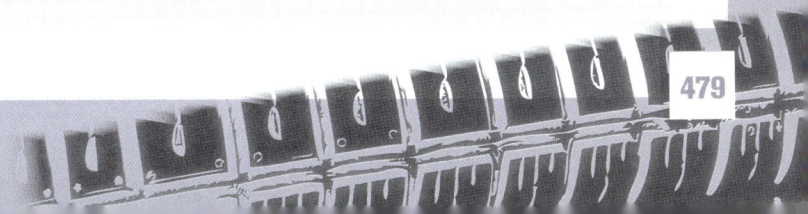

Level 2 Light Vehicle Maintenance & Repair

GETTING READY FOR ASSESSMENT

The information contained in this chapter, as well as continued practical assignments in your centre or workplace, will help you to prepare for both the end-of-unit tests and diploma multiple-choice tests.

By working through this chapter, you will have gained the knowledge and skills you need to carry out a range of inspection and servicing activities and to recommend service action on light vehicles.

You will need to be familiar with:

- Working safely when carrying out light vehicle inspection and servicing
- The importance of carrying out light vehicle inspection and servicing
- The understanding and skills needed to carry out a range of inspections on light vehicles
- Using a variety of prescribed testing and inspection methods
- Using relevant information to carry out light vehicle inspection and servicing
- Using the appropriate tools and equipment for light vehicle inspection and servicing
- How to conduct routine servicing and inspection, adjustment and replacement activities as part of the periodic servicing of light vehicles
- How to record information and make suitable recommendations following servicing and inspection activities

This chapter has given you an introduction and overview of light vehicle inspection and servicing, providing the basic knowledge that will help you with both theory and practical assessments.

You now need to apply the knowledge you have gained in this chapter in your day-to-day working activities. For example, you can follow the procedures given in this chapter to practise using the prescribed testing and inspection methods.

Before you try a theory end-of-unit test or multiple-choice test, make sure you have reviewed and revised any key terms that relate to the topics in that unit. You will need to read all the questions carefully. Take time to digest the information so that you are confident about what the question is asking you. With multiple-choice tests, it is very important that you read all of the answers carefully, as it is common for two of the answers to be very similar, which may lead to confusion.

For practical assessment tasks, it is important that you have had enough practice and that you feel that you are capable of passing. Always check that you are working safely throughout the test. Before you begin a task make sure you have the correct PPE, VPE, tools and equipment to hand and that you have a plan to follow, along with the equipment and information to complete the task.

For example, you could be asked to carry out a safety inspection. For this you will need to:

- Wear boots, overalls, a helmet and latex gloves.
- Protect the vehicle using wing covers, seat covers, steering wheel cover and foot mat covers.
- Assemble the tools and equipment you need – a hoist, tyre pressure gauge, tread depth gauge, replenishing fluids such as coolant, washer fluid, engine oil, and brake fluid and tyre pressure setting information.

Make sure that you observe all health and safety requirements and that you use tools correctly and safely. Always communicate effectively with customers and colleagues.

Re-read the chapter to confirm you have completed and understood all the tasks. This will help you to be sure that you are working correctly and to avoid problems developing as you work.

Good luck!

Index

Key terms are indicated by **bold** type.

1, 2, 3 system 56–57
abrasive wheels 11–12
accidents 33
  bleeding 38–39
  burns 40
  electrical injuries 38
  electricity 40–42
  emergency services 36–37
  first aid 36
  loss of consciousness 39–40
  objects in the eye 40
  prevention 33–35
  recovery position 37
  shock 39
Ackerman system 140–41
actuator 156, 269
acute 48
adhesives 111
aerodynamics 128–29
aftermarket 74
air charge 286
air conditioning 241–42
air cooling systems 235
air suspension systems 156
air systems 261, 274–79,
  301–305, 309, 452
air temperature sensors 279
airbag labels 476
airflow meters 275–76
alcohol 49–50
all terrain vehicles (4 x 4) 125
alloys 114–15
alternating current (AC) 371
alternators 368, 398–400
aluminium 114–15
amperes 371
amps draw 394
angle gauges 98
antifreeze 237
anti-lock braking systems 180–84
anti-theft devices 410–11
API (American Petroleum Institute) classification 232–33
aquaplaning 157
armature 275, 381
aspect ratio 193–94, 195
atomise 265
atoms 367–68
authorisation 76
Autodata 431
automatic gearbox 336–38
automating 44
auxiliary air valves 278
axle shaft arrangements 147
axles 350–51

backbone chassis 127
backlash 133
baffles 267
balancing wheels 197
ball joints 130
bank 207
bar 12

base, collector and emitter 296
batteries 32, 372–76, 378, 397–400, 471–72
baulk ring 336
bead 187
bearing race 144
bedding in 169
bevel gears 346, 347
bleeding
  braking systems 180
  fuel systems 310
bleeding, dealing with casualties of 38–9
block tester 250
blow-by gas 227
bodies, car
  aerodynamics 128–29
  inspection and servicing 469
  materials used for 126
  types of 124–25
body shop 70–71
boiling point of liquids 243
bore 220
bottom dead centre (BDC) 218
boundary lubrication 224
brake discs 91
brake fade 167
brake linings 164
brake pads 168–69, 457
brake shoes 165
braking systems
  anti-lock systems 180–84
  bleeding 180, 461
  brake fluid 179–80, 460–61
  brake lights 184
  brake pads 168–69, 457
  calliper assemblies 169–71
  disc brakes 167–68, 456
  drum brakes 164–67, 458–59
  efficiency of 177–78
  electronic parking brakes (EPB) 171
  faults 185
  hydraulic 172–78
  inspection and servicing 456–61
  regenerative braking 223
  requirements of 164
  servicing requirements 178–80
  testing 178
  warning systems 184
brushes 381, 398
bulbs 380–81, 415
bullying 50
burns 40
bus bar 379
butterfly 265
bypass lubrication system 229
by-products 227

cables 376–77
calibration 86
calliper assemblies 169–71
callipers 89–90, 94

calorific value 263
cam 91
cam and peg steering boxes 133
cam ring 283
cam timing 218
camber angle 137–38
capacitors 294, 295, 413
carburettors 264–66
carcinogenic 48
carcinogens 30
case hardening 130
cast alloy wheels 187
caster angle 138
catalytic convertors 304–305
central door locking 410
centre link 134
centrifugal forces 329
cetane value 288
chafed 392
charging 374
charging system 397–400
chassis units and components
  aerodynamics 128–29
  bodies 124–25, 126
  collision safety 128
  construction of 126–29
  fabricated sections of chassis 127
  health and safety 120–22
  inspection and servicing 470
  layout of vehicles 125–26
  personal protective equipment (PPE) 120
  tools 123
  See also braking systems; steering systems; suspension systems; tyre systems; wheel systems
check sheets
  pre-delivery 433–35
  pre-MOT 435
  pre-work 432–33
  service checklist 436
  service inspection 312
chemicals, control of 24–30
chisels 99
chronic 48
circuit relays 379–80
circuits 92, 368, 368–69, 382–83, 391, 393
cleaning 64–65
clearance 247
clearance volume 220
clevis pins 101
closed loops 305
clutch 321
clutch systems
  automatic transmission torque convertors 329–30
  bleeding 327
  coil spring clutch 324
  components of 322
  construction of 325–25
  diaphragm spring clutch 324
  differences in design 323–24
  dog clutches 328–29

drag and slip 323, 353
engaging the clutch 322
engineering principles 328
friction clutch 321–22
functions of 321
hydraulic operation 326–27
inspection and servicing 468–69
mechanical operation 325–26
slip and drag 353
testing and maintaining 352–54
coefficient of friction 169
coil spring clutch 324
coil spring suspension 152
cold starting 285–86, 310
collision safety 128
combined unit 136
combustible 52
combustion 206, 449
combustion chambers 217–18, 263, 284–85
combustion process 289–90
common rail diesel 267
communication
  barriers to 81
  effective 78
  methods of 78
  non-verbal 79–80
  sequence for 82
  team work 82–83
  technology for 81
  verbal 79
  written 80 (See also documentation)
commutator 381
compression ignition (CI) engines 209, 212–15
compression ratios 220–21
compression (squeeze) 210, 212, 213, 216
compression testing 244–45
compressor 414
compressor vanes 286
computer-controlled electronic ignition systems 298
condensers/capacitors 294, 295
conduction 293
conductors 368
connectors 377
consciousness, loss of 39–40
constant mesh 335
consumables 66
consumer 371
continuously variable transmission (CVT) 338
contractor 30
contracts 76
control measure 48, 57–58
controlled waste 30
convection currents 240–241
convertibles 124
cooling systems
  air 235
  air conditioning 241–42

481

convection currents (cont.)
  checking and replacing coolant 444–45
  engineering principles 242–43
  expansion tanks 238
  fans 240–41
  faults 251–52, 253
  liquid 235–40
  pressure testing 248
  purpose of 234
  radiators 236–7
  strength of coolant/antifreeze 249
  thermosiphon process 239–40
  thermostats 238
copper 367-8
core 237
corrosion inhibitor 155
corrosive 29
COSHH risk assessments 25–27
coupés 124
coupling point 330
cranked lever 167
crankshaft 207
  crankshaft rotation 211, 212, 214, 215, 216
cross flow 237
cross-contamination 65
crown wheel and pinion 346
crumple zone 128
current 92, 368
customer satisfaction 477
CV joints 345
cycles of operation 209–16
cylinder block 207
cylinder bore 209
cylinder capacity 220
cylinder head 86
cylinder head gasket failure 251

dampers 150, 157, 162
data from manufacturers 431
dead short to earth 392
deflection 447
deionised water 373
demisting 412
density 279
detent 333
detergent 65
diagnostic equipment 93
diagnostic trouble codes (DTCs) 305
diagrams, wiring 385, 409
dial gauges 90–91
diameter 89
diaphragm 155, 323
diaphragm clutch 323
dies 102
diesel engines 212–15, 266–67, 281–86, 303, 309–11
differential 342
differential units 347–48, 358
diodes 401
direct current (DC) 369
direction indicators 408–409
directional tyres 196
disc brakes 167–68, 456
discharging 374
disengaged 321
dissipate 237
dissipation 293
distributor caps 295

distributorless ignition systems 297–98
dividers 94
documentation 74–77, 430–36
dog clutches 328–29
drag 128, 323, 353
drills 103
drive 125–26
drive range 330
drive shaft 145
driveline systems 320
  bevel gears 347
  components of 344–46
  CV joints 345–46
  differential unit 347–48, 358
  drive shafts 345, 356–57
  engineering principles 352
  final drive unit 346–51, 358
  four wheel drive 343
  front wheel drive 341–42, 347
  limited slip differentials (LSD) 348–49
  propeller shafts 344–45, 357
  purpose of 341
  rear wheel drive 342–43, 350
  reverse lamp switch 350
  speedometer drive 349–50
  testing and maintaining 356–58
  transfer box 349
  wheel bearing arrangements 350–51
driveshaft 126
drugs 50
drum brakes 164–67, 458–59
dry sump lubrication systems 230
dwell 294
dynamos 398

earth 377
earth return 377, 378
ecotoxic 29
EGR lift value 303
electrical vehicles (EV) 41, 222–23, 370, 376, 388, 397
electrical continuity 377
electrical systems
  amperes 371
  anti-theft devices 410–11
  batteries 372–76, 378, 397–400, 471–72
  bulbs 380–81, 415
  cables 376–77
  central door locking 410
  charging system 397–400
  circuit relays 379–80
  circuits 368–69, 382–83, 391
  demisting 412
  electricity 366–72
  fans 381
  faults 390–95
  fuses 382
  health and safety 362–64
  heating system 381, 413–14
  immobilisers 411
  inspection and servicing 471–75
  inspection sheets 418
  lighting 405–409, 412, 415–17, 472–74
  manufacturer's equipment and tools 396

mirrors 412
multimeters 385–90
oscilloscopes 394–95
parasitic drain 394
protective equipment 362–63
resistance 371–72, 389–90, 392
service and maintenance 414–18
short circuit 393
starter motor 400–405, 414–15
sunroofs 413
switch 379
symbols 384
terminals and connectors 377
tools 365
volts 370, 388–89
watts 372
windows 411–12
wiring diagrams 385, 409
electrical tools, care of 105
electricity
  injuries 38, 40–42
  legislation 13
  measuring 91–93
  principles of 366–72
  use of 66–67
electrode (spark plug) gap 449
electrodes 290
electrolyte 373
electromagnet 379
electromotive force (EMF) 371
electronic control unit (ECU) 269–70
electronic parking brakes (EPB) 171
electrons 92, 367–68
elements 367
emergencies
  fire 51–55
  See also accidents
emergency services 36–37
engaged 321
engine bay 267
engine coolant temperature sensors (ECT) 279
engine mechanical systems
  air conditioning 241–42
  combustion chambers 217–18
  components and layout 207–209
  compression ignition (CI) engines 209, 212–15
  compression ratios 220–21
  compression testing 244–45
  crankshaft rotation 211, 212, 214, 215, 216
  cycles of operation 209–16
  cylinder capacity 220
  diesel engines 212–15
  engineering principles involved 217–24
  faults 250–51
  four-stroke engines 209–11
  health and safety 202–205
  heating system 241
  hybrid fuel engines 222–23
  inspection and servicing 439–50
  internal/external combustion 206
  naturally aspirated engines 222

petrol engines 209–12, 215–16
piston design 219
power and torque 221–22
rotary engines 215–16
safety labels 476
spark ignition (SI) engines 209–12
thermal efficiency 217–18
tools 205
turbocharged engines 222
two-stroke engines 211–12
types of engines 206–207
valve timing 218–19
See also cooling systems; lubrication systems
engine speed sensors 280
engine sump 225
engineering principles
  clutch systems 328
  cooling, heating and ventilation 242–43
  cooling system 242–43
  engine mechanical systems 217–24
  gearbox 339–41
  ignition systems 299–300
  lubrication systems 230–33
environmental protection 9–10
EPC software 73–74
epicyclic gear train 337–38
equipment
  care of 85, 104–105
  extraction 106
  fire-fighting 53–55
  health and safety 3–6
  hoist and lifts 105–106
  instructions for use 106–107
  lifting 14–15, 46–47
  manufacturer's 396
  measuring 85–95
  safe use of 107
  See also hand tools; tools
escape routes 53
estate cars 124
evacuation routes 53
exhaust (blow) 210–11, 212, 213, 216
exhaust systems
  catalytic convertors 304–305
  checking emissions 306–307
  cold starting 310
  gases 289–90
  inspection and servicing 453–55
  recirculation of gas 302–303
  secondary air injection 303
  silencers 301
  tools 261
expansion, linear/cubical 243
expansion tanks 238
external combustion 206
extinguishers, fire 53–55
extraction equipment 106
extruded 114

fabrication 84
fan belts 447–48
fans 240–41, 381
fastening devices 109–11
faults
  braking systems 185
  cooling system 251–52, 253
  electrical systems 390–96

Index

engine mechanical systems 250–51
fuel systems 305–307
lubrication systems 252
service inspection check sheet 312
steering systems 143
suspension systems 162
tyre systems 198
feathering 143
feeler gauges 93
ferrous metals 112–14
field coil 401
files 101–102
final drive 124, **342**, **346**–51
final drive unit 358
final gear reduction 346
fires 51–55
firing impulse 207
first aid 8, 36
fit, limits of 94–95
fixing devices 107–109
flame travel 299
flammable 29
  liquids 51–52
flash point 51
flasher unit 408
float chamber 265
flywheel 207
fog lamps 407–408, 409
four wheel drive 343
four-stroke engines 209–11
free play 141, 352
frequency meter 447
friction 321, 328
friction clutch 321–22
front wheel drive 341–42, 347
fuel injector 269
fuel systems
  air intake 274–79, 301–305, 309
  bleeding 310
  carburettors 264–66
  cetane value 288
  cold starting 310
  combustion process 289–90
  components of 266–67
  diagnosing faults 305–307
  diesel engines 266–67, 281–86, 309–11
  electronic control unit (ECU) 269–70
  filters 273, 309
  injection systems 267–80
  injectors 274
  inspection and servicing 451
  octane value 288
  petrol engines 263, 264–280, 311
  petrol/diesel comparison 287
  pressure regulators 273–74
  properties of fuels 262–63
  pumps 271–73
  replacement of components 308
  supply of fuel 270–74, 281–86
  tanks 270–71
  tools 261
  volatility 287–88
fulcrum 323
full-flow lubrication system 227–28
functional test 195

fuses 382, **383**

gas containers 12
gaskets 111
gearbox 331
  automatic 336–38
  baulk ring 336
  continuously variable transmission (CVT) 338
  engineering principles 339–41
  epicyclic gear train 337–38
  gear ratios 339–40, 352
  gears 332–34
  helical gears 334
  interlock mechanism 335
  manual transmission 331–32
  purpose of 331
  spur gears 333
  synchromesh 335
  testing and maintaining 355
  torque multiplication 339, 340
generators 368
geometry, steering 137–42
governor 283
graduation 87

hacksaws 100
hall effects sensors 297
halogens 30, **381**
hammers 99
hand tools
  angle gauges 98
  care of 104
  chisels 99
  drills 103
  files 101–102
  hacksaws 100
  hammers 99
  pliers 96
  punches 100
  screw thread cutting sets 102
  screwdrivers 95
  socket sets 97–98
  spanners 96–97
  spark plug thread chaser 103
  stud extraction 103
  thread restorer 103
  vices 104
  See also equipment; tools
handbooks, vehicle 73, 430
handbrake 459
handbrake travel 459
harassment 50
hatchbacks 124
hazards and risks
  COSHH risk assessments 25–27
  disposal of waste 29–30
  housekeeping 61–67
  legislation 10–11
  occupational 48–50
  routes of entry 24
  storage of substances 28, 29
  toxic substances 24–25
  See also health and safety
headlamps 406–408, 409, 473–74
health and safety
  abrasive wheels regulations 11–12
  chassis units, working on 120–22
  collision safety 128

consultation with employees 8
electrical systems 362–64
electricity 13, 40–42
employer's responsibilities 6–7
environmental protection 9–10
equipment 3–6
fires 51–55
hazardous substances 10–11
hazards and risks 23–30
housekeeping 61–67
incidents/accidents 9
inspection and servicing 422–42
labels 476
legislation 2–16
management of 6
manual handling 7–8, 42–45
mechanical systems, working on 202–205
multimeters 387
noise and vibrations 13
occupational health 48–50
personal protective equipment 5–6, 17–21, 120, 202, 258, 316, 362, 422
pressure systems 12
recycling 30–32
risk assessments 6, 55–59
signs and signals 9, 58, 60
use of equipment 107
waste disposal 32
working at height 14
working practices 122, 260, 261, 318, 364, 424
See also accidents
heat exchanger units 230
heat transfer 242
heater matrix 413
heating system 241, 242–43, 381, 413–14
helical 152
helical gears 334, **335**
helix 402
high tension 294
hoists and lifts 105–106
hot wire mass airflow sensors (MAF) 277
housekeeping 61–67
hub 138
hunting and surging 454
hybrid fuel engines 41, 222–23, 370, 376, 388, 397
hydragas suspension systems 155–56
hydraulic braking systems 172–78
hydrodynamic lubrication 224
hydrometers 397
hygroscopic 179
hypoid 346

identification codes 74
idler arm 134
ignition distributor 447
ignition systems
  cold-starting 285–86
  components 290–94
  computer-controlled electronic 298
  conventional 294–95
  distributorless 297–98
  electronic 295–99
  engineering principles 299–300

flame travel 299
hall effects sensors 297
ignition amplifier 295–96
ignition coil 293–94
ignition timing 299–300
inductive sensors 296–97
optical pickup system (OPUS) 297
tools 261
wasted spark 298–99
imbalance, wheel 196–97
immobilisers 411
impact drivers 95
improvement notices 3
included angle 138
independent suspension systems 151
induction (suck) 209, 211, 213, 215
inductive sensors 296–97
inert gas 381
inertia 332
inflation 191
information
  documentation 74–77, 430–36
  sources of 72–74
infrared spectrum 242
inhibitor switch 330
injection pump 447
injection systems
  diesel 281–86
  petrol 267–80
inline fuel injection pump 281–82
inspection and servicing
  air intake systems 452
  assessment methods 439
  bodies, car 469
  braking systems 456–61
  chassis units and components 470
  checks following 477
  clutch systems 468–69
  customer satisfaction 477
  documentation 477
  documents and information needed for 430–36
  electrical systems 471–75
  engine mechanical systems 439–50
  exhaust systems 453–55
  final checks 438
  fuel system 451
  health and safety 422–42
  importance of 426–27
  inspection sheets 437–38
  law and vehicle safety 428–29
  personal protective equipment (PPE) 422
  road tests 426–27
  service checklist 436
  steering systems 463–64
  suspension systems 462
  tools 425
  transmission system 468–69
  types of service 436
  tyres 464–67
  warranty requirements 426
  wheel systems 464–67
inspection sheets 75, 418
insulators 368
intake manifold 267
integrated circuits 297

483

intercoolers 286
interlock 335
internal combustion 206
International System of Units 84
internet 73
invoices 77
isocyanate-based paints 29

job cards 74, 430
juddering 321
jugs, measuring 94

keys 109
kick back 135
kinetic energy 66
kinetic lifting 43–44
king lead 294
king pin inclination (KPI) 138
Kitemark pledge 70–71

ladder chassis 126, 127
lambda window 288
laminated 151
layout of vehicles 125–26
leaf spring suspension 151–52
leaks 64–65
legislation 2
   abrasive wheels regulations 11–12
   consultation with employees 8
   electricity 13
   employer's responsibilities 6–7
   environmental protection 9–10
   equipment 3–6
   hazardous substances 10–11
   health and safety 2–16
   incidents/accidents 9
   manual handling 7–8
   noise and vibrations 13
   personal protective equipment 5–6
   pressure systems 12
   signs and signals 9
   tyre systems 191, 195
   and vehicle safety 428–29
   waste disposal 32
   working at height 14
levers, principles of 328
lifting
   equipment 46–47, 105–106
   manual handling 7–8, 42–45
   operations and equipment 14–15
light-emitting diodes (LEDs) 297
lighting 184, 405–409, 412, 415–17, 472–74
limit switch 413
limited operating strategy (LOS) 305
limited slip differentials (LSD) 348–49
limits of fit 94–95
liquid cooling systems 235–40
live axle 126
load index 194
longitudinal 158
low tension 294
lubrication systems
   boundary 224
   bypass 229
   classification of lubricants 230–33

components of 225–27
construction and operation of 227–30
dry sump 230
engineering principles involved 230–33
faults 252
final drive 349
full-flow 227–28
hydrodynamic 224
need for lubrication 224
oil coolers 230
oil filters 227
oil pressure test 246–47
positive crankcase ventilation (PCV) 226–27
pressure relief valve 227
properties of lubricants 233–34
wet sump 229

MacPherson strut 154–55, 163
main jet 265
malfunction indicator lamps (MILs) 305
manifold 210
manifold absolute pressure sensors (MAP) 276
manual handling 7–8, 42–45
manual transmission gearbox 331–32
master cylinders 172–74
materials
   bodies of vehicles 126
   ferrous metals 112–14
   forces applied to 115–16
   non-ferrous metals 114–15
   plastics 115
   properties of 112
measuring
   calibration 86
   care of equipment 85
   diagnostic equipment 93
   electricity 91–93
   limits of fit 94–95
   parallax error 86
   standards 84–85
   types of equipment 86–91, 93–94
mechanical advantage 326
mechanical systems. *See* engine mechanical systems
mesh 133, 333
metals
   ferrous 112–14
   non-ferrous metals 114–15
micrometers 87–88
microswitch 411
miniaturise 366
mirrors 412
monocoque chassis 126, 127
MOT tester's handbook 73
motion 128
MOTs 429, 435
multimeters 92, 385–90
multi-plate clutch 323
multipoint injection 269
multi-purpose vehicles (MPVs) 125
mutual inductance 293

naturally aspirated engines 222

nearside 188
negative offset 139
neutral 321
Newton metres 339
noise and vibrations 13
non-ferrous metals 114–15
non-independent suspension systems 150
non-verbal communication 79–80
nyloc nut 145

obligatory 416
occupational health 48–50
octane value 288
offside 188
ohms 371, 389
oil
   additives 233–34
   coolers 230
   filters 227, 442–43
   galleries 226
   grades 442
   level 440–42
   pressure relief valve failure 252
   pressure test 246–47
   pumps 225–26, 252
   purpose of 224
   recycling 31
   replacement 442–43
   See also lubrication systems
open circuits 379, 391
open loops 305
optical pickup system (OPUS) 297
oscillation 152
oscilloscopes 394–95
Otto cycles 209–16
ovality 458–459
oversteer 141

parallax error 86
parallel circuits 383
parallelogram 134
parasitic drain 394
parts, replacement 85
   See also inspection and servicing
parts department 70
passive sensors 297
pawl 167
pedal travel 167
periodic table 367
personal protective equipment (PPE) 5–6, 17–21, 120, 202, 258, 316, 362
petrol engines 209–12, 215–16, 263, 264–280, 303, 311
phase 215, 399
pickup 296
piezoelectric crystal 283
pinion gear 402–403
pinking 300
piston 207
piston crown 218
piston design 219
piston rings, worn 250
pitman arm 134
plastics 115
plenum chambers 279
pliers 96
plies 190–91
pneumatic tools, care of 104

polarity 368
polytetrafluoroethylene (PTFE) 104
ports 211
positive crankcase ventilation (PCV) 226–27
positive offset 139
potential difference (Pd) 371
potential energy 366
potentiometer 275
power 221–22, **320**
power (bang) 210, 213, 216
power parasite 303
power train 124
power-assisted steering 135–36, 464
pre-delivery check sheets 433–35
pre-focus 407
pressed steel wheels 187
pressure cap/radiator cap 237
pressure relief valves 227
pressure systems 12
pre-work check sheets 432–33
prism 407
processes and procedures, need for 71–72
profile 189
prohibition notices 3
propeller shafts 344–45, **345**, 357
propulsion 235
protective equipment 5–6, 17–22, 120–21, 202–203, 258–59, 316–17, 362–63, 422–23
punches 100

rack and pinion steering systems 129–31
radiators 236–37
ratchet 87
rear wheel drive 342–43, 350
reception 69
reciprocating 209
recirculating ball steering boxes 132
records. *See* documentation
recovery position 37
rectifiers 400, **401**
recycling 30–32
reference guides 74
refrigerant 241
regenerative braking 223
regulators 400, **401**
relay 379
reluctor ring 296
replenishments 436
resistance 92, 371–72, 389–90, 392
resistors 275
retail repairs 76
revolution 207
ring gear 402–403
risk assessments 6, 25–27, 47, 55–59
rivets 110
Road Fund Licence 428
road tests 426–27
roadworthy 426
rocker switch 411
roll pins 101
roller bearings 144
rotary engines 215–16

484

Index

rotary fuel injection pump 282–83
rotor arms 295
rotors 215, 398–99
rubber components in suspension systems 153–54
run-out 91, 456

SAE (Society of Automotive Engineers) classification 230–31
safe working load (SWL) 15
safety. *See* health and safety
safety cage 128
safety data sheet 27
sales department 69
saloons 124
scan tools 306
scavenging 212
screw and nut steering boxes 133
screw thread cutting sets 102
screwdrivers 95
scrub radius 139
seals 111
secondary air injection 303
selector fork 333
self-aligning torque 142
self-centering action 138
semiconductors 401
semi-elliptic 151
sensible heat 242
series circuits 382–83
service department 70
service indicator 477
service manuals 72–73
servicing
    PAS systems 136
    records 430
    schedules 74
    See also inspection and servicing
servo 135
set up 137
shimmy 150, 196
shims 133
shock 39
short circuits 373, 393
signs and symbols 9, 58, 60, 384
silencers 301
simulation 439
single point injection 268
slip 323, 353
slip angle 141, 141
slip rings 398
socket sets 97–98
solenoid 269, 403
solvent 65
spacers 89
spanners 96–97
spark ignition (SI) engines 209–12
spark plug thread chaser 103
spark plugs 291–93, 449–50
specific gravity 249, 397
specific heat capacity 243
specification 430
speedometer drive 349
sphere 144
spillages 64–65
splash feed 336
splayed 137
splined 133
splines 323

split pins 101
spoilers 128
spot lamps 407–408, 409
spot welding 126
sprockets 338
spur gears 333
squish 218
standards 84–85
starter motors 400–405, 414–15
stators 399
steel manufacturing 113–14
steel rules 86
steering geometry 137–42
steering systems
    axle shaft arrangements 147
    faults 143
    geometry 137–42
    inspection and servicing 463–64
    power-assisted 135–36, 464
    purpose of 129
    rack and pinion 129–31
    steering boxes 132–33
    steering gear ratio 130
    steering linkage 134
    wheel bearings 143–46
stepless 338
stoichiometric 279, 288
storage of hazardous substances 28, 29
stress
    applied to materials 115–16
    in driveline systems 352
    personal 50
stroke 220
structure, organisational
    body shop 70–71
    office 71
    parts department 70
    reception 69
    roles in 68
    sales 69
    service 70
    and vehicle repair process 68
stub axles 130
stud extraction 103
sunroofs 413
supercharged 222
superchargers/compressors 215
supercharging 301–302
suspension systems
    air suspension 156
    coil spring 152
    dampers 157, 162
    faults 162
    hydragas 155–56
    independent/non-independent 150–51
    inspection and servicing 462
    inspection of 161–62
    layout and components 150–51
    layouts 157–61
    leaf spring 151–52
    MacPherson strut 154–55, 163
    purpose of 148
    rubber components 153–54
    stability of vehicle 161
    terms for 149–50
    torsion bar 152–53
    unsprung weight 148

swarf 101
swept volume 220
switch 379
swivel axis inclination (SAI) 138
symbols. *See* signs and symbols
synchromesh 335
synchronised 321
synergy 82

tape measures 86
tappet clearance 250
taps 102
team work 82–83
technology for communication 81
tension 87
terminals 377
tester's handbook 73
thermal efficiency 217–18
thermal energy 263
thermistor 279
thermosiphon process 239–40
thermostats 238, 239, 253
thread designation 108–109
thread restorer 103
three-phase 399
throttle body 269
throttle butterfly 274–75
throttle position sensors 277–78
thrust line 140
tickover/idle speed 279
timing advance 300
timing belt 446–47
toe-in/toe-out 140–1
tolerance 91, 187, 273, 416
tools
    chassis units and components 123
    electrical systems 365
    engine mechanical systems 205
    fuel, ignition, air and exhaust systems 261
    inspection and servicing 425
    instructions for use of 106
    manufacturer's 396
    safe use of 107
    scan tools 306
    transmission/driveline systems 319
    See also equipment; hand tools
tooth pitch 101
top dead centre (TDC) 218
torque 135, 142, 216, 222, 323, 328
torque converter 329
torque multiplication 331, 339, 340, 402
torsion bar suspension 152–53
toxic substances 24–25
track rod assemblies 134
track rod ends, replacing 131
track rods 130, 141
traction 182, 342
trade plates 428
tramp or hop 196
transaxle 342
transfer box 349
transfer port 211
transistor 294
transmission 320
transmission systems 126
    clutch systems 321–30
    components 320

gearbox 331–41
    inspection and servicing 468–69
    purpose of 320
    testing and maintaining 352–58
    tools 319
trapezoidal 158
tread 190
tread contact patch 189
tri square 94
trim clips 110
trouble codes 93
true rolling 139
true vertical 138
trunking 21
tubed/tubeless tyres 191–92
turbines 286, 329
turbo boost 286
turbocharged engines 222
turbochargers 286, 302, 311
two-stroke engines 211–12
tyre shuffle 191
tyre systems
    aspect ratio 193–94, 195
    faults 198
    functions of 189
    imbalance 196–97
    inspection and servicing 196–98, 464–67
    laws 191, 195
    load index 194
    parts of tyres 189–90
    plies 190–91
    pressure 195
    recycling 30–31
    side wall markings 192–95
    speed rating 194–95
    tread 190, 195
    tubed/tubeless 191–92

understeer 142
universal joints (UJ) 345
unsprung weight 148
upright 237

vacuum 239
valve clearance 250–51, 446
valve lead/lag/overlap 218
valve stem seals 250
valve timing 218–19
valve train 208
vans 125
vaporise 51
vapour lock 309
variable resistors/ potentiometer 373
variable speed 107
vee blocks 91
vehicle identification number (VIN) 431
vehicle protective equipment 22, 121, 203, 259, 317, 363, 423
vehicle track 158
venturi 264–65
verbal communication 79
Vernier callipers 89–90
vices 104
violence 50
viscosity 179, 230–31, 442
viscous coupling 241
volatile 51
volatility of fuel 287–88

485

volt drop 381
voltage 92, 388–89
volts 370
volumetric efficiency 218, 301

warning systems 184
warranty repairs 76
warranty requirements 426
waste
   hazardous 29–30
   legislation 32

wasted spark ignition systems 298–99
water pumps 235
watts 372
welfare facilities 49
WELs 29
wet sump lubrication systems 229
wheel bearings 143–46, 350–51
wheel systems
   cast alloy wheels 187
   detachable rims 188
   inspection and servicing 464–67

pressed steel wheels 187
   requirements of 186
   retention of wheels 188
   wire (spoked) 188
winding 275
windows, electric 411–12
wiper system 475
wire (spoked) wheels 188
wires 376–77
wiring diagrams 385, 409
working at height 14
workshop manuals 73, 430

workshop operation. *See* communication; health and safety; materials; measuring; processes and procedures; structure, organisational; team work
worm and roller steering boxes 132, 133
worm gear 349

Zener diodes 401